Linow

Quantitative Evaluation of Energy Systems

Sven Linow

Quantitative Evaluation of Energy Systems

Methods – Solutions – Applications

HANSER

Print-ISBN: 979-8-950184-03-1
E-Book-ISBN: 9798950184307

Bibliographic information of the German National Library:
The German National Library lists this publication in the German National Bibliography; detailed bibliographic data are available on the Internet at http://dnb.d-nb.de.

AI-based systems were used for the translation of this work as well as for the creation of alternative text descriptions for images and illustrations. Responsibility for the content, as well as the final editorial review and approval, rests solely with the author(s) and the responsible editors.

Vilshofener Straße 10 | 81679 München | *info@hanser.de*
www.hanser-fachbuch.de
Production management: Frauke Schafft
Cover concept: Marc Müller-Bremer, www.rebranding.de, München,
Cover design: Max Kostopoulos
Cover picture: © shutterstock.com/Summit Art Creations
Typesetter: in-house
Print: CPI Books GmbH, Leck
Printed in Germany

Table of Contents

Preface

Do we want to shape our future within the realm of the possible, or do we simply let it happen to us? Do we want to use the actual state of our living Earth — in which our future will unfold — and the possibilities that our living Earth allows us, as the basis for shaping our future? What spaces of possibility for active shaping do we have today? These questions are at the center of our current societal debates.

The societal discussion on this matter is in a process of finding its way: On one side, voices dominate that are firmly anchored in trust in the magic of technological progress. This progress is supposed to simply solve all problems on its own in due course. This event will occur once progress, entirely without active support, guidance, or direction from the state or society, but perhaps through the action of market forces, produces the solution. On the other side, there are demands for rapid and proactive shaping of the societal framework conditions, in which just solutions for a shared future are actively developed together. Many arguments in this discourse are characterized by a lack of verifiable factual reasoning and quantitative evidence.

The belief in and hope for purely technical solutions addresses today's and future engineers directly. We are the *they* in "they will find a solution." The goal of good engineering education is to enable the consistent use of quantitative methods and factual reasoning. Core competencies of good engineers are the handling of numbers, the quantification of causes and effects, and the ability to make decisions on the basis of factually and quantitatively justifiable arguments. Who among us would want to board a flying taxi whose design was carried out without regard for (technical) rules and is based solely on trust in progress, i.e., on the assumption that a solution would spontaneously emerge from nothing if a problem occurs during the flight.

This is where the present book comes in. It introduces the current state of the spaces of possibility for our future energy supply on our living Earth. For this purpose, it presents quantitative methods that allow us to argue with reliable numbers and verifiable facts. It is understood that the framework within which a future technology is feasible is constrained by the laws of nature, by the earth system and the resources available within it, as well as by the planetary boundaries within which a technically advanced and just human society remains possible. Technical development can only be shaped by society within this framework.

I hope that we begin to address our pressing problems with the necessary seriousness of thought and the appropriateness of action. In this way, we can shape our future within our real possibilities ourselves. The alternative — a solution through disaster, which quickly becomes more probable through simply waiting for "progress" — seems to me entirely inappropriate. Let us help the new to be born, and let us be willing to let the old go.

Darmstadt, March 2025

Sven Linow

1 Introduction

The sufficient availability of well-usable energy is a fundamental prerequisite for the continued existence of our society. Only through energy do the production, transport, use, and disposal of all our technology become possible. Only through energy available in the future can we manage the accumulating effects of our previous technical actions in the world.

At the same time, the framework conditions for our energy technology are changing. Many other decision-making spaces of our society, and thus those of technology, are also shifting. This is the subject of the present book. The aim of the book is to help you recognize the situation and its requirements, to provide tools and convey skills that enable you, in your role as engineer, to describe your assignments as clearly as possible, to recognize your scope for action and decision-making, and thereby to make the best and most robust decisions possible.

The book is aimed at representatives of all technical disciplines, since every discipline requires not only disciplinary expertise but also fundamental shared competencies: engineers need deep disciplinary knowledge, paired with a broad interdisciplinary foundation on which to work. Therefore, this book provides broad expertise that extends beyond the boundaries of technology, in order to be capable of action and decision-making in the subject areas covered here.

1.1 Core Theses

The pressing questions are: How do we enable ourselves to live a good life and maintain human civilization in the earth system over the long term? How can we respect the natural limits of our living earth and ensure that we live and act within a range of conditions that is good for us?

With regard to our energy system, this can be formulated more concretely:

1. How do we ensure the long-term adequate supply of our society with necessary energy, without excessively adversely affecting the earth system?

2. What resources are available for the future and for the development of our society and our lives over the long term?
3. How do we ensure in the short term that the stable climate regime of the last 10,000 years does not collapse in a way that brings major or existentially threatening disadvantages for us and our society?

These are not purely technical questions. Interwoven with them are ethical questions and questions of equitable distribution that go far beyond technical considerations. We must keep these in mind. The core theses of this book are:

- Time is pressing if we ourselves want to proactively shape a good and just future.
- The pressing problems are poorly defined. They are very wicked problems (see Chapter 2). We must understand this and accept that there are no simple solutions, only alternatives among poor options.
- Solutions to these wicked problems are never purely technical or purely social. They emerge from a combination of both approaches.
- We have long known enough to be able to move immediately to the necessary actions. Many current literature references are intended to provide direct access to this knowledge.
- Waiting now, only to have to chase after the very dynamic developments in the earth system in the near future without being able to shape them ourselves, is an exceptionally unattractive idea.
- Many problems can be described with simple means at sufficient accuracy to arrive at relevant conclusions. Essential for our decisions and actions are human engagement and the assumption of responsibility. This we cannot delegate to our machines.

The focus is on energy literacy, with which we can think, describe, plan, or implement what is relevant. Energy literacy means the competency to grasp energy and energy-related questions in a systemic form, i.e., with their effects and in their orders of magnitude.

1.2 Focus

This textbook focuses on the relationship between the factors considered central to the future of our society, future energy technology itself, the provision of resources for this energy technology, and the ongoing global warming and biodiversity losses. The major changes in these factors will strongly influence the requirements for and the space of possibilities for technology in our future development:

- In the future, regenerative technologies will have to enable the provision of energy for society. This brings major changes for the entire technological base and for the forms of socially possible use of energy. This has massive implications for industrial uses and the economy. For engineers who design this technology or work with it, it is therefore important to understand the changes and new requirements in order to be able to respond appropriately in their work.

- The effort required for the provision of raw materials is increasing. Many raw materials are not available in the quantities needed for the global technical implementation of all new ideas. Some are seriously scarce. At the same time, modern technology often requires very special materials from partly rare raw materials. Engineers need fundamental ideas about how this may develop in the future in order to be able to respond appropriately.
- The currently rapidly accelerating shift in climate globally and locally is changing the space of possibilities for technical design. Limiting climate warming and necessary climate adaptation require new social and technical approaches. They influence the availability and use of all technically accessible forms of energy globally and locally.
- Our living earth needs space and resources to thrive, and we are completely dependent on it and its flourishing. Here exist limits and requirements for our human appropriation of resources.

These areas are intensively intertwined. The systemic understanding of the dependencies in depth exceeds the possibilities of a classical engineering degree program. At the same time, these framework conditions cannot be ignored. For this reason, key performance indicators are used in the following as a means of reducing complexity. Key performance indicators are today an essential element in many organizations for analysis, planning, and control. They are essential for the execution of projects, for shaping the future of the organization, and for day-to-day and long-term management. Good key performance indicators have high informational value from a technical perspective and enable the structured evaluation and conscious selection of solutions. For this reason, working with key performance indicators occupies an important place — while being fully aware that key performance indicators always represent a simplification.

1.3 Competencies

This book aims to convey the competencies that enable and encourage you to act actively in the world (Biggs/Tang 2011). In order to be able to apply these competencies with confidence, we need a broad base of knowledge. This knowledge can only be accommodated to a limited extent in a book. Therefore, in this book you will find many references from the field of technology into further relevant areas of knowledge. These references are an invitation.

The connectivity to relevant competencies of education for sustainable development has been deliberately incorporated (Brundiers et al. 2021). Systems thinking, strategic thinking, and futures thinking are essential for addressing the questions mentioned above. At the same time, a clear value orientation is taken as a basis here.

All important current social issues are closely intertwined with the secure availability of energy and resources. This is not always directly visible and is often ignored, much to the detriment of the discussion. It is necessary that our ideas and actions for shaping our future are grounded in a fundamental knowledge of energy and resources, energy inputs for actions, and future real availability (energy literacy). Many meaningful technical or scientific statements on resource use, the provision of energy, and system effects are only possible on the basis of a fundamental understanding of energy. The application of fundamental knowledge and methods of thermodynamics in the discussion of energy-related topics is a sign of seriousness of thought and effort.

Evidence-based decision-making processes — i.e., decisions in which substantive arguments count — require robust technical facts and methods. As engineers, we take on the role of bringing discussions of technical problems back to what is actually technically feasible. For this we need a robust framework of skills and knowledge. For the methods and examples presented here, a small part of the concepts, definitions, and formulas from an introductory thermodynamics lecture is required as a toolkit.

The goal of good teaching/learning situations is to convey to students genuinely applicable competencies, whereby these competencies can only be applied if, alongside them, a base of usable knowledge about interrelationships is available. We learners must acquire these competencies ourselves in our very individual learning process. The hope associated with this book is that real, current examples of high technical and social relevance for one's own further life provide additional motivation that carries you through the not always easy material.

Specifically, the following competencies are to be conveyed:

1. The fundamental methods from an introductory thermodynamics lecture are applied. This is intended to encourage you to apply the systemic perspective, which is so central to the thermodynamic method, to a wide variety of problems.
2. The ability to gain orientation in poorly defined problems or unclear situations is very helpful in professional life. In addition, it is necessary to first roughly estimate numbers in order to better understand and narrow down the problem and the task. This situation differs greatly from studying, in which usually only clearly and completely specified tasks need to be solved. The approaches used here are suitable for initial estimates. They are ideal for narrowing down a problem, comparing solution variants, and thus laying the foundation for good communication.
3. Many of the methods and approaches presented are based on key performance indicators. The construction of the key performance indicator is often the focus. The competency conveyed here is the construction of good and informative key

performance indicators as well as the assessment of whether a key performance indicator carries a relevant statement. In the best case, you yourself take influence on the design of key performance indicators and can raise inappropriate, imprecise, or poorly defined key performance indicators to a level appropriate for the problem.

4. The scientific method is briefly introduced and pitfalls in one's own research are discussed. The literature references in this book are kept (as far as possible) up to date and make use of specialized scientific literature from various disciplines. This is intended to encourage you to include all forms of literature while at the same time evaluating sources with the appropriate critical distance.
5. The examples and the numbers calculated therein are deliberately chosen to be large. In this way we practice dealing with initially unimaginably large numbers. It represents its own challenge to repeatedly bring these large numbers back to a tangible basis.
6. Many of the results and values determined here are not very precise, but always sufficiently robust to proceed to further evaluations and conclusions. The ability to work and argue with values that are not absolutely correct but sufficiently accurate represents one of the fundamental competencies for engineers.
7. Dealing with deliberately global problems is intended to help overcome a common obstacle in the discussion of solutions to large problems: in public discussion, the smallest solutions are often celebrated in detail, which upon systemic examination make no contribution due to their changes being minuscule compared to the global problem. Recognizing this is a competency in itself.
8. The problems discussed are at least of regional and often of global relevance. Many of the problems addressed are not small independent tasks, but global problem structures in which there are feedbacks and interactions that make everything confusing and complicated. Recognizing problems as global and interconnected is an essential competency when it comes to shaping the future.
9. At many points, relevant cross-references to other topics and aspects are inserted. This is intended to show the nonlinear feedbacks through which many global problems are so difficult to grasp and address.
10. The connection between technical and social aspects is of fundamental importance for engineering education. Good engineers possess a clear attitude and ethics. Developing and enabling this gift can only succeed if the connection between the technical and the social sphere is regularly and vividly presented.
11. The image of the living world or the earth system is developed from a natural science perspective. This illustrates the applicability of thermodynamics to natural and living systems. A fundamental understanding of the essential processes and sequences in the earth system is the basis for the further analysis of the effects

of human action. Only from an understanding of the earth system is it possible to grasp many particular aspects of the problems addressed here.

12. Real and large problems are never simple tasks. They are wicked problems that resist simple approaches. It is an essential step in understanding to recognize a wicked problem as such, rather than believing in simple solutions.
13. Many problems can be approached with relatively simple tools. Often rough calculations and simple models in combination with thorough reflection and acceptance of uncomfortable possibilities are sufficient to obtain the important insights.

1.4 Structure

Part I: Fundamentals, Methods, Key Figures, and Framework Conditions

Part I provides many fundamentals that together create a basis with which technical and societal questions of the transformation of our energy system can be addressed.

Chapter 2 is not technical. It conveys essential concepts for understanding large problems such as the transformation of our energy system. For this purpose, an established classification of problems is introduced. This makes clear that this book focuses on wicked problems that resist simple, purely technical solutions. To this end, it contains essential foundations for the constructive, evidence-based, and solution-oriented handling of problems and the procurement of reliable facts. We want to approach our problem in an evidence-based and scientific manner. This requires a critical approach to sources and information.

In Chapter 3, the essential contents of thermodynamics necessary for understanding the subsequent chapters are summarized in brief form. The chapter cannot replace a thermodynamics course. All contents are presented in a didactically more accessible and technically more in-depth form in my textbook *Angewandte technische Thermodynamik* (Linow 2022). There, a large number of examples with a direct reference to the transformation of our energy system can be found.

Chapter 4 offers an introduction to the life cycle of products. Further central concepts are the availability of resources and the technology readiness level of solution ideas.

In Chapter 5, central energetic key performance indicators are introduced. Key performance indicators allow a first good comparison when they have been determined cleanly and reliably.

Chapter 6 summarizes descriptions with which a growing or shrinking quantity and its behavior over time can be described. For this purpose, aspects of growth for a number of required resources are examined. The understanding gained here is intended to help describe and evaluate the speed and limits of transformation processes.

In Chapter 7, energetically fundamental relationships of our earth system are introduced. We want to live and operate economically in this system in the long term. Therefore, the focus is on relevant subsystems and their limits.

Part II: Shaping the Transformation

With wicked problems, the problem description is part of the solution attempt. Therefore, the problem description cannot take place at the beginning, but waits until here.

Chapter 8 introduces hard framework conditions within which future civilizations and their energy systems will operate. To this end, it discusses important concepts and classifies their possible contributions to a transformation.

Against this framework of possibilities and requirements, the relevant technical concepts for the energy infrastructure are then introduced:

In Chapter 9, the focus is on the provision of heat from fossil fuels, biomass, and other sources.

The potential of regeneratively produced hydrogen and synthetic fuels produced from it is the subject of Chapter 10. Hydrogen and its derivatives have a special role and are therefore deliberately presented separately from fossil fuels.

Chapter 11 focuses on hydropower and wind as direct conversion of mechanical power into electricity, as well as their potential and requirements.

In Chapter 12, the main focus is photovoltaics. For this purpose, further fundamentals on solar irradiance are first introduced.

Every energy system contains storage. In Chapter 13, storage options for electricity are the focus. Synthetic fuels as storage are the subject of Chapter 10.

Chapter 14 addresses the topic of reducing harmful effects on the earth system and the associated energetic costs. First, it deals with flue gas treatment as it is common today. The second focus is then on dealing with carbon dioxide and other greenhouse gases.

Chapter 15: Outlook and Spaces of Possibility

Chapter 15 deals with the question of how quickly we can shape the transformation of our energy system and what we need for it. Chapter 2, Chapter 7, and Chapter 8 deliberately go beyond the engineering framework and introduce societal aspects that are essential to the problem. In Chapter 15, we pick up some of these threads and also some technical threads and bring them together.

Chapter 16 (List of Symbols) and Chapter 17 (Appendix)

Chapter 16 contains a list of the constants, symbols, and SI prefixes used in the book. Chapter 17 (Appendix) contains many useful data – useful for the worked examples, but also for one's own work and analyses.

Examples

The examples illustrate the application of many introduced concepts. At the same time, they serve to present concrete environmental impacts. No technology works without problems. Concrete examples are chosen as a place to illustrate this. The numerical values used in the examples have been determined to the best of knowledge; nevertheless, they are sometimes imprecise, outdated, or there are differing values from other sources. Often they are average values that must be verified upon closer examination or during analysis. The results of the examples are therefore also estimates; however, they are sufficiently robust to allow conclusions to be drawn.

1.5 Acknowledgments

Even the first edition of this book was the result of a long personal process. Since then, much has happened in our world, so that a new edition with a thorough revision of all contents became worthwhile. I am very grateful that Hanser Verlag opened this path and I am delighted by the outstanding support, in particular from Julia Stepp and Tim Borck.

Many of the topics addressed in the book have since become significantly clearer and more sharply defined. At the same time, I have had the opportunity to discuss numerous topics with a wide variety of people. I am very grateful for the many opportunities as well as for the patience and ideas of all those involved. At Hochschule Darmstadt, these include, among many others, Anna Schilling, Carsten Barth-Schulmeyer, Sonja Kleinod, Ulrich Klüh, Martina Mohrbacher, Alexander Landfester, Iris Steinberg, Tom Phillips, Martin Führ, Silke Kleihauer, Anja Hentschel, Nicole Saenger, Jens Hoffmann, and Leon Pezzica. In Darmstadt, these include, among many others, our climate office, Patrick Voos, Oliver Ottinger, Oliver Kah, as well as many people from civil society such as Sigita Urdse and Heike Böhler. In Hesse, these include the climate departments in the Ministry of Agriculture, my contacts with the HLNUG, the LEA, people from initiatives, associations, parties, and other municipalities such as Dietzenbach, as well as my fellow campaigners on the scientific climate advisory council of the state of Hesse.

Important impulses came from the Lehre^{n}-annual program 2020, the StIL annual program 2022, and the Lehre^{n}-conferences. Great thanks go to all organizers and participants of these spaces of possibility, which are very important to me.

In addition, there are important technical discussions within the framework of the For-Future organizations here in Hesse and at the federal level. Special thanks are due to my co-authors for our projects and to the authors of the S4F impulses on nuclear energy, which I was able to co-edit, as well as to Jens Clausen and the co-authors of the policy paper on the municipal heat transition.

Special thanks go to all my students, who have learned the subject of this book with me and whose many questions made this project possible in the first place. Their questions, their sometimes unexpected approaches, their discussions and learning processes can be found in very many places. In addition, there are a large number of student theses that I was able to supervise. The discussions associated with them have provided me with new insights and ideas, some of which can be found in the book.

My brave family has once again endured a father and husband who was at times absent for months. They put up with an enormous drifting dune of literature in their living room and a completely covered dining table. Yet they always encouraged me to tackle and complete this project – awesome!

References

Biggs, J./Tang, C.: Teaching for Quality Learning at University. McGraw Hill, Maidenhead 2011

Brundiers, K./Barth, M./Cebrián, G./Cohen, M./Díaz, L./Doucette-Remington, S./Dripps, W./Habron, G./Harré, N./Jarchow, M./Losch, K./Michel, J./Mochizuki, Y./Rieckmann, M./Parnell, R./Walker, P./Zint, M.: Key competencies in sustainability in higher education—toward an agreed-upon reference framework. In: Sustainability Science 16, 2021, pp. 13–29

Linow, S.: Angewandte technische Thermodynamik. Carl Hanser Verlag, München 2022

Quaschning, V.: Regenerative Energiesysteme. Technologie – Berechnung – Klimaschutz. 11. Auflage. Carl Hanser Verlag, München 2021

Smil, V.: Energy and Civilisation. A History. The MIT Press, Cambridge 2017

Part I – Fundamentals

How do we as a society want to address the pressing social and technical problems associated with the transformation of our energy systems? Part I provides you with the foundational knowledge needed to think more deeply about this system. The complexity of our historically grown energy system permeates large parts of our society and our planet. In Part I, we approach the description of the system, with each chapter opening its own perspective. The energy system is subject to constant change, but in order to make it fit for the future, we will need to strongly accelerate this change and adapt to the possibilities of life on this planet. This is the subject of Part II.

The aim of this book is to present important tools for evaluating energy technology and to illustrate their application. In particular, methods are to be provided that estimate, from a thermodynamic perspective, the extent to which an energy technology achieves important goals and what environmental impacts it generates in the process. A systemic perspective is consistently required for this purpose. This systemic perspective is established through clearly chosen system boundaries and methods that describe the life cycle of a product as completely as possible. In this context, the goal definition from Chapter 8 already has a retroactive effect, meaning that the focus here is primarily on those fundamentals that will later help us on the path toward the goals yet to be developed.

Chapter 2 examines what exactly characterizes these relevant, very difficult problems. It offers guidance on the question of how we can succeed in establishing reliable facts, and it addresses the question of how we can generate good solutions. This also gives us a way to scrutinize existing solution approaches and find out how good they are.

Chapter 3 provides a brief introduction to thermodynamics. Here, important terms used in the following chapters are defined and fundamental relationships are introduced. The chapter cannot replace a genuine introduction to thermodynamics. It is rather to be understood as a refresher or an invitation to engage more intensively with this method and description.

Chapter 4 delves deeper into the systemic perspective. The focus is on the life cycle of energy and of components of the energy system. The foundation of technology is the availability of resources. The concept is therefore introduced in a very general way and then explored in more depth specifically with regard to metals and their future supply. With a view to technological developments and the question of when new technologies can solve today's problems, the technology readiness level is introduced as a perspective.

In Chapter 5, the fundamentals of energy and life cycle are brought together to introduce important key figures for energy systems. These key figures are extremely helpful for describing and comparing different technical options.

Transformation and change always involve growth and contraction. Chapter 6 provides tools with which we can describe this. In doing so, we cannot avoid the question of whether long-term economic growth is planetarily possible.

Chapter 7 offers an introduction to the description and magnitude of energy flows in the earth system and to the relevant human influence on these. Our aim is to understand processes and to be able to estimate orders of magnitude. This too is an invitation to engage more deeply with these topics.

The goal is to make sound, robust decisions that are positive for as many people as possible, in order to address the major future challenges and manage them as effectively as possible. To do this, we must not only generate possible solutions, but also analyze them with regard to the problem itself, i.e., with regard to the possible outcomes, results, and consequences. We do this generally on the basis of facts, taking into account applicable regulations, laws, standards, and rules, and using the scientific method. In doing so, we also orient ourselves toward engineering ethics as most recently formulated by the VDI in 2021. We begin with this aspect.

2 Problems and Solutions

The subject of this book is major challenges that at first appear to be technical in nature. They are major challenges because, on closer inspection, their solution becomes very difficult, because technology alone is not sufficient for a solution, because technology may be part of the problem, because there is often no good solution at all. Such challenges are difficult to grasp and even more difficult to solve.

Problems can be divided into tame problems and into wicked problems. In English, *tame problem* and *wicked problem* are used as established terms (Rittel/Webber 1973). They have been translated here.

- *tame*: The term *tame* is deliberately intended to evoke associations such as friendly, obvious, cuddly or nice, and to suggest that this type of problem has completely submitted to human design.
- *wicked*: On the other hand, *wicked* illustrates the character of those problems that elude all human control in a devious, unpleasant and unpredictable manner, and unfold their effects unexpectedly, differently than planned and with force, regardless of whether we ignore or deny them.

In VDI 2221-1, the terms "task" and "problem" are used: tame problems or tasks in the sense of VDI 2221-1 are fully described by the problem statement, a solution exists, and there are unambiguous criteria for whether a proposed solution is right or wrong. This is the type of assignment we are primarily confronted with during our studies, but which engineers encounter rather rarely in their professional daily lives. VDI 2221-1 understands as a problem those technical challenges or specific design tasks for which there is no simple solution. Even these problems require their own systematic approach, but they are often not wicked problems in the sense of the definition used here.

Wicked Problems

The problem areas focused on here — energy supply, climate, biodiversity and resources — give rise to wicked problems, as they as challenges mostly go well beyond the problem definition of VDI 2221-1. One of the goals of this book is to illustrate this.

Wicked problems are characterized by the following ten criteria (Rittel/Webber 1973, slightly revised, see also Table 2.1):

1. There is no complete or correct description of the problem. The description is itself part of the solution, since the solution inherently contains a specific definition of the framework for action, the actors, the causes, etc.
2. There is no unambiguous criterion for when the solution has been reached. The solution is frequently the state that is arrived at after a certain amount of effort and that is now implemented after the resources provided for finding the solution have been exhausted.
3. Solutions are neither right nor wrong — they are good or bad. Most often they are better or worse. Criteria for good and bad frequently arise from a normative or ethical framework.
4. There is no unambiguous evaluation of how good a solution is. With tame problems, deviations from the ideal solution can be determined and individual proposed solutions can thus be classified. With wicked problems, the consequences of the solution can only be recognized in hindsight, since the effects propagate in complex ways throughout the entire affected system. Frequently, the implemented solution cannot be compared with other proposed solutions.
5. Every implementation counts in full. There is no possibility of approaching the solution through trial and error. Every implemented solution or measure leaves traces in the world that cannot be undone and that influence and distort every subsequent attempt.
6. There is no countable or fully describable set of potential solutions. There is no well-describable set of operations by which a solution is reached.
7. Every wicked problem is unique. There are no classes of similar or similarly solvable wicked problems.
8. Every wicked problem can be regarded as the symptom of another wicked problem. It follows that improving one problem creates difficulties in addressing the next.
9. If we describe a wicked problem as a deviation from a desired different state, we thereby already determine solution paths or exclude possible solution paths. This happens solely through the framing of the problem.
10. Errors in the solution often cannot be tolerated, as they have direct effects on the circumstances and lives of people. Those seeking solutions bear a high responsibility that their solution works.

Thus the definition of the wicked problem goes significantly beyond the problem definition of VDI 2221-1. In particular, wicked problems always contain social effects or repercussions that cannot simply be resolved. The central thesis of this book is that all relevant technical and social problems are wicked, at least with regard to a transformation of our energy supply. Constraining boundary conditions arise, among other things,

- From the second law of thermodynamics, which sets natural limits on many possibilities (see Chapter 3),
- From the limited availability of resources (see Chapter 4, Chapter 5 and Chapter 6),
- The necessity of remaining within planetary boundaries (see Chapter 8), as well as
- From social and ethical limits.

The magnitude of the pending required changes and the complexity of the links between social and technical aspects means that all problems of the Great Transformation of our energy system are wicked (this sounds almost like a version of the second law of thermodynamics, and perhaps it also follows from it in some convoluted way).

Table 2.1 Classification of Problems

Aspect	Tame problems (*tame*), tasks	Wicked problems (*wicked*)
Problem description	unambiguous closed complete	An unambiguous or complete description is impossible. The description depends on the perspective, the attitude, and the chosen approach to a solution. The problem can only be fully described as part of a solution.
Solution	unambiguous closed	Often there is no solution, only better or worse ways of dealing with the problem. There are any number of possible solutions.
Deviation from the solution	Clearly determined The error can be calculated immediately.	It is unclear when a different solution begins. It is unclear from what degree of deviation the proposal is no longer a possible solution.
(Numerical) simulation	Well suited	A simulation is never complete and the model is inadequate: missing aspects, consciously or unconsciously chosen assumptions, and simplifications used distort the simulation. The result may be irrelevant.
Example	Number, mass, volume	What is the right way to raise my child? What education should I choose?

Capturing Wicked Problems

In many political or technical discussions, complex wicked problems are presented as simple tame tasks for which there is a correspondingly simple solution or product.

The aim of this book is to convey a fundamental, careful, and ethically appropriate approach to wicked problems and to examine proposed solutions with the necessary attention. For this purpose, we can ask ourselves these questions:

- Is the problem meaningfully and completely described? Is the cause clearly identified? Is the symptom a cause or a consequence of the problem?
- Is the proposed solution appropriate to the problem? What purpose does this proposed solution serve? For whom is this proposed solution useful?
- Is the problem really a simple tame problem for which there is only this one solution, or is it in reality a wicked problem?

AI So-called artificial intelligence (AI) cannot help us here: AI has no complete description of the problem. AI has no model that maps all interrelationships and feedback effects. AI has no *skin in the game.* AI inevitably produces bullshit. AI has no consciousness and therefore cannot deliver decisions on ethical issues (VDI, 2021; Bridle 2019; Townsen et al. 2024; Zhou et al. 2024). More on digital technology can be found in Chapter 8.

2.1 The Scientific Method

In order not to be confused or overwhelmed by wicked problems, it is necessary to approach them in a structured manner and in this way to accumulate the necessary knowledge about the problem, the entanglements of the problem with other areas of the world, and about possible solutions. The scientific method is particularly well suited for this, since unlike other approaches it is fundamentally systematic and error-oriented, and since in the scientific approach the result is not predetermined. The goal of every university education is to be able to think and work scientifically (even if this sometimes feels more like vocational training at the bachelor's level).

Essential in this regard, alongside solid professional competencies, are the critical handling of sources (Section 2.2) and the capacity for personal reflection. We need critical distance from our own emotions and needs in order to prevent them from unconsciously influencing our thinking and anticipating results. At the same time, we need our emotions as an ethical compass.

In science, a fundamental assumption is that the physical world exists and has a very concrete, unambiguous form and shape, entirely independent of how we ourselves wish to see or are able to see it. To recognize this real world ever more clearly — that is something we can and should attempt.

If we want to solve the wicked problems ahead of us, then we must understand the world as it is, with all its challenges and inconveniences, as well as possible. The only reasonably reliable path to this that we know of so far is the scientific method. The direct application and consequence of this thesis is that one's own opinion cannot be the starting point or the center of one's own theory of the world. The world does not orient itself according to my own opinion; rather, I must grasp the world as it is in order to develop a current theory of the world based on a sound scientific analysis of reliable findings. My individual theory is thus dependent on my current understanding of the world, and I must change my theory of the world whenever I receive new reliable findings. Always.

Scientific method

The scientific method is clearly defined by a fundamental attitude and general approach:

- It seeks answers to questions from nature and technology, always looking to nature or technology itself for those answers.
- It presents preliminary answers to its questions — so-called hypotheses — and tests, through careful observation or through repeatable experiments, whether and to what extent these hypotheses are reliable answers. Only this form of confirmation of hypotheses is recognized.
- All results that build up into a theory as a structure of verified hypotheses are always regarded as provisional.

Every theory is valid only until it is refuted by new observations or experiments. As long as it is valid, we should use it — even at the risk of contradictions arising. From these contradictions we can then further improve our scientific description of the world by first understanding where the contradictions come from. Afterwards we can adjust our theory so that these contradictions diminish or disappear.

The aim of this book is to approach questions always from a scientific perspective. This often already requires courage. In many discussions, fundamental assumptions are taken as given: there is often a felt consensus about what may still be discussed and what is already settled. Questioning this consensus can already provoke massive rejection within a social group or organization. Many social groups and organizations have belief systems that predetermine decision-making processes. Particularly within groups, acceptance of the fundamental belief systems is often a prerequisite for participation. A scientific approach, however, requires that we ensure all our assumptions are correct. We must verify and question. This is sometimes uncomfortable or difficult.

Any theory that cannot be confirmed or refuted by experiments or observations in the real world is not scientific. In the subject area of this book, we can confirm or refute everything through experiments or observations.

Any theory that is also based on supernatural forces or magic is not scientific. Magic keeps creeping into our world and into our thinking. It is a laborious process to keep reassuring oneself that one's own thinking is not guided by ultimately magical expectations or hopes. In social discourse, magical hopes and expectations are always present. Identifying these elements is already a major and strenuous step.

2.2 Recognizing Reliable Sources

It can be difficult, especially at the outset, to obtain reliable information on a topic. Scientific specialist literature would first have to be found, and it can quickly become too complicated as an entry point. This section therefore describes further possible sources and is intended to help you strengthen your judgment.

Many communication channels follow the rule: "*Whoever shouts loudest is right.*" This is consistently applied particularly in the realm of the attention economy of digital platforms, in social networks, in politics, and in forms of popular reporting: texts often also target emotions such as anger and outrage directly, which are intended to prevent any critical distance from the text (Törnberg/Chueri 2025). In most cases, simple solutions are simultaneously presented with which the problem addressed would be directly solved. The solution (or the truth), this form of reporting implies, is being prevented by evil, stupid other groups or media. In the process, the information about the actual occasion is already heavily filtered. Particularly within opinion bubbles and echo chambers in social media, these emotions then quickly amplify — and every network in social media has the tendency to be such an echo chamber of like-minded people, since any form of deviation from the general outrage is immediately and emphatically punished there with social ostracism (Bridle 2019). That is not scientific.

Social media Common social media are provided by commercially oriented actors with a strong political agenda of their own. Here the primary goal is to bind users as intensively as possible to the medium in order to show them as much lucrative advertising as possible. To achieve this, the hidden algorithms that suggest content to us are designed to deliver content that matches the interests and emotional needs of the individual as closely as possible. Very often these are contents with a high sensational or arousal component. In particular, the algorithms are indifferent to the truth content of individual pieces of content — there is deliberately no distinction between bullshit and facts (on bullshit see Frankfurt 2020). What is essential is to clearly identify users and to know them as precisely as possible. Through selection and use, a user profile is thus created. This user profile then shapes what users are able to see of the world at all. As users, we thereby shape ourselves and our world, and live in an information bubble without any contact with the real physical world (Bridle 2019).

Original source or hearsay The scientific method demands an approach that runs diametrically counter to this description of the world and this form of information gathering. In particular, a scientific approach requires first of all knowing and evaluating the original sources. Before I as a reader allow myself to be overwhelmed by the expected outrage over a terrible new piece of draft legislation, I first read the original draft and begin to analyze it using my own criteria. If I lack information to understand the text, I research it, where possible using additional scientific sources. I maintain a critical distance toward every source. This critical distance is reflected in the fact that

I do not blindly believe what is stated in the source, but wish to be convinced by arguments.

Critically examining sources

When critically examining sources, we should ask ourselves the following questions:

- Is the source internally consistent?
- Does it accurately and sufficiently completely reproduce other sources, contributions, or opinions?
- Does it place other contributions sufficiently clearly and correctly in the necessary context?
- Are the methods used here appropriate for addressing the topic?
- Can the results be correct?
- Are the conclusions logically derived from the results?
- Are the motivations of the authors comprehensible and sufficiently clearly disclosed?
- Are analysis and opinion distinguishable from each other?

Studies Good studies are reliable scientific representations. This applies when they are conducted by an independent research institute and when the client has not predetermined the result. Good studies follow the logic of scientific texts and disclose all framework conditions. Good studies are complete, i.e., they also contain information about deviating results and interpretations, failed experiments, and difficulties in implementation.

Journalistic Texts

These texts follow a different logic and are therefore not always suitable as reliable sources: Authors often attempt to document their "objectivity" by presenting the subject they are reporting on as a dispute between two poles (Dean 2012). This is a form of representation that originates from political reporting and is already unsuitable there for describing problems or developments (as false balance). This perspective must be evaluated very critically with regard to the topics of this book: If the position that is actually supposed to be reported on is a scientific consensus, then the authors, if they want a contrast, have no choice but to find a second position that lies far outside the scientific consensus. Such positions lying far outside the scientific consensus can easily be found today via the internet: self-promoters, quacks, magicians, merchants of doubt (Conway/Oreskes 2012), or cast pseudo-scientists offer themselves there. While scientists must speak of *probability* (see box), the opposing side is allowed to be *absolutely convinced.* Unlike scientists, they also do not have to report on their conflicts of interest, or even obvious self-interests are simply ignored. What aspires to be objective reporting quickly becomes a vehicle for untruth or creates the space for spreading actually nonsensical positions. In addition, journalists are sometimes forced to report on topics to which they have no specialist access of their own. They are therefore sometimes unable to recognize obvious falsehoods, untruths, bullshit, or nonsense. This applies all the more when it comes to rapid reporting.

Good journalistic texts draw a sharp distinction between reporting and opinion. The previous paragraph highlights the danger of an apparently "objective" presentation that reduces topics to opposing poles. This can also appear in an otherwise clean and fact-based text that clearly contains no opinion of its own. The opinion might then be found in a text marked as a commentary elsewhere. However, this clean approach is no longer always maintained even in quality media, and opinion and reporting then merge into one another. Many media outlets already abandon minimum standards out of consideration for revenue or the political agenda of their owners and begin with emotion (Törnberg/Chueri 2025).

Describing probability and certainty

Scientific texts have an established form for expressing how certain a statement is or how significant a result is. The formulation "very likely" may colloquially be rated at a 20% probability of occurrence on a good day, but in a scientific text it means a value well above 90%.

The fact that scientific texts couch their statements in probabilities is not an indication that the situation might be entirely different, but follows from the inherent logic of science, which always represents an approximation to the world and accepts that things may be otherwise. This makes statements from scientific articles considerably more reliable, truer, and more useful than so-called ultimate truths from other sources.

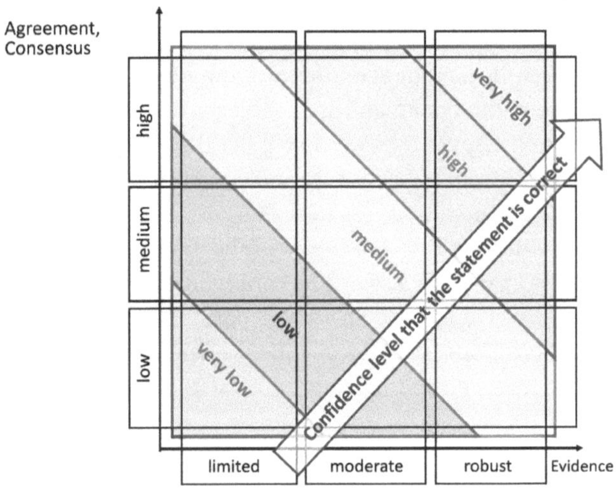

Figure 2.1 The confidence level for a statement depends on the aspects of agreement (within the panel) and evidence (which, how good, how reliable, how consistent) (based on Mastrandrea et al. 2010).

In scientific publications, the probability and the certainty of statements are often expressed verbally. The reports of the Intergovernmental Panel on Climate Change (IPCC) are a good example here. Table 2.2 lists the terms used there and their definition for probability along with a translation. Based on two independent parameters — the existing evidence for a statement on the one hand and the degree of consensus among the experts involved on the other (Figure 2.1) — the confidence or certainty with which a statement is made in the IPCC reports is additionally defined. This clarity and traceability is what distinguishes good scientific sources.

Table 2.2 Standardized Terms of the IPCC to Express the (Likelihood of the Outcome) (Source: Mastrandrea et al. 2010)

Term (EN)	Probability of occurrence
Virtually certain	> 99 %
Extremely likely	> 95 %
Very likely	> 90 %
Likely	> 66 %
More likely than not	> 50 %
About as likely as not	33 % to 66 %
Unlikely	< 33 %
Very unlikely	< 10 %
Extremely unlikely	< 5 %
Exceptionally unlikely	< 1 %

Science and technology journalism A particular difficulty arises in reports on new technical developments: in order for scientists to gain attention, they must present their results as directly relevant. Journalists, in turn, find it extremely difficult to report on a development that is not directly and concretely applicable. This creates the risk that the potential of an invention or discovery is significantly exaggerated in their report. This effect explains the frequent appearance of ultimate solutions and silver bullets in the press. This need is additionally exploited by start-ups to push themselves forward and raise capital: unfounded promises are gratefully received (and the massive limitations in the fine print are simply overlooked).

If we read uncritically through the technology section, we will all soon be traveling to the moon in autonomous taxis and sourcing our raw materials from Neptune. The impression is created that there are technical solutions to all problems, which could practically be rolled out tomorrow morning. Some readers feel confirmed in their optimistic view of the future, while others become cynical, as these wonders soon dissolve into thin air. Such reports always have a kernel of truth (a study or a planned experiment in a first demonstrator), but everything has simply been presented more optimistically and much more simply than it actually is. The only remedy here is to seek out the original source and evaluate it yourself (Dean 2012).

Value of journalistic contributions Criteria for good journalistic work are presented, for example, in the press code of the German Press Council. These requirements can be used directly as a benchmark to assess whether we can use a journalistic article, a blog entry, or a video as a source. Good journalistic sources have very high value for tracking current developments, such as the political implementation of measures. However, day-to-day reporting is at the same time hardly able to identify long-term trends or causes, so that both perspectives become important here: it is only scientific texts or long-term journalistic contributions that provide important perspectives.

Journalistic contributions offer, at best, the significant opportunity to get to know personal perspectives and experiences. Scientific literature must, for methodological reasons, largely forgo this. Only through the personal experience and perspective of key actors and those affected can a complete picture emerge. We need these perspectives in particular to prevent a lack of imagination, which characterizes many contributions to the topics of this book.

You can find the German press code in its current form here: *https://www.presserat.de/pressekodex.html*

The core is respect for truth and human dignity: *"Respect for truth, preservation of human dignity, and truthful information for the public are the supreme commandments of the press. Every person working in the press upholds on this basis the reputation and credibility of the media."*

Other important aspects of the code are thoroughness of research, separation of activities, and the separation of advertising and editorial content. Not all media succeed in this and in other requirements set out therein. Some media deliberately position themselves clearly outside of it.

Politics

In politics, a worldview is widespread that assumes all facts are negotiable. Negotiable means that facts are the result of a negotiation in which the more powerful determine the "facts" for everyone through money, influence, power, threats, or coercion.

Accordingly, in politics opinions are often fixed, or opinions are not oriented toward facts in the sense of scientific findings. Potential facts are then adapted to the respective prevailing opinion. Autocratic regimes display this behavior particularly clearly and distinctly. A hallmark of autocratic regimes is the weakening or destruction of scientific institutions. At the same time, we find opinionated politicians with little respect for scientific knowledge and little capacity to question their own position everywhere. This opinionatedness and the associated mercilessly self-centered determination to prevail are often regarded as strength of character, although they would be better described as ignorance, stupidity, or in terms of psychiatric diagnoses. This is not a helpful position or preferred method for addressing the challenges of future sustainable development, but rather endangers our future (DeLay 2024).

Structure of scientific journal articles

Scientific articles follow an established structure and argue with logic. For this purpose, they use a specific language. Both clearly distinguish them from other forms of written communication. The structure of a scientific article is clearly defined so that readers can understand the argumentation as precisely as possible and follow the arguments in detail (if they wish to do so). At the core, it is about being able to understand how the authors arrived at their conclusions:

1. The research question and the motivation for the investigation are clearly presented at the beginning: what is being analyzed and for what purpose?
2. An overview of the relevant literature and the state of research is provided. The focus is on fundamentals that are considered common consensus, as well as on important recent contributions. With regard to the problem being analyzed, gaps in knowledge, open questions, and also points of contention are indicated where relevant.
3. The methods used are presented (often with reference to other literature) in such a way that readers can apply these methods themselves – so that they can ultimately repeat the investigation independently.
4. The results of the analysis are presented clearly. This is a purely factual presentation of data, figures, calculations, and simulations.
5. In the evaluation, the results are analyzed with regard to the research question. Only here is an attempt made to provide an answer to the scientific question. Here, recommendations for action can be developed from the hypotheses thus generated. The investigations of others are incorporated here, confirmed, extended, or rejected.
6. At the end of the article are the conclusions, which should develop logically from what has preceded them.

Enforcing this logical structure and ensuring that the chain of argumentation is coherent and comprehensible is the task of the peer review process, to which the vast majority of scientific articles are subjected. Peer review means that specialist colleagues critically read the article before its publication and assess whether it is suitable for publication based on the criteria mentioned, or how it would need to be revised in order to be published.

The language of scientific literature is very objective and restrained. The goal of the presentation should be to avoid confusing critical readers with emotions, but rather to make the technical content as accessible as possible. Respect for readers is expressed in the fact that they are encouraged to draw their own conclusions.

2.3 Social Aspects of Technical Change

Wicked problems and their solution variants have the property of creating affected people and violating interests. This makes early communication of this aspect important. Not In My Backyard (NIMBY) and also Build Absolutely Nothing Anywhere Near Anything (BANANA) describe current social trends. There is always a group that argues why something may actually be good in principle, just not here or not now. These groups usually find a receptive audience; they perform well in media systems or serve the expectations and interests of others. Sometimes they raise aspects that are worth considering but have either been overlooked or not communicated so far. Sometimes they are instrumentalized by other interests. Sometimes they retreat into their own social echo chamber and are no longer receptive to arguments from outside. Sometimes something is simply instrumentalized.

Due to the low power density of natural energy flows, it is not possible to make infrastructure for the provision of energy and for the generation of electricity from sustainable sources small. Due to the power losses when transporting electricity over longer distances, it is appropriate to spatially couple generation and demand. This represents a fundamental conflict that in most cases cannot be resolved purely technically but must be negotiated socially.

Figure 2.2 Energy landscape Eemshaven, Netherlands

Figure 2.2 shows Eemshaven in the Netherlands from a distance of 30 km. Eemshaven is located far from all population centers in the Netherlands. There is a coal-fired power plant (rated capacity 1560 MW, allegedly CCS-ready, see Chapter 14), a total of eight combined cycle gas turbine power plants (CCGT), three of which with a combined rated capacity of 1311 MW are visible in Figure 2.2 (the five additional CCGT power plants have a further combined rated capacity of 1750 MW), and approximately 50 wind turbines (combined rated capacity 156 MW). Figure 2.2 illustrates the size ratios and the space requirements of the individual technologies well. It also illustrates how a society pushes its energy generation to the sparsely populated periphery and out of the population's sight. There in the periphery, sacrifice landscapes then emerge that we would not want close to us in the center (Brandt/Wissem 2017).

In our society there is a lack of understanding of the magnitude of energy quantities or energy demand. (Populist) politics shamelessly exploits this. Many of the proposals being discussed show that the magnitude of one's own demand, the demand of one's own city or one's own country is not understood. The only way forward here is to calculate these numbers, state them, and illustrate them well. The following is intended to provide you with helpful data and methods for this task.

Energy demand

A person requires approximately 100 W or 10 MJ per day for their supply of energy through food in order to stay alive and healthy. This corresponds roughly to one kilogram of bread (Figure 2.3). The annual per capita primary energy consumption in Germany is 45,000 kWh. This corresponds to the energy content of 45 such loaves of bread for each of us per day. Our technology requires 45 times as much as we do. For the annual per capita final energy consumption of 31,000 kWh, we would need approximately 30 energy slaves for each of us to do this work for us.

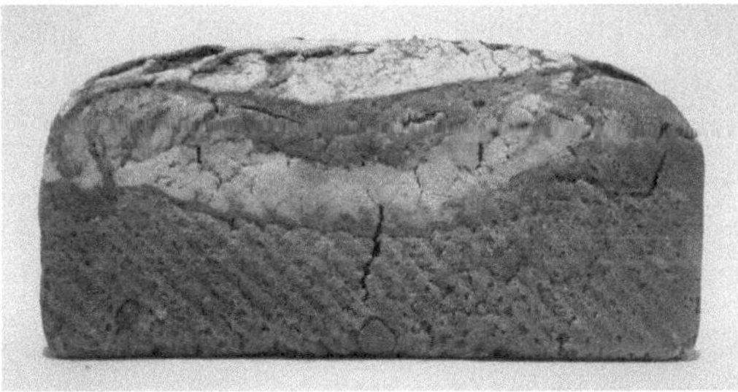

Figure 2.3 Daily energy demand of a person

2.4 What Do We Know?

In a simple representation, four categories of knowledge can be identified (Figure 2.4). Here it is two axes that determine what we know:

1. The simple axis of what is actually known: What knowledge and insight is really available and usable?
2. The difficult axis of what we know and want to know about our state of knowledge

From this, the four categories develop in Figure 2.4:

- **Known knowns:** The first category is established, secure knowledge, whereby this knowledge is continuously changed, expanded and adapted, at least from a curious or scientific basic attitude. New findings must be integrated, old certainties dissolve and some competencies are lost over time.

- **Known unknowns:** The second category consists of known knowledge gaps. In the best case, we make an effort to fill these, or we turn to people who can take on this task for us. Of course, every gap we fill will cause our established knowledge to change. Therefore, the first defensive battles are already fought here, intended to prevent learning.
- **Unknown unknowns:** The third category consists of all those things that we objectively cannot know that we do not know. This source of genuine surprises may currently be estimated as rather small.
- **The repressed:** The fourth category is the knowledge that we have, indeed must have, but which we consciously or unconsciously and always at our own risk ignore. Shifting as much of our more difficult or critical knowledge here as possible is the result of our own inertia and the goal of trolls, populists and merchants of doubt.

In Chapter 8 we look at what possibilities we have to shape a good future. It will be essential that we cannot afford ignorance if we want to shape a humane and good future. Keeping one's own repressed knowledge to a minimum requires reflective capacity, humility and a very low bullshit tolerance in all of us.

Figure 2.4 A portfolio of our knowledge

2.5 Environment and Technology

We humans, our environment and our technology – these are not separate spheres, but all three exist only together in the earth system and interact closely. Therefore, the scientific method uses the same theories for their description and analysis. The same laws of nature apply to all three.

Long since dismissed as false, but still powerful systems of thought in the Western world separate between humans and their technology, between body and mind, and between humans and nature as their environment. If we view the world from a thermodynamic perspective, it is possible to describe humans or humans together with their technology as a system. However, these are open systems that are in direct and very close exchange with their (living) environment:

- The three systems are in constant and very close exchange of substances. Nothing remains confined to one area.
- Energy permeates all three domains and flows across all their system boundaries.
- Through the exchange of energy and entropy, the exchange of effects takes place within the common earth system. At the same time, the exchange of entropy can be used to describe the impact of one's own actions in the world.

The common system is the critical zone of our earth (the area of the earth full of life) and we are absolutely dependent on the state of this critical zone. A meaningful hierarchy of the systems cannot be achieved; however, we humans with our technology and our economy are by now a very large and powerful part of the earth system.

Anthropocene Environment and technology interact with each other at all their interfaces and in an unmanageable diversity of processes. Due to the mass, size, and extent of energetic flows of the earth that pass through technology, these interactions and their repercussions for the earth system are overwhelming. The environmental impacts of human technology on the limited and by now comparatively small earth system have short- and long-term consequences for our earth system: the term Anthropocene has since been coined for this. Anthropocene describes the action of us humans as an ongoing geological event with massive global impacts in all areas of the (for us) critical zone of the earth. What is fascinating is the scientific discussion on the question of when humans have exerted so much influence on the earth system that we can regard this as a distinct epoch in earth history (Zalasiewicz et al. 2024).

The goal of this book is to provide methods with which we can evaluate possible solutions for a future energy technology under the conditions of the Anthropocene and within planetary boundaries. To do so, the environmental impacts of human action must be part of this evaluation. In order to do this, it is necessary to understand them, to be able to estimate and assess them.

Impacts

The impacts of our human actions, which are always actions involving technology, can be subdivided as follows:

1. There is the intended impact, i.e., what was actually planned and envisioned.
2. In addition, there are impacts that, while not desired, are nonetheless foreseen and knowingly accepted (collateral damage such as, e.g., climate warming as a consequence of the use of fossil fuels).
3. There are also direct unintended impacts. These are unforeseen consequences of actions that arise because something was overlooked during planning, because certain interrelationships were not considered, or because interrelationships within the earth system were not known
4. Unintended systemic impacts arise from overlooked interactions between the deliberately altered subsystem and other systems. Very often these possible interactions have not been taken into account or were misjudged.
5. Furthermore, there are future impacts of the intended action that may only come into effect through scale effects. This includes technology lock-in, i.e., the long-term commitment to a particular technology or standard through initial implementations. One example is the 1435-mm standard gauge of railways, which cannot simply be changed today. Also included here are impacts that only arise because something accumulates over long periods of time and thus only gradually becomes noticeable (toxic substances, see Adam 2024), as well as slow responses of complex nonlinear systems (oceanic currents).
6. And finally, there are fundamentally unknown impacts. While all other effects would have been foreseeable with sufficient care in planning, or at least become apparent with increasing knowledge, these effects occur suddenly and unexpectedly. These must not be confused with disavowed interrelationships, which are already included further above.

Perceiving Common societal ways of dealing with unintended environmental impacts, especially the unconsidered, long-term, difficult, or demanding ones, are mostly:

- **Denying:**"*No, such a thing does not exist*", "*That is all scaremongering by crazy fanatics*", "*99.9% of all scientists and experts are wrong*", ...
- **Ignoring (as a form of denial, DeLay 2024):**"*That is not my area of expertise*", "*That is not my responsibility*", "*There are truly more important things*", "*pragmatic and measured*", ...
- **Hoping and believing (as another form of denial, DeLay 2024):**"*Someone else will solve it*", "*It won't be that bad*", "*I am a technology optimist*", "*Entrepreneurs will invent something*", ...

  There are a number of current offerings for belief. Many ideologies have imaginative ideas about why the problem does not exist, why it is irrelevant, or why it will disappear on its own if we follow the right prophets.

- **Lack:** Relevant are the lack of imagination (Ghosh 2017), the lack of attention or focused thinking, and the lack of humility.
- **Giving up:** This includes despair, apathy, and cynicism. The age-old theme of the incorrigibility of humans or of youth belongs here, as does a complacency in one's own thinking and acting.
- **Accepting:** Actively and consciously acknowledging the situation and dealing with it. What is meant is coming to act and to carry on, i.e., in a certain sense the normal engineering approach: (i) understand the problem → (ii) develop solution approaches → (iii) define evaluation criteria → (iv) evaluate and select solution approaches → (v) implement → (vi) respond appropriately when things do not work out as planned.

This book focuses on the last approach. It contains many approaches and methods to grasp problems and possible solutions and to make them describable. The emphasis is on evaluation criteria and thus on assessing and comparing possible solutions. Only the last option is based on genuine strength of character, as it requires consciously facing and confronting the truly unpleasant and difficult facts.

2.6 Finding Solutions Methodically

Decisions can be made for very different reasons. As engineers, you learn very specific perspectives and methods to arrive at good decisions. Good engineers possess a certain humility that allows them to prefer working systematically and using these methods. This collection of tools is intended to help analyze solutions in an appropriate and comparable manner. It is meant to help weigh and evaluate solutions against each other. In this way, it aims to contribute to keeping a clear head in difficult situations and finding the way even in the half-dark. The fundamental VDI guideline on design methodology, VDI 2221-1, discusses exactly this and seeks to protect practicing engineers from insufficiently thought-through, foolish or dangerous snap decisions through a methodical approach. As users of the guideline, we are to be forced, through a deliberate slowing down of the decision-making process, to analyze robustly in a first step and only later, and then for comprehensible reasons, to make a decision. The same purpose is served by core elements of management standards such as ISO 9001 or ISO 14001. From the social sciences come methods such as the delta analysis (Bizer/Führ 2014), which also pursue this goal. They all follow the same logic and are directly transferable to one another.

The relevant steps, based on VDI 2221-1 as well as Bizer and Führ, are presented in Table 2.3 (Bizer/Führ 2014). For practical application, it is essential to note that this is not a linear process of sequential execution, but that each step also has repercussions on previous steps and results. Thus, each individual work step can influence the preceding steps again. In particular, the definition of the goal must be adjusted with the process and the description of the current state must be expanded when new findings or new criteria emerge. The approach requires intellectual flexibility from all involved, goodwill, and often a negotiation process. Other representations of this methodology may name more or fewer steps, but are comparable in essence.

Further prerequisites for this approach are always:

- The group or organization proceeding in this way is honest with itself.
- All goals are clearly named or are artfully co-developed.
- The underlying values are conscious and named.
- The group or organization and its members are capable of learning, discourse, and consensus. This applies in particular to leadership as well.
- The process is open-ended.
- Goals and criteria are adjusted and optimized in the sense of a learning curve. Goals and criteria are not distorted in order to fabricate a preferred solution.

All notes and boundary conditions for wicked problems apply here (see the beginning of Chapter 2).

Table 2.3 The General Approach to Evidence-Based Solution Finding: Solutions Can Be Technical, Organizational, or Behavior-Based

No.	Step	Key question
1	Describe the initial situation	Where do we stand today? What are fixed, predetermined framework conditions and values? What competencies are already available?
2	Describe the goal	Where do we want or need to go? What should our solution be able to do, accomplish, or achieve?
3	Describe the current situation	Why is it the way it is today? What is not good about the current state and should therefore be changed? What is good about the current state and should remain as it is? Which criteria are important and informative enough to assess whether a solution idea or another situation or design is better than the current one?

No.	Step	Key question
4	Describe the difference between the current state and the goal	What needs to be done to achieve the goal? What prevents the actors from pursuing or achieving the goal? What functions must our suitable solution have in order to achieve the goal? How should people use a suitable solution?
5	Find possible solutions	Here, if possible, all relevant possible solutions should be found. At a minimum, however, relevant and fundamentally different options must be found in order to compare them with one another: Which approaches can achieve the goal? Which approaches change the situation in an appropriate way? Which solutions enable the required functions?
6	Evaluate the possible solutions based on the criteria	All possible solutions are compared and evaluated using the same criteria and without bias: Which goals are achieved by which solution? What are the inputs required for the individual solution options? What are the unintended effects of the individual solution options? Which solution, based on current knowledge, offers the best opportunity to achieve the goal without jeopardizing other values?
7	Implementation	The solution option selected in the previous step is implemented.
8	Evaluating the implementation	Accompanying the implementation, the change is to be evaluated: Is the solution as we planned it? Does it function appropriately? Are its other effects tolerable? Do we have undesired effects that we had not thought of beforehand? Do we need to correct or improve the solution? Do we need to abort and start over?
9	Learning	What feedback do the experiences gained have for the organization or society? Are the methods appropriate? Should we change them? Are the fundamental assumptions, including those about the goals, correct?

Making poor decisions

Wicked problems in particular lead to poor decisions in all areas of life. This has several reasons, which ultimately all stem from the effort required:

- There is no correct, unambiguous, or easily accessible solution, but only many more or less poor ways of dealing with the problem.
- It quickly becomes apparent that there is no simple solution, but that the problem requires particularly great attention or resources.
- Nobody already has the necessary resources, experience, knowledge, and competencies to adequately address the problem. Learning is compulsory.
- Every initial analysis already reveals that all proposed solutions have specific disadvantages. Different solution approaches may be disadvantageous for different groups.
- No solution delivers even a satisfactory answer to the more pressing problems right away; it will be necessary to weigh disadvantages against one another.
- Not all consequences of a decision are obvious. Not even after a thorough analysis.
- Therefore, gut decisions, group hysteria, belief systems, charlatans, or magicians can quickly prevail, as they all promise a simple solution without effort and without weighing of interests.

2.7 Decision Spaces

The examples and tasks in this book are designed as if we were allowed to develop a solution to the respective problems. This is not always the case in real life. In everyday life, this situation often presents itself differently. Engineers have a more or less clearly defined decision space within which they are expected to deliver good results. The art here is to make good use of one's own decision space:

- If I am assigned a clearly and completely defined task, it is expected that I make full use of the given decision space: anything else is an indication of decision weakness, lack of capability, or fear. Superiors notice this immediately, and this can close off any opportunity for more responsibility, higher pay, or advancement.
- Often, assigned tasks are not clearly defined. In this case, defining one's own decision space is already part of the necessary clarification of the task. On the one hand, there are boundary conditions that must be adhered to, or certain decisions that must be made by others. This must always be taken into account. On the other hand, the courage to interpret one's own decision space generously is often rewarded, as it demonstrates that I have the big picture in view. This may require either responding flexibly to decisions that were ultimately made elsewhere, or initiating decision-making processes in other places.
- Sometimes it may be necessary to also provide an assessment of problems within one's own organization or of solution approaches by other members of the organi-

zation that was not requested. This is difficult terrain and must be communicated with the necessary sensitivity and flexibility.

- *"Don't ask for permission, ask for forgiveness"* is a motto that in some situations is consciously associated with a personal risk. Asking for permission may take too long, or the answer may initially be fixed and would make one's own action, deemed necessary, impossible. This approach, when applied with the necessary sense of proportion, is a prerequisite for one's own proactive and shaping action.

As citizens, interest representatives, journalists, consultants, non-governmental organizations (NGOs), but also as employees of a company, we often find ourselves in the situation of having to define our own decision space for the benefit of society or the organization. In that case, it is one's own values, one's own competence and knowledge, and one's own motivation that determine this within the framework of the rules of society or the organization.

Attitude influences outcomes

In an organization, it is important to recognize which attitude is fundamentally compatible, which attitude dominates, and which attitudes are not desired. Attitude here refers to a fundamental set of values, expectations, belief systems, but also bases for action that influence decisions without having to be explicitly stated each time.

Classification of attitudes

The attitude with which the observance of planetary boundaries, the transformation of the energy system, or more specifically the handling of greenhouse gases is approached from the perspective of engineers can be classified. Of course, the transitions are fluid and an organization can have different attitudes in different areas (Seager et al. 2012).

Business as usual In this attitude, techno-optimism is at the forefront: the starting point of thinking is the assumption that there will always be a technical solution or that others will solve the problem. More specifically, techno-optimism can be understood as an attitude that already sees the core questions of this book as answered. For techno-optimists, there would always be enough resources, energy would appear when needed, entrepreneurs would deliver solutions at the right time and unconstrained by (natural) law limitations, and infinite growth would represent not a problem but the solution. Complex and wicked problems are in this worldview mostly reduced and simplified to the point where they appear as a purely technical development task for which others are then responsible. From this attitude, it is often already difficult to understand the concept of the wicked problem. Thus, no contribution to solving wicked problems is possible from this stance (Morozov 2013; DeLay 2024).

Systemic approach This is a pragmatic view in which wicked problems are recognized as such. This fundamentally allows interdisciplinary approaches to finding solutions. From this attitude, an approach follows in which previous goals continue to be pursued, but are additionally realized within a sustainability constraint. Financial aspects thus continue to be the drivers of development, with the boundary condition that additional regulatory requirements are now to be met by the sought solution. Green Growth falls into this category.

A somewhat more far-reaching second variant consists of methods within the framework of a Triple Bottom Line (as a three-pillar model: profit, people, planet). Within this framework, in addition to purely financial drivers, social aspects and aspects of interaction with the (natural) environment are also included. At its core, this involves a negotiation process between these three aspects, which can be structured, for example, in the form of a risk assessment. The result is products that are manufactured or used more efficiently or are less harmful with regard to their environmental impact, and which individually can indeed contribute to sustainable development goals, as they have less harmful environmental impacts. Scale effects, which are not taken into account in the analysis, can then reduce or even reverse this contribution. Even if Triple Bottom Line will not be sufficient to solve the major problems of sustainable development (see Chapter 8), it is nonetheless a common and attractive attitude for companies today, from which at least an attempt is made to provide one's own serious and relevant contribution.

Sustainable Engineering Science This approach is characterized by a fundamental skepticism toward purely technical solutions. This skepticism relates in particular to elaborate, material- or energy-intensive solution proposals. Transformation to sustainable development is fundamentally understood as a wicked problem, for which purely technical solutions can rarely exist. At the center of engagement with wicked problems are feedback effects and scale effects.

The methods in this book support a systemic approach and sustainable engineering science.

Where do I want to work?

I can choose the organization in which I want to work based on its attitude. Attempting to implement a fundamentally different attitude within an organization is very difficult, time-consuming, and often fails. In a company, it is the task of top management to develop the organization's attitude and then also enforce it internally (ISO 14001). I can help company management with this goal. However, a change in attitude cannot be enforced against management without at least considering a change in that management.

Compliance A typical and important example is the question of how an organization deals with legal requirements. The attitude toward this is understood by everyone within the organization, even if this attitude is not always clearly formulated in writing or the official statement on it may differ. Compliance means adhering to existing legal requirements. For organizations, compliance in this sense involves considerable effort, as the current status must be regularly researched. This must be communicated within the organization; when changes occur, procedures, processes, and products must be adapted and, if necessary, extensively modified, and compliance must be verified internally. Compliance is one of the essential elements of risk management within an organization. Compliance is essentially law-abidingness and thus self-evident, but this can be viewed and presented differently within an organization. It is then my personal decision whether I want to be part of such an organization and am willing to bear the associated personal risks.

2.8 Risk

Dealing with risks is a high art. In many areas of life, risks can be relevant or significant. Risks and thus the sensible and necessary approach differ for different areas or sectors. Companies must know their business risks and develop appropriate strategies for dealing with them. Insurance companies must additionally identify, assess, and quantify the risks of their customers. States must, within the framework of public services, recognize the risks of their citizens and determine the appropriate way to deal with them. In all cases, complex trade-offs between different types of information, legal interests, and claims are necessary.

Organizations ISO 31000 describes the method of risk assessment for risks affecting an organization. It is therefore the perspective of the organization looking outward and inward that is considered here. This standard provides a good introduction to methods for dealing with risks, and the concept of risk used there is relevant here.

Term risk

A helpful definition of the term risk according to ISO 31000 is as follows:

"Effect of uncertainty on objectives"

→ *An effect is a deviation from expectations – in a positive and/or negative direction*

→ *The objectives can cover various aspects (e.g., finance, health and safety, and environment) and apply at different levels (e.g., strategic, organization-wide, project-, product-, and process-related objectives)*

→ *Risks are often characterized by reference to potential events and consequences or a combination thereof. Risks are often described in terms of the consequences of an event (including changes in circumstances) in combination with the likelihood of its occurrence*

→ *Uncertainty is the state, even partial, of deficiency of information related to understanding or knowledge of an event, its consequence, or likelihood"*

A different perspective on the concept of risk and forms of risk reduction is brought by product safety, as fundamentally introduced in ISO 12100, already through the definition. Here, a risk is the

"combination of the probability of occurrence of harm and the severity of that harm."

This standard refers very specifically to the placing on the market of products for which the respective CE directives apply within the EU. The fundamental method for risk reduction for products is presented in ISO 12100. This concerns concrete physical risks to persons and, where applicable, to domestic animals that arise directly from products. Long-term, systemic, psychological, or generally societal risks are explicitly not covered.

Management The goal of risk management is to develop an orderly and systematic procedure for dealing with risks and then to apply it consistently. The resulting risk assessment and ultimately risk reduction aims to identify all relevant risks and to define measures by which a possible impact, in particular a possible harm, can be limited to an acceptable level.

A risk assessment is at least in the European legal system, in which the precautionary principle applies, fundamentally necessary before a measure is taken that may contain risks. As an important requirement for risk management and thus for risk assessment, ISO 31000 states:

- The goal is to protect or create value in all areas of life. Specifically mentioned are human health and safety, compliance with legal and regulatory requirements (compliance), public acceptance of measures or products, protection of the environment from environmental impacts, ensuring product quality, aspects of project management, and good governance.
- Risk management should be an integral part of all decision-making processes and part of the decision-making itself.
- Risk management explicitly addresses uncertainty, i.e., uncertainty must not be a reason to neglect an aspect, but is first and foremost an indicator of a relevant aspect.
- The method is systematic, structured, and timely, i.e., scientific. It is based on currently available information. It uses orderly procedures where possible. The purpose of the standard is thus to effectively prevent naivety, belief, intellectual laziness, entrenched thinking patterns, excessive optimism, and other obstacles to an evidence-based engagement with one's own objectives and uncertainties.
- Each organization must find its own procedures and structures that adequately cover its risk portfolio.

Time frame A fundamentally difficult question is the time frame over which risks are considered. Here, any legal requirements are only one aspect. A very long planning horizon is necessary especially for governments. For long-term companies in the area of raw material extraction or basic industries, planning horizons of 30 years are rather common. A cost-benefit perspective is often used (see Chapter 6). This implicitly uses a time frame, but does not answer the underlying question, merely shifting it.

Methods The general method for dealing with risks consists of the steps indicated in Table 2.4. Independently of this, the following considerations are relevant for establishing the criteria for the assessment of risks:

- Are the causes of change one-time or long-term, unexpected or systemic?

- What are the possible impacts of deviations on the organization or society? What is their duration and intensity?
- Are there ways to measure or otherwise quantify the causes and their effects?
- How high is the probability of occurrence? This assessment depends on the method used to capture the event, on the quality of the data available, and on the time frame within which the change can occur.
- Which method is used to determine the level of risk? Which aspects are taken into account? What obligations exist for the organization and which laws apply?
- Within what framework are risks tolerable or accepted for the organization or society? Which views of the affected parties must be taken into account?
- Which affected parties were considered and in what form?
- To what depth of possible combinations of events or probabilities should risks be taken into account?
- How is it ensured within society or the organization that arguments are evidence-based?

Connectivity A society or an organization cannot perceive all the risks that affect it. This follows from its own logics, ways of interaction, and communication channels. Not all topics are connectable within an organization, meaning they cannot actually be discussed within it (Luhmann 1986). Some topics are consciously or unconsciously ignored and thus represent blind spots.

Responses A society or organization always has several options for dealing with an identified and assessed risk, in particular:

- Understanding the risk as an opportunity: consciously accepting the risk in order to seize the opportunity
- Avoiding the risk by not starting or discontinuing the activity associated with the risk
- Voluntarily accepting the risk without critical discussion
- Ignoring the risk, since the level of risk (probability of occurrence, impact on the organization) has been classified as sufficiently low and tolerable
- Eliminating the cause of the risk itself or taking out insurance against it
- Taking measures that favorably influence the probability of occurrence
- Taking measures that ensure the impacts become significantly more favorable, i.e., for example, remain tolerable
- Sharing or distributing the risk through consolidation, insurance, contractual safeguards, etc.

Table 2.4 General Approach for Risk Management (Based on ISO 31000)

No.	Step	Task
0	Communication and consultation	Early engagement with the expectations, information, requirements, and perceptions of the interested parties of the organization: i.e., communication inward and outward The goal is to enable a truthful, relevant, accurate, and understandable exchange of information. For this purpose, background information required in all subsequent steps and persons with the relevant knowledge are involved.
1	Establishing the context	Defining the system boundary of the activity and its framework conditions, identifying the external and internal contexts This also concerns cultural and organizational framework conditions, contractual obligations and relationships. Furthermore, objectives and responsibilities as well as the scope of research activities and depth of information are to be defined.
2	Risk identification	Identifying risk sources, events, and developments that bring about changes, as well as affected areas, causes, and potential impacts, with the aim of drawing up a comprehensive list of risks In addition to the obvious and direct risks, this includes in particular follow-on effects, cascade and cumulative effects. With regard to impacts, all relevant points should be taken into consideration, even if the risk source or cause is not obvious. For this purpose, it may be necessary to work with scenarios in order to understand the impacts of more complex situations.
3	Risk analysis	The goal is to develop an understanding of each risk. This is the basis for the subsequent steps and for the development of strategies for risk treatment. For this purpose, the criteria developed previously (see preceding explanations) are applied. This process is iterative and can lead to a changing view of individual risks or risk groups. The result is the most robust and accurate description of the risks possible.
4	Risk evaluation	The goal of risk evaluation is to classify the risks with regard to urgency. Here it is prepared to what extent and with what priority the risk is then further treated. The focus is on the level of risk, the resources of the organization, legal requirements, the broader context of the risk, and its external effects.

No.	Step	Task
5	Risk treatment	In risk treatment, the steps are defined and implemented by which the organization influences the risks. This is a classic management activity that can be handled in accordance with the Plan-Do-Act-Control cycle according to ISO 9001. The starting point is the data from the risk evaluation. Measures are now developed and defined to treat the risks. The success of the measures must be evaluated on a regular basis. In the event of deviations from the target, new or different measures must be taken.
6	Monitoring and review	The fundamental process presented up to this point must be reviewed at regular intervals to determine whether it is being carried out as planned and whether it is sufficient in the intended form or needs to be adapted.
7	Documentation	Essential steps must be documented. There may be legal regulations for this.

References

Adam, B.: Tempering the not-yet. Towards a social theory for the Anthropocene. In: European Journal of Social Theory 27, 2024

Bizer, K./Führ, M.: Praktisches Vorgehen in der interdisziplinären Institutionenanalyse. Ein Kompaktleitfaden. sofia-Diskussionsbeiträge 14-7. Darmstadt 2014.*https://www.sofia-darmstadt.de/fileadmin/Dokumente/Diskussion/2014/Netzversion_Stufenheuristik.pdf*

Brandt, U./Wissem, M.: Imperiale Lebensweise. Oekom, München 2017

Bridle, J.: New dark age. Der Sieg der Technologie und das Ende der Zukunft. C. H. Beck, München 2019

Conway, E. M./Oreskes, N.: Merchants of Doubt. How a Handful of Scientists Obscured the Truth on Issues from Tobacco Smoke to Global Warming. Bloomsbury 2012

Dean, C.: Am I Making Myself Clear? A Scientist's Guide to Talking to the Public. Harvard University Press 2012

DeLay, T.: Future of denial. The ideologies of climate change. Verso 2024

DIN EN ISO 9001: Qualitätsmanagementsysteme – Anforderungen

DIN EN ISO 14001: Umweltmanagementsysteme – Anforderungen mit Anleitung zur Anwendung. **Hinweis:** Die ISO-14000-Serie wendet sich an Unternehmensführungen oder Manager. Sie soll Hilfestellungen bieten, wie Umweltschutz von oben in eine Organisation eingebaut werden kann. Einen Überblick gibt *http://www.iso.org/iso/theiso14000family_2009.pdf* (siehe auch die Einführung unter *http://www.iso.org/iso/introduction_to_iso_14001.pdf*)

DIN EN 31000: Risikomanagement – Leitlinien

Frankfurt, H. G.: Bullshit. Suhrkamp, Berlin 2020

Ghosh, A.: Die große Verblendung. Der Klimawandel als das Undenkbare. Blessing 2017

Luhmann, N.: Ökologische Kommunikation. Kann die moderne Gesellschaft sich auf ökologische Gefährdungen einstellen? Verlag für Sozialwissenschaften 1986

Mastrandrea, M. D./Field, C. B./Stocker, T. F./Edenhofer, O./Ebi, K. L./Frame, D. J./Held, H./Kriegler, E./ Mach, K. J./Matschoss, P. R./Plattner, G.-K./Yohe, G. W./Zwiers, F. W.: Guidance Note for Lead Authors of the IPCC Fifth Assessment Report on Consistent Treatment of Uncertainties. Intergovernmental Panel on Climate Change (IPCC), Geneva 2010

Morozov, E.: To save everything, click here. The folly of technological solutionism. Public Affairs 2013

Rittel, H. W. J./Webber, M. M.: Dilemmas in a General Theory of Planning. In: Policy Sciences, 4, 1973, 155–169

Seager, T./Selinger, E./Wiek, A.: Sustainable Engineering Science for Resolving Wicked Problems. In: J. Agric. Environ. Ethics 25, 2012, pp. 467–484

Törnberg, P./Chueri, J.: When do parties lie? Misinformation and radical-right populism across 26 countries. In: The International Journal of Press/Politics 2025

Townsen Hicks, M./Humphries, J./Slater, J.: ChatGPT is bullshit. In: Ethics and Information Technology 26, 2024

VDI – Verein Deutscher Ingenieure: Ethische Grundsätze des Ingenieurberufs. Düsseldorf 2021

VDI 2221-1:2018: Entwicklung technischer Produkte und Systeme – Modell der Produktentwicklung

VDI 2221-2:2018: Entwicklung technischer Produkte und Systeme – Gestaltung individueller Produktentwicklungsprozesse

Zalasiewicz, J./Thomas, J. A./Waters, C. N./Turner, S./Head, M. J.: What should the Anthropocene mean? In: Nature 632, 2024, pp. 980–984

Zhou, L./Schellaert, W./Martínez-Plumed, F./Moros-Daval, Y./Ferri1, C./Hernández-Orallo, J.: Larger and more instructable language models become less reliable. In: Nature 634, 2024, pp. 61–68

3 Energy Fundamentals

The topics discussed in the following require from us a certain set of concepts, definitions, and formulas from thermodynamics. At the center are an understanding of energy and its conversion, unavoidable losses in energy conversion, and dissipation. The most important points are summarized in this chapter. For details, derivations, and further depth, reference is made to my book *Applied Technical Thermodynamics* (Linow 2022).

3.1 Systems

The concept of the system and the resulting methods of description and analysis are a central core of the thermodynamic method. The systemic perspective runs through the methods presented here and extends well beyond thermodynamics. The system is the domain that is to be examined in more detail. This can involve very simple or extremely complex objects. In thermodynamics, systems are defined as simply as possible for good reason. This is quite different, for example, in life cycle assessment (LCA) and comparable methods. There, the goal is to appropriately link many individually interacting components with one another and to determine meaningful connections and contributions. The system can become impressively complex.

System Methodically, a fundamental distinction is first made between the analyzed system and its environment. The system can then either be examined in isolation, or all interactions between the system as defined and its environment are included in the further analysis. The system under consideration is always the domain that is specifically examined and thus given particular attention. It is delimited from the surroundings by a defined and closed system boundary. A system can be broken down into further subsystems. Individual, spatially non-contiguous objects can together form a system. These may have formed a common system at an earlier point in time or may merge at a later point in time.

System boundary The system is separated from its environment by a system boundary. This system boundary encloses the system in an unambiguous manner:

- The system boundary is a completely closed surface, i.e., there are no back exits through which something can disappear.
- Objects can be cut by the system boundary, but no object can be part of the system boundary.
- System boundaries can change over time.
- Surfaces of objects are well suited as system boundaries, but this need not be the case.

Balance boundary The system boundary is a balance boundary, i.e., if something crosses the system boundary, it must be counted. In thermodynamics, these are matter, energy, and entropy. Time-dependent balances of all flows for the system that cross the system boundary are established — typically as a mass balance, a molar balance, or an energy balance. The sign convention used here is that every flow into the system is counted as positive and every flow out of the system is counted as negative, regardless of which physical quantity is being considered. The system boundary as a balance boundary makes it possible to account for all flows that cross the system boundary. In this way, all contributions are completely compiled and their influence on the system can be analyzed.

State A system can be described unambiguously by its state: the state is defined by state variables such as pressure p, temperature T, volume V, mass m, and amount of substance n. For simple homogeneous systems, two or a few state variables are sufficient for a complete description; for very complex systems, we can always capture only individual aspects through state variables, even if we decompose the system into simpler subsystems. In Section 3.4 tools are identified with which we can describe simple systems.

A system can be in thermodynamic equilibrium. In this state, energy is uniformly distributed throughout the system, the entropy of the system is thus at its maximum, and no work can be performed within the system. The system has reached the endpoint of its thermodynamic evolution. It is dead. Simple systems that serve as building blocks of thermodynamics are often in this state. Figure 3.1 illustrates the evolution of an isolated system in which energy becomes ever more uniformly distributed throughout the system. The living Earth is far from equilibrium.

Flow system Systems that are in exchange with their surroundings and that can release not only energy or mass but also entropy to the surroundings are flow systems. A special case is represented by steady-state flow processes, in which relevant state variables of the system under consideration remain constant while energy or mass simultaneously flows through the system. Complex flow systems are far from thermodynamic equilibrium: the entropy is not at its maximum, the energy is not evenly distributed, and a multitude of processes can take place within the system. Typical examples are the earth or a living organism.

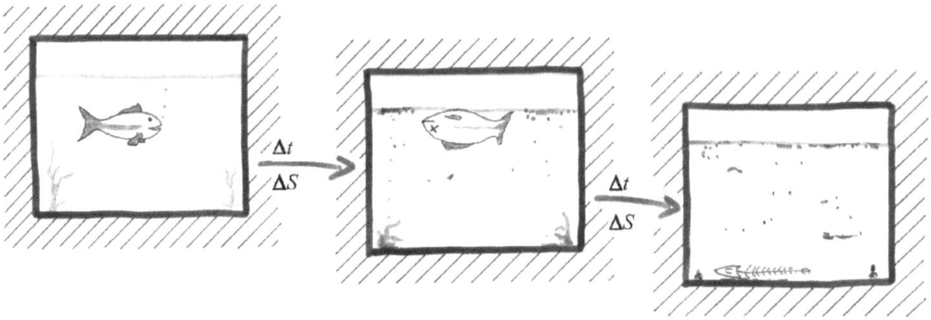

Figure 3.1 Time evolution of a closed system: over time it transitions into an equilibrium state in which the entropy *S* is at its maximum.

State Variables

The most important state variables, with which we describe simple systems, are the following:

- Pressure p and temperature T of the system: in large systems both can vary across the volume of the system – for many calculations, however, it is useful to limit the size of the system so that both are homogeneous
- The mass m and the volume V of the system
- The internal energy U, the enthalpy H and the entropy S
- Material properties such as specific heat capacity, viscosity, and many others

Specific state variables are often used. These are the ratio of the extensive state variable and the mass

$$v = \frac{V}{m}$$

or

$$h = \frac{H}{m}$$

The specific volume v is the volume occupied by 1 kg and the specific enthalpy h is the enthalpy of 1 kg of the system.

3.2 Energy and Changes of State

We want to describe how energy flows through systems, what effects energy has on these systems, how the conversion of energy from one form to another takes place and what losses are associated with this. The core of the description is the introduction of an energy of the system as a balance quantity: transferred heat Q and work performed

W change this energy. A positive sign consistently means that the system has absorbed energy, and a negative sign means that it has released energy. Transferred heat and work performed are then process quantities: they have the unit of an energy, but describe a change – the process that changes the state of the system and its surroundings.

First we look at the state variables, i.e. how energy can be present in the system. We then describe the two energetic process quantities.

3.2.1 Energetic State Variables

Energy can be contained in the system as an external form of energy:

Kinetic energy This is the energy that a system of mass m has due to its motion at velocity c relative to a reference system:

$$E_{kin} = m \cdot \frac{c^2}{2}$$

The kinetic energy thus depends on the reference system. Often the reference is the earth's surface.

Potential energy This is the energy that a mass has due to its difference in height Δz relative to a reference system in the gravitational field of the earth:

$$E_{pot} = m \cdot g \cdot \Delta z$$

The potential energy also depends on the reference height. This can be sea level.

A second group consists of forms of energy that become available through reactions:

Chemical energy When a system changes through chemical reactions taking place within it, the energy released or absorbed in the process can be exchanged between the system and its surroundings. Typical is the chemical energy of a fuel, which is released by combustion with air and transferred as heat to the surroundings. The chemical energy contained in a fuel is described by the heating value (more on this in Chapter 9).

Nuclear energy When the reactions take place in atomic nuclei, this is nuclear energy (more on this also in Chapter 9).

Electrical energy Electric and magnetic fields or electric charges can store energy. This can be expressed, e.g., by the voltage U or the charge C. Included here are batteries in which the charge is stored as ions.

When we say energy, we mean ...

The concept of energy is often used rather loosely in discussions:

- At a first level, the concept of energy as a state variable is mixed with the process variables work and heat. The clean separation between these concepts is relevant in order to be able to talk or think thermodynamically about systems, energy and in particular the exchange of energy between systems through processes. Unambigouosly separating these three concepts is a learning objective of thermodynamics.

- Sometimes potentials such as voltage or pressure are also referred to as energy. Potentials are important independent concepts, because they enable energy to flow (via electrical voltage, via pressure differences, etc.). They are not themselves energy.
- In everyday language, the physical meaning of the term energy is sometimes barely discernible, and the transition into magical thinking is reflected linguistically.

If we want to think reliably about the transformation of our energy system, we need clearly defined terms so that we can think with certainty and so that others can understand us well.

The third group comprises the two energetic balance quantities – the energy of the system in the narrower sense:

Internal energy This balance quantity describes the energy content of a system solely as a function of its absolute temperature $U = U(T)$. For homogeneous substances, the internal energy can be determined via the specific heat capacity $c_v(T)$ and taking phase transitions into account.

The zero point of the internal energy can be chosen freely. This zero point can be the zero point of the absolute temperature. For many substances and questions there are common conventions. For the determination of the internal energy in heterogeneous systems, a precise knowledge of the composition and the temperature distribution in the system is necessary. Heterogeneous systems are those in which different phases or substances are present simultaneously. The knowledge required for the determination of the internal energy is often only obtained for chemically simple systems of homogeneous temperature.

Enthalpy The sum of internal energy and the product of pressure and volume of the system is the enthalpy:

$$H = U(T) + p \cdot V$$

For homogeneous substances, the enthalpy can be determined via the specific heat capacity $c_p(T)$ and taking phase transitions into account. The reference point of the enthalpy can be chosen freely and is oriented to that of the internal energy.

The product of pressure and volume is the displacement work. This describes the work required to displace the atmosphere so that space exists for the system: ultimately, the atmosphere must be shifted upward in accordance with the volume of the system in order to place the system under consideration. For this, the additional energy $p \cdot V$ is required.

3.2.2 Energy Conservation and Process Variables

Transferred heat Q and work performed W change a system through processes. A change of state occurs. The first law of thermodynamics states that energy is a conserved quantity: it can be converted from one form to another and it can be transported, but it cannot disappear and it cannot be created from nothing. This law is one of the foundations of the scientific description of the world. It follows that any description of the world in which energy behaves differently is a magical theory.

Energy Conversion

All forms of energy and process quantities can be converted into one another. To obtain kinetic or potential energy, work must be performed. Chemical and nuclear energy are results of processes in which heat was transferred at the macroscopic level. Electricity is a form of work.

Work can, at least theoretically, be completely converted into another form of work. Work can be completely converted into heat, either as an intended process, e.g. in electric heating, or as a loss of work. In practice, losses always occur when converting from one form of work to another, even in technical processes. Relevant technical losses are the following:

- First, there is the friction of flowing fluids on walls in pipes, in units, compressors, turbines and when flowing around vehicles. Here, friction is closely related to the viscosity of the fluid and the velocity gradient near the surface.
- Mechanical friction in bearings is, in the intended case, friction in the lubricant, not direct friction between the shaft and the bearing.
- In addition, there is the direct electrical resistance due to the flowing current, as well as indirect losses in which the electric field of a current interacts with matter in the vicinity of the conductor and, for example, induces currents there.

In this process, part of the work is converted into heat by friction. The loss of work is equal to the waste heat generated in this way. This is dissipation, i.e. dispersion, because the actually well-concentrated usable energy is dispersed into the surroundings by dissipation. Energy dispersed in this way is lost as work for the actual process.

Heat can also never be completely converted into work, even theoretically. Part of the heat is lost as unusable waste heat. This is a consequence of the second law of thermodynamics (section 3.3). Only because of this are friction losses and dissipation actual losses of usable work.

The energetic process quantities and the energetic state quantities are linked by the first law. Any change in the energy of the system can only occur either through mass transport across the system boundary or, if we can exclude this, through the process quantities work and heat. The internal energy U changes from state 1 to state 2 via the transferred heat Q and the performed volume work:

$$\Delta U_{12} = Q_{12} + W_{v,12} \tag{3.1}$$

The enthalpy H of the system changes via the transferred heat and the performed pressure work:

$$\Delta H_{12} = Q_{12} + W_{p,12} \tag{3.2}$$

The process quantities effect a change in the system via the change in internal energy ΔU_{12} and the enthalpy ΔH_{12}. We need two ways to describe work. On the one hand, there is the volume work, which causes a change in the closed system when the volume of the system changes:

$$W_{V,12} = -\int_1^2 p(V) \cdot dV = -m \cdot \int_1^2 p(v) \cdot dv \tag{3.3}$$

On the other hand, there is the pressure work in open systems:

$$W_{p,12} = \int_1^2 V(p) \cdot dp = m \cdot \int_1^2 v(p) \cdot dp \tag{3.4}$$

The pressure work can be extended by contributions of kinetic and potential energy. This is then the technical work:

$$W_{tech} = W_p + W_{kin} + W_{pot} \tag{3.5}$$

Volume work is intuitive for piston machines; otherwise, enthalpy and technical work dominate in applications (Linow 2022).

3.2.3 Heat

Heat is energy that flows due to a temperature difference. Heat flows only from a system of higher temperature to a system of lower temperature.

Sensible heat If only the temperature of a system changes, then the following applies

$$Q_{12} = m_{system} \cdot c_{system} \cdot \Delta T_{12} \tag{3.6}$$

The transferred heat results in a measurable temperature change. Therefore, this is sensible heat. The material property that describes how strongly the temperature changes with the transferred heat is the specific heat capacity c_{system}. This heat capacity takes different values depending on the form of the change of state. In most technically relevant cases, the pressure of the system does not change during heating. Then the specific heat capacity at constant pressure is used:

$$Q_{12}\big|_{p=const} = m \cdot c\big|_{p=const} \cdot \Delta T_{12} = m \cdot c_p \cdot \Delta T_{12} \tag{3.7}$$

Latent heat If the state of aggregation of the system changes with the transferred heat, the temperature remains constant during the phase transition,

$$Q_{12} = m \cdot \Delta h_{phase} \tag{3.8}$$

where Δh_{phase} denotes the specific melting or vaporization enthalpy. Phase transitions can also occur in solids. In that case, the crystal structure transforms.

Another form of latent heat is chemical energy. Heat is stored in the system by chemical reactions taking place there that require energy (heat): the system thus absorbs heat. Through the chemical reactions, substances are formed that store this energy. Later, the energy chemically stored in the system can be released again through chemical reactions.

Most phase transitions proceed spontaneously in both directions when the temperature and pressure are appropriate and heat can be absorbed or released. For chemical reactions, this is more complex and depends on many individual details. Thus, there are chemical reactions that proceed spontaneously and reversibly, such as reactions in chemical solutions. Other chemical reactions require elaborate technical installations, such as the production of hydrogen or synthetic natural gas. This is explored in more depth in Chapter 9 and Chapter 10.

3.2.4 Flow Processes

Work performed and heat transferred are finite quantities. Often, however, we want to describe flow processes in which a heat flow $\dot{Q}$ flows over a long period of time or a work stream is performed (Figure 3.2). The power P denotes the current and time-dependent flow of work, i.e., the work flowing across the system boundary per unit time:

$$P(t) = \frac{dW_p}{dt} \tag{3.9}$$

The heat flow is correspondingly

$$\dot{Q}(t) = \frac{dQ}{dt} \tag{3.10}$$

Conversely, work performed is a power integrated over a period of time:

$$W_{p,12} = \int_1^2 P(t) \cdot dt \tag{3.11}$$

Heat transferred is a heat flow integrated over a period of time:

$$Q_{12} = \int_1^2 \dot{Q}(t) \cdot dt \tag{3.12}$$

Power always refers to pressure work or, more generally, to technical work.

Work or heat thus accumulate over a period of time. It therefore makes sense to specify a time period or process in addition to the work or heat. Conversely, power or heat flow can change over time. It is then important to also specify whether the system is stationary, i.e., power and heat flow are constant, or to specify the point in time at which the power or heat flow occurs. If power or heat flow do not change with time, this is a stationary process. In this case,

$$P = \frac{W_{12}}{\Delta t_{12}} \text{ and } \dot{Q} = \frac{Q_{12}}{\Delta t_{12}} \tag{3.13}$$

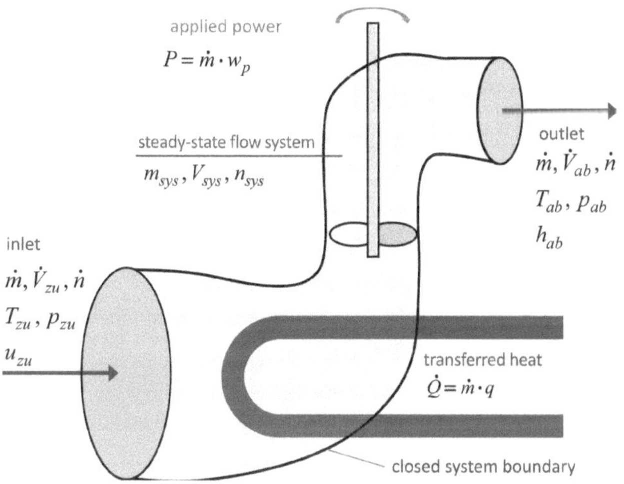

Figure 3.2 Transferred heat flow, power performed, and the change in state variables in a stationary flow process (source: Linow 2022)

Pump as a system

An electrically driven pump performs work by pumping water into an elevated tank. Potential energy is ultimately supplied to the water, and the amount of water in the tank is a measure of the work the pump has performed over the time it has been pumping water into the tank. Power is the current flow of work that moves the current mass flow of water. The system boundary is decisive for the value of the work:

- If the water is the system, the work supplied to the system corresponds to the change in the potential energy of the water.

- If the pump is the system, the work transferred from the drive shaft into the pump is greater than the work absorbed by the water due to friction losses in the pump. Dissipation occurs in the pump and in the water.
- If the electric drive is additionally included, further conversion losses must be taken into account.

Work and power of galley slaves

The concept of the energy slave is found in the literature and serves to illustrate how much work we have machines do for us today and what value this work could have. In the following example, the term slave is taken literally, making the direct comparison between engine and human labor directly tangible.

The galley was at the time often a death sentence, in which the condemned had to suffer for several months before his death. The slave was chained to the galley and thrown overboard upon his death or final exhaustion. A rowing slave requires at least 10 MJ of food per day to function, and in reality significantly more. He can then deliver a typical power of 250 W during a sprint over 10 min and a continuous power of 100 W over 8 hours. A typical galley is manned by 144 rowers.

What is the work performed in a sprint?

The work of a sprint is

$$W_{sprint} = N_{rower} \cdot P_{sprint} \cdot \Delta t = 144 \cdot 250 \text{ W} \cdot 600 \text{ s} = 21.60 \text{ MJ}$$

The work of a day's labor is

$$W_{day} = N_{rower} \cdot P_{rower} \cdot \Delta t = 144 \cdot 100 \text{ W} \cdot 8 \text{ h} = 115.2 \text{ kWh} = 414.7 \text{ MJ}$$

What is the thermal efficiency of the rowers?

First, the daily energy consumption of the rowers is needed

$$Q_{day} = N_{rower} \cdot Q_{person} = 144 \cdot 10 \text{ MJ} = 1440 \text{ MJ}$$

and thus the energy efficiency related to the day's work of

$$\eta_{rower} = \frac{W_{day}}{Q_{day}} = \frac{414.7 \text{ MJ}}{1440 \text{ MJ}} = 0.2880$$

Thus, the efficiency of the rowers is comparable to that of internal combustion engines.

What fuel requirement does an outboard motor have to replace the performance of the rowers?

The power requirement for the galley can today be met by a commercially available outboard motor. Many manufacturers offer suitable motors that can also deliver the sprint power. We assume the same efficiency as for the rowers: this is a realistic estimate for such a motor. The gasoline required at an efficiency η_{motor} = 0.2880 amounts to the work of a day's work of 414.7 MJ

$$Q_{gasoline} = \frac{W_{day}}{\eta_{motor}} = \frac{414.7\ \text{MJ}}{0.2880} = 1440\ \text{MJ}$$

and thus

$$V_{gasoline} = \frac{m_{gasoline}}{\rho_{gasoline}} = \frac{Q_{gasoline}}{H_{I,V,gasoline}} = \frac{1440\ \text{MJ}}{30.5\ \frac{\text{MJ}}{\text{l}}} = 47.2\ \text{l}$$

What is the daily food requirement of the 144 rowers?

The typical calorific value of wheat as a food is 1260 kJ per 100 g.

The daily grain quantity is calculated as

$$m_{grain} = \frac{Q_{day}}{H_{S,grain}} = \frac{1440\ \text{MJ}}{12.60\ \frac{\text{MJ}}{\text{kg}}} = 114.3\ \text{kg}$$

What agricultural area is required?

How large is the area for wheat cultivation to feed the rowers or to operate the motor? A good yield of wheat under sustainable farming is 3000 kg ha^{-1}. A typical yield when converting wheat into fuel under sustainable agriculture is 670 l ha^{-1}. Since no time period was specified, figures per day or per year are meaningful. The area requirement for grain to supply the rowers is per day

$$A_{grain} = \frac{m_{grain}}{q_{grain}} = \frac{114.3\ \text{kg}}{0.300\ \frac{\text{kg}}{\text{m}^2}} = 381\ \text{m}^2$$

The area requirement for gasoline to supply the motor amounts to per day's work

$$A_{gasoline} = \frac{V_{gasoline}}{q_{gasoline}} = \frac{47.2\ \text{l}}{0.067\ \frac{\text{kg}}{\text{m}^2}} = 705\ \text{m}^2$$

When switching from human labor to biofuels for propelling the galley, the demand for agricultural land increases by approximately a factor of 2. The difference becomes significantly more favorable for the motor if the boat is essentially not in use, since the motor only consumes fuel when operating, whereas the slaves consume regardless of their work.

The value of the daily work of 144 rowing slaves, depending on the current gasoline price, is today approximately €100. This is roughly the value of pure physical labor today. Only through special skill or craftsmanship does human labor become valuable. The task illustrates how easily we can replace impressive human labor with small machines, and conversely, what a large number of people (or draft animals) we would have to employ if modern engine technology were not available.

3.3 Entropy and the Second Law

The second law is the technical core of thermodynamics. Every scientific description of processes or events must, in order to have relevance for technical or natural systems, always be consistent with the second law. The second law and the closely related state variable of entropy are not easy to understand.

One formulation of the second law is that the entropy S in a closed system either remains constant or increases until it reaches a maximum. From the perspective of engineers, this makes the law manageable, since it concerns the use and interpretation of a calculable state variable: entropy S is a non-conserved state variable of a system. It cannot decrease in a closed system:

- Entropy can remain constant in a change of state. These are ideal reversible processes. Here, reversible means that the process can be completely reversed at any time, so that the system and its surroundings return to their initial state. Such changes of state cannot be observed. They are only important as a theoretical limiting case.
- In all natural and technical systems, the result of a change of state is always a permanent change in the system or in its surroundings. This also applies when a process has been completely reversed. This is accompanied by an increase in entropy in the system or the surroundings in which a change remains.

heat transport The change in entropy can be attributed to certain processes. Entropy is generated when heat Q flows across a system boundary in a process:

$$\Delta S_{Q,12} = \int_1^2 \frac{dQ}{T} \tag{3.14}$$

This equation includes heat losses of the system as well as the mixing of two streams of a substance that have different temperatures. Radiation is a form of heat flow. Therefore, radiation also transports entropy. Important relationships can be found in Section 3.6 on radiation.

Dissipation Entropy is generated during the dissipation of work. Dissipation is the irreversible conversion of work into heat, e.g. by friction:

$$\Delta S_{diss,12} = \int_1^2 \frac{dW_{diss}}{T} \tag{3.15}$$

Here the dissipated work is exactly the part of the work that is lost. Typical examples are pressure losses in pipes and bearing friction.

Mixing Entropy is generated when substances are mixed:

$$\Delta S_{mix} = -R \cdot \sum_i n_i \cdot \ln y_i \tag{3.16}$$

Here $R = 8314\ \mathrm{J\ kmol^{-1}\ K^{-1}}$ is the universal gas constant, n_i the amount of substance of component i and y_i the mole fraction of substance i of the total amount of substance n_{mix}. For the mole fractions the following applies

$$\sum_i y_i = \sum_i \frac{n_i}{n_{mix}} = 1 \tag{3.17}$$

The remarkable thing about the sum in Equation 3.16 is that it is dominated by the small fractions. The specific mixing entropy is

$$\Delta s_{mix} = \frac{\Delta S_{mix}}{m_{mix}} = -\frac{R}{n_{mix} \cdot M_{mix}} \cdot \sum_i n_i \cdot \ln y_i = -R_{mix} \cdot \sum_i y_i \cdot \ln y_i \quad (3.18)$$

Here R_{mix} is the specific gas constant of the mixture. Figure 3.3 illustrates the development of the contributions. The product $y_i \cdot \ln y_i$ is more strongly dominated by the logarithm. As a result, although the entropy of mixing for a binary system as in Figure 3.3 is maximum when two equal molar fractions are mixed together, the sum is quickly and significantly dominated by the smaller fraction when the molar fractions are unequal. It is therefore sufficient, when a small fraction is added to a very large reservoir, to calculate this small fraction. The question of what exactly the large reservoir is cannot always be answered unambiguously.

Material property Entropy is a state variable, i.e., it is transported across a system boundary by means of mass. Every substance has a specific entropy s_m that depends on temperature and phase state.

What is entropy from a macroscopic perspective?

Entropy is macroscopically seen a measure of how concentrated energy is in the system and thus how available it is for use. The application of entropy to purely macroscopically described systems requires its own interpretation, which differs from the stochastic microscopic interpretation.

If the system has low entropy, there is still a potential within the system to perform work. As entropy increases, the potential for work decreases accordingly. If the energy in the system is completely and uniformly dispersed and thus every potential for work is exhausted, the system has reached its maximum entropy: in thermodynamic equilibrium, the energy is dispersed throughout the system such that no more work can be performed within it. It is dead (Figure 3.1).

A system far from thermodynamic equilibrium can perform work. Differences in potentials such as pressure, chemical composition, or temperature serve as prerequisites. The work performed is converted into heat through losses or due to the process itself. Over time, in a closed system, all energetic possibilities for work are depleted and dissipated through the performance of work. In this process, entropy is generated and the entropy in the system increases. Thus, the entropy generated in a process is a measure of the work irreversibly dissipated in that process. This is summarized by the Gouy Stodola theorem :

$$W_{diss} = T_0 \cdot \sum_i \Delta S_i \quad (3.19)$$

Here the dissipated work is the product of the ambient temperature T_0 and the sum over the total entropy generated in all processes i.

This equation can also be used to determine the minimum work required for a process. As an example, the minimum work for separating an impurity from a material stream is exactly the work dissipated during the corresponding mixing. However, separation processes in practice typically require considerably more work than this minimum.

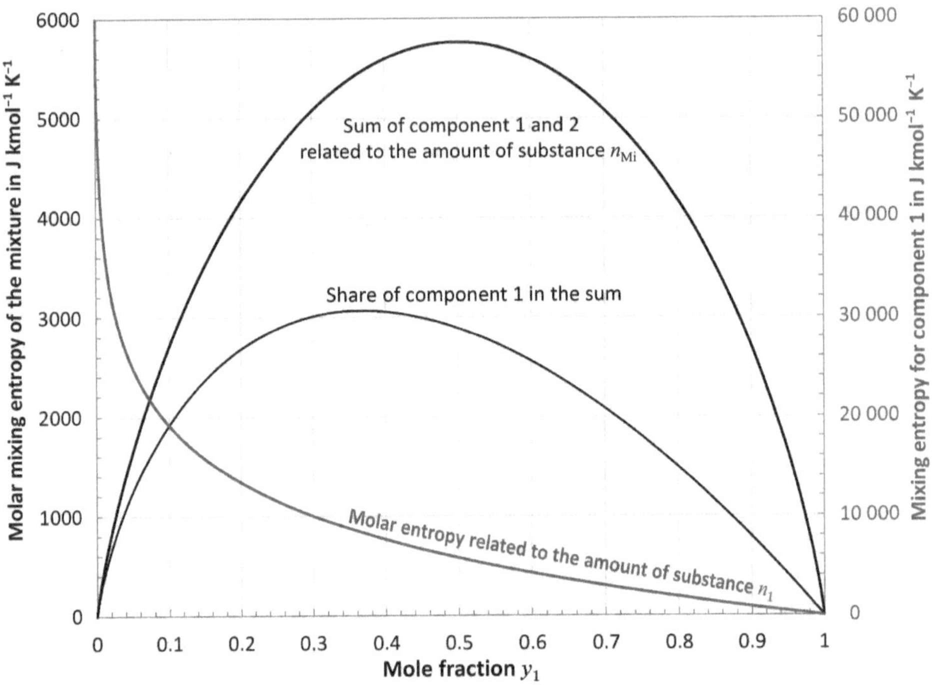

Figure 3.3 Influence of molar fractions on the terms in Equation 3.18

3.4 Describing Fluids

For many estimates or calculations it is necessary to describe the substances or material streams involved by their specific material properties. Often a macroscopic description of the substances involved is sufficient. Figure 3.4 shows how the phases of a homogeneous substance are arranged. Carbon dioxide is used as an example (and this diagram is used in Chapter 14).

For a technical description, the phases in which substances can exist are relevant, since each phase has particular properties and different phases are described using different models. In thermodynamics and fluid dynamics, fluids are the primary focus. Fluids include gas, liquid, supercritical fluid, and plasma. The specific models discussed below are each valid only in small regions of this diagram (Figure 3.4)

- The ideal gas is found in the lower right area of the diagram, far away from all phase boundaries, at low pressure and high temperature. The limit up to which this model applies depends on the accuracy that the calculation is intended to yield. It is possible to make estimates even into the supercritical region.
- Liquid carbon dioxide behaves almost like an ideal liquid at low temperature.
- For the other regions, diagrams or software are then needed in order to extract the state values from the models available for many pure substances (Linow 2022).

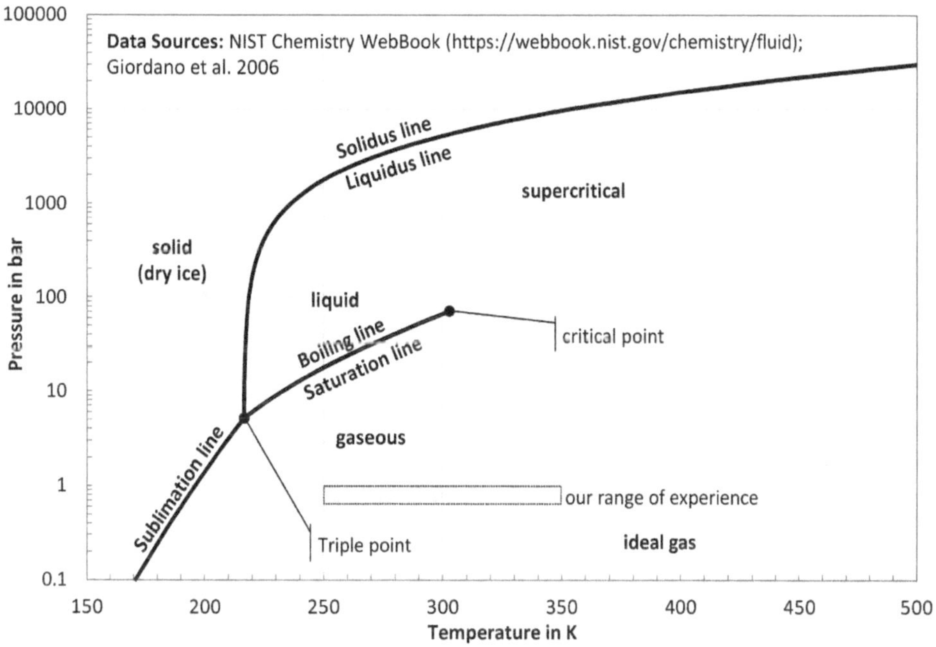

Figure 3.4 The phases of carbon dioxide (data sources: NIST Chemistry WebBook, Giordano et al. 2006)

Amount of Substance

Substances are composed at the microscopic level of atoms or molecules. For the description of chemical reactions, focusing on this level is necessary. The connection between the macroscopic view and chemical reactions is made via the amount of substance n. The amount of substance indicates the number of particles in a system:

- In a gas the gas particles are present as stable molecules. Only at very high temperatures, such as those observed in a flame or in a plasma, does a breaking apart of molecules occur and significant proportions of reactive radicals or atoms can be present. In the gas, the amount of substance thus refers to molecules. Exceptions to this rule are the noble gases, as these do not form molecules. They remain atoms.

- In technically relevant liquids the particles are mostly present as stable molecules. In liquid metals they are atoms.
- Solids can often be better described atomically. This applies to all crystalline materials such as metals and ceramics. Most glasses are also still well described atomically, but can already be represented via their oxides. Important examples of a molecular solid are water and carbon dioxide, which as ice remain a crystal of molecules.

The unit of the amount of substance is the mole and it denotes a number of $N_A = 6.022 \cdot 10^{23}$ pieces. This is the Avogadro constant. This constant is defined by the number of ^{12}C carbon atoms that together yield a mass of 12 g. The mass of a homogeneous substance is directly related to its amount of substance, with the number of particles N.

$$m(\text{substance}) = N \cdot m(\text{particle}) \tag{3.20}$$

Since the absolute number of particles N is unwieldily large and the mass of a single particle is unwieldily small, we instead calculate with the amount of substance and the mass that one mole of the particles has, the molar mass M. Thus

$$m = \frac{N}{N_A} \cdot M = n \cdot M$$

Chemical reaction equations refer to the specified number of atoms and molecules, but they can also be read as equations in terms of amounts of substance. In that case, the numbers in a reaction equation are, for example, 1 mol or a multiple thereof.

Ideal Gas

For many calculations, gases can be treated as an ideal gas This applies in particular to air, but also to many gaseous fuels such as natural gas at sufficiently high temperature and low pressure, as well as hydrogen or carbon dioxide. A gas behaves ideally when the equation of state of the ideal gas applies:

$$p \cdot V = m \cdot R_i \cdot T \tag{3.21}$$

In this equation of state of the ideal gas, R_i is the specific gas constant of the gas. This contains the properties of a specific gas. It is a substance-specific quantity that can be found in tables (e.g., Table 17.2 in Section 17.2). In molar quantities (amount of substance), the equation of state of the ideal gas reads

$$p \cdot V = m \cdot R_i \cdot T = n \cdot \underbrace{M_i \cdot R_i}_{=R} \cdot T = n \cdot R \cdot T \tag{3.22}$$

The universal gas constant $R = 8.314\ \text{kJ kmol}^{-1}\ \text{K}^{-1}$ is independent of the substance. This means that at the same pressure and the same temperature, any ideal gas contains the same number of particles in a given volume. Thus, for each substance i the specific gas constant R_i can be determined easily:

$$R_i = \frac{R}{M_i} \tag{3.23}$$

Figure 3.5 shows the real gas behavior of some gases. The compressibility factor is a measure of how accurately the ideal gas approximation is applicable. Carbon dioxide illustrates how the ideal gas approximation becomes more accurate at higher temperatures.

Gases at standard conditions

The volume of gases is not suitable for specifying their properties or for trading quantities of fuel gases. In some cases, masses or mass flow rates are therefore used for gases as well. For natural gas and some other gases, however, it is common to use the volume at standard conditions. This is particularly helpful when calculations are to be performed in terms of volume in all cases, e.g., for turbomachinery, ventilation ducts, or pressure losses. The standard condition is defined as T_N = 273.15 K (= 0 °C) and 101 325 Pa.

The standard condition is the legally defined reference condition for many applications. For ideal gases, any volume can be converted to standard conditions:

$$V_N = V(p,T) \cdot \frac{T_N}{T} \cdot \frac{p}{p_N} \tag{3.24}$$

Conversely, the volume at other pressure and temperature can be determined from the volume at standard conditions:

$$V(p,T) = V_N \cdot \frac{T}{T_N} \cdot \frac{p_N}{p} \tag{3.25}$$

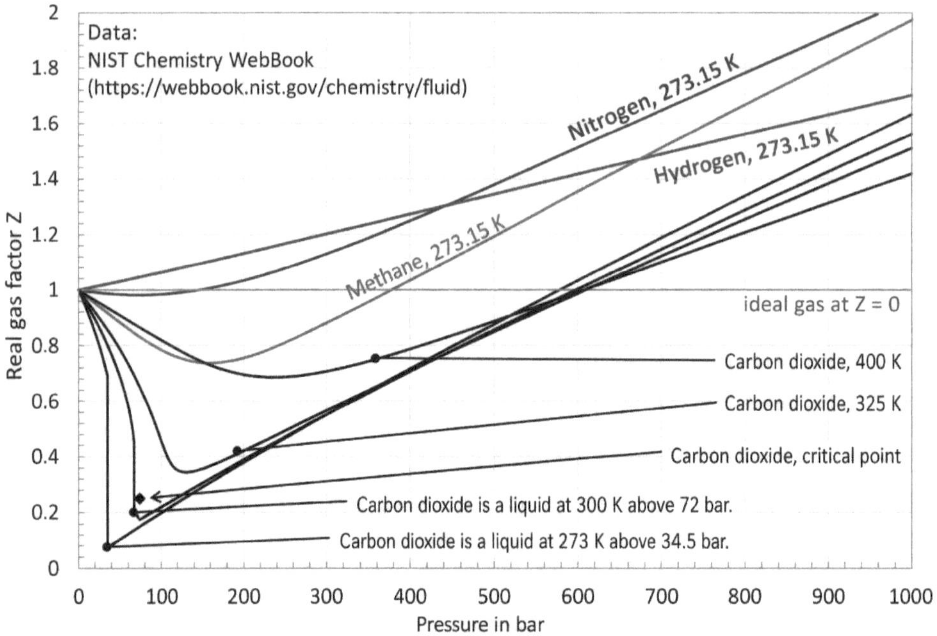

Figure 3.5 Real gas factor for typical gases: carbon dioxide illustrates how the ideal gas only becomes an acceptable model for technically relevant pressures at a greater distance from the critical point (data source: NIST Chemistry WebBook).

Ideal Fluid

The ideal fluid is incompressible, i.e., a change in pressure would not lead to any change in other state variables. This approximation describes many technical fluids such as water well. The state variables remain temperature-dependent. In particular, the specific volume and density change with temperature, so that buoyancy is included in the model.

3.5 Energy Conversion

Energy conversion refers to the conversion of one form of energy into another. Energy conversion is at the center of technical processes and plants as well as natural processes in the earth system: on earth, the radiation of the sun (a heat flux) is repeatedly absorbed, converted, and thus drives the incredible diversity of living and non-living processes and things.

3.5.1 Efficiency

The central key figure of energy conversion is the efficiency – a measure of the efficiency with which energy is converted into a new form. In the specific thermodynamic sense, the efficiency indicates what fraction of the supplied energy is converted from the considered input energy E_{input} into the considered target energy. In technical systems, a distinction is made between the useful energy E_{use}, i.e., the converted form of energy that is desired, and energy losses. In this case, the efficiency is

$$\eta = \frac{E_{use}}{E_{input}} \tag{3.26}$$

The efficiency always has a value between 0 and 1. An efficiency of 1 means complete dissipation-free and thus reversible conversion. This is not possible in technical and natural systems. Here, the quality of energy (see Chapter 5) is not taken into account. The non-useful remainder of the input energy is – from the perspective of the considered process – dissipated. In most cases, the dissipated energy is then heat, which is released to the surroundings at low temperature.

Central to the calculation and discussion of meaningful efficiencies is to clearly define the system boundaries, i.e., to unambiguously identify what benefit is specifically measured where and which inputs are included in the calculation. This is discussed in Chapter 4 and Chapter 5.

3.5.2 Converting Heat into Work

A central consequence of the second law of thermodynamics is that the conversion of transferred heat into performed work is fundamentally never complete. The theoretical efficiency of this conversion is always much less than 1. The maximum possible efficiency is the Carnot efficiency. This theoretical limiting case is relevant, as this is the maximum of conversion that engineers can strive for. If a heat flux at temperature T_{in} is converted into power, a waste heat flux at temperature T_{out} remains, and the Carnot efficiency is

$$\eta_{Carnot} = \frac{P_{out}}{\dot{Q}_{in}} = \frac{\dot{Q}_{in} - \dot{Q}_{out}}{\dot{Q}_{in}} = 1 - \frac{T_{out}}{T_{in}} \tag{3.27}$$

This ideal case is a reversible process in which no entropy is generated. In the most favorable case, T_{out} is the ambient temperature. This maximum efficiency is a direct consequence of the second law of thermodynamics and the conservation of entropy. Thus, for power that can be obtained from a heat flux, it always holds that

$$P < \eta_{Carnot} \cdot \dot{Q} = \left(1 - \frac{T_{out}}{T_{in}}\right) \cdot \dot{Q} \tag{3.28}$$

3.5.3 Converting Work into Work

Unlike the conversion of heat into work, there is no theoretical limit to the efficiency for the conversion of one form of work into another. This is due to the fact that work itself does not generate entropy when it is performed across a system boundary. Work fundamentally does not transport entropy. This applies in particular to electricity, which can be transferred from one system to another without mass transfer. Common forms of work are the already briefly introduced volume work, pressure work and technical work associated with flowing fluids. In addition, there are types of work:

- **Shaft work:** This is mechanical work that is transmitted across a system boundary by means of a shaft and through the rotation of the shaft. A shaft running idle therefore transmits no work. The work must be supplied to the rotating shaft at one end and extracted at the other end.
- **Mechanical work:** This is any form of work performed by mechanical forces such as technical work, volume change work or shaft work.
- **Electrical work:** This refers to forms of work performed by means of electric current or electric fields. Here the power is

$$P_{el} = U \cdot I$$

The voltage difference U is the potential that enables the current I to flow. A voltage difference can also be induced in a body by an externally applied electromagnetic field. A current then flows there.

Electrothermal heating

The conversion of work into heat can be intentional in a technical process. This is particularly the case with electrical heating. The electrical energy is converted into heat that is further utilized; examples of conversion processes in electrothermal heating are summarized in Table 3.1 (Pfeifer et al. 2011).

For many specialized technical processes, electrothermal heating is today the only way to provide heat at a sufficient temperature. In vacuum and in high-purity atmospheres, only extremely clean forms of heat may be used. This is only possible electrically. Very high temperatures, such as those required for melting tungsten, molybdenum, niobium or tantalum, can only be generated electrically. Very high power in a small space and very fast processes with very high power are often only possible electrically. Such processes are overall very energy-intensive due to the long energy conversion chain, but are unavoidable for many technologies such as the manufacture of digital devices.

Table 3.1 Technically Relevant Processes of Electric Heating

Process	Applications
Joule effect, resistance heating	Flowing electric current is converted into heat due to the electrical resistance in the conductor itself. Heating conductors with a nearly temperature-independent specific resistance are preferably used. Typical for common applications are heating conductors made of AlCrFe or NiCr alloys. In special applications, heating conductors with a more strongly temperature-dependent resistance are also used, such as tungsten, silicon carbide (SiC) or molybdenum disilicide (MoSi).
Inductive heating	Electric current is moved in another conductor solely through the effect of an electromagnetic field. The electrical power in the other conductor can then be further utilized. One application is the heating and melting of metals. The induced current in the workpiece or material generates heat through the Joule effect. Induction works at low frequencies of the field.
Dielectric heating	An electromagnetic field of high frequency is used to heat non-conductors. The process is similar to induction. For example, silicon is remelted and purified in vacuum in this way.
Microwave	Electronic components generate radiation in the microwave range, which is used for further applications or processes, as it can be absorbed by the material.
Light, optical radiation	Heating conductors (incandescent lamps) or electronic elements (LEDs) can directly emit light, or light is generated in a gas discharge. In this case, the real efficiency of the conversion is limited. The generated light is thermal radiation.
Plasma	A plasma is generated by high voltage. Plasma means that atoms or molecules are ionized, i.e., they release electrons. The kinetic energy of the charged particles can be used for targeted heating of the material. Plasma can be used for further processes such as coatings or cleaning.
Electron beam	Electrons are released, accelerated in a field, and then melt, e.g., metals in a vacuum when they strike them.

3.5.4 Pressure Loss As Dissipation

A concrete and technically relevant example of dissipation is the pressure loss. In a flowing fluid, the pressure decreases along the flow path. In general, the following applies to the pressure loss in a component

$$\Delta p = \zeta \cdot \rho_{fluid} \cdot \frac{\bar{c}^2}{2} \tag{3.29}$$

Here ζ is the pressure loss coefficient, ρ the density and $\bar{c}$ the mean velocity of the flow. The pressure loss coefficient can be calculated for pipelines; for almost all other components, measured values are available (Stephan et al. 2019; Idelchik 2005). The pressure loss to be overcome increases quadratically with the velocity of the fluid.

Pressure loss is the result of friction in the fluid as well as between the fluid and the wall, i.e., dissipation. The heat generated in the process is dissipated via the pipeline or, in insulated pipelines, leads to heating of the fluid. For round pipes, the volumetric flow rate is

$$\dot{V} = A_{pipe} \cdot \bar{c} \tag{3.30}$$

Thus the pressure loss for a round pipe ($A = \pi/4 \cdot d^2$)

$$\Delta p = \zeta \cdot \rho_{fluid} \cdot \frac{8}{\pi^2} \cdot \frac{\dot{V}^2}{2} \tag{3.31}$$

This equation shows that even small changes in the pipe diameter d have a large influence on the pressure loss.

The pressure loss in pipes additionally depends on the shape of the flow in the pipe and on the surface of the pipe. Very smooth pipes have the lowest pressure loss, fast turbulent flows in pipes with roughness elements extending far into the flow have a high pressure loss. In analogy to the treatment of electrical resistance, a pipe friction factor λ is introduced, which relates the potential (here the pressure instead of the voltage) to the flow. The pipe friction factor is thus analogous to the specific electrical resistance. Then, for round pipes (Marek/Nitsche 2015)

$$\Delta p = \frac{l_{pipe}}{d_{pipe}} \cdot \lambda\left(\mathrm{Re}, \frac{k}{d}\right) \cdot \rho_{fluid} \cdot \frac{\bar{c}^2}{2} \tag{3.32}$$

The pipe friction factor λ depends on the flow regime, defined by the Reynolds number Re, and the ratio of wall roughness k to the diameter of the pipe d. The Reynolds number depends on the mean velocity of the flow $\bar{c}$, a typical dimension of the geometry L^* and the kinematic viscosity ν :

$$\mathrm{Re} = \frac{\bar{c} \cdot L^*}{\upsilon_{fluid}} = \frac{\bar{c} \cdot d_{pipe}}{\upsilon_{fluid}} \tag{3.33}$$

The viscosity is a temperature-dependent material property of the fluid. For a pipe, the typical dimension L^* is the diameter of the pipe.

If the pipe is hydraulically smooth, i.e., the irregularities of the wall are so small that they lie entirely within a laminar boundary layer of the fluid, an approximation is (Marek/Nitsche 2015)

$$\lambda(\mathrm{Re}) \cong \frac{0.3164}{\mathrm{Re}^{0.25}} = 0.3164 \cdot \left(\frac{\upsilon_{fluid}}{\bar{c} \cdot d_{pipe}}\right)^{\frac{1}{4}} \tag{3.34}$$

For rough pipes, where wall irregularities protrude into the turbulent flow, the following applies

$$\lambda\left(\mathrm{Re}, \frac{k}{d}\right) \cong \left(\frac{1}{2 \cdot \log_{10}\left(3.71 \cdot \frac{d}{k}\right)}\right)^2 \tag{3.35}$$

Transport of fluids

Optimal solutions are sought for the transport of fluids and in particular fossil energy carriers over long distances: natural gas and crude oil as well as, in the future, possibly hydrogen and carbon dioxide must be transported over long distances. The energy consumption should be as low as possible without the expenditure for the pipes becoming too great: For this purpose, pipes with the smoothest possible interior surfaces are manufactured and attention is also paid to the quality of the surface at weld seams. Furthermore, an optimum is sought between manufacturing costs (pipes as small as possible) and low operating costs through low velocity (pipes as large as possible).

For natural gas, it is sufficient – depending on the pressure – to use a compressor approximately every 100 km to 1000 km to rebuild the pressure in the pipeline and thus convey the gas (Cerbe/Lendt 2016). The situation is different for crude oil. Crude oil has a much higher viscosity and a density that is 10 to 100 times higher than that of natural gas under pressure. The viscosity depends on the composition of the crude oil (its grade) and strongly on the temperature. As a result, the pressure drop at the same throughput is fundamentally much higher than for natural gas: significantly more pumping stations are required. In cold regions, the pipeline must additionally be heated to prevent the viscosity from becoming too high.

3.5.5 Heat Transfer

Heat transfer is essential in technical processes and energy conversion. Either the transfer of heat should occur with as little resistance as possible and at nearly constant temperature. This applies to heat exchangers and thermal processes. Alternatively, the heat should ideally not flow at all. This applies, for example, to insulation of pipelines, buildings, or refrigerators. However, heat always finds a way from a system into its surroundings. Technically, only the magnitude of the heat flux can be influenced.

Heat is transferred via three fundamental processes:

Thermal conduction Here heat flows within a material (i.e., solid, liquid, or gas). In this process, it is passed on, so to speak, from one molecule or atom to the next. This describes the material property of thermal conduction, a temperature-dependent material property.

Convection Currents transport heat:

- In fluids, temperature differences lead to density differences. This causes buoyancy. Heat is carried along with the flow. This is free convection.
- When a fluid flows over the surface of a body at a different temperature, heat flows depending on the temperature difference between the fluid and the body. The higher the flow velocity at the surface, the more heat can be exchanged. Due to the predetermined flow velocity of the fluid, this is forced convection.
- Often a mixture of forced and free convection is present. In addition, convection, like heat conduction, is bound to matter. Heat conduction is a prerequisite for convection.

Radiation Heat is transferred by electromagnetic radiation directly and without matter. Every surface and every gas emits heat and can absorb thermal radiation.

Methods for describing heat transfer can be found in the technical literature (Marek/Nitsche 2015).

3.6 Optical Radiation

The earth is supplied with energy by the sun and, conversely, emits heat into space. These heat flows occur essentially as optical radiation. Every surface emits radiation. The material, its phase, its temperature and its surface properties influence what radiation the surface emits and also what radiation it absorbs, and what is reflected or simply transmitted by it (Howell et al. 2015, Marek/Nitsche 2015). Radiation can be described as a wave, so that wavelength λ, frequency v and propagation velocity c_0 of the radiation are related:

$$c_0 = \lambda \cdot v \tag{3.36}$$

In a vacuum, the propagation velocity of light is the vacuum speed of light c_0. In matter, this propagation velocity is slightly reduced. The vacuum speed of light is a natural constant.

Radiation can also be described as a stream of particles. The particle of radiation is the photon. Photons have energy, entropy, momentum and angular momentum, but no mass. For the energy of the photon, the following applies

$$E_{photon} = h \cdot \upsilon = h \cdot c_0 \cdot \frac{1}{\lambda} \tag{3.37}$$

The energy of a photon decreases with its wavelength or increases with its frequency. The factor h, which maps the correlation between frequency and energy, is Planck's quantum of action. Here, h has the value of $6.626 \cdot 10^{-34}$ J s. This is the famous and often invoked quantum leap: the smallest physically still relevant energetic unit. There is nothing smaller and more insignificant than a quantum leap.

The Radiation Spectrum

The energy of individual light quanta leads to the different effects of electromagnetic radiation depending on the wavelength or frequency. Electromagnetic radiation is subdivided according to these different effects. Different regions of radiation have different effects on matter. An overview is given in Table 3.2. Optical radiation is mostly described in terms of wavelength. For radio waves and electromagnetic fields, on the other hand, the frequency is preferably specified.

Radio and microwave Electromagnetic radiation of low frequency is transmitted by a very large number of photons of very low energy; therefore, the particle effect of the radiation is not relevant here. The radiation is generated by electrical or electronic components and is well and intuitively described by its frequency.

Optical radiation This borders on hard particle radiation (X-rays) at 10 nm and on microwaves, i.e., the radio range, at 1 mm. The further subdivision of optical radiation is shown in Figure 3.6.

Infrared The range of classical thermal radiation is further divided between IR-A, IR-B and IR-C, whereby this is motivated by the effect on the human body:

- IR-A reaches the retina and can damage it at very high intensity.
- IR-B contributes to burns of the skin and cornea of the eye.
- IR-C is physiologically mostly harmless. The effect is overheating up to burns.

Visible light This extends from 380 nm to approximately 780 nm. The relationship between spectrum and color perception is complex and is described, e.g., via the CIE standard color space. Monochromatic light, i.e., light of a single wavelength, has clearly assignable colors:

- Violet corresponds to 380 nm.
- Blue light corresponds to approximately 470 nm.
- Green is at 520 nm (the maximum of solar radiation).
- Yellow is at 570 nm.

- Red is above 620 nm.

Ultraviolet This range is again subdivided according to physiological effect:

- UV-A can cause sunburn at a sufficient dose.
- UV-B will cause sunburn at a very low dose.
- Upon irradiation with UV-C, the upper layer of skin dies.
- Vacuum UV (VUV) is optical radiation for which air is no longer transparent.
- EUV was previously classified as X-ray radiation and today is included in the UV range, since state-of-the-art exposure machines for the production of microchips operate here. With very great effort it is still possible to construct optical components for this range.

Table 3.2 Spectral Ranges of Electromagnetic Radiation

Range	Wavelength and frequency	Properties
Radio waves	$\infty > \lambda > 1$ mm $0 < \nu < 300$ GHz	Radio waves penetrate most non-conducting solids. Radio radiation can be emitted by electrical circuits, as it influences free charges.
Microwaves	1 m $> \lambda >$ 1 mm 0.3 GHz $< \nu <$ 300 GHz	Microwaves are a special part of the radio range. They are technically generated by special electrical components.
Optical radiation: infrared	1 mm $> \lambda >$ 780 nm IR-A 780 nm to 1400 nm IR-B 1400 nm to 3000 nm IR-C 3000 nm to 1 mm	Infrared radiation belongs to optical radiation. It is absorbed and emitted by most substances. Infrared radiation is technically emitted preferably by warm surfaces or substances. There are also LEDs that emit infrared radiation.
Optical radiation: visible light	780 nm $> \lambda >$ 380 nm	This is optical radiation that can be perceived by the human eye. Visible light can be used for chemical or electrical reactions (photosynthesis or photovoltaics). Visible light can be emitted by hot surfaces and gas discharges or by semiconductor components (LEDs).
Optical radiation: ultraviolet	380 nm $> \lambda >$ 10 nm	This is optical radiation that can destroy chemical bonds. Below 180 nm, air is opaque. This range is called vacuum UV.
X-ray radiation	10 nm $> \lambda >$ 0.01 nm	This is radiation that can release electrons from atoms and molecules and destroy chemical bonds.

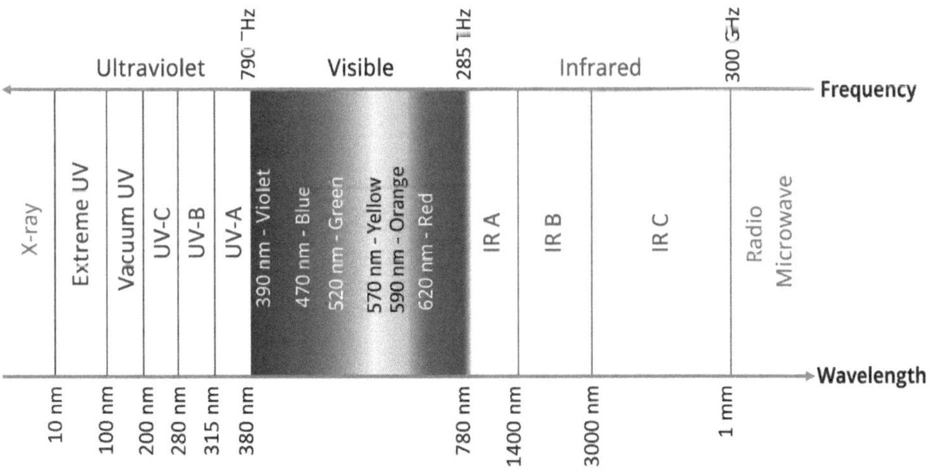

Figure 3.6 Essential ranges of optical radiation (explanation in text)

Spectral Emission – Planck's Law

The spectral emission of real surfaces can often be approximated with a simple model. The model of the black body describes a surface that, depending on its temperature T, emits the maximum possible energy. The emitted heat flux related to the area of the body depends only on the temperature and is a function of the wavelength:

$$M_{black}(\lambda,T) = \frac{2 \cdot \pi \cdot h \cdot c_0^2}{\lambda^5} \cdot \frac{1}{\exp\left(\frac{h \cdot c_0}{k_B \cdot \lambda \cdot T}\right) - 1} \tag{3.38}$$

It has become established practice to combine the constant factors and thus simplify the equation:

$$M_{black}(\lambda,T) = \frac{c_1}{\lambda^5} \cdot \frac{1}{\exp\left(\frac{c_2}{\lambda \cdot T}\right) - 1} \tag{3.39}$$

The constants are

$$c_1 = 2 \cdot \pi \cdot h \cdot c_0^2 = 3.142 \cdot 10^{-16}\ \text{W m}^2$$

$$c_2 = \frac{h \cdot c_0}{k_B} = 0.014387\ \text{m K}$$

Equation 3.39 is Planck's law, c_0 the speed of light in vacuum and k_B the Boltzmann constant. The Boltzmann constant is related to the universal gas constant R and Avogadro's number N_A as follows:

$$R = N_A \cdot k_B \tag{3.40}$$

Equation 3.39 describes the distribution of the emitted energy over the wavelength as a function of temperature. This function is shown in Chapter 7 for 5776 K. The shape of the function always remains similar, only the position of the maximum and the enclosed area change significantly.

Real surfaces emit less power for each wavelength than the ideal black body radiator. This can be described by a temperature- and wavelength-dependent material property, the spectral temperature-dependent emissivity $\varepsilon(\lambda,T)$ of the surface of the body:

$$M(\lambda,T) = \frac{c_1}{\lambda^5} \cdot \frac{\varepsilon(\lambda,T)}{\exp\left(\frac{c_2}{\lambda \cdot T}\right) - 1} \tag{3.41}$$

This spectral emissivity $\varepsilon(\lambda,T)$ contains all material-dependent data and their temperature dependence. It is a function with numerical values between 0 and 1.

The Total Emission – Stefan-Boltzmann Law

If the heat flux emitted by a surface is to be determined, the spectral details of the radiation are not of immediate interest. The heat flux leaving a surface of homogeneous optical properties and uniform temperature is

$$\dot{Q}_{radiation}\left(T_{surface}\right) = A \cdot \varepsilon(T) \cdot \sigma \cdot T^4 \tag{3.42}$$

In this equation, $\sigma = 5.67 \cdot 10^{-8}$ W m^{-2} K^{-4} is the Stefan-Boltzmann constant and the emissivity used here $\varepsilon(T)$ is a function with numerical values between 0 and 1 that usually depends only weakly on temperature. Values for $\varepsilon(T)$ and for a large number of typical technical surfaces can be found in the literature (Stephan et al. 2019, Section K; Howell et al. 2015). Often only numerical values are given in the literature, without any indication of a temperature dependence.

The Stefan-Boltzmann law in Equation 3.42 applies to the emission of a surface. If the heat flux is to be calculated as the sum of emitted and absorbed heat, then for surfaces that absorb and emit radiation well overall and that are located in an environment with approximately homogeneous (radiation) temperature,

$$\dot{Q}_{radiation} = A \cdot \varepsilon(T) \cdot \sigma \cdot \left(T^4 - T_0^4\right) \tag{3.43}$$

In Table 3.3 the gross emission and the net heat flux are given for some common surface temperatures.

Table 3.3 Numerical Values for the Emission of Surfaces

Temperature		Emission, Equation 3.42	Maximum, Equation 3.44	Heat flux, Equation 3.43, T_0 = 25 °C	Examples
255 K	-18 °C	239 W m^{-2}	11 360 nm	-207 W m^{-2}	Emission of the earth
288 K	15 °C	390 W m^{-2}	10 060 nm	-57 W m^{-2}	Mean temperature of the earth's surface
300 K	27 °C	459 W m^{-2}	9659 nm	12 W m^{-2}	
360 K	87 °C	952 W m^{-2}	8049 nm	505 W m^{-2}	Solar thermal without concentrator
700 K	427 °C	13 614 W m^{-2}	4140 nm	13 167 W m^{-2}	Solar thermal with parabolic trough concentrator
1200 K	927 °C	117 573 W m^{-2}	2415 nm	117 126 W m^{-2}	Fire
1811 K	1538 °C	609 900 W m^{-2}	1600 nm	609 450 W m^{-2}	Melting point of iron
5776 K	5503 °C	63 109 000 W m^{-2}	502 nm	63 108 625 W m^{-2}	Surface temperature of the sun

Maximum – Wien's Displacement Law

The total information content of the Planck function is too extensive for many purposes. In many technical applications, the fundamental information required is the wavelength of the maximum of the spectral emission. This is described by Wien's displacement law. In its specific form it states where the maximum of the Planck function for black body radiators lies:

$$\lambda_{max} = \frac{c_{Wien}}{T} \tag{3.44}$$

The constant is $c_{Wien} = 2.89777 \cdot 10^{-3}$ m K. This also applies to so-called gray emitters, where $\varepsilon(\lambda,T)$ does not depend on the wavelength. Many ceramics are well described as gray emitters in the technically relevant infrared wavelength range, so that this relationship can be applied particularly well to ceramic surfaces. In Table 3.3 the wavelengths of the maxima of the emission are also indicated.

3.7 Heat Transfer by Radiation

Intensity of Radiation

When a surface emits radiation, the spectrum does not change as long as this radiation can propagate undisturbed – i.e., in a vacuum. However, the energy density of the radiation changes with the distance from the source. For any closed surface enclosing the entire source, the power flowing through this surface remains constant due to conservation of energy. For nearly point-like sources, the intensity decreases quadratically with distance, because the area of a sphere enclosing this source is

$$O_{sphere} = 4 \cdot \pi \cdot r^2 = \pi \cdot d^2 \tag{3.45}$$

Thus the intensity – i.e., the heat flux related to a surface perpendicular to the direction of propagation of the radiation – decreases accordingly:

$$I(r) = \frac{\dot{Q}_{emission}}{O_{sphere}} = \frac{\dot{Q}_{emission}}{4 \cdot \pi \cdot r^2} \tag{3.46}$$

Sources can be regarded as point-like when the observer is at a distance of approximately 50 times the source diameter from the source.

Radiation Striking Matter

When radiation strikes matter, three processes occur, which can be described by corresponding material properties:

reflection A portion of the radiation is reflected directly at the surface. The fraction of energy that is reflected is ρ. The reflectivity of a surface ρ depends on the angle at which the radiation strikes, the wavelength of the radiation, the temperature of the surface, as well as the composition of the body near the surface and the structure of the surface. In principle, every interface at which the optical properties change sufficiently fast exhibits reflection.

absorption A portion of the radiation, or the energy contained in it, is absorbed by the body. The absorption rate α depends on the direction from which the light strikes the surface of the body, on the wavelength of the radiation, and on the temperature of the body, as well as of course on the composition and structure of the body.

transmission When a portion of the radiation passes through the body, this portion is transmitted. This transmission τ depends on the angle, the wavelength of the radiation, the temperature of the body, its composition and structure.

conservation of energy From conservation of energy, Kirchhoff's law follows:

$$\rho + \alpha + \tau = 1 \tag{3.47}$$

This Kirchhoff's law applies to radiation that (i) is incident at a specific angle β, (ii) for a specific wavelength λ, and (iii) for a temperature. Strictly speaking, only

$$\rho(\lambda,T,\beta) + \alpha(\lambda,T,\beta) + \tau(\lambda,T,\beta) = 1 \tag{3.48}$$

Additionally, the following applies to the material properties of emission and absorption

$$\alpha(\lambda,T) = \varepsilon(\lambda,T) \tag{3.49}$$

Reflection depends on the refractive index

The reflection at surfaces is described by the Fresnel equations. These describe the reflectivity at this boundary when radiation passes from one medium through a clearly defined phase boundary into another medium. The equations depend on the angle, the optical properties of the two media, the wavelength, and the polarization of the light. For the special case of normal incidence and non-absorbing media,

$$\rho_{\perp}(\gamma) = \frac{\left(\frac{n_2}{n_1} - 1\right)^2}{\left(\frac{n_2}{n_1} + 1\right)^2}$$

Here n_1 denotes the refractive index of the medium from which the radiation passes into medium 2 with refractive index n_2. Often medium 1 is air, for which $n_{air} = 1$ is a good approximation. Glass typically has a refractive index of about 1.5, so the reflectivity of normally incident radiation at a glass surface is about 4%. Refractive indices depend on the wavelength.

Metals are opaque to optical radiation. They absorb strongly. For this case, for normal incidence and air as the surrounding medium,

$$\rho_{\perp}(\lambda) = \frac{(n_{metal} - 1)^2 + n_{metal}^2 \cdot k_{metal}^2}{(n_{metal} + 1)^2 + n_{metal}^2 \cdot k_{metal}^2}$$

Here k is the extinction coefficient. Values for k or for the imaginary refractive index $n_i(\lambda) = n(\lambda) \cdot (1 - i \cdot k(\lambda))$ can be found in the literature (Touloukian/Dewitt 1970, Touloukian/Dewitt 1972). For very thin layers of metals, it can be observed that they become transparent to optical radiation. Very thin gold or silver films, as used as reflector layers, can become greenish translucent below a layer thickness of approximately 100 nm.

Absorption and Scattering

The absorption of radiation of a specific wavelength in a homogeneous medium can be described using the Lambert-Beer law as follows:

$$I(x) = I_0 \cdot \exp(-a(\lambda) \cdot x) \tag{3.50}$$

The intensity of the radiation decreases exponentially with the distance traveled x in the medium. The absorption coefficient a depends in solutions or glasses on the concentration of the absorbing medium and the wavelength. The absorption coefficient can be calculated from the imaginary part of the refractive index:

$$a(\lambda) = \frac{4 \cdot \pi}{\lambda} \cdot n(\lambda) \cdot k(\lambda) \quad (3.51)$$

The absorption of gases – e.g. in the atmosphere – is exceptionally complex, since gases exhibit a very well-defined line spectrum. These lines, their position, their shape and their absorption strength depend on the concentration of the gas, on pressure and on temperature. Therefore, for example, the absorption behavior of radiation in the atmosphere cannot be described in general terms, but requires precise information about the altitude-dependent data on temperature and pressure and their effect on absorption and emission by atmospheric gases (Howell et al. 2015).

Transparency of glasses

Many technical glasses are non-absorbing or only weakly absorbing in the visible spectral range. Pure fused silica can transmit radiation over several 100 km in the near infrared. Normal architectural glass and borosilicate glass are also transparent over many meters. The absorption of glasses in the visible spectral range is caused by impurities. Architectural glass typically shows a greenish discoloration. This is attributable to iron (green glass). In normal applications this is acceptable. Glasses used for high-quality photovoltaic modules have a significantly reduced iron content, since this allows the transmission of the glass to be increased by a few percentage points (white glass).

Scattering

Radiation can also be altered in an otherwise radiation-transparent medium by scattering. The direct transmission is reduced for the simple case of low scattering by

$$I(x) = I_0 \cdot \exp(-s(\lambda) \cdot x)$$

Here the scattering depends on the size of the particles, their shape and their composition. This is described by the factor $s(\lambda)$.

A simple example is the scattering of sunlight in a blue sky: the scattering depends strongly on the size of the particles relative to the wavelength of light. Blue light has a shorter wavelength than red light, so blue light is preferentially scattered by the air molecules. The intensity of this scattered radiation is not insignificant. During the day, all stars and planets are outshone by this scattered radiation.

Scattering also occurs at water droplets or dust particles in the atmosphere. This scattering has different spectral properties, so that clouds made of water often appear white, but clouds made of other substances do not necessarily.

References

Cerbe, G./Lendt, B.: Grundlagen der Gastechnik: Gasbeschaffung. Gasverteilung. Gasverwendung. 8. Auflage. Carl Hanser Verlag, München 2016

Chase, M. W.: NIST-JANAF Thermochemical Tables. Springer, 1998 (siehe auch *https://janaf.nist.gov*)

Delgado-Bonal, A.: Entropy of radiation. The unseen side of light. In: Nature Scientific Reports 7, 2017, p. 1642. *https://www.nature.com/articles/s41598-017-01622-6*

Giordano, V. M./Datchi, F./Dewaele, A.: Melting curve and fluid equation of state of carbon dioxide at high pressure and high temperature. In: J. Chem. Phys. 125, 2006, 054504

Howell, J. R./Menguc, M. P./Siegel, R.: Thermal Radiation Heat Transfer. CRC, Boca Rocca 2015

Idelchik, I. E.: Handbook of Hydraulic Resistance. Jaico Publishing, Mumbay 2005

Linow, S.: Angewandte technische Thermodynamik. Carl Hanser Verlag, München 2022

Marek, R./Nitsche, K.: Praxis der Wärmeübertragung. Grundlagen – Anwendungen – Übungsaufgaben. Carl Hanser Verlag, München 2015

Maurer, T.: Kältetechnik für Ingenieure. VDE Verlag, Berlin 2016

NIST Chemistry WebBook, SRD 69. Thermophysical Properties of Fluid Systems. *https://webbook.nist.gov/chemistry/fluid*

Palik, E. D. (Hrsg.): Handbook of Optical Constants of Solids. 5 Bände. Elsevier, 1997

Pfeifer, H./Nacke, B./Beneke, F.: Praxishandbuch Thermoprozesstechnik. 2 Bände. Vulkan Verlag, Essen 2011

Stephan, P./Kabelac, S./Kind, M./Mewes, D./Schaber, K./Wetzel, T. (Hrsg): VDI Wärmeatlas. 12. Auflage. Springer, Berlin 2019

Touloukian, Y. S./Dewitt, D. P.: Thermal radiative properties. Metallic elements and alloys. In: Thermophysical Properties of Matter, Volume 7, NASA 1970

Touloukian, Y. S./Dewitt, D. P.: Thermal radiative properties: Nonmetallic solids. In: Thermophysical Properties of Matter, Volume 8, NASA 1972

4

Describe Infrastructure and Technology

4.1 Life Cycle Assessment

For many of the methods and metrics introduced below, such as cumulative energy demand (CED) and Energy Returned on Energy Invested (EROI), as well as life cycle assessment (LCA) in general, an analysis of the life cycle of products is fundamental. It is often intended or required for the application of these methods. It is extremely helpful in the environmentally conscious design of products. The life cycle of a product is defined in ISO 14040 as

> *"consecutive and interlinked stages of a product system, from raw material acquisition or generation from natural resources to final disposal"*

The concept seems straightforward at first glance, but it often creates difficulties in concrete application and implementation. One of the main focuses of this book is therefore a systematic overview of contributions to the life cycle with a focus on energy infrastructure.

A product can never stand alone. Products exist in a network of mutual relationships and dependencies. Means of production, means of transport, components and spare parts, services and rights are themselves products that are required for the life cycle. The life cycle assessment of a single product is always a complex task. This begins with the definition of system boundaries. Completely describing a life cycle is barely achievable even for simple products and quickly becomes more demanding for more complex products or services.

Products are goods and services

ISO 14001 defines products as goods and services. Goods are physical products that can be touched (tangible). Services are initially non-physical (intangible), but they always require physical goods of a technology in order to be provided.

Ownership is understood here in a legal sense as current power of disposal. This represents an important concept, because it means that brands and patents also fall under the term products. Thus the term product encompasses in particular

- Objects such as parts, components, tools, furniture and household goods,
- Buildings and infrastructure,
- Means of production, machines, plants and factories,
- Raw material sources such as crude oil and gas fields, mines, open-cast mines, forests and fields,
- Extraction equipment such as conveying systems, excavators or hoisting machines,
- Means of transport such as infrastructure, ships, vehicles, railway networks, pipelines, power lines and conveyor belts,
- Processes, procedures and methods, algorithms and digital platforms,
- Services such as transport, brokerage, consulting, legal services, sales, catering, care and supply,
- Rights such as contracts, patents and trademarks as well as rights for mining, collecting, utilizing or distributing (This applies at least insofar as these rights are restricted or licensed and can thus be traded. In this sense, fields, farmland, meadows and forests also become products.),
- Financial services such as loans and insurance, as well as
- Disposal and its services such as waste bins, landfills, waste incineration plants, recycling processes and facilities, means of transport and rights.

Life Cycle

The fundamental idea of the life cycle is illustrated in Table 4.1 for a physical product. For a life cycle assessment, all relevant activities of a product's life cycle must be taken into account as far as possible. All direct contributions, i.e. all components used in the product, all transport routes and all manufacturing steps must be considered. All indirect shares also belong to the system, i.e. all auxiliary means used and all proportional expenditures for manufacturing processes and services. All required resources and all environmental impacts across the stages of the life cycle must be included in this assessment. This is therefore an assessment of a system with initially clearly defined but difficult-to-identify system boundaries in detail.

Termination criteria The assessment must be terminated at a suitable point in order to make it manageable at all. There are established procedures for this in a life cycle assessment. Below a (to be estimated) minor impact, a contribution can be neglected. This must be documented and justified appropriately in each case.

Goods The life cycle of physical products can be divided into individual phases. The most important phases are summarized in Table 4.1.

Since physical products are almost always composed of a large number of individual components, these components each go through their own life cycle. In order to define the system boundary for the life cycle of a specific product, the network of further products that enable the product's life cycle must also be taken into account:

- For a simple product such as a screw manufactured from the iron of an iron mine, this may at first glance seem entirely feasible. However, once the blast furnace and the steelworks are included in all their complexity, this possibility of a complete description already breaks down.
- For a complex product such as a mobile phone, which is composed of nearly all elements of the periodic table, whose supply chain is global and frequently changing, completeness is hardly achievable even without including the means of production.
- In many supply chains there are transitions that are deliberately withheld from close scrutiny, such as raw materials from crisis regions, inferior or counterfeit products from dubious sources, secret processes, procedures and compositions, as well as other trade secrets. Here, an external assessment is not possible.

In summary, a truly complete inclusion of all real contributions in a life cycle assessment is impossible. In order to nevertheless arrive at a meaningful result, the LCA method has been developed accordingly.

Functional unit All results are reported with reference to a functional unit specified. Defining this unit is straightforward for some products (refrigerator). For raw materials and semi-finished products, it is common practice to report results relative to mass (specific) or relative to volume. For many services, the concept can be transferred appropriately. In the field of information technology, it becomes extremely difficult to define suitable units.

Table 4.1 Life Cycle of a Physical Product

No.	Step	Activities
1	Extraction 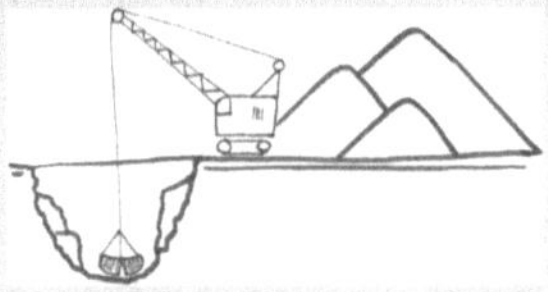	Extraction of the required raw materials from a raw material deposit or harvesting of renewable raw materials
		Transport from extraction to raw material processing, conversion and primary shaping 
2	Primary shaping 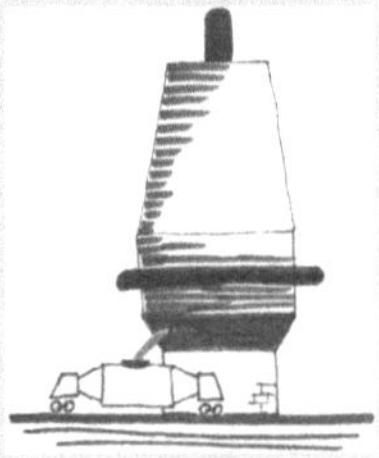	Conversion of raw materials into the required starting material (primary shaping): A blast furnace is shown here as an example.
		Transport of semi-finished products from primary shaping to further processing 
3	Manufacturing 	Manufacturing of the product: Manufacturing can comprise several stages. First, individual parts are produced, next assemblies or components are assembled, and finally the finished product is manufactured. Each step may take place in a different facility at a different location.
		Transport of intermediate products from one production site to the next Transport of finished products to the market 

No.	Step	Activities
4	Distribution 	The distribution often involves several transport steps until the goods are available in the shop or reach the customer via distribution centers and parcel service providers. Additional activities such as marketing and advertising are also included. Transport of products to the point of use 
5	Use 	The use of the product can take place in one location or in various locations. It may be necessary to provide additional products for use. Transport of products at end of life for disposal 
6	Disposal 	Disposal can occur as a direct route to the landfill, or energy and raw materials can be returned to use. In that case, they are suitable for evaluation. 

Life cycle assessment (LCA)

The ISO 14040 series of standards describes the method of life cycle assessment (LCA). It is a structured method for evaluating the environmental impacts of products or processes. It can most easily be implemented using suitable software and additional databases.

In principle, an LCA is a rather time-consuming method. The major challenge today is often less the data basis, since accepted data can be obtained from databases if necessary. The major challenge is defining well-balanced system boundaries: if the boundaries are drawn too broadly, the effort involved in the assessment can become too extensive. If the system boundaries are drawn too narrowly, relevant and substantial contributions are neglected. This also makes it clear that the clean documentation of the system boundary and the assumed effects outside the defined system boundary is an important component of an LCA.

ISO 14040 describes a procedure in which cutoff criteria can be used. For this purpose, the system boundary itself becomes a subject of investigation and is continuously refined during the LCA process (Ausberg et al. 2015). Key parameters for investigating the system boundary are the availability of data, the relevance of the data to the result, and the objective of the LCA. If the goal is a comparison of different variants, a comparable system boundary should be drawn for all variants.

Relevant key figures determined by an LCA are:

- The cumulative energy demand (CED, see Chapter 5)
- The CO_2footprint, i.e., the total amount of CO_2 or other greenhouse gases emitted over the life cycle
- The total water consumption or the mass of water used and polluted
- The total sulfur emissions and associated effects such as acidification of soils and water bodies
- The total emissions of reactive nitrogen and associated health effects and effects on ecosystems such as eutrophication of soils and water bodies

An LCA represents too great a time investment in many situations. This limitation applies in particular when different solutions need to be quickly evaluated as an intermediate step in a design or engineering process. Therefore, in product design, instead of a complete LCA, it is proposed to estimate the CED approximately (Ashby 2009).

Cumulative energy demand (CED)

If no LCA is carried out, VDI 4600-1 offers an independent procedure for determining the cumulative energy demand (CED) for a product, and already provides typical CED values for common materials and semi-finished products from power plants and energy infrastructure, which can then be used instead of a complete individual analysis. This can also be used for a first rough estimate (Ashby 2009).

In principle, this data can also be taken from an LCA database or from the GEMIS software. By using such typical values from databases, the comparability between different products or production routes is significantly improved, since systematic errors affect all variants equally. In addition, the use of average or typical values from the literature provides a safeguard against one's own errors. At the very least, one's own values should be critically compared with common and established figures.

System Boundary of the Investigation

Defining the system boundary is decisive for the informative value of studies or analyses in which the life cycle of a product forms the data basis. Conversely, it is possible to optimize the result of a life cycle assessment toward the expectations of the client through a skillful choice of system boundaries. Only if a study clearly and completely describes its system boundary and clearly justifies why the system boundaries were defined in exactly that way can it be credible. The background is that every single aspect and every stage of the life cycle requires energy and resources and causes impacts. These impacts and resource requirements are essential

- For the provision of energy across all life phases of the product,
- For the provision of resources for the product,
- For production and transport,
- For (illegal) disposal, recycling and (illegal) landfills,
- For social aspects that influence or change the product and the life cycle,
- For health aspects, namely the direct aspects of occupational safety and the health of the user as well as conditions of extraction and disposal of products and emissions along global supply chains,
- For health aspects, namely the indirect health consequences of emissions, as well as
- For digitalization including questions of accessibility, equity and surveillance, for the handling of fundamental rights and for the impacts on governance systems.

All these impacts are simultaneously the responsibility of the manufacturer. This responsibility is now enshrined in a large number of laws (e.g., in the German Circular Economy Act, KrWG § 23).

Implementation of the EU REACH Regulation and ingredients

Regulation (EC) No. 1907/2006 (REACH) of the European Union stipulates in Article 33 the obligation to pass on information about substances in articles:

> *"1. Any supplier of an article containing a substance (...) in a concentration above 0,1 % weight by weight (w/w) shall provide the recipient of the article with sufficient information, available to the supplier, to allow safe use of the article including, as a minimum, the name of that substance.*
>
> *2. On request by a consumer any supplier of an article containing a substance (...) in a concentration above 0,1 % weight by weight (w/w) shall provide the consumer with sufficient information, available to the supplier, to allow safe use of the article including, as a minimum, the name of that substance. The relevant information shall be provided, free of charge, within 45 days of receipt of the request."*

This means that manufacturers of products are obliged to know which ingredients are present in which weight fractions (w/w) in all parts and components of their products. This information must therefore be transmitted correctly and completely along the supply chain. In addition, in global supply chains it is necessary to know and comply with all applicable regulations on ingredients for as many countries as possible that are traversed by the product and its life cycle. For this purpose, current and complete information is required about which substances are regulated or prohibited where.

The automotive industry has developed a digital tool for itself that can provide information on ingredients completely and as accurately as possible. This International Material Data System (IMDS, *https://www.mdsystem.com*) is a database that maps the entire global supply chain. Specific rules apply as to what information is provided at which point, by whom and how, without disclosing trade secrets where applicable. The system is voluntary, but does exert considerable pressure within supply chains to participate.

In addition, there is a binding list, the Global Automotive Declarable Substance List (GADSL, *https://www.gadsl.org*), which lists substances that are relevant in the automotive sector, whose use is prohibited or restricted, or where this is foreseeable. For all participants in the supply chain, it is thus directly apparent when they are no longer permitted to use which substances in their products.

The administrative and thus energetic input for these measures is considerable for such a large and globally connected sector. However, the input is small compared to the alternative of all automotive manufacturers having to test and monitor every component they use themselves.

Manufacturing processes A major challenge for the analysis is often already the understanding of the manufacturing processes of raw materials and components. For central processes such as aluminum or steel, we may have a rough idea of the procedures, but there is an impressive variety of processes and in some cases also dependencies on other processes or raw materials. It quickly becomes necessary to seek out experts who can show the way here.

Use phase The use phase of a product frequently, but not always, determines resource and energy consumption. The use phase can vary with respect to many aspects and the product can change or remain the same over the use phase (Table 4.2). During the use phase, operating supplies, fuels, auxiliaries, spare parts, new components, software, energy and much more may be necessary to use, maintain or modify the product. The use phase may or may not determine the energy demand of the product in the life cycle. This often depends on the system boundary.

Products can be disposed of at the end of their first use by their owners or passed on for further use. Examples of further use are

- The sale of a used product on the primary market,
- The lending of a product that is not currently being used,
- The gifting of a used product,
- The sale of the used product on a market with lower requirements regarding safety, environmental impacts or similar, and
- The sale of components from the product for further use.

This determines the actual period of use and intensity of use and thus influences the resource consumption of the use phase. Particularly in the case of capital goods, this can represent a significant aspect.

Smartphone

The energy consumption of a modern smartphone is surprisingly low in terms of computing performance. The electricity required to operate the smartphone is on average significantly lower than the energy expended for its manufacture.

This statement is only valid as long as the smartphone alone is to be the system. It changes in an impressive way through use: the smartphone is worthless without the system of cell towers, data transmission lines, data centers, servers, algorithms, programmers, click workers in low-wage countries, etc. If we take this entire infrastructure into account, we arrive at a completely different result. The results of energy and resource consumption therefore always depend on the chosen system boundary and on the actual use. Especially for networked systems, it is therefore extremely challenging to define a suitable functional unit and meaningful system boundaries.

Table 4.2 Relevant Aspects of the Use Phase of Goods

Aspect	Explanation
Service life	The service life can vary from a single short-term use (toilet paper, paper tissue, disposable razor) to very long periods of time (buildings, dams).
Intensity of use	A product can be used continuously over its lifetime (blast furnace, glass melting furnace, residential building, locomotive) or it may not be used at all (knick-knack, sports car in the garage, mistaken purchase at the boutique). This is described by the utilization rate. The purpose of a product may be to be used continuously (bed, table, chair) or never (fire insurance).
Maintenance and upkeep	A product can be well maintained and lovingly cared for, but it can also be neglected. There are products for which regular maintenance is planned and necessary (machine tool, agricultural machine). There are products that manage without any maintenance, either because it is truly not necessary or because it is not really intended (smartphone).
Conversion	A product can be modified repeatedly over its service life (residential building) or it can remain unchanged (painting).
Upgrade and expansion	A product can be expanded with new functions (reflex camera, new software or hardware for a machine). There are products for which this is not possible (cutlery).
Downgrade	A product can be simplified by removing functions that have become obsolete.

Use phases

A German freight forwarder operates a fleet of new trucks (first life). He regularly passes on the trucks after a few years. The trucks then first go to another freight forwarder in Germany (second life), then, as it becomes too costly to keep the trucks roadworthy according to German regulations, to Eastern Europe (third life). From there they end up in Africa a few years later (fourth life). At some point they fall apart, but the engines can still drive pumps (fifth life).

End of life The end of life of the product encompasses all aspects of the product's further life cycle after its use phase. It is important to distinguish between the end of use and the end of life:

- End of use: The product can continue to be used for its intended purpose. It is nevertheless disposed of.
- End of life (EOL): The product can no longer be used for its intended purpose and is therefore disposed of.

The end of life can be prevented or delayed by repair, upgrade or downgrade. The end of use could also be prevented by these measures. The decision not to continue using a product depends on a number of factors and on whether it is a commercial or industrial use or purely private use.

Once the decision has been made to dispose of the product, the end of life of the product begins. The actual end of life of products is determined by a multitude of very different aspects. Legal requirements, their actual implementation or enforcement by the legislator, the possibilities and willingness of society, existing business models and actors define the processes. In a life cycle assessment, it is therefore less the plan than the reality that needs to be considered.

When is the end of life reached?

Commercial or industrial

The reasons for an end of life are in particular:

- A repair is no longer worthwhile: The costs exceed the benefit.
- Repair is no longer possible, in particular because the required spare parts are no longer available.
- The product itself has become obsolete. Often the process for which the product was intended is no longer needed and the product can no longer be used for any other process either.
- **Automation:** The product cannot be profitably integrated into new or stronger automation.
- **Scaling:** The product cannot be profitably integrated into a larger, faster, or even smaller system.

- **Process costs:** A significant reduction in energy or material consumption makes replacing the product with a significantly better product sensible and worthwhile.
- **Safety and occupational health:** The product no longer meets the occupational health and safety requirements.
- **Legal aspects:** The product or process is no longer permitted.

Based on economic reasons, it is examined whether the product can then be relocated to another site, whether it can be sold or reused as a whole or in part, or whether it actually has to be disposed of (as a last resort).

In the private sector

The reasons for end of use in wealthy regions and households, in particular:

- The repair seems rather tedious: It involves effort to find out who, where, how and when the product can be repaired.
- A new expression is desired, because there is a new design or a new fashion.
- The product no longer conveys sufficient status or the neighbors or friends have a newer product ("have what the Joneses have").
- The product no longer satisfies current needs. The needs have shifted.
- The product is no longer needed.
- No repair is worthwhile: The ability for one's own repair is lacking. The effort seems too great. The effort is too great.
- Repair is no longer possible: The spare parts are not available. No suitable craftspeople can be found for the repair.

In wealthy households with storage space, these products then accumulate. Households that are poor, have scarce resources, or are located in regions where there is a market will try to resell the product at the end of use or end of life.

When does the life of a product end energetically?

This is a question of system boundaries. The energy expenditures for transport and further treatment steps can be included without difficulty. The volume reduction in a waste incineration plant or the placement in a landfill can also be included.

With smartphones it becomes difficult when these, declared as economic goods, leave the EU in order to be "disposed of" in poor countries. The concrete technical steps that are carried out, for example in Korle (Ghana, 5° 31′ 30″ North and 0° 15′ 10″ East), to extract precious metals using backyard technology, are not described. What efforts will be necessary in the long term to limit the social and ecological consequences of e-waste there, or at least to secure the landfills there, is unknown. The city is partly built on e-waste. E-waste recycling and disposal take place directly within the settlement.

For hard coal from Germany, so-called eternal burdens exist. This means that the water management of the shafts and tunnels will be carried out for all eternity, since impressive quantities of toxins are stored there. Independently of this, the polders created by ground subsidence will have to be drained for all time. If these activities are included within the system boundary, the energy expenditure for this disposal over geologically long periods of time increases impressively. A similar situation applies to radioactive waste from nuclear energy. Here too, security measures are planned and necessary over thousands of generations. The benefit occurred within a few decades. From now on, energy debts are being repaid.

4.2 Resources

Resources are always needed to provide products or services. Resources are fundamentally limited foundations for economic action. As part of a life cycle assessment, it is always necessary to identify these resources. Essential resources are in Table 4.3 listed, together with a generally formulated assessment of their availability. In particular, the provision of socially usable energy always requires resources. Dealing with the impacts of a product life cycle requires additional resources for the treatment, minimisation, adaptation or avoidance of these impacts. This is impressively illustrated by the increasingly necessary adaptation to the consequences of global heating. Here we are only at the very beginning of the foreseeable efforts that will be required (Kotz et al. 2024).

Table 4.3 defines groups of resources that are closely interconnected with each other. In particular, available energy is a link between resources. It is only through available energy that the other resources can be accessed. Conversely, many other resources are needed to access energy.

Table 4.3 Resources As the Basis for Technological Human Action

Resource	Definition	Limits
Inorganic substances	This refers to any material resource, in particular raw materials, semi-finished products, preliminary materials, all process gases as well as all liquids including water and air, insofar as it originates from inorganic sources. This includes fossil fuels.	The quantity of mineral raw materials is fundamentally limited. The quantity of mineral raw materials available at any given time and the flow of mineral raw materials are limited already due to the energy required for their provision and the machinery for extraction.

Resource	Definition	Limits
Organic substances	This refers to any material resource, in particular raw materials, semi-finished products, preliminary materials, all process gases as well as all liquids including water and air, insofar as it originates from organic sources – i.e. living plants, animals or fungi.	The available mass of renewable raw materials is limited at any given time. The mass flow of available renewable raw materials is limited due to the energy required for cultivation, the machinery, the people, the land area required for cultivation and the soil required for cultivation.
Energy (work and heat)	These are all forms of work and heat that are required over the entire life cycle of a product or service or that enable this life cycle.	The work and heat usable in the earth system over a period of time are limited. The total available amount of energy from fossil fuels is limited, because the quantity of available raw materials itself is limited. The available power and heat flow at any given time are limited, since the mass flow of raw materials and the solar incoming radiation are limited. The power and heat flow from renewable raw materials are limited, because the flow of renewable raw materials for this purpose is limited.
Capital	Capital is the resource that is needed by society to make things possible.	Capital can be in private hands or in public hands (village, town, district, state, country), with different motivations for the use of capital. The amount of available capital over a period of time and the flow of capital are fundamentally limited. Private capital usually flows to activities that enable a good return at low risk. This further restricts the availability of capital for activities with high risk or low yield in principle.

Resource	Definition	Limits
Land area (*Ricardian Land*)	This is the area on the earth's surface that is occupied for a specific activity and thus cannot be available for another.	The land area on earth is limited. Land area can often only be occupied for one use at a given time. One use generally excludes all others. Thus possible uses compete with one another. This concerns uses such as agriculture and food, housing, transport, infrastructure, recreation, industry, energy generation, extraction of mineral raw materials, living and functioning ecosystems, as well as the shifting and migration of ecosystems due to climate change.
Solar incoming radiation	This refers to the solar incoming radiation at ground level, depending on weather, climate and time of day and year.	The solar incoming radiation available at a location and the total incoming radiation into the earth system are limited and determined by the earth system.
Water	This is the water directly available at a location at a given time for an activity.	This means sufficiently clean water with a low salt content and a low content of other contaminants or pathogens. The total mass of fresh water is limited. The flow of fresh water is limited. Seasons, weather, climate influence availability.
Soil	The soil is a living substrate in which plants can grow well and productively.	Soil is, alongside land area and water, another necessary prerequisite for agriculture. The requirements for good soil are extensive. They include the composition, the particle size distribution of the solid inorganic components, the quantity and chemical representation of nutrients for plants, the type and composition of organic material, existing organisms, and the absence of toxic substances. The amount of soil is limited and is decreasing rapidly. At the same time, soil is the prerequisite for terrestrial ecosystems, including agricultural and forestry ecosystems (Gomiero 2016).
Organisms	These are individual living beings of a species (plants, animals, fungi, bacteria, viruses, etc.) as a self-renewing resource.	The number of organisms is finite and countable, both spatially and temporally.

Resource	Definition	Limits
Species	This is the sum of all organisms of a species that can collectively ensure the continued existence of the species.	Every existing species is a potential for current or future solutions. The number of species is limited in the earth system. It is currently decreasing very rapidly (*6th mass-extinction*). The rate at which new species can develop and occupy vacated ecological niches is high. Often just a few 10,000 years are sufficient for higher life forms. However, this is negligible for the time frame in which human societies act. Relevant exceptions are pathogens and pests.
Ecosystems	Ecosystems are functional units in which a large number of species coexist together, thereby ensuring the existence and dynamics of the ecosystem.	Individual ecosystems are the basis of ecosystem services. These are services that would otherwise have to be provided by technology. Furthermore, ecosystems and in particular their diversity offer a potential for adaptation to changing environmental conditions or future problem-solving. The number and quantity of valuable ecosystems are limited in the earth system. Both values are currently decreasing rapidly.
People	These are sufficiently physically and mentally healthy, sufficiently capable, and sufficiently trained people for a specific task.	Industrial global warming threatens the health and cognitive abilities of all people through a multitude of mechanisms.
Law	This is the reliable order of a society.	Law encompasses the ability to enforce rules, agreements, and contracts. Law is the foundation for trust in mutual dealings and in action.
Stable, plannable conditions	These are the earth system within planetary boundaries.	What is meant is a tolerable risk of life within the human ecological niche. Within this, there are limited risks in the area of basic human needs (food, shelter, water, ...).

4.3 Raw Materials

In this section we look specifically at inorganic resources. While many of the other resources from Table 4.3 are renewable or remain available only within a certain framework, inorganic substances have the special characteristic that they are also absolutely finite in availability as long as they are not completely reused.

Reserves and Resources of Finite Raw Materials

For all raw materials, which are extracted from geological deposits, clear definitions exist for relevant subsets. These are summarized and explained in Table 4.4. The terms from Table 4.4 are illustrated in Figure 4.1. These subsets refer, for a raw material, either to a specific deposit, to a region, or to the entire earth system. Unless otherwise stated, these figures always refer to the earth system, i.e., to global quantities. In public discourse, these terms are very frequently conflated. Resources located at depths inaccessible today are readily counted among the reserves, as are possible potentials. In addition, outside of scientific consensus there are wild theories about the formation of raw materials. These theories are romantic or surprising, but lack any scientific basis.

Table 4.4 Terms Relating to Raw Material Deposits

Term	Definition	Relevance for the future
Extracted	This is the total amount of the raw material that has already been extracted from geological deposits.	Extraction lies in the past. Deposits still exist in human artifacts and in waste.
Reserve	This is the total amount of the raw material that can be safely extracted under today's conditions.	For reserves, all necessary prerequisites for their extraction are in place today: ▪ The deposits are known. ▪ The quantity of the raw material present in the deposits and its grade are well known. ▪ The costs of extraction are known and the required market price for economically viable extraction is given for these raw materials. A change in the market price thus leads to a change in the quantity of reserves.

Term	Definition	Relevance for the future
Resource	These are the known deposits of the raw material that cannot be economically extracted under today's conditions.	The necessary economic prerequisites for extraction from known deposits are not given: ▪ The deposits of the reserves are known. ▪ The quantity of the raw material present in the deposits and its grade are well known. ▪ The costs of extraction are known, but the required market price or a significantly more cost-effective technology for economically viable extraction is not available for these raw materials today. A change in the market price or the technologies used thus leads to a change in the quantity of reserves.
Potential	These are unproven but geologically possible raw material deposits.	Specifically, these are suspected deposits: ▪ It is not known whether the raw material is actually present at the suspected location, or the raw material has in principle already been found at the location, but the quantity present there and the grade in the rock are unknown. ▪ The exact shape of the deposit is unknown. ▪ The costs of extraction are not yet known.

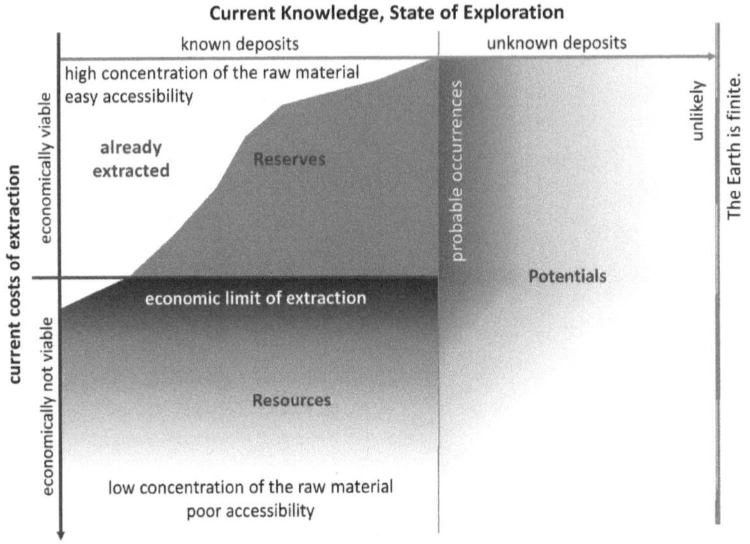

Figure 4.1 Definition of the terms reserve, resource and potential

Range

Essential for reliable planning is the range of the reserves of a raw material in conjunction with the market price. The range of minerals is often given as the static range. This is the time that results when the raw material is extracted from the available reserves at the current extraction rate: The range increases when the market price for the raw material rises and when new reserves become known. The range decreases when the extraction rate grows over time.

For a known or assumed reserve, different figures for the range are given, such as Figure 4.2 illustrated for silver. The range depends on a very large number of factors and changes due to market conditions and technological development. One measure is the static range. This is the range at constant consumption (Narbel et al. 2014). A second, somewhat more complex option to calculate is the application of a bell curve of annual extraction. This function takes into account the relevant information about historical consumption, reserves and resources and is then fitted to the historical consumption based on these boundary conditions. An example of a bell curve can be found in (Lundaev et al. 2023).

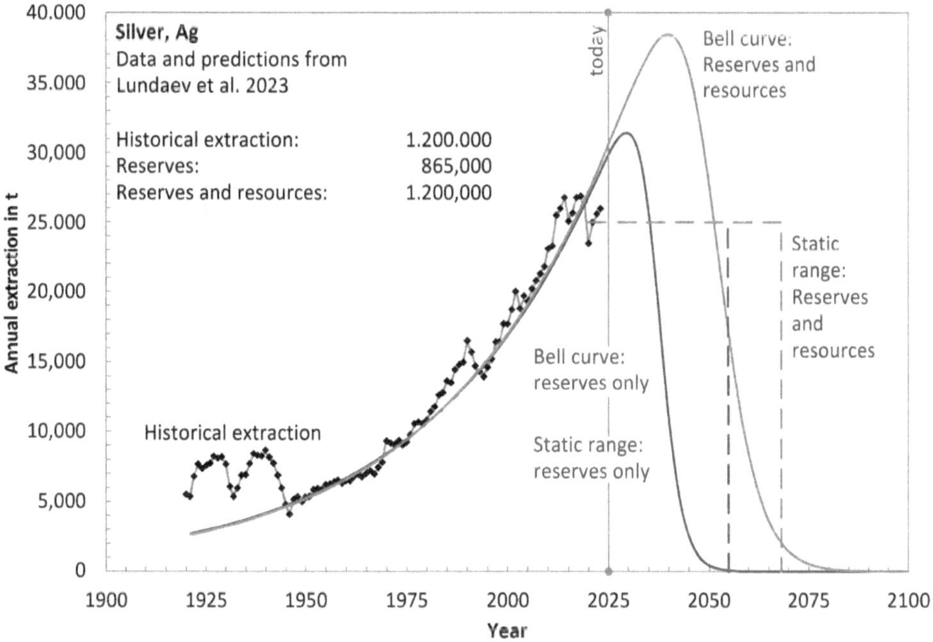

Figure 4.2 Ranges for known reserves and resources of a raw material using silver as an example (data and predictions from Lundaev et al. 2023)

Technical and economic limits for the extraction of raw materials are closely interrelated. Important aspects are:

Concentration This refers to the typical content of the desired raw material in the deposit: the higher the concentration in the rock, the less total mass needs to be moved. This also makes the separation of the raw material from the gangue easier. As a result, the technical, energetic and financial efforts for processing are lower overall. For this reason, these high-grade deposits are preferentially mined first.

- **Mass – the size of the deposit:** Large deposits that can be extracted over long periods of time allow significantly higher on-site investments than small or scattered ones.
- **Access – the accessibility of the deposit:** Proximity to trade routes, existing infrastructure, railways, ports and the electrical grid reduces the necessary own investment in infrastructure and transportation costs.
- **Depth – the proximity of the deposit to the earth's surface:** With the depth of the deposit, all costs increase significantly. Depending on the geological conditions, depths of more than a few thousand meters are no longer accessible due to geothermal heat alone. In general, deposits below 3000 m depth are considered unreachable today, although this limit should be set at 1000 m for many substances.

An exception is the gold and platinum mines in South Africa, which reach depths of over 4000 m. Here, the market price justifies the associated effort.

- **Geology – the composition of the rock:** Different types of rock require or allow differently complex methods for mining and separating the raw material from the gangue. This directly influences costs.
- **Regulations:** These are the local requirements for safety, environmental protection, working conditions, etc., or their absence. These factors determine further costs of extraction, as they define or allow the avoidance of measures that themselves generate costs.

The enumeration directly illustrates that large, easily accessible deposits of high concentration are mined first, and that over time mining must shift to more remote, more difficult, smaller and more expensive deposits. The financial and energetic costs of extraction therefore inevitably increase over time. Technical developments and efficiency improvements of the required processes can influence this effect to a certain extent. In the long term, it is to be expected that the price at which the raw material is traded on the market will increase.

Data – How long will a raw material last?

Predictions about ranges are difficult to make. They are not very accurate and always contain many assumptions about the future. The assumptions about future developments are not always explicitly stated or are based on estimates.

An overview of the challenges involved in developing reliable statements on this topic is provided by König (using the example of predictions for lithium and copper, König 2023). For critical metals of the energy transition, Lundaev et al. provide an overview based on historical data, reserves and resources (Lundaev et al. 2023). Figure 4.3 provides an overview of the reserves and resources of important metals, Figure 4.4 specifically for the metals of the platinum group and rare earths. The figure is always related to the mass already extracted and gives an impression of how many times we can still extract this mass. Some raw materials are close to exhaustion.

Mineral raw materials For many raw materials, the static range was roughly 30 years. This period is the planning horizon within which new potentials are explored in order to be able to develop them in time. This applies to raw materials with extensive potential, where potential deposits have not yet been developed for various reasons:

- The potential deposit is far from suitable infrastructure such as ports, railways or roads. This case is becoming increasingly rare, as even the most remote regions of the earth are now being developed, such as Baffin Island for iron ore, Arlit in the Sahara for uranium, or Dikson on the Siberian Arctic Ocean for coal.
- The deposits are technically inaccessible today. The occurrences lie too deep in the earth, under the ice of Antarctica or in Greenland.
- The ore concentration is so low that mining is not worthwhile.

- The deposits are legally or socially inaccessible. They are located in a national park or a nature reserve, under cities, or the local community does not consent to mining.
- The deposits are politically inaccessible today due to civil war or war. A counterexample is the mining of tantalum and cobalt in the Congo.

Fossil energy carriers Reliable information about the quantity of reserves and resources of fossil primary energy carriers is actually necessary for planning. In this area in particular, however, the data situation is especially difficult. On the one hand, there is a recurring confusion of concepts, so that many reports must be read with great caution; on the other hand, there are large reserves that have remained constant for many years despite intensive extraction and dramatic price fluctuations. This is due to the fact that oil-producing nations in particular do not want to provide precise figures for political reasons (McGlade/Ekins 2015).

Data Relevant sources providing – as far as possible – reliable statements are provided by, for example, the following institutions:

- German Energy Agency (*https://www.dena.de*)
- German Mineral Resources Agency (*https://www.deutsche-rohstoffagentur.de*)
- Federal Institute for Geosciences and Natural Resources (*https://www.bgr.bund.de*)
- US Geological Survey (*https://www.usgs.gov* – whether this will remain so in the future remains to be seen)
- Raw Materials Information System of the Joint Research Center of the EU (*https://rmis.jrc.ec.europa.eu*)
- World Fact-Book of the CIA (*https://www.cia.gov/the-world-factbook* – whether this will remain so in the future remains to be seen)
- International Energy Agency (*https://www.iea.org/energy-system*)
- BP Energy Outlook (*https://www.bp.com/en/global/corporate/energy-economics/energy-outlook.html*)

In addition, there are publications by major management consultancies on specific topics.

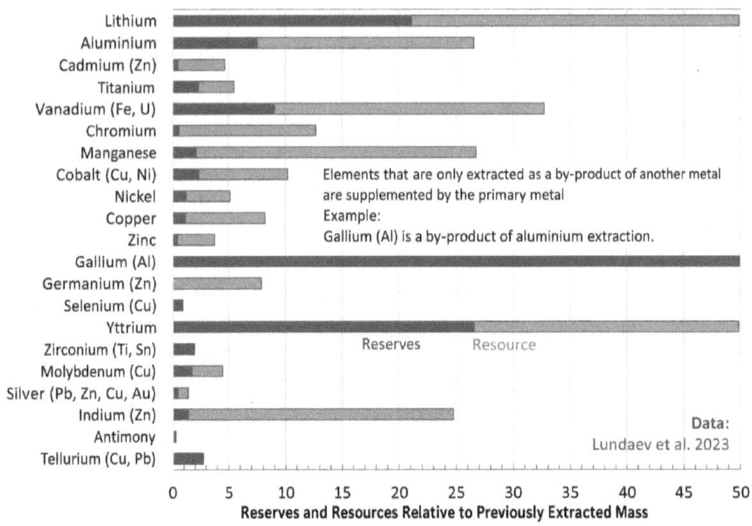

Figure 4.3 Long-term availability of critical metals (data source: Lundaev et al. 2023)

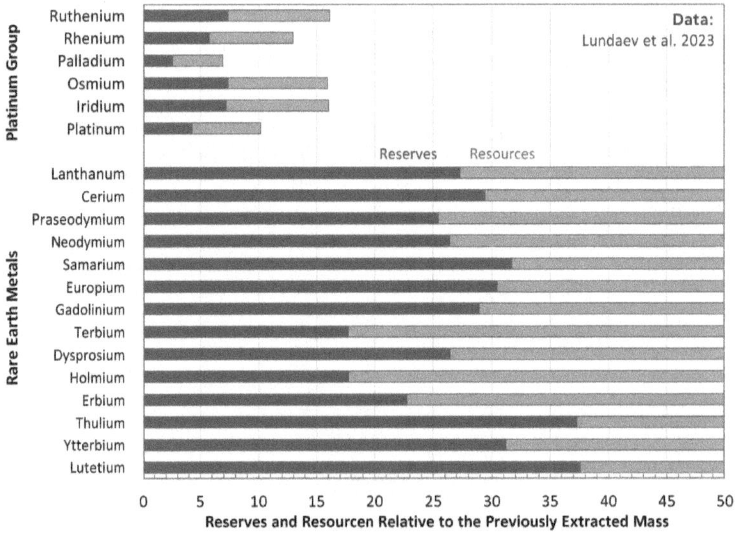

Figure 4.4 Long-term availability for platinum group metals and rare earth metals (data source: Lundaev et al. 2023)

Peak-X

Every raw material in the earth system is finite. The limit is not the total mass of a raw material available in the earth's crust (see Table 17.1 in Section 17.1). The balance between the financial and energy input for extraction versus the maximum price achievable on the market determines how much of the raw material can be extracted in total.

For every extractable raw material, there is inevitably a point of maximum extraction: before this point, annual extraction increases on average, and after this point, this quantity declines. It has become common practice to call this the peak point (Narbel et al. 2014).

For some raw materials, the point of maximum production lies in the past, such as for mercury or asbestos. For most raw materials, it lies in the future from today's perspective. A retrospective analysis of the production volume of a raw material, either locally from a mine, for a national economy, or globally, is possible for most raw materials. Based on this data, it is then possible to determine retrospectively when the point of maximum production occurred, provided it lies some time in the past. Looking ahead – predicting when the production volume of a raw material will reach the point of maximum extraction and how the annually extracted quantity will subsequently decline – is more difficult.

The incomplete overview in the box shows that simple linear assumptions for predictions are not valid. There are many feedback loops within the system as well as through the achievable price and energy costs into the market. Price signals often have very delayed effects, since from the start of exploration to the start of production, even in the fastest cases, years pass. A currently low price will only lead to a shortage with a very significant delay.

What determines the timing of the peak?

Relevant factors, which influence the further course of extraction, are:

- The available reserves increase when the price of the raw material rises: in that case, the higher extraction costs of more difficult, remote, or smaller deposits become economically viable.
- The available reserves increase when new deposits are found through exploration. The prerequisites are that (i) sufficient capital is available for exploration, (ii) new undiscovered deposits of relevant size still exist, and (iii) the new deposits can be extracted at normal market prices, i.e., they become reserves.
- The sector must continuously receive sufficient capital to advance the exploration and development of new deposits. The capital market must have confidence that the capital invested will yield an adequate return. This is fundamentally not straightforward in the raw materials market, since decades typically pass from exploration to production.
- For raw materials whose extraction or processing is particularly energy-intensive, the market and thus the volume of reserves can change significantly when energy prices rise or fall.

- Technological developments in extraction and processing can influence costs, so that resources become reserves through new technology. Such developments take place over long periods of time and are generally known within the sector, so they can be factored into planning.
- Some raw materials can be replaced by other materials. Especially for rare and expensive raw materials, industrial users make great efforts to replace them with cheaper and readily available alternatives, thereby rendering them obsolete.
- Users want and can, in the long term, only pay a certain maximum price for a raw material: the resulting cost pressure on the companies extracting the raw material can lead to the premature cessation of mining at more expensive deposits. This typically prevents investment in the exploration and development of new deposits. Scarce capital can also prevent new investments in better processes. This initiates a vicious cycle that can lead to a sudden severe shortage of the raw material.
- Many smaller mines or sources depend on shared market access. When the first mines are exhausted, the transport costs of the others increase. This can lead to an early end of production well before all deposits are exhausted.
- Regulatory requirements can reduce or end the use of certain substances, and they can influence the costs of extraction at certain locations. They can ensure that other raw materials are needed in greater quantities.
- Regulatory requirements regarding emissions or health protection can ensure that reserves in certain regions can no longer be extracted profitably.
- If, for example, a closed recycling loop develops for the raw material as a result of regulatory requirements, this can lead to new extraction being needed only to a minor extent. This is the goal of a circular economy.
- Many elements that are used only in small quantities or are present only in low concentrations are not mined as raw materials in their own right. They are extracted as a by-product of another raw material, or they occur in worthwhile concentrations only as an admixture of another raw material. In that case, the production volume and availability depend on that other raw material. This applies to some critical raw materials for electronic components (Figure 4.3).

Peak Oil/Peak Gas/Peak Coal

The discussion on the peak for fossil energy carriers has particular significance: On the one hand, the extraction of all other raw materials depends on the availability of affordable fossil energy carriers. Diesel plays a special role here, as it powers many of the machines and vehicles used for extraction and cannot easily or cost-neutrally be replaced by other forms of energy. Thus, the price of fossil energy carriers has a direct impact on the price of all other raw materials through extraction costs and transport costs. On the other hand, the extraction of fossil energy carriers itself requires large quantities of energy and other raw materials. As a result, price increases for raw materials also affect the costs of extracting fossil energy carriers.

The consumption of fossil raw materials is increasing, while at the same time old, easily extractable large deposits, particularly for crude oil, are slowly drying up. Therefore, development is moving toward sources that are more difficult to develop (fracking, oil sands, deep sea, Arctic). To use these sources, more equipment, more complex infrastructure, and more energy are required. All of this increases extraction costs. The following impacts are possible:

- The market price rises sufficiently in the future. This means that increasingly difficult deposits become resources and the costs of extraction can always be covered. This leads, via the increase in energy costs of fossil fuels, to a rise in the cost of all other raw materials. Conversely, this would also lead to a further increase in the costs of extracting fossil fuels. This would be a (positive) feedback loop. Positive is meant in the sense of self-reinforcing.
- The price achievable on the market does not rise sufficiently in the future to be able to count the more difficult deposits as resources. This means that these raw materials remain in the ground. This first leads to a long-term scarcity up to the cessation of production of the affected fossil fuels. As a consequence, the extraction of all other dependent raw materials will be switched to new, possibly more expensive technologies and energy carriers.

Both possibilities are limiting cases that do not occur in this way. They illustrate the challenge for the market, the industrial sector, and society: In the interest of a long-term supply of fossil fuels, a rising price must be accepted. At the same time, a rising price for energy will have negative effects on all other areas of the economy that require this energy directly or indirectly. If these costs are increasingly borne by society as a result of political decisions, these subsidies distort or obscure important signals and may thereby jeopardize necessary transformation processes. Options include direct subsidies or allowing the externalization of costs by reducing safety requirements, easier expropriation, or greater environmental destruction.

For the extraction of fossil fuels, the peak point in time is of central societal importance:

- If the decline in production occurs now due to declining demand, this means, in the best case, that the global energy transition has worked and the shift to the use of non-fossil energy carriers has occurred in time.
- If the decline in production occurs due to the actual foreseeable exhaustion of all deposits, all parties involved can prepare for this and attempt to transition their energy supply in time. This succeeds if it is recognized as a fact for decision-making processes.
- If the decline in production only occurs when the market can no longer bear the increased extraction costs, and this happens in a society that has not made the transition to a non-fossil energy supply, then this will cause a severe economic crisis or a systemic collapse. When the supply of energy to industry and society in the form still required, expected, and solely usable, and at costs considered bearable, no longer functions, then even essential products and services can no longer be provided. The inability to continue massively subsidizing costs can also be the trigger.

This book focuses on evaluation methods for analyses and not on the societal consequences of Peak-X. However, the systemic feedbacks and societal consequences are relevant and it is worthwhile to know and understand them. Predictions that assume a long-term supply of fossil energy at today's prices, as well as expectations that demand this, would be irresponsible and naive. Any political or societal decision-making that is oriented toward the longer term and that takes into account evidence-based relevant risks must incorporate Peak-X early on into all raw material- and energy-relevant aspects of its actions.

4.4 Manufacturing of Products

In order to give the energy indicators in Chapter 5 a solid foundation, this section discusses energy-relevant activities for the production of products. Our view also includes fossil fuels.

Extraction of Raw Materials

Often the life cycle begins with the extraction of new raw materials. This extraction involves a whole range of processes and invested energy. Important individual activities are:

1. The starting point is the acquisition of permits or licenses to search for and explore deposits in a region.
2. Financiers and investors must be found for the search. They must be presented with a sufficiently detailed plan that makes the possible yield and the financial risk recognizable in particular.
3. The actual exploration must be carried out with due thoroughness. Whereas historically it was often chance finds that led to deposits, this has since become its own science, which uses impressive machines to visit remote parts of the world in order to find new deposits there.
4. Once the deposit is known, the mining license must now be acquired. Further permits for the operating site or for emissions may be required.
5. The infrastructure must be adapted, expanded, or built entirely from scratch. Infrastructure refers in particular to the required facilities on site, transport routes, ports, accommodation for workers, supply of people, supply of energy and materials, workshops, and intermediate storage.
6. The actual production access to the deposit must be built: for oil and gas, the first exploratory wells must be expanded or new production wells must be drilled so that production can begin. For the extraction of solid raw materials in open-cast mining, overburden must be cleared away. For the extraction of solid raw materials underground, shafts must be drilled and tunnels driven. Supply infrastructure must be installed in the shafts and tunnels.

7. For production, the required infrastructure must be maintained and kept running. It must be relocated if necessary and adapted to the progress of extraction.
8. Extracted raw materials must be prepared for transport.
9. Separated residual materials such as waste rock and processing residues (tailings) must be disposed of. They are often simply tipped on site as a spoil heap or directed into tailings ponds.
10. At the end of extraction, when the deposit is exhausted, the wells must be sealed and the shafts secured. Open-cast mines must be backfilled depending on applicable regulations. If long-term impacts remain, long-term remediation measures must be carried out. Mines must continue to be dewatered and contaminated wastewater and soils must be cleaned. These can become perpetual burdens.

Particular energy expenditure is associated with the construction and expansion of structures such as boreholes, tunnels, and shafts. During operation, the expenditures for pumps, ventilation, and ore processing in particular can be high.

The extraction of raw materials requires, directly and indirectly, a large number of the resources from Table 4.3. The directly required resources are unavoidable for the purpose of extraction. They are often used efficiently. The indirect and often externalized impacts could be minimized with careful procedures: it is actually of great importance for every society that there are no toxic deposits, long-term destroyed ecosystems and soils, chronic diseases, etc. in its midst. In reality, these perpetual burdens are accepted, as for example in the Ruhr region. There, for example, it is PCB-contaminated wastewater from old coal mines that must be expensively treated for all eternity. The associated non-discountable energy expenditures and losses of further resources must be attributed to extraction.

Processing of Raw Materials

Preparation Usually still at the site of the mine or the well, the raw materials are processed for the first time. The goal is to separate everything that does not necessarily need to be transported: waste rock, unwanted admixtures, toxic substances, etc. are separated from the valuable part with the expenditure of energy and other resources and remain on site as spoil heaps or in tailings ponds.

It is these often poorly secured deposits that have long-term impacts on the surroundings of the mine or well as well as on ecosystems and communities: through toxic substances that slowly and steadily, or in the event of failure suddenly and violently, bring disease and death. The associated expenditures and losses of resources are part of the costs.

Reduction The vast majority of ores occur as oxides, sulfides, or in other compounds. Only precious metals are extracted as native metal. Sulfides are often roasted: the sulfur reacts to form gaseous SO_2 (this often creates acid wastelands around such facilities, as historically in Røros or currently around Nikel). The oxide is then reduced: for iron in blast furnaces, for aluminum in electric furnaces, for many other ores in electric arc furnaces. Almost always the reducing agent is carbon, so that carbon dioxide is released.

Primary shaping In many processes of the basic materials industry, the liquid metal can be taken directly from the reduction furnace. For some important metals, downstream processes are still required: the steelworks converts pig iron into steel. For silicon, a whole series of further purification processes are carried out downstream until it has the required purity.

The specific energy and resource requirements for the provision of semi-finished products (ingots, bars, slabs, ...) depend on these processes. In Table 17.10 and Table 17.11 (see Section 17.6), typical values are compiled for technically significant basic materials, giving an impression of how much energy is expended in each case for their provision as semi-finished products. The externalized and long-term expenditures are not taken into account here.

Manufacturing Products

For the manufacture of products, a whole series of processes is usually necessary. For simple products, it is often possible to survey the entire manufacturing process and describe it with a manageable number of individual steps. For simple products that are manufactured in large quantities, a few process steps sometimes dominate, so that a nearly complete description can be achieved with reasonable effort. For complex products, on the other hand, the number of individual components and manufacturing steps is difficult to survey. In many value chains, it is not possible today to reliably specify the composition of finished products, since this information is already lost along the value chain. It is all the more difficult to identify the individual concrete processes in the life cycle or to know the input of specific individual steps.

For products in which one process dominates the manufacturing, or if different forming processes are to be weighed against each other, Table 17.12 in Section 17.6 provides typical data for common process steps.

Fossil energy carriers

Processing fossil energy carriers

The raw materials are cleaned and components that can already be problematic during transport are separated:

Natural gas In the case of natural gas this means at least the separation of long-chain hydrocarbons that can condense in the pipelines. Desulfurization is advisable in order to reduce the mass to be transported and to prevent corrosion by sulfuric acid in the pipelines. The gas must also be dried to avoid condensation of water. The separation of carbon dioxide is also useful, since carbonic acid has a corrosive effect. These processes each require energy.

Once natural gas has been dried, desulfurized, separated from condensable long-chain hydrocarbons, dedusted, and freed from other possibly undesirable impurities, it is almost ready for use. The last step is the adjustment of the heating value. The heating value is either legally specified or guaranteed by the energy suppliers within a fairly narrow window. For this purpose, either an inert gas or ethane and propane is added to the natural gas. After that, the natural gas can be fed into the general supply network.

Natural gas for pipelines is compressed. The work for compression is comparatively low for natural gas (see Chapter 10). Natural gas can also be transported as liquefied natural gas (LNG) in tankers. For this purpose, the natural gas is cooled to approximately –162 °C, the boiling temperature of natural gas at a pressure of 1 bar.

Crude oil Crude oil can, depending on its composition, which influences viscosity, be transported either in pipelines or in tankers by road, rail, or ocean. Some grades must have a sufficient temperature for transport and pumping, so that some pipelines are heated.

Coal In the case of coal separation of coal and rock is carried out based on density. For this purpose, the coal is already broken to a well-suited size. This process is energetically very favorable. Lignite as a primary energy carrier is mostly not further processed but goes directly to power plants. Hard coal for energy generation usually does not need to be further processed either.

Oil sands For the processing of oil sands and other special raw materials, many additional processes are required, some of which require considerable additional energy (Carroll 2015).

Refinery

Crude oil is a mixture of a wide variety of hydrocarbons. This mixture must first be broken down in such a way that common, market-accepted fractions are produced. This is done in refineries. Important energy products produced there are:

- **Gasoline:** This is a mixture of relatively short-chain hydrocarbons with a low boiling point that evaporate easily. Therefore, gasoline forms an ignitable mixture in air.
- **Kerosene:** This is a mixture of relatively short-chain hydrocarbons that does not form an ignitable mixture.
- **Diesel:** This encompasses many very different supply forms. In the passenger car sector, diesel is more comparable to kerosene. For other applications, diesel can consist of much longer chain hydrocarbons with a high boiling point and be significantly more viscous.
- **Liquefied gas:** This consists primarily of propane (C_3H_8) and butane (C_4H_{10}).

In addition, there is an incredible variety of solvents, waxes, greases, oils and other products that are not used as fuel but are also separated or produced.

4.5 Transport of Goods

Transport is part of every life cycle. Raw materials, semi-finished products, finished goods and waste must be transported from the raw material storage to processing, from one production facility to the next, from the manufacturer to users and from users to the landfill or waste recovery facility. Transport requires energy, in particular:

- The direct energy input to move the mass or volume: Typical values for this are summarized in Table 17.13 in Section 17.7. Actual inputs may deviate from these.
- The input for packaging, i.e. packaging material, machines and equipment as well as the disposal of the packaging after use
- The input for the provision of fuels or energy for the means of transport such as filling stations, bunkering facilities, fuel tanks, electric overhead lines and power plants
- The input for the provision and maintenance of the built infrastructure for the means of transport, i.e. roads, ports and airports, traffic signs, lighthouses and signal boxes
- The input for the special infrastructure in the supply chain such as warehouses, distribution centers and logistics centers
- The input for data centers, digital infrastructure and digital aspects of logistics
- The input for the required financial services and insurance
- The socially borne input for emergency preparedness such as weather services, fire brigades and ocean-going tugs
- The input for the supply of the required personnel, who either directly operate the means of transport or necessary other facilities and indirectly build and maintain the required infrastructure

This incomplete list illustrates that transport is a complex system in which many components and activities must interlock. Energy inputs for transport therefore depend on the extent to which the individual aspects are taken into account. The values in Table 17.13 (see Section 17.7) are therefore lower bounds.

4.6 State of Development of a Technology

In many discussions and publications, new technologies at every stage of their technical development are presented as ready-made solutions to problems. It is then assumed that everything that is conceivable will also be feasible on a large scale within the shortest possible time. Important political decisions are now being prepared and made on the basis of such bets. In the discussion, important categories become confused. This section provides a basic framework for assessing the current state and the chances of implementation.

Start Many interesting ideas are first developed in science fiction and sometimes their consequences are also mentally explored. An important first step is then to check whether the idea is compatible with our laws of nature. Only people who know our laws of nature and have the competence to assess this can do so. A second source is tinkerers who try things out. In this process of trying things out, the incorporation of our natural possibilities happens automatically.

Goal Ultimately, a problem can only be solved with an established technology that is available on the market, for which all resources are available and which is affordable. The path from a fundamentally possible idea in this world to such a product is long. The vast majority of ideas die on the way from the first sketch to market maturity, e.g. because they are not feasible or not affordable or because they tie up too many resources (Narbel et al. 2014).

So that we can clearly assess how far a technology has matured and when we might (if all goes well) expect to use it, there is the robust tool of the technology readiness level (Technology Readiness Level, TRL, see Figure 4.5). Table 4.5 defines the individual levels. Two terms are essential for orientation within these levels:

- A **demonstrator** is a preliminary version of the technology at laboratory scale. Its goal is to demonstrate the effect once. Scale effects and challenges of future technical production are not taken into account here. The second goal of a demonstrator is to gather initial experience and identify key technical challenges to be solved in further implementation. The result of working with a demonstrator can and may be that the technology is not pursued further. Demonstrators are typically operated by people in white lab coats.
- A **prototype** is a version of the technology manufactured and then also operated under conditions already comparable to the intended industrial production. A prototype should already have a size that is fundamentally comparable to the later version. In particular, endurance tests under realistic conditions should be carried out with the prototype. If the prototype can be operated successfully and the experience gained with the prototype is positive, then no further critical scale effects are expected and the fundamental technical challenges of industrial production should essentially have been mastered. The result of working with a prototype can and may be that the technology is not pursued further. Prototypes are typically operated by people in blue overalls.

Photovoltaics

A very good example of the differences between demonstrator and prototype are photovoltaic cells. The best efficiency achieved for such cells has so far been reported for research demonstrators of 1 cm^2 area and for real modules by NREL (*https://www.nrel.gov/pv*). Demonstrator cells are the result of scientific research facilities. Prototypes are produced on sufficiently industrially scaled machines. A closer comparison shows that some technical variants never progressed from the demonstrator to the prototype stage and that some of the prototypes never reached market maturity.

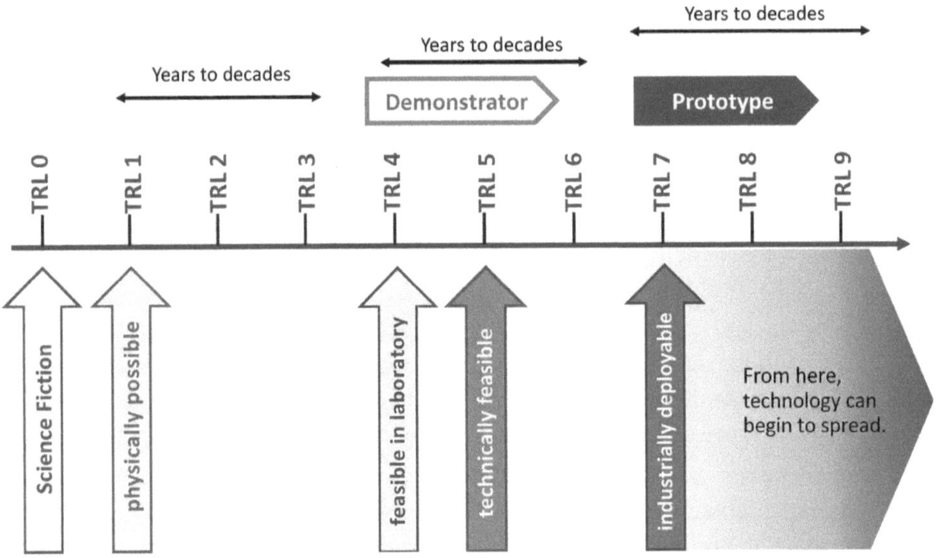

Figure 4.5 Development steps and important hurdles of a technological development

New technology must, in order to keep the financial and technical risk for all parties involved within limits, be developed step by step and scaled in the process. The individual stages are also to be understood as decision points: it must be decided whether further development is possible, economically viable, and desirable.

At a successful TRL 4 of all required components, initial pre-developments are completed and it can be assumed that a technological implementation can succeed, provided that the technological challenges for large-scale implementation identified up to this point can be solved and a market can be found.

At a successful TRL 6, planning on an industrial scale can begin, although the issues of social and market acceptance still need to be clarified.

At TRL 8, important societal decisions can assume that this technology is capable of actually taking on the role assigned to it under the now clearly identified framework conditions. Not before.

Chapter 9 discusses nuclear fusion – a technology in which essential core questions of future power plants are still being addressed in the TRL 0 to TRL 1 range. Chapter 14 looks at geoengineering, where small-scale experiments do not always measurably demonstrate the effects that are to be investigated. Large-scale experiments are immediately very costly and the associated unknown risks might not be socially acceptable.

Table 4.5 Technology Readiness Levels (TRL)

EU 2020	Description	Target/Status
TRL 0	Science fiction: this involves an untested idea or an untested proposal. The concept exists only on paper. In particular, no analysis has been carried out and no initial experiments have been undertaken.	Dreaming
TRL 1	The basic physical or chemical feasibility has been demonstrated by the following: ■ By an analysis (on paper) ■ By references to properties that are sufficiently similar in other existing technologies ■ By initial experiments on sub-processes or sub-systems	This step is intended to demonstrate that the technology can take place within the laws of nature.
TRL 2	A technology concept has been formulated.	Concept phase, in which the first necessary experiments, calculations, and simulations are defined
TRL 3	The concept has been confirmed. A first design of the concept or novel properties of a construction have been confirmed in their function in a model or on a small scale in a laboratory.	This is intended to demonstrate that the technology can be expected to meet specified criteria if additional experiments are carried out.
TRL 4	The new technology has been tested. A **demonstrator** has been built and the function demonstrated through experiments that already took place close to possible operating conditions.	Basic technical prerequisites are in place. Implementation in a larger facility can be prepared.
TRL 5	A full-scale prototype has been built. With the prototype, the technology has been qualified through experiments, with the intended process and environmental conditions being real or simulated.	Initial experience is available at relevant scales and under realistic conditions.

EU 2020	Description	Target/Status
TRL 6	The integration of the technology has been tested. A full-scale **prototype** has been deployed in the intended operating environment and the functions and interaction with the environment have been fully tested.	Technical feasibility has been demonstrated under future conditions.
TRL 7	The technology has been installed. A full-scale prototype has been integrated into the intended process environment and a complete test program of its function and interaction in the intended environment has been carried out.	The technology has demonstrated acceptable performance and reliability over a certain period of time.
TRL 8	The system is complete and qualified.	
TRL 9	This is proven technology integrated into an operational system. The technology has functioned successfully, has demonstrated acceptable performance, and is reliable within the predefined acceptance and qualification criteria.	

We will soon conquer space and harvest raw materials from asteroids

Let us apply this framework to current future ideas: where exactly do we stand with the colonization of Mars by billionaires and mining companies starting in 2033 and the commercialization of raw materials from Mars on Earth?

The television series "MARS" (*https://www.nationalgeographic.com/tv/mars*) provides the focus of the current discussion here. It is based on the state of public debate and a childlike faith in technology. The film "The Martian" with its quite obvious errors can also serve as a reference here: on Mars there is no shielding atmosphere, so humans and potatoes are exposed to intense high-energy radiation and quickly succumb to radiation sickness.

The answer to this question depends on the TRL of all required components: only from TRL 9 onwards can a company send employees whose lives depend on this technology. This means that all components must be at TRL 9. Heroes may risk their lives with considerably less reliable technology, but below TRL 7 they simply will not reach their goal.

It may well make sense to define additional TRL levels for space requirements that indicate a higher degree of reliability and redundancy than applies to technology on Earth: it is relatively straightforward to place a replacement on the seabed within a few days or weeks. Spare parts take years to reach Mars – and until then, things must manage without them.

References

Ashby, M. F.: Materials and the Environment: Eco-Informed Material Choice. Elsevier, Amsterdam 2009

Ausberg, L./Ciroth, A./Feifel, S./Franze, J./Kaltschmitt, M./Klemmayer, I./Meyer, K./Saling, P./Schebek, L./ Weinberg, J./Wulf, C.: Lebenszyklusanalysen. In: *Kaltschmitt, M./Schebek, L.:* Umweltbewertung für Ingenieure. Methoden und Verfahren. Springer, Wiesbaden 2015

Ayres, R. U.: On the Practical Limits to Substitution. IIASA Interim Report. IIASA, Laxenburg 2005

Carroll, A. R.: Geofuels. Energy and the Earth. Cambridge University Press, 2015

DIN EN ISO 12100:2011: Sicherheit von Maschinen – Allgemeine Gestaltungsleitsätze – Risikobeurteilung und Risikominderung

DIN EN ISO 14001:2015: Umweltmanagementsysteme – Anforderungen mit Anleitung zur Anwendung

DIN EN ISO 14040:2009: Umweltmanagement – Ökobilanz – Grundsätze und Rahmenbedingungen

DIN ISO 31000:2018: Risikomanagement – Leitlinien

Gomiero, T.: Soil degradation, Land Scarcity and Food Security. Reviewing a Complex Challenge. In: Sustainability 8, No. 3, 2016, p. 281

Graedel, T./Allenby, B. R.: Industrial Ecology and Sustainable Engineering. Prentice Hall, 2009

König, M.: Entwicklung der Versorgungssituation und des Energiebedarfes mineralischer Rohstoffe im Kontext der globalen Energiewende. Masterarbeit. Darmstadt 2023. *https://doi.org/10.48444/h_docs-pub-394*

Kotz, M./Levermann, A./Wenz, L.: The economic commitment of climate change. In: Nature 628, 2024, pp. 551–557

Lundaev, V./Solomon, A. A./Le, T./Lohrmann, A./Breyer, C.: Review of critical materials for the energy transition, an analysis of global resources and production databases and the state of material circularity. In: Minerals Engineering 203, 2023, 108282

McGlade, C./Ekins, P.: The geographical distribution of fossil fuels unused when limiting global warming to 2 °C. In: Nature 517, 2015, pp. 187–190

Narbel, P. A./Hansen, J. P./Lien, J. R.: Energy Technologies and Economics. Springer, Heidelberg 2014

5 Energy Performance Indicators

Meaningful performance indicators help us evaluate technologies for energy supply and communicate this assessment in an understandable way. This chapter explains relevant indicators and how they can be used for the comparison and evaluation of power plants and energy infrastructure. General concepts are presented first. The specific indicators follow the framework of ISO 13602-1.

5.1 Quality of Energy

Are all forms of energy equivalent? Do all forms of energy have the same value? Can I convert all forms into one another? The question itself suggests the answer, and the brief introduction to thermodynamics in Chapter 3 shows that there are relevant differences here. Concentrated energy is valuable. Energy is more valuable when it can provide more usable work. From the Carnot efficiency, a direct statement follows regarding the quality of transferred heat compared to work performed. Table 5.1 provides an orientation to forms of energy in ascending quality.

This question is of high relevance when, later in the chapter, different forms of energy are added together or divided by one another: results that are unreliable or even incorrect arise when the quality of energy is not taken into account. There is currently no consensus on this, only proposals for how it could be appropriately resolved (Kraan et al. 2019; Solomon et al. 2023). Here, the use of the same unit (joule or watt) creates the impression of comparability, which is however not given, as Table 5.1 shows. For energies to be added, the unit must have an identical meaning. Ratios such as an efficiency only become dimensionless when both energies refer to the same form of energy.

This problem can be solved if all energies refer to a truly comparable form, or are converted to a comparable unit via correction factors (Solomon et al. 2023). Therefore, in the following, electricity is always meant by energy unless explicitly stated otherwise. Other forms of energy are then converted into electricity using the values given in Table 5.1 – as then comparable joules J_{el}.

Table 5.1 Quality of Energy Forms: the Efficiency of Conversion Always Refers to the Conversion of This Form of Energy into Work Performed.

Form of energy	Use	Conversion	Quality
Heat or cold near ambient temperature (ΔT small)	Heating, cooling, drying Use only in the intended form	Not intended	Very low
Cold at approximately 5 °C	Cooling, air conditioning, storage technical processes	Not intended	Medium
Cold at < –18 °C	Long-term storage of food drying of food technical processes	Not intended	High
Heat at approximately 40 °C to 80 °C	Heating, hot water	Not intended	Low
Heat at approximately 100 °C to 150 °C	Preparing food, drying, long-term preservation of food first thermotechnical processes	Rarely intended, approx. 10 % η = 0.10 for electricity from geothermal energy	Medium
Heat at 500 °C to 800 °C coal, gas, oil, concentrated solar thermal energy	Steam turbine generates work many thermotechnical processes	η approx. 20 % to 40 % η = 0.38 for electricity	Medium
Nuclear fuels	Complex provision of heat at < 500 °C	η approx. 20 % to 40 % η = 0.33 for electricity	Medium
Heat at approximately 700 °C to 1400 °C	Combustion engine and gas turbine generate work	η approx. 20 % to 40 % η = 0.33 for electricity	Medium
Chemical fuels	Heat over a wide temperature range	Via heat	High
Mechanical work, shaft work	Conversion into any other form of work	η close to 1	Very high
Potential energy, kinetic energy of large masses of fluids	Conversion into any other form of work	η close to 1	Very high
Very intense directed electromagnetic radiation (> GW m^{-2})	Almost every technical process cutting, welding, separating		Very high
Electricity	Conversion into any other form of energy	η close to 1	Maximum

Exergy One attempt to capture this quality of energy is the concept of exergy. Exergy is defined as the maximum work that a system can deliver to its surroundings when it transitions from its state to the state of the surroundings, or conversely the minimum work that is theoretically required to bring a system from the defined conditions of the surroundings to a new target state. Exergy refers to a reference state (environment). The atmosphere at standard conditions, seawater at 4 °C, or the average composition of the earth's crust are often used for this purpose. Exergy is an overall cumbersome concept that will not be pursued or applied further here. In Szargut you will find a compact introduction, important data and relevant applications (Szargut 2005).

Based on the Gouy-Stodola theorem the loss of work and thus the change in the quality of energy can also be made tangible without using exergy.

5.2 Maximum Power or Maximum Efficiency?

Power plants should have the highest possible output at maximum efficiency. Both cannot be achieved simultaneously. Therefore, the trade-off between efficiency and power output is a necessary part of the design of energy systems. It is noteworthy that natural processes also frequently operate at the extremes of high power or high efficiency. When the natural system (such as living organisms or a hurricane) can convert a very large amount of energy, it operates close to maximum power (Kleidon et al. 2016). Conversely, living organisms and self-organized structures operate close to maximum efficiency when the system has been settled for a long time, is therefore close to thermodynamic equilibrium, and little free power is available. Many systems can be well understood and modeled through the maximization of the power of energy flows and thus with maximum dissipation and generation of entropy (Hall 2017; Kleidon et al. 2016).

Gas turbines

Gas turbines are described in simplified form by the Joule reference process (Linow 2022). The system is then the air flowing through the gas turbine. The heat released from the fuel via combustion and the work performed in the compressor and in the turbine are related to this flowing air.

A key characteristic of a gas turbine is the pressure ratio of the compressor, i.e., the ratio of the current pressure after the compressor p_2 to the ambient pressure p_1:

$$\Pi_V = \frac{p_2}{p_1} = \left(\frac{T_2}{T_1}\right)^{\frac{\kappa}{\kappa-1}}$$

The theoretical efficiency increases with the pressure ratio Π_V. The more the compressor compresses the incoming air, the more of the supplied heat can be converted into work. The efficiency of the gas turbine is always lower than the Carnot efficiency:

$$\eta_{Joule} = 1 - \frac{T_1}{T_2} = 1 - \frac{1}{\Pi^{\frac{\kappa-1}{\kappa}}} < \eta_{Carnot} = 1 - \frac{T_1}{T_3}$$

In the Joule reference process, the heat is supplied isobarically between the compressor and the turbine. The flowing air is heated from the temperature after the compressor T_2 to the temperature before the turbine T_3. The heat supplied in this way decreases with increasing pressure ratio of the compressor, since the temperature T_2 increases with the pressure ratio. The specific technical work of an ideal gas turbine is therefore theoretically

$$w_{pp} = \eta_J \cdot q_{in} = \eta_J \cdot q_{23} = \left(1 - \frac{T_1}{T_2}\right) \cdot c_{p,air} \cdot (T_3 - T_2)$$

This function has a maximum at

$$T_{2.\max} = \sqrt{T_1 \cdot T_3}$$

and the maximum specific work of the reference process at the maximum is

$$w_{GT,\max} = c_{p,air} \cdot (T_1 + T_3) - 2 \cdot c_{p,air} \cdot \sqrt{T_1 \cdot T_3}$$

Figure 5.1 illustrates the curves of efficiency, transferred heat, and work performed as functions of the compressor outlet temperature T_2. At a low pressure ratio and thus low temperature T_2, the turbine can only perform a small amount of work. This work initially increases together with the efficiency. As the pressure ratio increases, the heat flow that can be supplied decreases, and with it the power that can be extracted from the turbine. The power is zero at a pressure ratio of 1 – no compression. Furthermore, the power is zero when the temperature after the compressor already equals the permissible inlet temperature of the turbine. Then no more air flows through the gas turbine, which however exhibits its maximum efficiency.

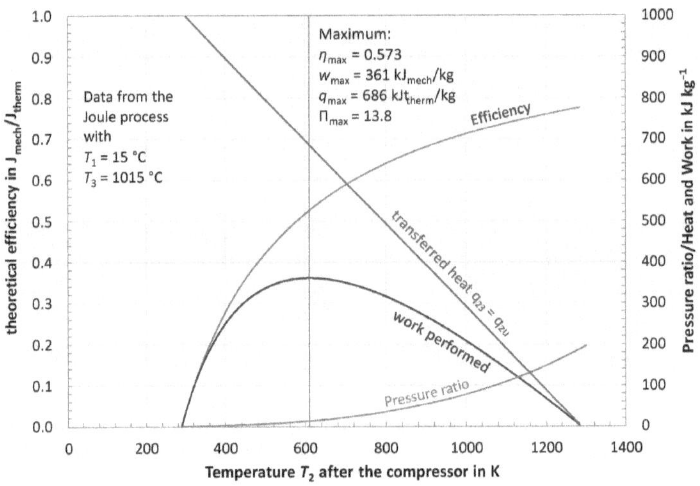

Figure 5.1 Dependence of efficiency, specific heat supplied, and specific work performed for the Joule reference process

When designing the gas turbine, an operating point must be defined. This is determined by the pressure ratio. A trade-off between power output and efficiency must always be made, since both cannot be optimized simultaneously. For stationary gas turbines, this operating point is usually only slightly above the point of maximum power. For aircraft engines, which are operated over a wide range of power and ambient pressure, the task is even more challenging.

5.3 Simple Key Figures

5.3.1 Rated Power

Every power plant has a rated power P_N. This is the maximum power that the power plant can provide on a long-term basis. In principle, a higher power can sometimes be generated for a short period. The rated power is the characteristic of the power plant or product that the manufacturer guarantees to users under a set of conditions:

- At the intended ambient conditions specified in the operating instructions, since the rated power often depends on ambient pressure and temperature
- During intended operation, i.e., operation of the plant within the intended limits and loads
- While maintaining the intended safety for the operators and for the plant
- With the intended fuel, operating materials, and spare parts
- Usually when the intended emission reduction measures and other auxiliary units are in operation

The rated power can be provided as mechanical work at a shaft, a chain, or a belt, or it is the electrical power generated by a generator. The rated power depends on the system boundary: if the power plant generates electrical energy, all electrical auxiliary consumptions are in principle part of the power plant and are therefore subtracted from the generated power. For the comparison of power plants, a comparable, if possible standardized, system boundary must therefore be used. This is often the plant fence.

In vehicles, the engine power is relevant; however, the power actually usable by the system is also the important figure here: that is, the power actually used by the vehicle, ship, or aircraft for propulsion, after all required or convenience-related auxiliary consumers have been supplied.

5.3.2 Efficiency

The efficiency is defined as the ratio of benefit to input. Therefore, both of these quantities must each be determined clearly and comparably. For this purpose, the definition of system boundaries and that of a functional unit are important. The efficiency always depends on the ambient conditions and on the current power output. Therefore, the efficiency is specified by the manufacturer at rated power and at specified ambient conditions. There are international standards for this, in which the relevant requirements are defined for specific plants (see Section 17.14).

Benefit The benefit is the provided (electrical or mechanical) power made available at the system boundary.

Input The input depends on the energy conversion used. When converting the chemical energy of a primary energy carrier (coal, oil, natural gas), this input is determined by the mass flow of the fuel crossing the system boundary and by the heating value of the fuel:

$$\dot{Q}_{fuel} = \dot{m}_{fuel} \cdot H_{I,fuel} = \dot{V}_{N,fuel} \cdot H_{I,N,fuel} \tag{5.1}$$

When utilizing energy flows of the earth, the maximum energy flow entering the power plant system can be used as the input:

- The mass flow of the air passing through the area swept by the rotors of a wind power plant, or the kinetic energy contained in this mass flow
- The kinetic and potential energy contained in the mass flow of the water flowing into a hydropower plant
- The total radiant power flowing onto the surface of a solar thermal or photovoltaic module – as module efficiency
- The total radiant power irradiating the property on which the solar thermal or photovoltaic system is located

Along energy conversion chains, efficiencies are linked in series: the usable energy inevitably decreases further with each conversion stage. The overall efficiency is the product of the individual efficiencies of the individual conversion steps. Such efficiencies can be found in Table 17.16 and Table 17.17 in Section 17.10.

Using and determining key figures for power plants

Rated power, efficiency, and many other properties of power plants are properties guaranteed by the manufacturer. In order for customers to rely on manufacturers' statements or to compare statements from different suppliers, industry standards are published that define as clearly and traceably as possible how these key figures are determined. Examples of IEC standards that describe essential key figures of power plants and general infrastructure for electricity generation are compiled in Section 17.14. In some cases, these standards go well beyond the power plant itself. They also partly include the methods by which sites can be evaluated. This makes them a valuable source for the detailed examination of a technology.

5.3.3 Capacity Factor

Infrastructure and power plants in particular are utilized with varying intensity. The reasons for this can be found in the market, in the purpose of the power plant, and finally in technical conditions. The capacity factor describes the intensity of utilization over time.

Power plant The capacity factor of a power plant is the ratio of the mean output power of the power plant to its rated power. The mean power is the power actually delivered by the power plant, averaged over a period of time Δt sufficiently long for a meaningful evaluation (Figure 5.2):

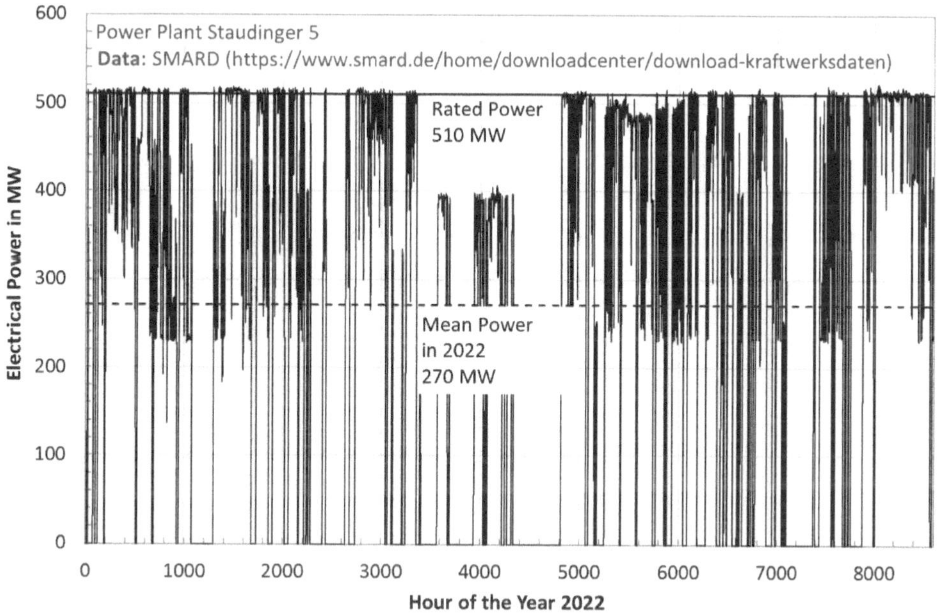

Figure 5.2 Illustration of the capacity factor using data from the Staudinger 5 power plant: the current output of the power plant varies between 0 and P_N = 510 MW. The mean power averaged over the period Δt = 1 year is $\overline{P}_{pp} = 270\,\mathrm{MW}$ (data source: SMARD).

$$\overline{P}_{pp} = \frac{1}{t_2 - t_1} \cdot \int_{t_1}^{t_2} P_{pp}(t) \cdot dt \tag{5.2}$$

This period becomes meaningful when it takes into account seasonal fluctuations in demand or in the available utilized energy. The capacity factor is the ratio of the energy actually provided to the maximum possible:

$$\psi = \frac{\overline{P}_{pp}}{P_N} = \frac{1}{P_N \cdot \Delta t} \cdot \int_{t_1}^{t_2} P_{pp}(t) \cdot dt \tag{5.3}$$

The product $P_N \cdot \Delta t$ is the maximum possible total energy that the power plant would have provided if it had operated at its rated power PN over the entire period Δt.

Ultimately, the capacity factor of a power plant can only be determined with certainty in retrospect, after the power plant has been decommissioned. Predictions are generally possible, but always subject to uncertainties.

Full load hours A frequently used variant is to express the capacity factor as full load hours, i.e., the time in a year over which the power plant would have to run at rated power to generate its average output delivered in that year:

$$flh = \psi \cdot 365 \cdot 24\ \text{h} = \psi \cdot 8760\ \text{h} \tag{5.4}$$

Influencing factors The capacity factor of a particular plant depends on a whole range of factors and can therefore vary considerably:

- Many power plants that are nuclear or coal-fired are intended to be operated only in a narrow power range close to their rated power and are designed so that they may be switched off only a few times during their planned service life. Such base-load power plants therefore inevitably achieve high capacity factors.
- Power plants with gas turbines exhibit their very good efficiency for fossil fuels only near rated power, but can be switched on and off quickly and frequently. Many gas turbines therefore have lower capacity factors.
- Photovoltaics and solar thermal collectors can achieve a maximum capacity factor of approximately 35%. Clouds, dust, shadows, and other effects reduce this further at many locations.
- Wind blows differently depending on the weather, and the output of wind power plants depends strongly on wind speed. Therefore, values of up to approximately 50% are achieved at favorable sites.
- Due to the response of power plants to actual electricity demand, their utilization can decrease significantly: The order in which power plants are taken off the grid depends, among other things, on the levelized cost of electricity, but also on the temporal responsiveness of the power plant.

The capacity factor is closely linked to economic benefit. The higher the capacity factor, the more energy the power plant provides. However, the economic assessment additionally takes into account the actual time-dependent price achieved for the generated energy.

Hydropower

Figure 5.3 shows for the Modau, a small river in Darmstadt-Eberstadt where many mills once stood, the annual variation of the volumetric flow rate (fluctuating gray line). For a mill or hydropower plant, this annual variation is not very intuitive. Typically, the data are therefore presented sorted by volumetric flow rate (steadily decreasing light gray line). For a fictitious hydropower plant at the gauge, a possible rated power is entered as the design discharge: this would be the maximum volumetric flow rate that a power plant could then utilize. Furthermore, it was assumed that the power plant ceases operation below a minimum value. The light gray shaded area is then a measure of the actual energy generated in that year, while the area spanned by the time and the design discharge represents the maximum value. Such a power plant would have achieved a capacity factor of approximately 50% in the drought year 2018. Such or comparable representations are the basis for the design of renewable power plants: based on the annual variation, the rated power is determined and at the same time the mean utilization factor is estimated.

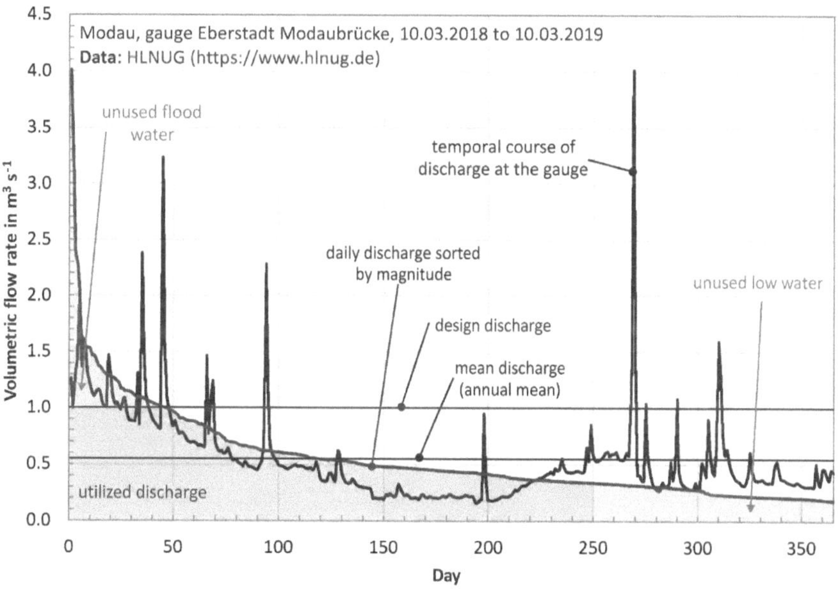

Figure 5.3 Typical annual variation of a watercourse, possible development for hydropower, and non-usable portions (data source: HLNUG)

Utilization of vehicles

Vehicles are designed for very different levels of intensive use. In addition, they are only very rarely and briefly operated at rated power. It therefore makes sense to describe their utilization differently. A purely time-based consideration of utilization is simpler: the utilization share is then the proportion of time during which the vehicle is in use. Typical numerical values are associated with this definition:

- Civil aircraft are in operation for well over 50% of their service life, depending on route length.
- Cargo and passenger ships achieve very high values. Idle times including shipyard stays can amount to less than 10%. Thus, a utilization share determined in this way will be around 90%.
- Railways, trams, and buses achieve very high utilization shares over their service life.
- Trucks are limited in their utilization share particularly by the rest periods of the drivers, but achieve values of around 50%.
- Motor vehicles for private use have typical utilization shares of approximately 2% to 4%. This is also why they are designed for an exceptionally short actual operating time of approximately 2000 h over their service life.
- As an extension, a function is suggested that includes the proportion of loading relative to the intended maximum loading, i.e., exemplarily

$$\psi_{load} = \frac{\overline{m}}{m_{\max}} = \frac{1}{m_{\max} \cdot \Delta t} \cdot \int_{0}^{\Delta t} m(t) \cdot dt \tag{5.5}$$

Here $m(t)$ the mass of the load and $m_{\max}$ the maximum permissible load; however, depending on the application, this could also be the number of passengers, the volume, or the value of the cargo. When this concept is applied to vehicles, the utilization shares decrease in all categories: for air freight typically by a factor of 0.7, in the area of privately used motor vehicles to values well below 1%.

5.4 Energy and Resources

5.4.1 Cumulative Energy Demand (CED)

The cumulative energy demand indicates the total energy input expended over the life cycle of the product. Either the entire life cycle of the product (see Chapter 4) or only a portion of it may be considered. The procedure for determination is specified in VDI 4600. Useful data and examples can be found in VDI 4600-1. The CED is usually a result of a life cycle assessment (LCA, see Chapter 4).

Since the cumulative energy demand for many products can already be estimated quite reasonably with relatively little effort, it also serves as a simplified LCA (Ashby

2009). This is particularly suitable for an initial comparison of solution approaches. The cumulative energy demand (CED) is defined in VDI 4600 as follows:

> *"The CED indicates the totality of the primary-energy-assessed input that arises in connection with the manufacture, use, and disposal of an economic good (product or service) or can be causally attributed to it. This energy input represents the sum of the cumulative energy demands for manufacture CED_H, use CED_N and disposal CED_E of the economic good, whereby for these partial sums it must be stated which upstream and ancillary stages are included."*

The CED is composed of the individual contributions:

$$\mathrm{CED} = \mathrm{CED}_H + \mathrm{CED}_N + \mathrm{CED}_E \tag{5.6}$$

For the determination of the CED and the analysis of energy flows, specific defined terms are used.

System boundary The system boundary results from the required life cycle assessment (see Chapter 4). For simpler comparative studies, other system boundaries may also be used. It is relevant that the CED does not include human or animal labor. This is due to the fact that this would cause impressive additional methodological and technical difficulties.

Primary energy carriers A distinction is made here between energy carriers,

1. That exist in nature and are harvested by means of power plants (renewable energy carriers), and those
2. That are first extracted using resources and energy (fossil energy carriers).

In every case, the lower heating value H_l of the energy carrier is included in the CED.

Natural physical energy carriers These are the material flows or energy flows in the earth system from which usable work can be obtained by means of suitable power plants.

Renewable energy carriers These are biomass from living plant material:

- Wood, wood pellets, and charcoal
- Harvest residues and manure as well as any other biomass
- Fuel gas obtained from biomass, such as methane
- Liquids obtained from biomass, such as ethanol or diesel

Additional processes are necessary to make these energy carriers usable: they are cultivated, harvested, or collected. Subsequently, they are cleaned, refined, processed, or converted. They must be transported so that they can be used in an engine or power plant. These processes are included in the calculation of the CED.

Fossil energy carriers Special processes are necessary to be able to use these. They are extracted. Subsequently, they are cleaned, refined, processed, or converted (see

Chapter 4) and transported to the point of use. These are energy carriers from dead plant material:

- Peat
- Lignite (brown coal)
- Coal (hard coal)
- Crude oil and petroleum products derived from it, such as gasoline, diesel, kerosene, and heavy fuel oil
- Natural gas
- Shale oil, i.e., crude oil variants accessed by fracking
- Oil shale, i.e., deposits mined conventionally and used as fuel or for the production of petroleum products
- Tar and oil sands, in which typical degradation products of crude oil are found
- Methane clathrates

In addition, there are nuclear energy carriers:

- Uranium and thorium as well as plutonium for nuclear fission
- Deuterium for nuclear fusion (tritium does not occur in earth system in quantities worth extracting and must be produced in nuclear reactors, e.g., from lithium)

Data on fuels can be found in Table 17.6 to Table 17.8 (see Section 17.5), Table 17.15 (see Section 17.9), and Table 17.16 (see Section 17.10).

Primary energy This is the energy contained in the respective energy carrier used. For fossil energy carriers and renewable energy carriers, it is the lower heating value H_I. For natural physical energy carriers, this is the energy flux contained in the medium, such as the kinetic power of the wind.

Secondary energy This is the energy content of the energy carrier that has been fully processed or converted for a specific use. Typical examples of secondary energy carriers are gasoline, diesel, purified natural gas, ethanol, coal (see Chapter 9). In addition, there is hydrogen and its derivatives (see Chapter 10). Electricity is assigned here.

The primary energy content of secondary energy carriers varies with the properties of the primary energy carrier, such as the source or mine, and the effort required for exploration, development, and extraction of the primary energy carrier. In addition, there are the required processing steps, transportation, and provision at the point of use.

Final energy These are the forms of energy that are actually used for a purpose:

- Mechanical energy for moving or driving machines or vehicles
- Electrical or magnetic energy in processes and devices
- Light or electromagnetic radiation
- Heat as process heat, including heat for cooking and space heating

- Cold

VDI 4600 introduces further definitions and subdivisions of the CED, each of which is useful for specific questions. This section is not a substitute for using the VDI guideline, but rather attempts to translate the inherently cumbersome presentation from the guideline into a more practical framework. It is intended to arouse curiosity to try out the method.

5.4.2 Determining the CED

The CED is the total primary energy actually required to provide a specific product, including that contained in the material of the product. The CED for a product is determined by summing up all energy inputs over the actual life cycle of the product, in particular for manufacturing, use, and disposal as well as for transport. The system boundary encompasses all activities required for this. This makes the method very close to that of a life cycle assessment (LCA, see Chapter 4). The functional unit for the CED is

- One piece for countable products,
- A unit of mass or a unit of volume for bulk goods, raw materials, and fluids, as well as
- An energy unit, a volume, or a mass for secondary energy carriers.

The system boundary is defined so that it contains all relevant inputs and processes. The considerations regarding cut-off criteria as in an LCA apply here. Table 5.2 describes the procedure for a determination of the CED where the target data are to be reliable (e.g., because a decision is based on them or they are to be published). The calculation of the CED using the method from Table 5.2 is comparable to creating an LCA and can therefore be very time-consuming. Conversely, a good LCA always yields a reliable CED as one of its results.

Simple estimation of the CED

To simplify the determination of the CED to keep it manageable and minimize the input, the method can be simplified for rough estimates. For this purpose, generally accessible, universally applicable values and assumptions can be used for the life cycle:

- For many semi-finished products or raw materials, established CED values exist (see the VDI 4600-1, GEMIS or PROBAS databases as well as further literature). Table 17.10 and Table 17.11 in Section 17.6 contain such values for relevant basic materials.
- Studies or reports summarize CED values for secondary energy carriers. Here too, an extensive body of literature exists.

- Mean values exist for transport routes and transport distances (see Table 17.13 in Section 17.7).
- For electricity, the current energy mix of the region can be used if no further data are available.

The use of such general values is often indicated, since precise data for a very specific product are not available. By using standardized values, good comparability can be achieved.

First estimate

Often the time or the necessity to determine a CED according to the procedure from Table 5.2 is lacking. In many questions, only an orientation value is needed. In particular for estimates or simple comparisons of solution variants, the resources for a CED study are often not available. In this case, the procedure can be significantly simplified. This strong simplification is what Ashby proposes as a substitute for an LCA (Ashby 2009): One application of the CED is not to determine the impacts of a product concretely and in detail via an LCA, but to use the cumulative energy demand as an aggregated measure for the impact. The core argument here is that impacts are very often directly and closely linked to the energy expended. At the same time, it is specifically about a design process for products and thus about a comparison of possible solutions.

The prerequisites for a simple, quick CED estimate that nevertheless delivers meaningful results are as follows:

- The essential contributions are to be identified.
- If solutions are to be compared, all small contributions that are similar or identical across all solutions can be neglected.
- The essential contributions must be traceable to suitable, i.e. good and reliable literature values. In Table 17.13 (see Section 17.7) for transport, in Table 17.10 and Table 17.11 (see Section 17.6) for basic materials, in Table 17.12 (see Section 17.6) for common forming processes, and in VDI 4600-1, a large number of reliable values can be found that can be used directly.

Table 5.2 Conducting a Determination of the CED

No.	Objective of the step	Key aspects
1	Define the purpose of the study	Documentation of the objective of the study and the key question to be answered The state of the literature should be given here if comparable studies already exist. The documentation of the method used and the definitions also takes place here.

No.	Objective of the step	Key aspects
2	Describe system boundaries and energy contributions Define the functional unit	The system boundaries document the scope of the study. Based on a process diagram or another suitable visual representation of the flows under consideration, the existing energy contributions are presented. In the course of the investigation, it often becomes clear (i) whether a particular contribution is relevant, (ii) whether reliable data are available for it, (iii) which further contributions are also important. This diagram and the associated documentation of the system boundary is the core of a thorough study. If no values can be obtained for certain contributions, this must be documented. If contributions are neglected, this must be justified.
3	Research and document energy contributions	As far as possible, all energy contributions, flows and values are compiled, obtained from the literature, queried from experts or estimated. Values taken from the literature are to be identified as such, and the literature used is to be cited. Where possible, the robustness, reliability and accuracy of all values are stated. Specifying error intervals can be helpful.
4	Use of financial data	If no energy data are available for relevant contributions, an attempt can be made to estimate these using financial data such as costs or prices. This is not considered in VDI 4600. It is an imprecise estimation method.
5	Calculate the contributions	Based on the researched data, the individual contributions can be compiled and summed up. Consistency of the data must be ensured during the calculation (quality of the energies).
6	Determine the CED	The CED is determined from all identified contributions. Here, the reliability of the CED value can be determined using the error intervals of the relevant values and the possible variants of the system boundary. Sensitivity analysis can be used to examine the influence of important variables on the result.
7	Take into account the quality of energy	Great care is required when adding different forms of energy. This refers to the addition of energy of different quality. If necessary, a common basis is to be defined and applied. In VDI 4600, the reference to primary energy is always intended. There are different approaches to how this can be implemented. What is essential is thorough documentation.

CED of power plants

In VDI 4600-1, concrete examples are used to show how the CED of power plants can be determined. The more detailed information on the method and examples given there can be directly transferred:

- The physical system boundary is drawn around the outside of the power plant fence.
- Machines used to manufacture the power plant or its components are not included in the balance.
- All manufacturing processes for the power plant and its components are included in the balance (example: the process by which concrete is produced belongs to the system, but not the cement plant itself).
- Operating materials including fuels are balanced by their mass.
- At the end of life, the system boundary for material that is reused is at the point of separation. For material that is not reused, landfill and final repository are proportionally part of the system.
- If material carries energy out of the system, this is taken into account as a credit accordingly (example steel: here, no pig iron needs to be smelted when reused).

5.4.3 Termination Conditions in the Calculation of the CED

In the literature, there are recurring discussions about which contributions are relevant. In particular, studies intended to confirm a result tend to focus on only a few aspects (see discussion in Hall 2017). In principle, however, the CED as a concept requires completeness: therefore, every contribution must first be considered and listed. The listings in Chapter 4 are to be understood in this sense. They provide a non-exhaustive list of relevant contributions. There are then two main reasons why not all contributions can be taken into account:

1. **Availability of data:** On the one hand, it is possible that no figures are available or can be researched. Confidentiality of processes or procedures is one possible obstacle here. Often, a reasonable estimate does not then appear possible. This is noted as an open point in a methodologically sound study. The data may become accessible at a later point in time. Then only the gap in the calculation needs to be identifiable in order to complete it.
2. **Volume of data:** Tracing along manufacturing chains very quickly leads to a very large number of individual branches and immense volumes of data. Here, the process can be terminated from a defined and documented value onwards.

One consequence of the termination conditions as well as the conscious or unconscious neglect of contributions is that the CED almost always yields a lower estimate, meaning the actual value is larger.

5.4.4 CED of Human Activities

The CED according to VDI 4600 explicitly excludes all human (and animal) activities from the calculation of the CED. In doing so, the method avoids major difficulties while at the same time creating technical comparability. The calculation of the CED of an animal or human activity or human labor raises major questions that arise from the definition of the system boundary:

- Draft animals such as oxen or horses are kept under fairly comparable conditions. They require sufficient feed and, if necessary, stables for their use. The system boundary would thus be well definable. At the same time, there are very detailed data on the performance and concrete benefits of animal labor (Smil 2017).
- Reducing humans to their work output and food consumption is only possible under extreme and inhumane conditions (see the example of galley slaves from Chapter 3). Only then can the data given there on energy costs be used for calculations.
- Only in rare cases is human activity purely work in the mechanical-thermodynamic sense. Examples are rickshaw drivers, laborers, and porters.
- Often, human activity is a mixture of action and deliberation. Many physical activities require a high degree of training before they can be performed. Blacksmiths who only after many years of practice can truly handle their hammer with confidence, and lawyers who only after years of study and concrete experience can reliably resolve their cases to the advantage of their clients, illustrate this. Thus, the training period, apprenticeship, or study belongs within the system boundary.
- We grant all people more than bare sustenance. A dignified human life includes unproductive phases (childhood, old age, illness, leisure) that go far beyond the minimum of physical recovery. Dignified human life encompasses further contributions in material terms and with regard to the required energy input. It would need to be determined whether and how these are to be evaluated.

Figure 5.4 provides an initial orientation as to what a model might look like in which only the CED of human labor is determined. It is assumed that the rest has already been determined from a conventional CED analysis and only this contribution is still missing. The individual elements in Figure 5.4 ultimately define the CED associated with an activity via income.

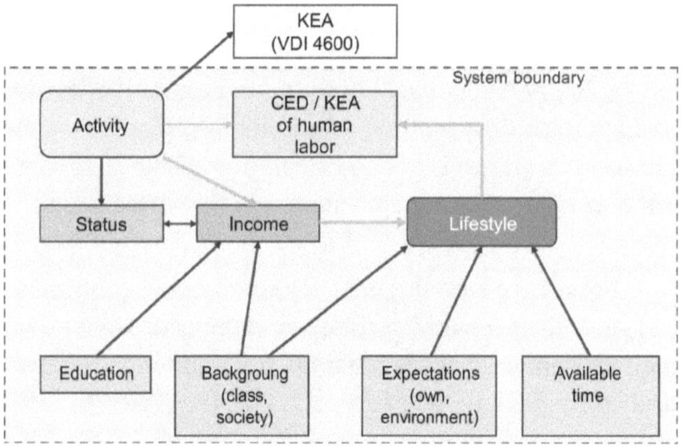

Figure 5.4 Aspects and interactions of the CED of human labor

In this context, a single self-contained activity is well suited as a functional unit. Activities are characteristic of the profession. They are understood as a definable unit, as countable and with a measurable time span. The profession practiced is one of the key factors that define income. Further factors are education, which influences which profession a person can practice, as well as background. Background is understood broadly here and refers to the place and society in which a person lives. Regional income differences are incorporated, for example, through this concept. Background understood as family history defines, through the education made possible, through financial means (assets), and through direct parental influence, which activities a person can perform and at what income. Background, available time, and societal expectations influence personal lifestyle. This lifestyle then determines what energy inputs a person has, and thus the CED of their activity.

A second level of consideration is the active lifetime, i.e., the proportion of lifetime actually used for valued activities. Farmers in subsistence agriculture and their families may quickly work 12 h every day. Children and the elderly are fully integrated within their capabilities. Thus, the active lifetime or (if one wishes to define it as such) the capacity factor of family members would be very high. This presents itself differently in a society with a focus on professions requiring long training periods, with regulated working hours and part-time work, as well as early retirement.

To date, there are virtually no reliable data in the literature on the CED of human labor. This makes it nearly impossible to specifically determine the energy costs of truck drivers, security personnel, engineers, or insurance brokers. This leaves only the option of using financial data: from the salary, the duration of an activity, and a suitable conversion of money into energy, a first approximation can be generated. Table 5.3 is more of a stimulus for reflection. It summarizes the main levels of need and a range of possible energy inputs between a simple form of society and a substantial way of life.

Table 5.3 Needs According to Maslow and Their Energy Input

Level of need	Specific needs	Energy input		
		min.	max.	Form of energy
Basic physical needs	Food	10 MJ d^{-1}	15 MJ d^{-1}	Food
	Air	–	Air conditioning	Electricity
	Water	–	In bottles	Fuel
	Excretions	–	Bathroom	Electricity
Protection and security	Physical integrity	0	Bodyguard, palace guard, personal physician	
	Protection from the elements	Hut/cave	Palaces	Building materials, fuels, electricity
	Heat (and cold)	10 MJ d^{-1}	Central heating	Fuel or electricity
Affection	Family	0	High	Transport
	Group membership	–	High	Transport
	Communication	–	Digital and direct	Transport, electricity
	Religion	–	Very high	Transport
Recognition	Respect	0	Unlimited	Nuclear weapons
	Status, influence, power	0	Unlimited	
	Expression, fashion, jewelry	0	Unlimited	
Self-actualization	Creativity	0	Very high	

Decent living standards

Closely intertwined with the question of a CED for human activities is the question of decent living standards. The CED initially provides a value. The magnitude of the value, however, immediately raises ethical questions. Starting from a minimum set of dignified conditions, the minimum energy requirement of a person can be determined as a function of regional boundary conditions such as climate (Millward-Hopkins et al.

2020). On a global average, the energy requirement of a decent lifestyle amounts to approximately 15.7 GJ per year and person.

CED of financial services

Products and services of the financial sector are an essential prerequisite for the provision of energy. Many activities must be initiated long in advance and thus pre-financed. For this purpose, investors and loans are required. The risks of many ventures are large and must be adequately hedged and insured. That this generates considerable effort is regularly documented in the cost structures of projects: there, and also elsewhere in the literature, concrete figures can be found on the share of financing costs and insurance in the total budget.

However, it is difficult to reliably infer energy costs from these financial costs. It can be concluded from the necessary activities and the infrastructure of the financial and insurance sector that relevant energy inputs exist, such as buildings, data centers, personnel structures, travel activities, trade fairs and events, research and studies, independent expert reports and on-site analyses.

5.4.5 Material and Energy Flow Analysis

One possible method for capturing the share of human labor and human needs in the CED or in energy inputs is offered by the material and energy flow analysis (MEFA). It is a further development of the material flow analysis (MFA). The MFA is a method with which the utilized material flows and stocks of material for a region or for an otherwise defined system can be examined. All materials flowing into the system and all flows leaving the system are evaluated. These flows are thus thermodynamic process variables (Eurostat 2001). Within the system itself, stocks and repositories remain as state variables that accumulate or decrease. The flows crossing the system boundary each include so-called hidden flows, which are to be regarded as a necessary part of the provision of the evaluated material flows. This includes overburden at mines, harvest residues, and waste (Haberl 2001).

A visual overview of the flows, systems, and stocks for a MEFA is given in Figure 5.5. Here, the flows associated with the provision of energy are considered, as well as all energy-rich materials, even if these are not intended for energetic use. The analysis regards society as something that, like a living organism, has a metabolism that must be nourished — i.e., supplied with energy — through the input and output of material. This approach can also be applied to other systems. The concept demonstrates the very close interlinking between available material and usable energy.

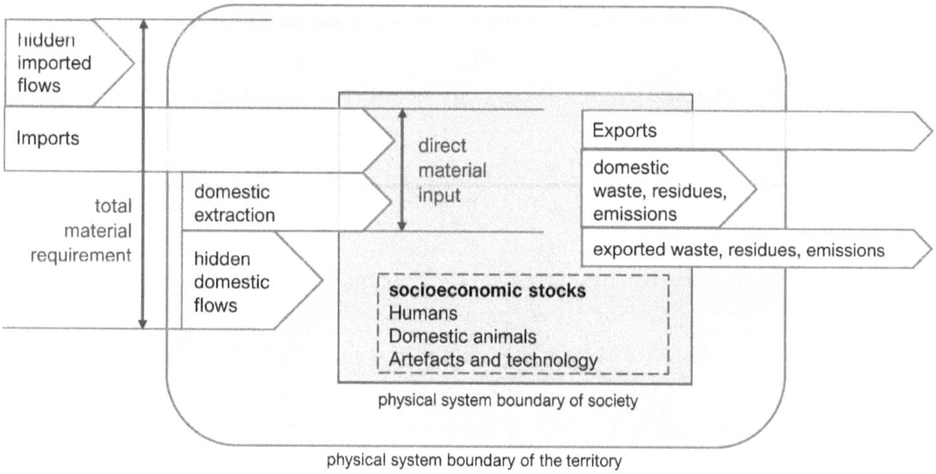

Figure 5.5 Essential flows of a MEFA analysis according to Haberl (source: Haberl 2001)

5.4.6 Reference Quantity Primary Energy

For the indicators CED and EROI, it is necessary to relate all energy data consistently to one form of energy. Different forms of energy have different quality and it makes little sense to mix them here. In particular, it makes a considerable difference in the numerical value of the CED whether the energy is reported as electricity or as the calorific value of a fuel.

To minimize this difficulty, primary energy is often used as the reference quantity. This means that all energy contributions must be stated as primary energy. This sounds simple, but is in fact quite complex in its implementation. In almost all processes and power plants, different forms of energy are used. The traceability back to true primary energy is more or less straightforward depending on the energy carrier and application.

In VDI 4600, the relationship between an energy and the CED for its provision is defined as the provision efficiency. The provision efficiency g of a fuel is the ratio of its lower heating value to the CED for the provision of the fuel at the point of use:

$$g_{fuel} = \frac{m_{Br} \cdot H_I}{\text{CED}_{fuel}} \tag{5.7}$$

Comparable is the supply efficiency for electricity:

$$g_{el} = \frac{W_{el}}{\text{CED}_{el}} \tag{5.8}$$

The numerical values given in VDI 4600-1 are to be considered outdated. They are consistent neither with rough calculations nor with more recent data from other sources.

For the conversion of electricity into (fossil) primary energy, (Hall 2017) proposes a blanket factor of 3. This factor is acceptable for purely fossil electricity generation. It follows from the efficiency of conventional coal power plants. The factor is overall set somewhat too low, since in addition to the efficiency, many other contributions are included in the CED.

A reliable method for converting electricity into primary energy is the use of current and well-researched figures for the regional energy mix. The argument for using these figures is that the electricity drawn from the grid in a given year corresponds on average to exactly this electricity mix and thus contains this share of primary energy. Conversely, the electricity fed into the grid by a single power plant in a given year will replace a portion of this electricity mix replace.

5.4.7 Net Energy

Net energy denotes the absolute amount of usable energy that an activity or a power plant provides. It is the difference between the yield of an energy-harvesting activity, i.e., the total energy provided by it, and the energy input required for it. Net energy is an extensive quantity, whereas the quantity EROI introduced in Section 5.1 is intensive. Both describe the same situation in different ways.

Net energy is the difference between the usable energy E_{use} delivered by the activity or the power plant and the energy E_{invest}expended for it:

$$E_{net} = E_{use} - E_{invest} \tag{5.9}$$

The term energy often refers to electricity or work, but net energy can also be calculated for the provision of fuel or food.

Under two prerequisites, net energy can also be calculated from the already introduced CED:

1. The CED includes, by its definition, the energy contained in a primary energy carrier. This may in some cases be included in E_{use} when it concerns the provision of an energy carrier, but it is certainly not included in E_{invest}. This needs to be adjusted.
2. Since the CED does not include human and animal labor, it can only be used for those applications in which these do not contribute significantly to the activity. This restriction applies in highly industrialized societies. It does not apply when evaluating subsistence economies.

For a fuel or a power plant that uses a fuel, this would give

$$E_{net} = E_{use} - E_{invest} = E_{use} - \left(\text{CED} - m_{fuel} \cdot H_{I,fuel}\right) \tag{5.10}$$

with the mass m_{fuel} of the fuel used for the energy E_{net} of the fuel used and the heating value $H_{I,fuel}$ of the fuel. In principle, the same requirements apply to the calculation as for the CED.

Social or systemic benefit only arises when net energy is positive. The amount of energy invested must be significantly smaller than the energy made usable by it. In principle, a power plant can also yield a positive financial benefit for the operator even when net energy is negative. This then depends on other sources of income such as government subsidies or high prices achieved for the energy provided. In energetic terms, and thus in social terms, this does not represent a benefit.

Values for net energy are unwieldy and non-intuitive. In particular, they are difficult to compare for power plants of different rated capacities. Therefore, they are rarely used.

5.4.8 Energy and Money

Sometimes, within the given time, no information can be found on the energetic effort required for something. In this case, the energy used E can be estimated via the capital K (King/Hall 2011):

$$E = K \cdot H_{€} \tag{5.11}$$

Here $H_{€}$ is, in analogy to the heating value, the energy value, i.e., energy per unit of money. This energy value is a quantity that is often not simply listed in databases. It must first be determined for the measure or activity under consideration, e.g., from the CED and the capital K expended:

$$H_{€} = \frac{E}{K} \simeq \frac{\text{CED}}{\sum_i K_i} \tag{5.12}$$

For fossil energy carriers, this can in principle also be used to estimate the EROI (King/Hall 2011):

$$\text{EROI}_{fossil} \simeq \frac{m_{fossil} \cdot H_{€,fossil}}{K_{invest} \cdot H_{€,invest}} \tag{5.13}$$

In fact, data can be found that make it possible to relate capital and energy:

- Current prices for energy raw materials are a good measure of what the provision of heat costs, and here heat or primary energy is directly linked to a monetary value.
- The current net price for electricity is a good measure of what the provision of work costs. Here too, work is directly linked to a monetary value.
- The gross domestic product (GDP) and the energy consumption of an economy yield an energy value that relates to the energy mix and the primary energy demand of the economy. Systemic aspects are included in these figures.
- Data on the energy and capital intensity of industrial sectors can be used for a specific application in this area.

A relevant error arises in the determination of the cumulative energy demand (CED) due to the high financial expenditure for capital costs and interest payments, which are not included here, as well as the fact that human labor is implicitly included in GDP but not in the CED. For this reason too, only approximate values can be determined here.

Literature in which money and energy are exchanged against each other and converted into one another must therefore be viewed critically. Additional errors from this conversion and from price elements not included are to be expected there. For initial estimates, however, these data are often sufficiently robust.

5.5 Energy Returned on Energy Invested (EROI)

Energy Returned on Energy Invested (EROI) is a fundamentally important concept for describing energy activities. It is another formulation of net energy. Both concepts can be applied not only to technology.

For every living organism it is essential that the return from invested energy is greater than the investment, otherwise it would starve. The sum of all leaves of a plant must return more energy to the plant than the plant has invested in building the leaves. A polar bear must take in more energy from the seals it kills than it invests in hunting them. Not every single hunt needs to be successful, but sufficiently many must be. The target value of the EROI is not an equilibrium in the sense of 'sufficient' (as energy expended = energy provided), nor a specific surplus, but a ratio: from the energy taken in, the basic needs of the organism (metabolism, social interaction, producing and raising offspring) must be met. For this, the energy taken in must be greater by a factor than the energy used for it. The exact factor depends on the organism. The surplus is needed as a reserve or used for reproduction, for physical growth, or for other activities, e.g., social ones. The same applies to technology for the provision of heat, work, and electricity: it must be energetically worthwhile and it must also enable further activities that make the long-term survival of society possible in the first place. The question of what value a society needs as the EROI of its energy system is addressed at the end of the section.

EROI is defined as

$$\text{EROI} = \frac{\textit{resulting energy from an activity}}{\textit{used energy for that activity}} \tag{5.14}$$

That is, at the center is an activity, an action, or a process by which energy is collected or converted. For this process, energy is invested, and from the process results the collected or converted energy. This can be a one-time activity or an ongoing process.

5.5.1 Determining EROI – the Method

A good, relevant, and scientifically sound EROI analysis is not straightforward. Data availability is often a major challenge. A further difficulty arises from the question of which contributions should be included. A methodical approach is essential in order to achieve the highest possible value and the most robust informative power. Table 5.4 describes a method, whereby the key points addressed there are discussed in further detail below. The documentation of an EROI analysis follows analogously to the LCA (see Chapter 4) or the CED (see Chapter 5).

The calculation of the EROI can be carried out in advance as a basis for planning. At this point in time, these calculations are valuable because they should reveal all the necessary and relevant assumptions. The forecast can then be continuously compared against the lifetime and operation of the infrastructure: data from past periods can be compared with the assumptions. This creates a valuable management tool with which critical developments quickly become visible. Correct and complete assessments only emerge in retrospect and once all activities have been completed. The retrospective calculation is important in order to derive requirements and conclusions for future projects.

Table 5.4 Procedure for an EROI Analysis (Based on King/Hall 2011; Murphy et al. 2011; Extended)

No.	Objective of the step	Key points
1	Define the purpose of the study	Documentation of the objective of the study and the key question to be answered The state of the literature should be presented here if comparable studies already exist. The documentation of the method used and the definitions is also provided here, either by reference to a published description in the literature or by explaining the extensions and modifications made to the method.
2	Describe system boundaries and process	The system boundaries document the scope of the study. Based on a process diagram or another suitable visual representation of the flows under consideration, it is shown which energy contributions exist and how these are connected to the environment or to one another. In the course of the investigation, it often only becomes clear (i) whether the contribution is relevant, (ii) whether there are reliable data on it, (iii) which further contributions are important. This diagram and the associated reliable documentation of the system boundary is the core of a rigorous study. If no values can be obtained for certain contributions, this must be documented. If contributions are neglected, this must be justified.
3	Research and document energy contributions	As far as possible, all energy contributions, flows, and values are compiled, sourced from the literature, obtained from experts, or estimated. All values are linked to the process. Where possible, the robustness, reliability, and accuracy of all values are stated. Specifying error intervals can be helpful.
4	Use of financial data	If no energy data are available for relevant contributions, an attempt can be made to estimate these using financial data such as costs or prices.
5	Calculate flows and contributions	Based on the researched data, the individual contributions can now be compiled and summed up. Consistency of the data (energy quality) should be ensured during the calculation.

No.	Objective of the step	Key points
6	Determine EROI	The EROI itself is determined. At the same time, the robustness of the figure can be established using the error intervals and the possible variants of the system boundaries. Sensitivity analysis can be used to examine the influence of key variables on the result.
7	Account for energy quality	Great care is required when adding different forms of energy. This refers to the addition of energy of different quality. Where necessary, a common basis must be defined and applied. There are different approaches to how this can be implemented; what is essential is rigorous documentation.

EROI for energy carriers

The calculation of the EROI of an energy carrier is at least at first glance quite simple. It is determined from the ratio of energetic benefit to energetic input for the provision of the benefit, with the energy provided at the defined point of transfer $E_{use,ec}$, which is potentially contained in the energy carrier, and the energy invested for this $E_{invest,ec}$.

$$\mathrm{EROI}_{ec} = \frac{E_{use,ec}}{E_{invest,ec}} \tag{5.15}$$

A fundamental stipulation is that the primary energy, i.e., the energy contained in the fossil energy carrier in its deposit, or the energy flow of the harvested form of energy (wind, sun, ...) is not included in the calculation, i.e., explicitly not included in the CED: Only all inputs required for extraction or harvesting are included in the denominator. Thus, for a fuel

$$\mathrm{EROI}_{ec} = \frac{E_{use,ec}}{E_{invest,ec}} = \frac{E_{use,ec}}{\mathrm{CED}_{ec} - m_{ec} \cdot H_{I,ec}} \tag{5.16}$$

Similar relationships, in which the primary energy is subtracted from the CED in each case, can also be established for other energy carriers.

In principle, the system boundary used, at which the generated work is transferred to the users, significantly influences the value. The EROI decreases with each step. These are typical variants of the system boundary:

- At the property boundary or the outlet of the infrastructure that extracts or harvests the energy carrier: In the English literature, fixed terms are used such as *from source to wellhead, ... to mine-mouth, ... to farmgate*. At this point, the primary energy carrier is still present.
- After the conversion of the extracted or harvested energy carrier into a usable form (in a refinery, a fermentation tank, etc.), i.e., specifically at the outlet of the facility: For fuels, this is the secondary energy.

- Where the usable energy carrier is made available to the user or end consumer: A fixed term used is, e.g.,*well to tank*, i.e., from extraction to the vehicle's tank.
- Where the final use of energy takes place: A fixed term used is, e.g., *well to wheel*, i.e., from extraction to driving the vehicle. This is the useful energy.

5.5.2 EROI for Power Plants

The calculation of the EROI for power plants is carried out similarly to that of an energy carrier (Kessides/Wade 2011):

$$\mathrm{EROI}_{pp} = \frac{W_{use,pp} + Q_{use,pp}}{E_{invest,pp}} \tag{5.17}$$

In this equation, $W_{use,pp}$ is the work provided by the power plant. If heat is provided as energy, it is $Q_{use,pp}$. The total energy invested over the life cycle of the power plant to generate the provided energy is $E_{Iinvest,K}$. Here too, the primary energy is not included in the calculation. In principle, the CED can continue to be used, but here too the primary energy contained in the energy carriers should always be factored out.

Discrepancies in the forms of energy can now be identified here: the delivered energy is mechanical work or electricity. In addition, usable heat can be delivered, e.g. as district heat or process heat. The energy expended often encompasses all possible forms of heat and work. To keep values comparable, documentation must be carried out with great care at this point, and if possible all energies should be reduced to a common quality. Methods are discussed in section 5.4.6 on primary energy.

The energy provided (the numerator) must be determined over a relevant period of time and ultimately over the lifetime of the power plant. What counts here is not the power possibly generated within the power plant, but the power delivered to society at a defined transfer point. This means that internal consumption must be taken into account appropriately. A relevant period of time is at least one year. Ultimately, these values can only be determined accurately in retrospect. The delivered power of all power plants varies over time, but this can be well described by the capacity factor. With the definition of the capacity factor,

$$W_{use,pp} = \int_0^{\Delta t} P_{use,pp}(t) \cdot dt = P_N \cdot \Delta t \cdot \int_0^{\Delta t} \frac{P_{use,pp}(t)}{P_N \cdot \Delta t} \cdot dt = P_N \cdot \Delta t \cdot \psi \tag{5.18}$$

In this equation, the current power output of the power plant $P_{use,pp}(t)$ at time t, the rated power of the power plant $P_{use,pp}$, the lifetime of the power plant Δt and the mean capacity factor of the power plant ψ are used.

The invested energy (the denominator) can often be divided into the energy bound in the power plant and its infrastructure E_{fix} for construction and disposal (this is a true

CED) and the required power $P_{invest,pp}(t)$, which is required to generate the delivered energy of the power plant during operation. This power $P_{invest,pp}(t)$ covers the internal consumption of the power plant. It is required, for example, for pumps, fans, lighting, control systems, or flue gas cleaning. In addition, all external consumption for the provision of the energy carrier and all other components and services must be taken into account (Kessides/Wade 2011).

$$E_{invest,pp} = E_{fix} + \overline{P}_{invest,pp} \cdot \Delta t + \overline{P}_{ext,pp} \cdot \Delta t = W_{fix} + \int_0^{\Delta t} P_{I,\mathrm{pp}}(t) \cdot dt + \int_0^{\Delta t} P_{E,\mathrm{pp}}(t) \cdot dt \quad (5.19)$$

This can be simplified:

$$E_{invest,pp} = E_{fix} + h_{intern} \cdot P_N \cdot \Delta t \cdot \Psi + h_{extern} \cdot P_N \cdot \Delta t \cdot \Psi \quad (5.20)$$

The factor h_{intern} contains all internal energetic expenditures of the power plant, the factor h_{extern} correspondingly the external energetic expenditures as a share of the delivered work. The benefit of these equations arises when such factors h_{intern} and h_{extern} are already known for certain power plant types and can be applied for initial estimates of similar projects. Both factors are not trivial to determine.

In order to obtain a true EROI that is comparable across different types of power plants, all proportional contributions for the extraction and processing of the required energy carriers must be included. The clear documentation and consideration of system boundaries is important for the significance of the result. Particular caution is advised when using values from older publications: for a clean general approach see (Murphy et al. 2016). A particularly glaring example of misallocation is discussed in (Raugel et al. 2017).

5.5.3 Derived Quantities

Net energy The net energy can be determined from the EROI:

$$E_{net} = E_{use,pp} - E_{invest,pp} = E_{invest,pp} \cdot (\mathrm{EROI} - 1) = E_{use,pp} \cdot \left(1 - \frac{1}{\mathrm{EROI}}\right) \quad (5.21)$$

Energy payback time The energy payback time Δt_{pb} is the time until the power plant has offset only the initially required investment of energy in construction, connection, and decommissioning. This is the point in time at which the following holds:

$$E_{use,pp}\left(\Delta t_{pb}\right) - E_{invest,pp}\left(\Delta t_{pb}\right) \quad (5.22)$$

This applies only if the invested and delivered energy are directly comparable. The energy payback time thus follows as

$$\Delta t_{pb} = \frac{E_{fix}}{\overline{P}_{pp} - P_{invest,pp}} \tag{5.23}$$

We can also express this, where applicable, using the factor h and the capacity factor:

$$\Delta t_{pp} = \frac{E_{fix}}{\psi \cdot P_N - (h_{intern} + h_{extern}) \cdot \psi \cdot P_N} \tag{5.24}$$

Only after this point in time does a new power plant deliver energy to society. Only then can the energy delivered by the power plant be used, for example, to manufacture additional power plants. This point is relevant for a technology transition. Either the old technology is still used to bring the new one into operation as quickly as possible, or the growth rate of the new technology based solely on itself remains limited by its energy payback time. This is one of the core problems with the rapid transition from fossil energy supply to renewable energy now envisaged. The speed of this transition remains limited, unless renewable energy technology is built using fossil energy (Sgourides et al. 2016).

EROI and curtailing

Curtailing is the technical term for shutting down wind or PV power plants that are currently capable of generating power but are prevented from feeding this power into the grid because their output is not needed at that moment. The EROI definition makes it clear that power plants can only achieve a high EROI if they provide a large amount of useful energy. There is a fundamental difference between (fossil) fuel-fired power plants and renewable power plants that harvest the earth's energy flows:

- The service life of power plants depends to varying degrees on the capacity factor: mechanical components are designed for operating durations and can remain idle for a long time with proper maintenance. Technology exposed to the weather or even to seawater ages primarily or, in the case of PV, exclusively through these influences.
- Fuel can be stored. If a power plant that uses fuel is not currently in operation, that fuel remains available at a later point in time.
- The earth's energy flows are either present or absent depending on the weather. They can only be harvested when they are available. If a wind or PV power plant does not harvest these currently available energy flows, this is directly a loss of the energy that could be generated over the service life $W_{use,pp}$.

Curtailing thus reduces the EROI of renewable power plants, but not the EROI of fossil fuel-fired power plants. Curtailing of renewable power plants is therefore socially disadvantageous and can be regarded as a subsidy for fossil fuel-fired power plants. Curtailing is detrimental with respect to all objectives of environmental policy and in particular for the reduction of emissions.

5.5.4 Interpretation of the EROI

At what point does the concept of EROI help to further analyze problems or evaluate possible solutions? The first step here is to illustrate the values. In Figure 5.6 this is shown from the perspective of the invested energy. Two other quantities then relate to this:

- The total energy provided is the sum of invested and delivered energy: $E_{use} = E_{invest} \cdot \text{EROI}$. This is the gross yield.
- The energy additionally delivered by the energy system under consideration is what this energy system generates as additional benefit. This is $E_{invest} \cdot (\text{EROI} - 1)$. That is the net yield. The activity or system is deployed for this additional benefit.

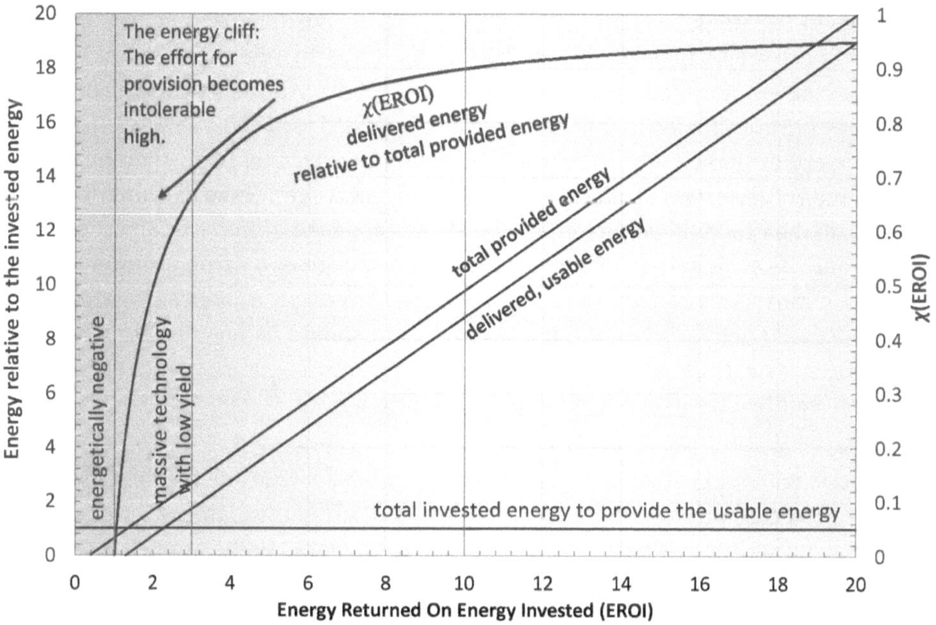

Figure 5.6 Total gross energy provided and net usable energy for constant invested energy as a function of EROI: the *energy cliff* is the steep drop of the function χ with decreasing EROI. The expenditure for net usable energy increases extremely.

That is, the gross benefit is reduced by what the system itself first requires for its own operation. The net benefit is what is usable. In Figure 5.6 an additional curve is therefore plotted. This is the ratio of net to gross energy:

$$\chi(\text{EROI}) = \frac{E_{use} - E_{invest}}{E_{use}} = 1 - \frac{E_{invest}}{E_{use}} = \frac{\text{EROI} - 1}{\text{EROI}} \tag{5.25}$$

This function is described in the literature as the *energy cliff* . Numerical values are contained in Table 17.20 in Section 17.11. What is socially or long-term relevant is the net yield of every energetic activity. This function illustrates what remains at the system boundary, after all actions have been completed, and can then still be used:

- At an EROI below 1, every energetic activity is an energetic loss-making venture, because more must be invested than is returned. Such activities are fundamentally to be avoided, unless a concrete and deliberate benefit arises that cannot be achieved otherwise, and it involves a defined and time-limited action. If a Savonius small wind turbine on the roof of an office building shows a negative energy yield over its service life, then it is the wrong strategy to install more of them. Even with more such rotors, the energy balance will not be reversed; the office building merely increases its energy intensity.
- Relevant for an assessment is the range of slightly positive EROI values, approximately between values of 1.0 and 3.0. In this range, the invested energy is somewhat greater than or comparable to the energetic gain. All energy carriers or power plants that are situated in this range and are intended to provide a relevant amount of energy over their service life are gigantic industries that consume incredible amounts of resources, from which only a modest energetic benefit is delivered relative to the expenditure, yet often with impressive further impacts. Since the EROI can only be estimated in advance with a certain degree of error, projects with an EROI in this value range can initially only be rejected: it is unclear whether they can actually deliver a benefit. In this context, it should be noted that many large projects are calculated on the basis of very optimistic assumptions, i.e., an additional significant safety margin must be required here, and all studies or projects in this range must be critically scrutinized.
- Projects with an EROI of approximately 10 or above lie outside the field of critical energy discussions. In this range, the returns clearly outweigh the input and the energy infrastructure begins to become small compared to the output.
- EROI values above approximately 100 may be considered of limited informative value. At best, such values are based on the assumption of a very long operating period. Here, 100 years as an upper limit for an analysis is reasonable even if the infrastructure itself is designed for a longer period. Over longer time periods, orderly or reliable predictions are often no longer possible, since rare events such as earthquakes or millennial floods already pose substantial risks. Only in retrospect may a higher value then appear.

The EROI can be determined for different system boundaries. Its value depends on the system boundary and must be interpreted differently with regard to the benefit in each case. Table 5.5 provides an overview for different system boundaries. It applies that each subsequent system boundary encompasses all previous ones (Hall 2017). Even the first simplest variant of pure extraction is extremely complex. This is because even this system boundary cannot be cleanly closed, since fuel and electricity flow back from broader system boundaries. The entire global infrastructure or technosphere is by now so interconnected that separations are always artificial. Nevertheless, the arrangement made here makes sense, as it clarifies fundamental aspects. In particular with regard to the statement of the EROI, it follows established representations and thus allows an answer to the question of from which EROI the activity can be meaningful. The values given in the right-hand column are taken from the literature and represent estimates.

Table 5.5 System Boundary As Well As Activities in the System and EROI (Based on Hall 2017, Smil 2017)

System and boundary	Activities	What does this mean?
Energy carrier 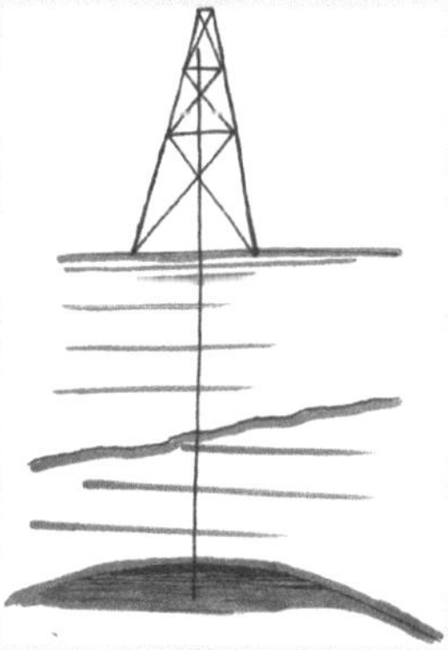 The system encompasses all extraction activities. The primary energy carrier is assessed as it is present at the fence of the infrastructure, i.e. at the wellhead/minemouth/farmgate.	Preparation: ■ Licenses for exploration and development ■ Business plan and risks ■ Capital and investors ■ Exploration ■ License for extraction, operating permit and emission ■ Construction or expansion of infrastructure for the plant and transport ■ Development of the mine/well ■ Commissioning Operation: ■ Maintenance and servicing ■ Extraction itself ■ Separation of unwanted additions ■ Accommodation and catering for workers ■ Disposal of waste and residual materials ■ End of operation ■ Decommissioning of the plant ■ Dismantling of the plant ■ Disposal of the plant and all other residues ■ Securing and sealing	EROI = 1.0 means that we can look at the energy carrier at the source, nothing more. From a social and technical perspective, it becomes interesting at an EROI of approximately 10. Conventional oil fields often have an EROI > 30. Unconventional oil often exhibits a surprisingly low EROI. Financially, different rules and expectations apply in this area, as the "fracking revolution" in the USA demonstrates.

System and boundary	Activities	What does this mean?
Processed energy carrier  The system encompasses the entire extraction as well as the transport to processing and the conversion or processing into a specific energy carrier itself. The system boundary is the fence of the refinery or purification plant.  	Additional activities are as follows: Transport: ■ Energy input to move the mass or volume ■ Packaging ■ Infrastructure for fuels or energy for the means of transport ■ Provision and maintenance of the infrastructure for the means of transport ■ Special infrastructure in the supply chain ■ Data centers and digital aspects of logistics ■ Financial services and insurance ■ Emergency preparedness ■ Supply of personnel Processing: ■ Operation of the plant or refinery ■ Flue gas cleaning ■ Consumables ■ Maintenance and upkeep ■ Cleaning work ■ Disposal of residual materials from operation ■ Security of the plant or refinery ■ Planning and permitting ■ Financial services and insurance ■ Construction of the plant ■ Required infrastructure for supply and disposal ■ Conversion or expansion ■ Decommissioning at end of life ■ Removal and disposal of all residual materials ■ Supply of personnel	EROI = 1.0 means that we can look at the processed energy carrier there, nothing more. For conventional fossil energy carriers, more than 10% of the primary energy is expended up to this point for the required life cycle. Then, from an EROI of 30 at the source, assuming that the required 10% is diverted from the energy carrier itself, an EROI of 6.75 results.

System and boundary	Activities	What does this mean?
Conversion of the energy carrier into electricity 	Additional activities at the power plant are: ■ Operation of the power plant ■ Internal consumption for pumps, fans, lighting, control and regulation ■ Flue gas cleaning ■ Consumables ■ Maintenance and upkeep ■ Cleaning work ■ Disposal of residual materials from operation of the power plant ■ Security of the power plant such as fences, surveillance electronics, plant security, as well as all indirect measures such as police activity, intelligence surveillance, data security and defense against hacker attacks, as well as military measures ■ Planning and permitting of the power plant ■ Financial services and insurance ■ Construction of the power plant ■ Required infrastructure for the power plant ■ Conversion or expansion of the power plant ■ Decommissioning of the power plant at end of life ■ Removal and disposal of all residual materials ■ Supply of personnel	EROI = 1.0 means that the lights on the power plant fence still illuminate the surroundings a little, nothing more. There is no energy available to distribute electricity or to create goods with which the electricity could be used.

System and boundary	Activities	What does this mean?
Delivered energy carrier  The system includes the entire extraction, processing and distribution up to the point of provisioning for consumption. The system boundary is the filling station, the household gas line or the household electrical connection. 	Additional activities are as follows: in the area of transport see previous explanations Electricity: ▪ The electrical distribution grid ▪ Substations ▪ Facilities for grid stabilization ▪ Control and regulation of the grid ▪ Maintenance and upkeep ▪ Construction and restructuring Fuels: ▪ Filling stations, coal dealers and wood dealers ▪ Roads and pathways to the transfer point ▪ Special means of transport	EROI = 1.0 means we can withdraw the processed energy carrier at the filling station, but without owning a tank or even a car to fill it into.
Use of the energy carrier  The system usually encompasses the entire technosphere.	Additional activities are as follows: ▪ Providing the product that can use the energy (car, washing machine, computer) ▪ The entire infrastructure to enable the life cycle of the product ▪ The social prerequisites to be able and permitted to use the product	EROI = 1.0 means that the system can be used as is and without modification. Even a repair is not provided for.
Change The energy infrastructure is subject to constant change and further development. This must be made energetically possible from within the infrastructure itself.	Additional activities are as follows: ▪ Changes to the energy infrastructure: conversion, expansion, upgrade and downgrade ▪ Technology change ▪ Research and development in all areas ▪ Training	

System and boundary	Activities	What does this mean?
Saving for times of need **Resilience in the system** Many fossil systems have storage facilities, for systemic or strategic reasons.	Additional activities are as follows: Storage: ▪ Construction of storage facilities and reservoirs ▪ Maintenance and upkeep of storage facilities and reservoirs ▪ Securing storage facilities and reservoirs against external influences ▪ Replenishing storage facilities and reservoirs Resilience: ▪ Robust infrastructure ▪ Local infrastructure ▪ Possibility of island operation ▪ Redundancies in the system	

Energy for Society

Noteworthy is the pyramid of energy needs of a modern energy-intensive society presented by Lambert et al. (Lambert et al. 2014). This pyramid is explicitly oriented at first toward Maslow's pyramid of human needs (see Table 17.22 in Section 17.13). It then transfers these to complex societies (Table 5.6). A comparison shows that from a societal perspective and with a pure focus on energy supply, the view of what constitutes essential basic needs and thus state responsibilities can shift considerably. The argumentation underlying this hierarchy follows directly from the analysis of EROI, here in connection with other indicators that are all closely linked to the energetic prosperity of a society. This fixation on the provision of energy seems alarming. What this perspective means concretely for a society must be negotiated socially and leads to wicked problems.

Table 5.6 Energetic Needs Hierarchy of a Society Linked to the Human Needs Hierarchy According to Maslow: the Minimum EROI for Conventional Crude Oil Follows from Hall and Klitgaard (Data: Hall/Klitgaard 2018). The EROI for Electricity Takes into Account the Quality of the Energy.

No.	Level of need	Specific needs	$EROI_{min}$ - ref crude oil	$EROI_{min}$ - ref electricity
1	Extract energy	Exploration extracting and mining energy resources harvesting renewable energy	1.1	1.1
2	Process energy	Cleaning and sorting refining and desulfurizing generating electricity	1.2	1.2
3	Transport	Transporting energy transporting energy infrastructure	3.0	2
4	Growing food	Feeding machine operators	5.0	2
5	Family	Producing offspring producing machine operators	8.0	3
6	Education	Training machine operators and explorers	10.0	3.5
7	Health	Maintenance of machine operators	12.0	4
8	Art	Entertainment of machine operators	14.0	4.5

5.5.5 Energy Cliff

In Figure 5.6 the function χ(EROI) is shown. In the literature it is common to represent the EROI decreasing from left to right, in which case the *energy cliff* runs in reading direction and an end is made clear. An average EROI of > 3.0 for energy carriers is considered the absolute lower limit at which a civilization or technical society can still function, an average EROI of > 10 as a target value for a society that wants to keep its environmental impacts stable below destructive levels in the longer term. In Table 17.21 (see Section 17.12), typical value ranges of achievable EROIs are compiled for relevant energy carriers and infrastructure. Values vary considerably depending on application, system boundary and source. Some of the values are not yet well established.

It becomes clear that all classical energy carriers lie well above the assumed limits. For nuclear fission, higher values are reported when ignoring the perpetual costs. The extremely low values of unconventional oil and fuel from plant material are relevant for their future use. That fracking, tar sands and oil sands cannot be energetically profitable is already indicated by the unfavorable cost structure of these energy carriers.

5.6 Energy Storage

Storing fuels is technically straightforward and energetically favorable for most relevant primary energy carriers. Generated electricity or mechanical work as well as hydrogen are less easy to store than fossil energy carriers. Since energy will also need to be retrieved from storage in the future in order to carry out important activities safely, suitable energy storage is an important building block of the energy transition (Palmer/Floyd 2020). A presentation with a focus on the technical implementation of special storage methods can be found in (Schmiegel 2019). This section specifically examines how storage systems can be characterized energetically as part of an energy infrastructure.

5.6.1 Fossil Energy Storage

Every infrastructure for the provision of final energy today contains many storage options in the transport and process chain. These storage systems are often an essential part of the infrastructure and serve many purposes:

- The deposit from which the primary energy carrier is extracted is a storage facility. The rate at which extraction takes place determines which mass flow is made available. Through the extracted mass flows, the primary energy carrier can be produced on demand or influence can be exerted on the market price.
- The transport infrastructure for crude oil and coal provides storage capacity at ports and loading stations. In particular, coal can be stored there for any length of time, while tanks are always required for crude oil.
- Fully loaded ships can travel faster or slower, or they can lie at anchor, so that they serve as intermediate storage.
- Refineries contain many tanks, as some products need to be stored temporarily. Many countries have strategic reserves for petroleum products, kerosene or diesel, in order to be able to continue operating the necessary infrastructure for some time even in the event of a supply failure.
- Coal-fired power plants have their own storage facilities, with a distinction sometimes additionally made between the operational storage and the strategic reserve.

- The transport infrastructure for natural gas is an immense storage facility. Due to the pressure, the mass in the pipelines, conduits and tanks can be varied over a very wide range. In addition, the natural gas infrastructure features special storage facilities that serve to absorb seasonal fluctuations.
- Harvested firewood is stored in piles or along the house wall and dried.
- Filled reservoirs and retention ponds are storage facilities for potential energy.

The capacity of the storage facilities for extracted but not yet combusted fossil fuels is large. It is sufficient to cover global demand for several months. These storage facilities stabilize the technical energy system and the raw material price. They bridge seasonal fluctuations and they are an insurance in case disruptions to the system occur due to unplanned outages, accidents, natural events or war.

EROI This storage capacity is already accounted for as a necessary part of the infrastructure in the EROI. Only a portion of these facilities is directly planned and built as storage. Much of it can simply be used alongside other functions. This makes it clear that for fossil energy carriers, the storage capacity represents only a very small share of the total energy invested in the infrastructure.

water In hydropower, the height of the dam often determines the usable potential energy and thus the rated capacity of the power plant. The large storage facility is then already part of the power plant and accounted for in the EROI. In large reservoirs, more than one year's worth of water for the power plant is usually impounded, also to be able to absorb strong seasonal or annual fluctuations. In hydropower, the dam represents the largest share of the project costs and the energy invested, although the project would not have been possible without the dam.

For these reasons, storage costs have so far not been visible as a relevant independent item of the infrastructure for energy supply. This will change when large quantities of electricity are to be stored.

5.6.2 Key Indicators for Storage

In the future, the storage of electrical energy will play a greater role in meeting the demand for electrical energy when little electricity can be generated, as well as in stabilizing the grid. Key performance indicators and options for an independent assessment of storage systems are summarized here.

ESOI The key figure Energy Stored on Energy Invested (ESOI) compares the energy input required for providing the storage system with the energy delivered by the storage system over its lifetime. In analogy to the EROI, the energetic benefit is the total energy delivered over the lifetime of the storage system $E_{use,s}$. The energy invested for this $E_{invest,s}$ encompasses all necessary activities for providing the storage system such as manufacturing, construction, maintenance, operation and disposal (Pellow et al. 2015):

$$\mathrm{ESOI} = \frac{E_{use,s}}{E_{invest,s}} \simeq \frac{E_{use,s}}{\mathrm{CED}_s} \tag{5.26}$$

As with the EROI, high values indicate a higher energetic benefit over the lifetime, since overall less energy and thus valuable raw materials are tied up in the storage system and can instead be used elsewhere in society. Regarding the use of the CED, the same boundary conditions apply as for the EROI: the primary energy of all raw materials used and the energy fed into the storage system are not part of the calculation and must be explicitly subtracted if necessary.

The same notes on the calculation of the delivered energy $E_{use,s}$ and the invested energy $E_{invest,s}$ apply here as for the EROI. Both can ultimately only be determined completely and correctly in retrospect. However, the calculation of both contributions and the resulting ESOI is also advantageous during the planning phase, as it reveals important requirements or assumptions that are necessary for advantageous operation of the storage system.

Storage cycles The number of usable cycles of a storage system is an energetically significant property. Here, a storage cycle is defined as (i) charging the storage system, (ii) storing the energy, (iii) discharging the storage system to make the stored energy available for use (Carbajales-Dale et al. 2014). The number of these storage cycles N_{cycle}, that the storage system can provide over its lifetime is counted.

Batteries typically have only limited cycle numbers that they can undergo at their intended capacity before they must be replaced due to aging processes. Pumped-storage hydropower plants can be used almost any number of times if they are adequately maintained.

Discharge The maximum discharge specifies the maximum permissible discharge of the storage system during normal operation without causing premature aging or damage. If a residual amount must be retained in the storage system, this directly reduces the ESOI, since additional energy input must be expended for storing this non-usable residual energy. Some storage systems can always be completely emptied, while others must retain a relevant residual energy:

- For many batteries, the intended withdrawal is limited to a portion of the contained energy. To avoid damage to the battery, a battery management system is required that controls the withdrawal and stops it if necessary.
- Hydropower plants can often only withdraw an upper portion of the water in the reservoir. The lower-lying portion of the water volume cannot be used energetically. Pumped-storage power plants, on the other hand, are optimized to be able to use as much of the upper water as possible.
- For gas and oil storage systems, it is important for safety reasons that there is always an overpressure relative to the surroundings in order to prevent the forma-

tion of flammable or explosive mixtures with infiltrating air inside the storage system.

Cycle efficiency The cycle efficiency of storage is a measure of what fraction of the energy required for charging the storage system $E_{in,S}$ can later be provided again as equivalent energy $E_{out,S}$:

$$\eta_s = \frac{E_{out,s}}{E_{in,s}} \tag{5.27}$$

This key figure completely maps the energy conversion chain of storage. As long as no heat is involved in the storage process, a high cycle efficiency of over 90% can be achievable. Storage via chemical energy and heat, on the other hand, leads to low cycle efficiency. The cycle efficiency can additionally depend on time if the storage discharges over time or the stored substance changes:

- Some battery types discharge over time.
- Pumped-storage and other reservoirs can lose water through evaporation in dry and hot regions. This is relevant when storage over weeks or months is intended.
- Fuels can change due to moisture or the loss of easily volatile components.
- Geological air and gas storage facilities can be leaky.
- Geological heat storage facilities can be depleted by groundwater flow.

Specific bound energy The specific bound energy ε_{cycle} is the ratio between the maximum energy delivered by the storage in one cycle $E_{use,cycle}$ and the energy required for manufacturing the storage $E_{invest,S}$. This quantity is directly related to the ESOI via the number of storage cycles N_{cycle}, when the storage is operated between full and empty in all cycles:

$$\mathrm{ESOI} = N_{cycle} \cdot \frac{E_{use,cycle}}{E_{invest,s}} = N_{cycle} \cdot \varepsilon_{cycle} \tag{5.28}$$

For the calculation of this figure ε_{cycle} the same reference energy must be used for both energies in order to keep the data internally consistent. This reference energy is usually defined as electrical energy. Thus, the invested energy $E_{invest,S}$ must be converted to electricity, e.g., via the energy mix. Other variants for a reference energy are possible and also meaningful, e.g., for the storage of fossil energy carriers.

Storage density This key figure is a measure of how large a storage unit must be in order to accommodate a given amount of energy. It can be expressed relative to the volume or the mass of the storage unit.

Resources The number of possible storage cycles of a storage unit N_{cycle} is a central influencing variable for the resource input related to the energy stored over the lifetime. A critical question is whether the raw materials contained in the storage unit are returned to use. Only if this cycle is closed does the use of this technology make sense for storing larger amounts of energy or over a longer period of time, and is feasible with regard to the resources employed.

In Table 17.19 in Section 17.9, average numerical values are compiled for concepts that can store electricity over a longer period of time. The data are taken from Barnhart et al. and the extensive compilation in the IEC Whitepaper (Barnhart et al. 2013; IEC 2011). In the field of batteries, great efforts are being made to optimize them with regard to these and other relevant key figures. Therefore, the values for the specific embodied energy and the ESOI indicate an order of magnitude, but may improve somewhat further with technological development.

5.6.3 Storage and EROI

When storage units are integrated into an energy supply system, this has an influence on the EROI of the system, since additional infrastructure must be built and cycle losses occur. On the one hand, the infrastructure required to provide the energy becomes more extensive and the invested energy E_{invest} must be extended by the invested energy tied up in storage $E_{invest,s}$:

$$E_{invest} = E_{invest,pp} + E_{invest,s} \tag{5.29}$$

At the same time, the energy provided from the storage unit is reduced compared to the energy originally provided by the power plants by the cycle efficiency

$$E_{use,s} = \eta_S \cdot E_{use,pp} \tag{5.30}$$

Only a fraction of the originally provided energy will ever be stored. This fraction is φ_S and depends on the specific energy system and its operation. Thus the EROI with storage can be estimated as

$$\text{EROI} = \frac{E_{use}}{E_{invest}} = \frac{(1-\varphi_s)\cdot E_{use,pp} + \varphi_s \cdot \eta_S \cdot E_{use,pp}}{E_{invest,pp} + E_{invest,s}} \tag{5.31}$$

This equation shows that it is advantageous to invest as little energy as possible into storage and to store only the absolutely necessary amount of energy. This can be supported by shifting from demand-oriented generation to generation-oriented consumption in society. However, this perspective changes when, without the storage unit, the system cannot provide a portion of the energy it is capable of generating. If the storage unit increases the total energy provided by the system, the EROI of the system also increases. Thus, for the assessment of the integration of storage units into an infrastructure, a trade-off between these effects becomes necessary.

References

Ashby, M. F.: Materials and the Environment: Eco-Informed Material Choice. Elsevier, Amsterdam 2009

Barnhart, C. J./Dale, M./Brandt, A. R./Benson, S. M.: The energetic implications of curtailing versus storing solar- and wind-generated electricity. In: Energy and Environmental Science 6, 2013, pp. 2804–2810

Carbajales-Dale, M./Barnhart, C. J./Benson, S. M.: Can we afford storage? A dynamic net energy analysis of renewable electricity generation supported by energy storage. In: Energy Environ. Sci. 7, 2014, p. 1538

Eurostat (Hrsg.): Economy-wide material flow accounts and derived indicators. A methodological guide. European Community, Bruxelles 2001 *https://ec.europa.eu/eurostat/web/products-manuals-and-guidelines/-/KS-34-00-536*

Fizaine, F./Court, V.: Energy Expenditure, economic growth, and the minimum EROI of society. In: Energy Policy 95, 2016, S. 172–186

Haberl, H.: The energetic Metabolism of Societies. Part I: Accounting Concepts. In: Journal of Industrial Ecology 5, 2001, S. 11–33

Hall, C. A. S.: Energy Return on Investment. A Unifying Principle for Biology, Economics, and Sustainability. Springer 2017

Hall, C. A. S./Klitgard, K.: Energy and the Wealth of Nations. Springer, Cham 2018

HLNUG – Hessisches Landesamt für Natur, Umwelt und Geologie. *https://www.hlnug.de*

IEC – International Electrotechnical Committee (Hrsg.): Electrical Energy Storage. White Paper. Genf 2011. *https://basecamp.iec.ch/download/iec-white-paper-electrical-energy-storage*

ISO 13602-1: Technical energy systems – Methods for analysis – Part 1: General

Kessides, I. N./Wade, D. C.: Deriving an Improved Dynamic EROI to Provide Better Information for Energy Planners. In: Sustainability 3, 2011, pp. 2339–2357

King, C. W./Hall, C. A. S.: Relating Financial and Energy Return on Investment. In: Sustainability 3, 2011, pp. 1810–1832

Kleidon, A.: Thermodynamic Foundations of the Earth System. Cambridge University Press 2016

Kraan, O./Chappin, E./Kramer, G. J./Nikolic, I.: The influence of the energy transition on the significance of key energy metrics. In: Renewable and Sustainable Energy Reviews 111, 2019, pp. 215–223

Lambert, J. G./Hall, C. A. S./Ralough, S./Gupta, A./Arnold, M.: Energy, EROI and quality of life. In: Energy Policy 64, 2014, pp. 153–167

Linow, S.: Angewandte technische Thermodynamik. Carl Hanser Verlag, München 2022

Millward-Hopkins, J./Steinberger, J. K./Rao, N. D./Oswald, Y.: Providing decent living with minimum energy: A global scenario. In: Global Environmental Change 65, 2020, 102168

Murphy, D. J./Carbajales-Dale, M./Moeller, D.: Comparing Apples to Apples: Why the Net Energy Analysis Community Needs to Adopt the Life-Cycle Analysis Framework. In: Energies 9 (2016), p. 917

Palmer, G./Floyd, J.: Energy storage and civilization. A systems approach. Springer, Cham 2020

Pellow, M. A./Emmott, C. J. M./Barnhart, C. J./Benson, S. M.: Hydrogen or batteries for grid storage? A net energy analysis. In: Energy and Environmental Science 8, 2015, pp. 1938–1952

Raugel, M./Sgouridis, S./Murphy, D./Fthenakis, V./Frischknecht, R./Breyer, C./Bardi, U./Barnhart, C./Buckley, A./Carbajales-Dale, M./Csala, D./Wild-Scholten, M. de/Heath, G./Jæger-Waldau, A./Jones, C./Keller, A./Leccisi, E./Mancarella, P./Pearsall, N./Siegel, A./Winke, S./Stolz, P.: Energy return on Energy Invested (ERoEI) for photovoltaic solar systems in regions of moderate insolation. A comprehensive response. In: Energy Policy 102 (2017), pp. 377–384

Schmiegel, A. U.: Energiespeicher für die Energiewende. Auslegung und Betrieb von Speichersystemen. Carl Hanser Verlag, München 2019

Sgouridis, S./Csala, D./Bardi, U.: The sower's way. Quantifying the narrowing net-energy pathways to a global energy transition. In: Environmental Research Letters 11, 2016

SMARD – Strommarktdaten für Deutschland. Kraftwerksdaten. *https://www.smard.de/home/downloadcenter/download-kraftwerksdaten*

Smil, V.: Energy and Civilization. A History. MIT Press, Cambridge, MA/USA 2017

Solomon, A. A./Manjong, N. B./Breyer, C.: The necessity to standardise primary energy quality in achieving a meaningful quantification of related indicators. In: Smart Energy 12, 2023, 100115

Sumper, A./Baggini, A.: Electrical Energy Efficiency, Technology and Applications. Wiley, Chichester 2012

Szargut, J.: Exergy Method. Technical and Ecological Applications. WIT Press, Southampton 2005

VDI 4600: Kumulierter Energieaufwand (KEA) – Begriffe, Berechnungsmethoden

VDI 4600-1: Kumulierter Energieaufwand – Beispiele

VDI 4661: Energiekenngrößen – Grundlagen – Methodik

VDI 4662: Bildung, Implementierung und Nutzung von Energiekennwerten

6 Growth

Growth is one of the core topics of our society (Hall/Klitgard 2018). This chapter provides the fundamental mathematical description for the growth of a quantity. To this end, it looks at consequences and effects of growth for relevant resources. For our understanding, the analogy with the concept of interest or compound interest can be helpful. Growth with a negative rate, i.e. shrinkage, is also described here.

If a value x changes from a point in time t_1 to a point in time t_2,

$$x(t_2) = x_2 = x(t_1) + \Delta x(t_2 - t_1) = x_1 + \Delta x_{12} \tag{6.1}$$

this can be described via a specific quantity z with

$$x_2 = x_1 + \Delta x_{12} = x_1 + z \cdot x_1 = x_1 \cdot (1 + z) \tag{6.2}$$

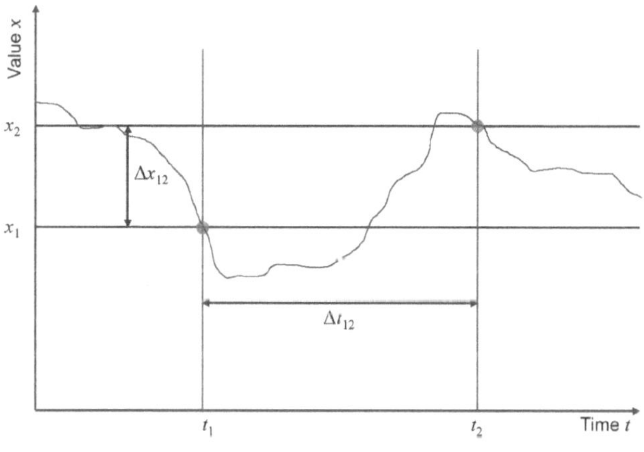

Figure 6.1 Change of a quantity x over time

In general, a quantity x will increase or decrease over time and z also varies with time (Figure 6.1). Figure 6.1 shows that the choice of the time interval can influence the result.

Linear Growth
A simple description is linear growth, i.e. the case where a quantity changes steadily by always the same value Δx. The time dependence of the quantity x is then described by a simple linear equation:

$$x(t) = \frac{x_2 - x_1}{t_2 - t_1} \cdot t + x(t = 0) = \frac{\Delta x}{\Delta t} \cdot t + x_0 \tag{6.3}$$

Linear growth is a good description for a whole range of applications.

6.1 Constant Growth Rate

A fundamentally different case arises when the growth rate z remains constant. The general description can be simplified for this case by defining the first point in time as a starting point t_0 is set (e.g., today) and the difference as a duration Δt, over which the change takes place:

$$x(t_0 + \Delta t) = x(t_0) \cdot (1 + z(\Delta t)) = x_0 \cdot (1 + z(\Delta t)) \tag{6.4}$$

Here the rate z is initially dependent on the time period. This becomes interesting when the duration Δt is regarded as fixed, i.e., as one day, one year, or the like. The quantity x is arbitrary and can be, for example, money in the bank, electricity consumption, population, gross domestic product, or per capita ownership of smartphones.

If a change occurs with a constant growth rate z, then for two equally long time periods Δt starting at different points in time, the same z is used. Positive z is growth in the strict sense, but negative z and thus shrinkage can of course also be described with the equations. Over the fixed time period Δt_0 and the constant rate of change z we obtain for the change

$$x(\Delta t_0) = x_0 \cdot (1 + z) \tag{6.5}$$

Over twice the time period it is

$$x(2 \cdot \Delta t_0) = x_0 \cdot (1 + z) \cdot (1 + z) \tag{6.6}$$

In general for the n-fold time period it is

$$x(n \cdot \Delta t_0) = x_0 \cdot (1 + z)^n \tag{6.7}$$

This also applies to time periods that are not integer multiples of the time over which the rate of change z remains constant.

It holds in general for an arbitrary time period Δt from the starting point

$$x(\Delta t) = x_0 \cdot (1 + z)^{\Delta t} \tag{6.8}$$

Figure 6.2 illustrates this behavior. This equation can be written as an exponential function:

$$x(\Delta t) = x_0 \cdot \exp\left(\ln (1 + z)^{\Delta t}\right) = x_0 \cdot \exp(\ln(1 + z) \cdot \Delta t) \tag{6.9}$$

Thus, any growth at a constant growth rate z is exponential growth. Human understanding struggles to grasp long-term exponential growth. The imagination tends to remain with linear growth over time. It is therefore extremely important to represent the behavior of exponentially (continuously) growing quantities over relevant longer time periods visually or in tabular form.

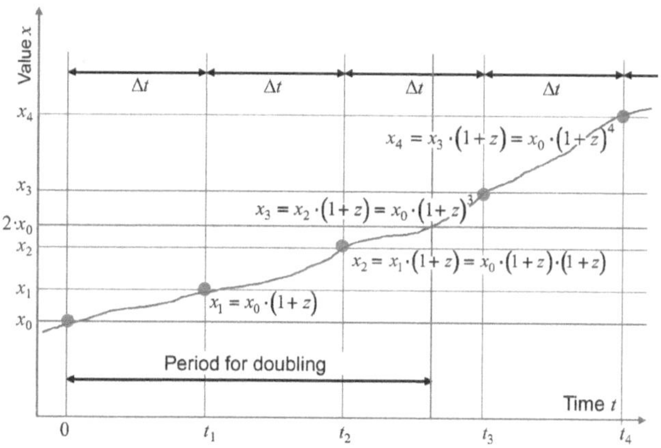

Figure 6.2 Growth at constant rate

Duration The last formula can be used, given a known constant rate, to calculate the required duration Δt for a change:

$$\Delta t = \ln\left(\frac{x(\Delta t)}{x_0}\right) \cdot \frac{1}{\ln(1 + z)} \tag{6.10}$$

Growth rate The growth rate can be determined from two data points:

$$z = \exp\left(\ln\left(\frac{x(\Delta t)}{x_0}\right) \cdot \frac{1}{\Delta t}\right) - 1 \tag{6.11}$$

Doubling The time for doubling of x is illustrative for many discussions. Examples include population development, the growth of bacteria, the annual consumption of resources, the value of an interest-bearing capital. Doubling means that $x(\Delta t_D) = 2 \cdot x_0$, and thus this time period is

$$\Delta t_d = \ln\left(\frac{x(\Delta t)}{x_0}\right) \cdot \frac{1}{\ln(1+z)} = \ln(2) \cdot \frac{1}{\ln(1+z)} \cong \frac{0.70}{\ln(1+z)} \tag{6.12}$$

Multiple Growth Rates

For more complex problems, it may be necessary to describe the temporal development of a quantity x as a function of more than one growth rate. Possible cases include a combination of several effects or a chain of dependencies. As an example: the mass of passenger cars is increasing and the number of passenger cars is also changing. What is then the total mass of all passenger cars? In this case the following applies

$$x(\Delta t) = x_0 \cdot (1+z_1)^{\Delta t} \cdot (1+z_2)^{\Delta t} \tag{6.13}$$

Using the exponential function, we obtain

$$\begin{aligned} x(\Delta t) &= x_0 \cdot \exp\left(\ln\left((1+z_1)^{\Delta t} \cdot (1+z_2)^{\Delta t}\right)\right) \\ x(\Delta t) &= x_0 \cdot \exp(\ln(1+z_1) \cdot \Delta t + \ln(1+z_2) \cdot \Delta t) \\ x(\Delta t) &= x_0 \cdot \exp((\ln(1+z_1) + \ln(1+z_2)) \cdot \Delta t) \\ x(\Delta t) &= x_0 \cdot \exp(\ln(1+z_1) \cdot \Delta t) \cdot \exp(\ln(1+z_2) \cdot \Delta t) \end{aligned} \tag{6.14}$$

A generalization to any number of growth rates is straightforward.

6.2 Integrals

Often not only the temporal evolution $x(t)$ is needed, but an integrated value $y(t)$: How much is consumed, required, or produced at a constant rate of change z over a given period of time? An example is the consumption of fossil resources with growing demand over a longer period of time. This integral is

$$y(t_1,t_2) = \int_{t_1}^{t_2} x_0 \cdot \exp(\ln(1+z) \cdot t) \cdot dt = x_0 \cdot \frac{\exp(\ln(1+z) \cdot t_2) - \exp(\ln(1+z) \cdot t_1)}{\ln(1+z)} \tag{6.15}$$

For the special case where integration starts from time 0, the integral simplifies:

$$y(\Delta t) = \int_0^{\Delta t} x_0 \cdot \exp(\ln(1+z) \cdot t) \cdot dt = x_0 \cdot \frac{\exp(\ln(1+z) \cdot \Delta t) - 1}{\ln(1+z)} \quad (6.16)$$

For the case where the time period Δt is sought, over which a quantity y was produced or consumed, we obtain in general

$$\Delta t = \frac{\ln\left(\frac{y(\Delta t)}{x_0} \cdot \ln(1+z) + 1\right)}{\ln(1+z)} \quad (6.17)$$

It also holds in general that

$$\int_{-\infty}^{t_1} \exp(a \cdot t)\, dt < \int_{t_1}^{t_1+\Delta t_D} \exp(a \cdot t)\, dt \quad (6.18)$$

This means that the value added or consumed during the last doubling period is greater than the total value that has been accumulated up to that point, or the consumption that has occurred up to that point.

6.3 Approximation for Small Rates of Change

For small rates or interest rates, i.e., $z \ll 1$, $\ln(1+z) \cong z$. The meaning of "small" ultimately depends on the error that can be tolerated in the calculations. If we now use the series expansion of the logarithm for $-1 < z < +1$

$$\ln(1+z) = \frac{z}{1} - \frac{z^2}{2} + \frac{z^3}{3} - \frac{z^4}{4} + \ldots \quad (6.19)$$

and truncate the series for $|z| \ll 1$ after the first term, we get for Equation 6.9

$$x(\Delta t) = x_0 \cdot \exp(\ln(1+z) \cdot \Delta t) \cong x_0 \cdot \exp(z \cdot \Delta t) \quad (6.20)$$

If the quantity grows as a function of more than one rate, this is

$$x(\Delta t) \cong x_0 \cdot \exp((z_1 + z_2 + \ldots) \cdot \Delta t) \quad (6.21)$$

The growth rates simply add up here.

The doubling period simplifies to

$$\Delta t_d \cong \frac{\ln 2}{z} \simeq \frac{0.7}{z} \quad (6.22)$$

If z is given in percent, the result is simply seventy divided by the growth rate. This doubling period is constant for a given growth rate. The integral over a time period simplifies to

$$y(t_1,t_2) \cong \int_{t_1}^{t_2} x_1 \cdot \exp(z \cdot t) \cdot dt = x_1 \cdot \frac{\exp(z \cdot t_2) - \exp(z \cdot t_1)}{z} \tag{6.23}$$

If integration starts specifically from time 0, the integral simplifies to

$$y(\Delta t) \cong x_0 \cdot \frac{\exp(z \cdot \Delta t) - 1}{z} \tag{6.24}$$

The period for generation or depletion is

$$\Delta t = \frac{1}{z} \cdot \ln\left(\frac{y(\Delta t)}{x_0} \cdot z + 1\right) \tag{6.25}$$

What does 1000 years of coal mean?

Donald Trump, as a US presidential candidate in 2016, argued that the USA would mine coal for the next 1000 years. We assume that he estimated the amount of coal mined annually will grow steadily in the future: we apply a growth rate of 2% here – comparable to the long-term increase in energy demand in the USA.

The amount of coal extracted in the USA in 2014 was approximately $900 \cdot 10^9$ kg. For simplicity, we assume that this coal consists only of carbon. We consider the environmental impact of the complete combustion of this coal, assuming that all other influences are negligible.

Determine the total mass of coal to be extracted in 1000 years, assuming a constant annual extraction rate.

This is simply a constant extraction without growth and thus

$$m_{C,stat} = m_{C,0} \cdot \Delta t = 900 \cdot 10^9 \ \frac{\text{kg}}{\text{a}} \cdot 1000 \text{ a} = 900 \cdot 10^{12} \text{ kg}$$

Determine the total mass of coal to be extracted in 1000 years.

The solution using the approximation of a small growth rate is

$$m_{C,w} = m_0 \cdot \left(\frac{\exp(z \cdot 1000) - \exp(z \cdot 0)}{z}\right) = 900 \cdot 10^9 \text{ kg} \cdot \left(\frac{\exp(0.02 \cdot 1000) - 1}{0.02}\right)$$
$$= 21.83 \cdot 10^{21} \text{ kg}$$

Using the exact function, a slightly lower value is obtained:

$$m_{C,w} = m_0 \cdot \left(\frac{(1+z)^{1000} - (1+z)^0}{\ln(1+z)}\right) = 900 \cdot 10^9 \text{ kg} \cdot \left(\frac{(1+0.02)^{1000} - 1}{\ln(1+0.02)}\right) = 18.10 \cdot 10^{21} \text{ kg}$$

How much oxygen mass is needed to combust 1 kg of coal?

The mass of oxygen is significantly greater than that of carbon:

$$m_{O_2} = m_C \cdot \frac{M_{O_2}}{M_C} = 1\ \text{kg} \cdot \frac{32.00\ \frac{\text{kg}}{\text{kmol}}}{12.01\ \frac{\text{kg}}{\text{kmol}}} = 2.664\ \text{kg}$$

By what factor does the oxygen mass of the earth's atmosphere decrease at a constant annual extraction rate over the 1000 years?

If we assume that the oxygen content of the earth's atmosphere changes over the next 1000 years only through the combustion of this coal, then we get

$$\Delta\mu_{stat} = \frac{m_{C,stat} \cdot \frac{M_{O_2}}{M_C}}{m_{O_2,atm}} = \frac{900 \cdot 10^{12}\ \text{kg} \cdot 2.664\ \frac{\text{kg}}{\text{kg}}}{0.23 \cdot 5.15 \cdot 10^{18}\ \text{kg}} = 2.2 \cdot 10^{-3} = 0.22\ \%$$

Used here is the Wikipedia value of the mass of the earth's atmosphere and the mass fraction of oxygen in the atmosphere of 23%. The increase in the carbon dioxide content of the earth's atmosphere would be quite substantial.

By what factor does the oxygen mass of the earth's atmosphere decrease for steady growth at 2%?

Substituting the corresponding values, we obtain

$$\Delta\mu_{stat} = \frac{m_{C,stat} \cdot \frac{M_{O_2}}{M_C}}{m_{O_2,atm}} = \frac{21.83 \cdot 10^{21}\ \text{kg} \cdot 2.664\ \frac{\text{kg}}{\text{kg}}}{0.23 \cdot 5.15 \cdot 10^{18}\ \text{kg}} = 49\ 100$$

The value means that we could only burn a small fraction of the coal before all the oxygen in the earth's atmosphere would be consumed: it is impossible to achieve the goal, and quite early on the majority of life on earth would perish (and higher mammals are among the first groups to go).

Determine the heat available in the thousandth year at a steady growth rate of 2%.

The mass of coal extracted at that point is

$$m_{C,w}(1000) = m_0 \cdot (1+z)^{\Delta t} = 900 \cdot 10^9\ \text{kg} \cdot (1+0.02)^{1000} = 358.4 \cdot 10^{18}\ \text{kg}$$

The amount of heat released by the complete combustion of this coal in that year is (provided sufficient oxygen is available for combustion)

$$Q(1000) = m_C \cdot H_{I,C} = 358.4 \cdot 10^{18}\ \text{kg} \cdot 30 \cdot 10^6\ \frac{\text{J}}{\text{kg}} = 10.75 \cdot 10^{27}\ \text{J}$$

By how much can this heat raise the temperature of the earth's atmosphere?

The calculated temperature is, assuming air as an ideal gas and constant specific heat capacity

$$\Delta T = \frac{Q}{m_{atm} \cdot c_p} = \frac{10.75 \cdot 10^{27}\ \text{J}}{5.15 \cdot 10^{18}\ \text{kg} \cdot 1004\ \frac{\text{J}}{\text{kg} \cdot \text{K}}} = 2\ 079\ 000\ \text{K}$$

This is a remarkably questionable result, since the assumptions used are not valid over this wide temperature range. Since iron melts at approximately 1600 °C, this figure merely illustrates once again that D. Trump's original assumption is not feasible.

Learning objective The example illustrates that it is often worthwhile to do the math for future visions: What do the ideas presented so convincingly actually mean in real terms? What is realistic and what is not? Do we want the result in all its consequences?

6.4 What Can Grow?

This section describes up to what point the use of relevant resources can grow.

Material The mass of material is limited by the fact that the earth system is finite. In addition, only a very small fraction of the earth's mass can ever be used for technology. The rate at which mass can be made available, whether through extraction and processing, through agriculture and harvesting, or through recycling of technology no longer in use, is also limited, since each of these activities requires other resources (e.g. Lundaev et al. 2023).

Chapter 7 provides an overview of the extent of the technosphere. The further incorporation of material into the technosphere depends on the availability of raw materials, i.e. reserves. Organic material cannot be accumulated infinitely in the technosphere, since this material is also needed in the living earth for plant growth.

Energy and power What is meant here is societally usable energy over longer periods of time and power at a given point in time.

The mass of fossil primary energy carriers on earth is finite. Therefore, the total heat obtainable from them is also limited. This is explored further in Chapter 9. The energy actually usable from fossil fuels also depends on decisions regarding climate policy (see Chapter 8) and fundamentally on prices achievable in the market. The price reflects the necessary inputs of energy and materials as well as rights and transport for the provision of fossil energy carriers, and these therefore act as additional limiting factors. The usable power from fossil energy carriers is limited by the fact that technology is required for each conversion process, which must first be provided at the expense of energy and other resources.

The usable power from renewable energy forms is limited by the energy flows available in the earth system and the material required for the technology used for conversion (more on this in Chapter 7).

Capital The current societal expectation arising from the growth paradigm is that capital multiplies exponentially at an appropriate rate of return. An appropriate rate of return is a matter of perspective. While in the time of Jane Austen and her heroine Elizabeth Bennet (Pride and Prejudice was published in 1813) an annual return of 4% was considered appropriate, this has shifted somewhat in modern companies with their internal rates of return of 30% to 50% per year for investments: these rates correlate with the typical expectation that an investment pays for itself within 12 to 18 months. The extent to which net capital can actually grow, whether capital can grow at all on its own or whether capital is merely a dependent variable of available energy, goes well beyond the scope of this book (Hall/Klitgard 2018). Whether it is possible for capital to grow steadily in an otherwise finite system, and whether it is desirable for capital to grow steadily, is addressed by Daly and Farley (Daly/Farley 2011).

The expectation of exponential growth of capital is also referred to in economics as "sustainable" growth: any capital that does not grow exponentially fast enough would thus grow "unsustainably." This is a peculiar and complete misunderstanding of the concept of a sustainable economy as described by von Carlowitz (1713) at the time: a sustainable economy is one that deals with a clearly limited and non-growing resource and uses it in such a way that it yields a good and constant return over the very long term. This is how the concept of "sustainability" is to be understood in this book.

At the same time, it is foreseeable that the economy and growth face ever new, self-generated problems that need to be managed and that tie up ever more capital and other resources.

Land Land area is finite in the earth system. Any growth for one use comes at the expense of another use. The use of land by natural or near-natural systems is included here as a form of use. Thus, any conversion of land should be well justified. With regard to an energy supply from biological sources, the sealing of agricultural land in particular poses a problem. This sealing often results in the loss not only of the area from agricultural use, but also of the soil through compaction. While in the global context it is mainly cities that occupy high-quality agricultural land with roads and buildings, in Germany this has now reversed. In urban areas, hardly any new land is being developed; rather, it is the rural municipalities that generously designate new residential areas and business parks. Agricultural products must then be produced elsewhere and often transported over long distances. This increases the energy intensity of agricultural products.

Soil This resource is only available in limited quantities. Since the formation of new soil takes a long time or occurs at a significantly lower growth rate than the expected return on capital, soil should be regarded as a non-renewable resource, at least with regard to economically common time horizons. Thus, the degradation and loss of soil through processes such as compaction, sealing, plowing, erosion, salinization, poisoning, and the use of pesticides simultaneously represent the destruction of non-regenerable capital. Soil cannot be discounted.

Entropy In a closed system, entropy increases with every process occurring within the system, because at the same time the system loses energy with every process: the ability to perform work within the system continually diminishes. The limiting case is a state in which no (significant) work can be performed anymore. This thermodynamic equilibrium is simultaneously the death of living things. If the closed system was still full of life at the beginning, this life has perished on the way to thermodynamic equilibrium. This begins first with the higher life forms and then continues down to the more undemanding survivors with ever lower metabolic rates.

An exponential increase in entropy in a system is difficult to imagine. It seems very atypical and very undesirable.

Endless economic growth

In this example, we assume that the basic prerequisite for economic growth is (i) the favorable availability of energy and that (ii) the available energy itself must continuously grow in order for the economy to grow. Thus, capital is understood as a quantity dependent on energy consumption (Hall/Klitgard 2018). Figure 6.3 shows the temporal development of the required energy for different growth rates of energy demand. The energy axis is logarithmic. As a result, the exponential functions appear as straight lines.

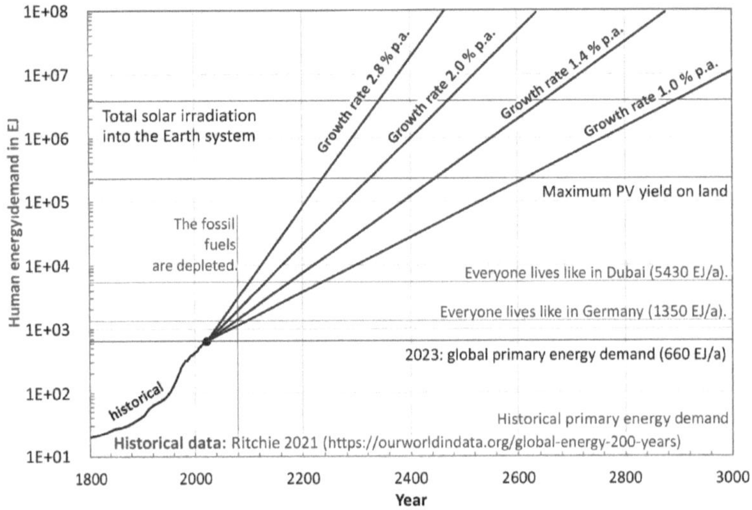

Figure 6.3 Increase in energy demand at different constant growth rates (data source: Ritchie 2021)

As the theoretically maximum energy available in the earth system, it was assumed that the entire land area of the earth is covered with photovoltaics.

If we take the demand for steady economic growth seriously and the previous growth in global energy demand continues, limits become apparent. Global energy demand has increased at approximately 2% per year since the industrial revolution. If this is to continue, then depending on the degree of consideration for ecosystems, the possibilities on earth will be exhausted in approximately 100 to 200 years and growth would come to an end. Therefore, Figure 6.4 examines extraterrestrial possibilities, which can rather be classified in the realm of speculative science fiction, since these technologies each have a TRL 0 or sometimes TRL 1:

- The sun provides $2 \cdot 10^{13}$ times today's demand. A prerequisite for harvesting is a suitable large-scale interplanetary spaceflight technology, which should be available within less than 100 years in order to keep our earth habitable. To avoid overheating of the planet Earth, moreover, major consumers of energy should no longer be located on Earth, and launching and landing of spacecraft on the Earth's surface should also be eliminated.
- The Milky Way provides $2 \cdot 10^{25}$ times today's demand. As a technical hurdle, spaceflight and energy transport at faster-than-light speed are relevant. Without this technology, even relatively nearby stars cannot be harvested in time to maintain growth. The technology for faster-than-light speed should therefore be available in approximately 1000 years so that the transition succeeds.
- The total energy flux available in the universe in the long term provides $2 \cdot 10^{36}$times today's demand. The required technology was already developed for the previous step. And with that, the limits of growth are finally reached.

This example is intended to show, by consistently thinking through the growth paradigm under the given conditions, that this paradigm is either not tenable, since it quickly leads to absurd results, or that its assumptions are wrong. To anticipate the argument: I cannot understand how economic growth without growth in energy demand or material consumption is possible. The mere addition of zeros on banknotes cannot be what is meant here. The increase in capital must manifest itself as a real value in the physical world through material and energy – and there is an upper limit to that.

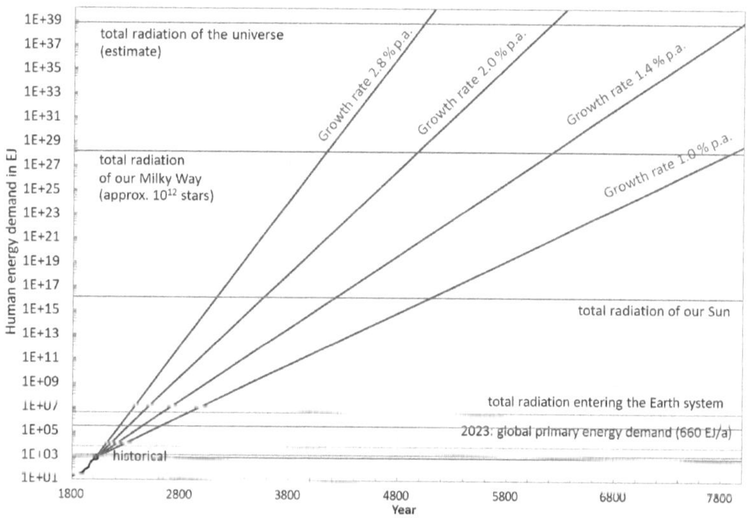

Figure 6.4 Increase in energy demand for different growth rates and extraterrestrial sources

6.5 Cost-Benefit Analysis

The cost-benefit analysis (CBA) is a method for the financial evaluation of an investment. It assesses whether an investment can generate the expected return on the capital employed. The method is thus directly linked to the topic of growth (of capital).

The value of an investment is determined as the sum of revenues minus expenditures over an expected period during which the investment is made and used. The value of each individual time period is discounted at the expected interest rate and then summed over all considered time periods (Carson/Roth-Tran 2009). Specifically for energy systems, the value is represented as follows (Narbel et al. 2014):

$$value = \sum_{t=1}^{n} \frac{B(t) - C(t)}{(1+z)^t} \tag{6.26}$$

Here z is the interest rate, $B(t)$ are revenues of period t (benefit) and $C(t)$ the expenditures of period t (cost). If the value determined in this way is positive, then the investment is financially worthwhile at the underlying interest rate. This equation is sometimes found with slight variations in the literature. The form reproduced here has no problems with the allocation of amounts, since these are always to be considered in the period in which they occur.

Externalizing Costs

The method immediately makes clear that costs must be minimized in order to make investments worthwhile. One option is efficient action (more on this in Chapter 8). A second powerful option is to externalize costs. The term means that costs associated with an economic activity are not attributed to it but are shifted externally. Examples of this are:

- **Subsidies:** Others pay for my investment. Subsidies can be direct or indirect, by providing required infrastructure.
- **Emissions:** All emissions have effects outside. If these effects then have to be borne or mitigated by affected third parties, these are externalized costs (on emissions see Chapter 14).
- Environmental legislation has the goal of minimizing the externalization of impacts. The non-application of existing laws, the weakening of laws, or the waiving of laws permits and promotes externalization.
- **Avoiding cleanup at end of use:** Mines or oil and gas wells are not reliably closed. Brownfields and contaminated soils – all of these are signs of the externalization of costs. Cleanup at the end is particularly critical, since no revenues are generated at that point, yet considerable costs continue to accrue. Insolvency is one way to externalize these costs.

The costs are borne, but by others. They can also be analyzed as energetic costs (Sheldon et al. 2015).

Interest Rate As a Critical Variable

Cost-benefit analysis is first of all a good method of economic evaluation, i.e., evaluation in a purely financial context. What must always be viewed critically in this context is the interest rate used, because the results are strongly influenced by the interest rate. At higher interest rates, it is nearly impossible to obtain a positive value from returns further in the future: by setting the interest rate, I define the time horizon over which I am willing to think and act into the future. Thus, an interest rate of 8% corresponds to a period of approximately 10 years from today, and an interest rate of 40% corresponds to a period of 1.5 years from today.

Reference Framework

Cost-benefit analysis has as its reference, and unless otherwise specified, *do nothing*, i.e., the decision not to act: no change to the system takes place and it is implied that in this case the world does not change. It is assumed that the product can continue to be manufactured at the same cost and sold at the same price, and that this also applies to all competitors, or that the *do nothing* decision generates no costs of its own. This assumption is already incorrect in the context of a purely profit-focused organization, and the reference scenario must always be defined with great care (Christensen et al. 2008): what needs to be clarified is how relevant indicators change if the organization does nothing.

This assumption fails entirely for questions of societal relevance, e.g., in a changing world, since the world as a *do nothing* scenario then generates its own costs for society that must be taken into account. This is strikingly illustrated by the decision for climate protection or the decision against climate protection (Kotz et al. 2024).

Strong Sustainability As a Boundary

Cost-benefit analysis relates to capital and thus not only to money, but also to all products that can be substituted by capital: resources that can also be replaced at any time in the future without difficulty by capital may be discounted, and cost-benefit analysis can be applied to them.

The basic assumption of "weak sustainability" assumes that all resources can be replaced at any time by capital, as long as the financial system remains stable. Since these two prerequisites – (i) all functions of the earth system and the biosphere can be replaced at any time by capital, and (ii) the financial system remains stable in the long term in the future – are not given, it is entirely irresponsible to rely on them. Only the principle of "strong sustainability" remains. The basic assumption of "strong sustainability" recognizes many resources that cannot be replaced by capital. These resources therefore cannot in principle be discounted: this applies in particular to energy, critical raw materials, soil, and land.

Applying the Method Safely

The cost-benefit analysis draws its appeal from being able to reduce a problem that initially appears complex to a single simple statement and recommendation for action. In this sense, it is most commonly used in the political context: as a way of not having to talk about wicked problems, but instead dressing the consequences of an action in a pseudo-rational argument. With regard to the wicked problems discussed here, cost-benefit analysis can usually make no contribution, because we must ask ourselves the following questions:

- Is capital the right unit for evaluating advantages and disadvantages?
- How do we assess illness, death, hunger, displacement, or trauma in monetary terms? What specific amount must be assigned to them?
- How do I assess creeping expropriation, the loss of home, the loss of security, or the loss of my familiar surroundings?
- Are insurance payments a good metric for assessing damages? What about damages that are not insured, about societies or people who are not insured or have no entitlement to insurance payments?
- If I set the value of a life today at €1 million and discount at 10%, the life in the year 2100 is worth approximately €500 in today's euros. Is this figure an objective assessment of the value of a human life at that point in time or at the present time?
- If I argue via insurance payments: is it true that the lives and health of different people are worth different amounts?
- Am I allowed to set the value of ecosystems and public goods at zero?
- Can I carry out my assessments with the required certainty? Are my figures sufficiently robust?
- Have I forgotten anything important? Can I be certain that I know everything and therefore cannot have forgotten anything important? Have I omitted anything important out of convenience or for personal advantage?
- What are the monetarily measurable benefits of a measure? Can I express all benefits in monetary terms?
- Can I directly offset the benefits and the input of a measure against each other for a given period of time?
- What happens to the objective advantages and disadvantages for which there is no monetary equivalent?
- Is *willingness to pay* an appropriate method for generating monetary value, or is it an expression of helpless fixation on numbers? Does *willingness to pay* mean that the opinion of people with less money is less important, since they can pay significantly less even if they wanted to, and that people without money have no say at all in decisions that affect them, since they cannot pay?

- What is an appropriate interest rate for discounting life, health, or the continued existence of my society?
- With long-term problems, the interest rate defines the decision. Is my theory robust enough to determine this interest rate with sufficient precision to also make decisions about life and death?
- Are personal interest rates, with which purchasing decisions are made, suitable for discounting long-term wicked problems (this is the usual approach for determining interest rates)?
- Has the theory of the world underlying the method been sufficiently scientifically tested, or is it rather the assertion of a theory that is kept alive solely by the personal financial interests of its beneficiaries?
- Is there a binding ethics, i.e., are there actions that I undertake even if they are not financially worthwhile for me?

Sen addresses some of these questions and others (Sen 2000). With reference to wicked problems, a considerable proportion of these questions must be answered in such a way that the use of cost-benefit analysis can be regarded as at least grossly negligent, and more often as deliberate, since the users — unlike those affected — are aware of the problems of the method and, through their authority to determine the interest rate, then force the outcome they desire.

Logistic growth

Are there other forms of growth that do not tend toward infinity at a constant rate? Are there ways to describe growth as it occurs in natural bounded systems? In biology, the growth of quantities in bounded systems plays a major role (as the number of animals in an ecosystem). A simple model for describing the development of the number of an animal or plant species is so-called logistic growth. In logistic growth, the temporal development of a variable quantity, which can only take meaningful values between $0 < x_n < 1$, is described using a series expansion. Starting from an initial value x_0, the following applies for each subsequent time step (i.e., as a time series)

$$x_{n+1} = x_n \cdot r \cdot (1 - x_n) \tag{6.27}$$

Here x_n is the value after n discrete time steps and r is a free parameter that influences the change. This quantity r describes the intensity of the interaction between the quantity x and the system. Figure 6.5 shows the typical S-shaped curve for two different initial values and two values of r (1.3 and 1.4). The influence of the parameter r is shown in Figure 6.6 :

- For $0 < r < 1$, x_n approaches the value 0.
- For $1 < r < 2$, x_n approaches the value $(r - 1)/r$. For $x_0 < (r - 1)/r$ this is monotonically increasing. For $x_0 > (r - 1)/r$ this is monotonically increasing after a sharp drop.

- For $2 < r < 3$, x_n approaches the value $(r - 1)/r$ to, where x_n alternates around the target value (oscillates back and forth).
- For $3 < r < 4$, there are always two or more solutions. Above 3.57, these are mostly chaotic solutions.
- For $4 < r$, the solution diverges. It grows without bound.

In natural systems, all forms of logistic growth are observed, from the final extinction of a population to the establishment of an equilibrium over a period of time to chaotic behavior. The more intensively or rapidly the quantity under consideration responds to impulses from its environment, the faster it tends toward an equilibrium. When a critical response rate is exceeded, the response is no longer monotonic. This can tip into chaos.

The more variables and influencing factors are interrelated, the more complex the behavior of the individual quantities. This behavior can, if necessary, be described by complex models that incorporate many feedback loops.

Logistic growth is a good robust description when an equilibrium is to be approached. It is also used in many forms in economics, e.g., when the utilization of a line is planned or it is otherwise assumed that a stable state is to be reached.

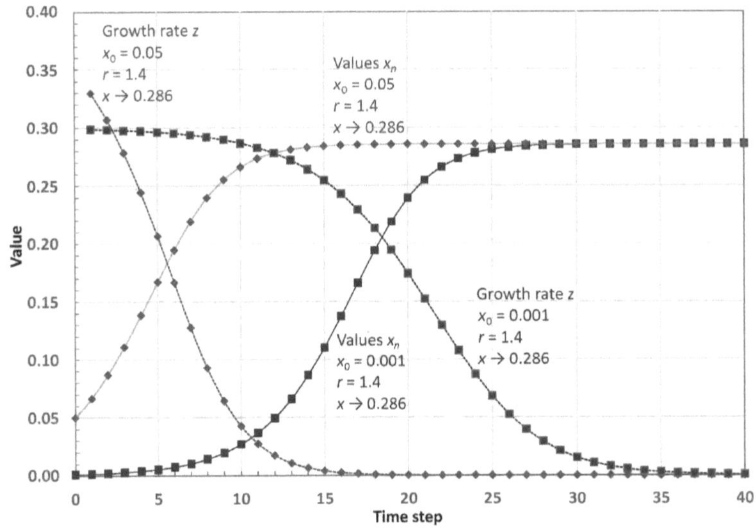

Figure 6.5 Logistic growth: solid lines show the development of the value and dashed lines show the growth rate for each time step.

The beginning of logistic growth cannot be distinguished from that of exponential growth, as Figure 6.5 shows. Concluding from only a few growth steps that the system would continue to grow at a constant rate can be wrong, since the growth rate in logistic growth remains nearly constant over the first steps and for $1 < (r - 1)/r$ remains nearly constant.

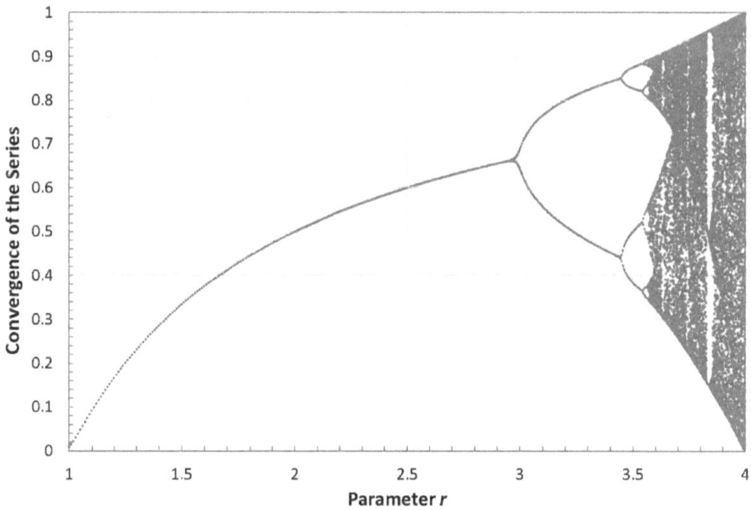

Figure 6.6 Development of the convergence of logistic growth as a function of the parameter *r*

Closed-form functions describe logistic growth

Describing logistic growth only via a series is not very practical for many tasks. For this reason, there are a whole range of equations that describe logistic growth for different purposes and degrees of complexity (Tsoularis/Wallace 2002). Common to these models is that they each yield an S-curve, comparable to the series expansions in Figure 6.5. Often an additional function for the relative or absolute growth is needed. This second function always has a bell shape.

As an example, the approach of Lundaev et al. (Lundaev et al. 2023) is used here. There, the absolute extraction of metals is approximated by an S-shaped function:

$$f(t) = A + \frac{K - A}{\left(1 + C \cdot \exp\left(-B \cdot \left(t - t_{peak}\right)\right)\right)^{1/v}}$$

This function is defined such that at the reference time t_{ref} it takes the value $f\left(t_{ref}\right) = 0$. For this to work, the lower asymptote -*A* is the value of the total mass extracted so far and *K* as the upper asymptote is either the reserve or the sum of reserve and resource of the metal at the reference period. The two asymptotes define the range of values between which the curve moves, with $f(-\infty) = -A$ and $f(+\infty) = K - A$. The remaining parameters define the course of the function. *B* is a growth rate. *C* is a free parameter that shifts the function. $1/v$ stretches the function along the y-axis. The maximum of growth, i.e., the maximum of the bell curve, is t_{peak}. To determine the free parameters, starting from historical data, one can use, for example, the "Solver" function built into some spreadsheet programs.

Strictly speaking, the growth rate is the derivative of the S-curve. Often the bell function sought is intended to describe an annual mass or rate. This can be well derived for many applications from

$$g(t) = \frac{df(t)}{dt} \cong \frac{f(t+\Delta t) - f(t)}{\Delta t}$$

calculate. Other equations for limited growth are discussed in the literature. Since every growth in a limited system is inevitably limited, only limited functions can describe real growth. This also applies to energy systems and thus to economic systems (Höök et al. 2011).

References

Carlowitz, H. C. von: Haußwirthliche Nachricht und Naturmäßige Anweisung Zur Wilden Baum-Zucht. Johann Friedrich Braun, Leipzig 1713

Carson, R. T./Roth-Tran, B.: Discounting behaviour and Environmental Decisions. In: Journal of Neuroscience, Psychology, and Economics 2, 2009, pp. 112–130

Christensen, C. M./Kaufman, S. P./Shih, W. C.: Innovation Killers, How Financial Tools Destroy Your Capacity to Do New Things. Harvard Business Review 2008

Coady, D./Parry, I./Sears, L./Shang, B.: How Large Are Global Energy Subsidies? In: IMF Working Paper 15/105, International Monetary Fund 2015.*https://www.imf.org/external/pubs/ft/wp/2015/wp15105.pdf*

Daly, H./Farley, J.: Ecological economics. Island Press 2011

Hall, C. A. S./Klitgard, K.: Energy and the Wealth of Nations. Springer, Cham 2018

Höök, M./Junchen, L./Oba, N./Snowden, S.: Descriptive and predictive growth curves in energy system analysis. In: Natural Resources Research 20, 2011, pp. 103–116

Kotz, M./Levermann, A./Wenz, L.: The economic commitment of climate change. In: Nature 628, 2024, pp. 551–557

Lundaev, V./Solomen, A. A./Le, T./Lohrmann, A./Breyer, C.: Review of critical materials for the energy transition, an analysis of global resources and production databases and the state of material circularity. In: Minerals Engineering 203, 2023, 108282

Narbel, P. A./Hansen, J. P./Lien, R. R.: Energy Technologies and Economics. Springer, Heidelberg 2014

Ritchie, H.: How have the world's energy sources changed over the last two centuries? December 1, 2021. *https://ourworldindata.org/global-energy-200-years*

Sen, A.: The Discipline of Cost Benefit Analysis. In: Journal of Legal Studies 29, 2000, pp. 931–952

Sheldon, S./Hadian, S./Zik, O.: Beyond carbon: quantifying environmental externalities as energy for hydroelectric and nuclear power. In: Energy 84, 2015, pp. 36–44

Smil, V.: Energy and Civilization. A History. MIT Press, Cambridge, MA/USA 2017

Tsoularis, A./Wallace, J.: Analysis of logistic growth models. In: Mathematical Biosciences 179, 2002, pp. 21–55

7
The Earth System

The utilization of the energy flows of our earth is the future possibility of supplying us humans and our civilizations with energy. This chapter is about developing relevant fundamentals of these energy flows: Which regenerative forms of energy are available in the earth system and what dependencies exist between important energy and material flows in the earth system? In doing so, this chapter can only focus on some central relationships.

7.1 The Earth As a Planet – Energy Balance

The earth system can be described as a stationary flow system, with a nearly constant energy and entropy flow across its boundary and with high dynamics within the system itself.

Earth mass The mass of the earth is constant. Almost no mass flows in (which is fortunate, as this would occur in the form of asteroids and the associated impacts). Mass losses are limited to the loss of hydrogen and helium from the uppermost layers of the atmosphere. Both gases are so light that, due to collisions with heavier molecules, they sometimes reach escape velocity and can thus leave the earth.

Planet The earth is, to a good approximation, a sphere with a diameter of 12,735 km. In fact, the diameter at the equator averages 12,756 km and between the poles 12,714 km. It moves on a nearly circular orbit around the sun. In doing so, it rotates about its own axis, so that the sun appears at the same position in the sky after every 24 h. The earth orbits the sun in 365.26 days.

Atmosphere The atmosphere is characterized by particularly high dynamics and very rapid mixing (Wallace/Hobbs 2006): Local deviations from the mean composition are equalized within weeks to months. The current composition of the air is given in Table 17.3 in Section 17.3.

Mass of the atmosphere

The mass of the atmosphere can be calculated surprisingly simply from the definition of pressure $p = F/a$ and the gravitational force $F = m \cdot g$, which the atmosphere exerts at sea level:

$$m_{atm} = O_{earth} \cdot \frac{p_{sea}}{g} = \pi \cdot d_{earth}^2 \cdot \frac{p_{sea}}{g} = \pi \cdot \left(12.74 \cdot 10^6\ \mathrm{m}\right)^2 \cdot \frac{100\,000\ \mathrm{Pa}}{9.81\ \frac{\mathrm{m}}{\mathrm{s}^2}} = 5.20 \cdot 10^{18}\ \mathrm{kg}$$

This value is close to the literature value of $5.148 \cdot 10^{18}$ kg. The deviation is due to mountains that extend above sea level.

Oceans Dynamic, but with significantly slower mixing times than the atmosphere, are oceans and seas. The exchange between near-surface layers and the deep ocean takes place over many hundreds of years (Czersky 2023).

Incoming radiation Due to the ratio between the diameter of the sun relative to the distance between the earth and the sun, the sun can be regarded to a good approximation as a point radiation source: the radiation of the sun reaches the earth nearly in parallel. The sun emits its energy in a spectrum that is well approximated by that of a black body radiator at 5776 K (Figure 7.6). Thus, the maximum intensity of the radiation lies in the green spectral range at approximately 520 nm. On the earth's orbit, the mean intensity of the sun is $I_{sol} = 1366\ \mathrm{W\ m^{-2}}$. This is the solar constant.

The mean irradiance by the sun

The mean intensity of the radiation of the sun at the position of the earth on its orbit amounts to $I_{sol} = 1366\ \mathrm{W\ m^{-2}}$. This value varies somewhat with the actual position of the earth on its orbit and thus with the season. As Figure 7.1 shows, the irradiated area of the earth as seen from the sun is a disk with the diameter of the earth. The earth is therefore irradiated with a heat flux of

$$\bar{\dot{Q}}_{in} = \frac{\pi}{4} \cdot d_{earth}^2 \cdot I_{sol} = \frac{\pi}{4} \cdot \left(12.74 \cdot 10^6\ \mathrm{m}\right)^2 \cdot 1366\ \frac{\mathrm{W}}{\mathrm{m}^2} = 174.1 \cdot 10^{15}\ \mathrm{W}$$

The intensity averaged temporally and spatially over the entire surface of the earth as a sphere amounts to

$$\bar{\dot{q}} = \frac{\bar{\dot{Q}}_{in}}{O_{earth}} = \frac{\frac{\pi}{4} \cdot d_{earth}^2 \cdot I_{sol}}{\pi \cdot d_{earth}^2} = \frac{1}{4} \cdot I_{sol} = 342\ \frac{\mathrm{W}}{\mathrm{m}^2}$$

This is the mean thermal radiation above the atmosphere.

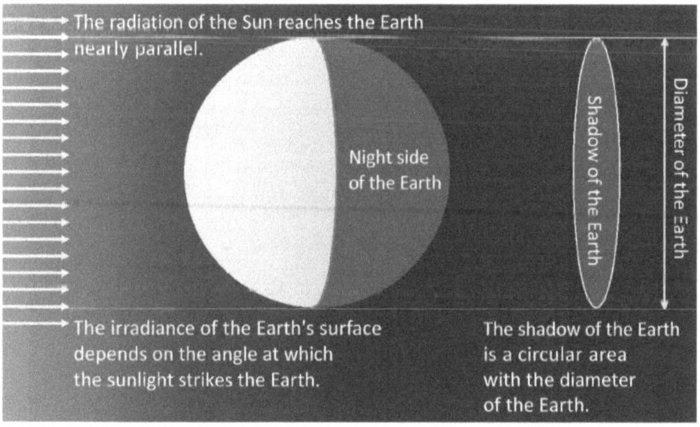

Figure 7.1 Geometry of the earth and the radiation of the sun: the sun always irradiates an area that corresponds exactly to the cross-section of the earth's shadow.

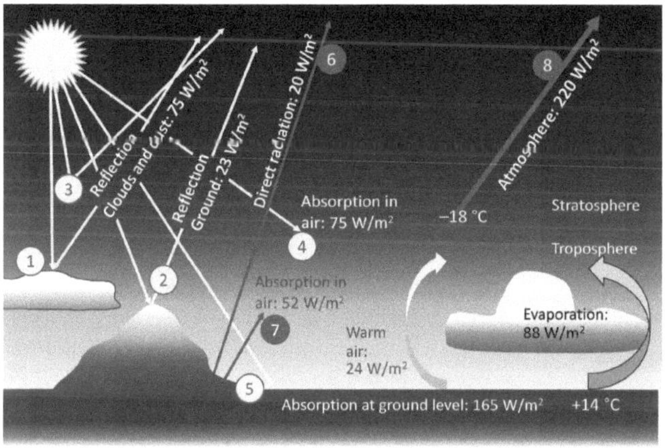

Figure 7.2 Mean net heat fluxes at the surface of the earth (explanation in Table 7.1)

Absorption Of the radiation from the sun, on average only about half reaches the ground. The other parts are scattered (i.e. reflected) or already absorbed in the atmosphere. Figure 7.2 illustrates the heat fluxes and Table 7.1 (see Section 17.1) gives an overview of the mean global values (Stephens et al. 2012). These values vary strongly with geographic latitude (i.e. the distance from the equator), the seasons, and the current cloud cover and particle loading of the local air (i.e. the weather and the climate).

Table 7.1 Mean Fraction of Heat Flows at the Surface of the Earth, Illustrated in Figure 7.2 (Based on Stephens et al. 2012)

No.	Radiation flux	Value in $W m^{-2}$	Fraction	Remark
	Solar energy input	342	100 %	Averaged over the earth's surface
1	Radiation from the sun reflected by clouds	47.5 ± 3	14 %	This fraction leaves the earth system unchanged.
2	Radiation from the sun that is directly reflected at the ground	23 ± 3	7 %	The scattering depends on the wavelength of the radiation.
3	Radiation from the sun that is reflected in the atmosphere (gases or dust)	27.2 ± 5	8 %	
	Total reflection of the earth	100 ± 2	29 %	Clouds, atmosphere and ground
	Radiation that is absorbed in the earth system	240	70 %	Atmosphere and ground
4	Radiation from the sun that is absorbed in the atmosphere	75	22 %	This fraction is amplified by greenhouse gases.
	Radiation that reaches the ground	192	78 %	
5	Radiation from the sun that is absorbed at the ground	165	48 %	Oceans and open water absorb solar radiation very well, ground and vegetation considerably less so.
6	Sensible heat that reaches the atmosphere from the ground by convection	24 ± 7	7 %	Dominant in deserts
7	Heat that reaches the atmosphere from the ground as latent heat	88 ± 10	26 %	When water condenses into clouds, this heat is released again in the atmosphere.
8	Net heat emitted as radiation from the earth's surface and re-absorbed by the atmosphere	51.8	15 %	This fraction is amplified by greenhouse gases.

No.	Radiation flux	Value in $W\,m^{-2}$	Fraction	Remark
9	Heat emitted as radiation from the earth's surface that directly leaves the earth system	20 ± 4	6 %	The spectrum lies in the far infrared. This fraction is reduced by greenhouse gases.
10	Heat that leaves the earth system as radiation from the atmosphere (upper layers)	240	70 %	The spectrum lies in the far infrared and is determined by the temperature of the stratosphere.

Emission As a stationary flow system, the earth must release as much energy as it absorbs. In doing so, we must still subtract the direct scattering, because the scattered radiation has not entered the earth system but leaves it directly. Measured values show that approximately 70 % of the radiation, or on average 240 W m^{-2} is absorbed. This absorbed heat flow must be released by the earth again so that the earth as a whole remains in a stationary state.

Emission temperature of the earth

The mean absorbed heat flow of the earth of 240 W m^{-2} is also re-emitted by it. From this heat flow we can determine a blackbody temperature:

$$T_{out} = \left(\frac{q_{out}}{\sigma}\right)^{\frac{1}{4}} = \left(\frac{240\,\frac{\mathrm{W}}{\mathrm{m}^2}}{5.67\cdot 10^{-8}\,\frac{\mathrm{W}}{\mathrm{m}^2\cdot \mathrm{K}^4}}\right)^{0.25} = 255\ \mathrm{K}$$

This is the temperature at which the earth system emits this very long-wave thermal radiation in the far infrared into space. The maximum of the radiation is found at approximately 10 000 nm. This temperature is significantly lower than the mean surface temperature of the earth of 288 K and is the mean temperature of the upper atmosphere (stratosphere). The difference is directly generated by the presence of the atmosphere, or more specifically by the greenhouse effect of the earth's atmosphere.

From Table 7.1 we can read off the mean heat flows. These vary depending on location and the prevailing weather there. In particular, the actual local heat flow depends on the latitude. The absorbed heat flow is distributed within the earth system and converted into other forms of energy (Kleidon 2016):

- Of the incident heat flow of $175 \cdot 10^{15}$ W, 70% is absorbed in the earth system. This amounts to $123 \cdot 10^{15}$ W.
- At the equator, more heat is absorbed than at the poles. This difference amounts to approximately 40% or $49 \cdot 10^{15}$ W. However, the efficiency of converting this

difference in the atmosphere into wind is only about 2%, so that approximately $1.0 \cdot 10^{15}$ W of wind power results from this.

- The conversion of wind into waves has an efficiency of approximately 6% and the conversion of wave motion into ocean currents is 8%. Therefore, the power of ocean currents is $5 \cdot 10^{12}$ W and thus much lower than that of wind.
- From the evaporation of water, a latent heat flux of $45 \cdot 10^{15}$ W is supplied to the atmosphere. The power of raindrops as potential energy resulting from the condensation of water amounts to $1 \cdot 10^{15}$ W.

This overview provides a first indication of how much of this power we can extract through technology.

Local solar irradiance The incoming radiation of the sun varies for every location on earth depending on these factors:

- The day-night cycle: The sun shines only during the day, i.e., on an annual average 12 h per day.

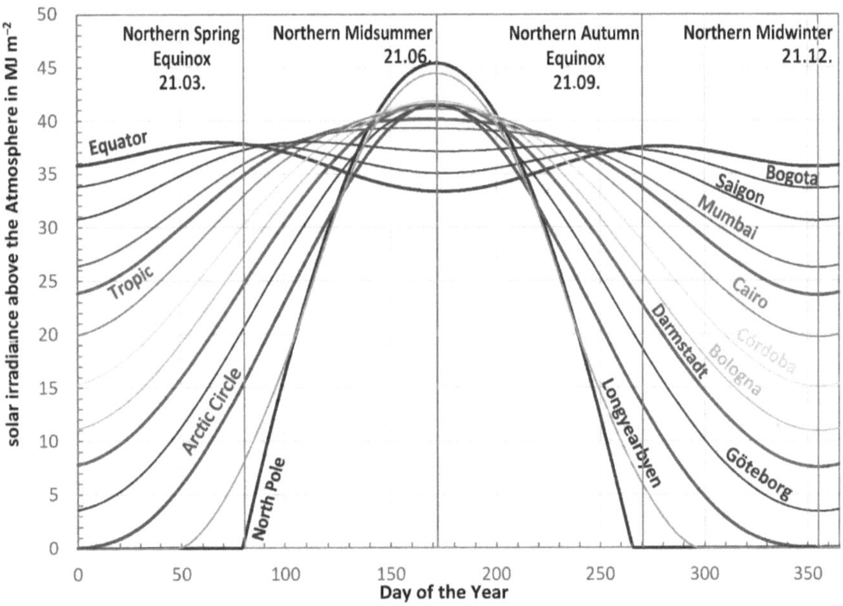

Figure 7.3 Annual cycle of solar irradiance as daily sum above the atmosphere for representative locations of the northern hemisphere

The season: The earth's axis is tilted by 23.5° and depending on the orbital position of the earth on its orbit, the orientation of this axis relative to the sun changes. This results in the variation of the day length and the angle at which the sun irradiates the respective location.

- The latitude of the location, i.e., the distance from the equator: Further away from the equator, the intensity relative to the surface is lower, since the radiation of the sun arrives at an ever shallower angle.
- Weather and climate: A cloud cover reduces the usable incoming radiation at the ground.
- Dust: Dust in the atmosphere scatters and absorbs sunlight.

These factors limit the use of solar radiation. Often, high energy demand and high incoming radiation do not occur at the same location or at the same time. Figure 7.3 shows the variation of daily incoming radiation with the season for the northern hemisphere and without atmosphere. Near the equator, the variation is very small. Near the poles, however, the available radiation is strongly limited to the summer half of the year.

Entropy Balance

The third relevant flux across the system boundary of the earth is the entropy transported with the radiation. Every photon transports the same amount of entropy regardless of its energy. Therefore, what matters is how many photons carry the incoming or outgoing energy.

The radiation of the sun consists of relatively high-energy photons streaming from a small solid angle: due to the large distance from the sun to the earth, the light particles arrive nearly in parallel. The radiated power of the earth is based on far more photons of low energy and occurs in all directions, so that this corresponds to a significantly larger entropy flux. Overall, the earth system is far from thermodynamic equilibrium, and thus many processes can take place that generate entropy. The entropy difference between emission and incoming radiation defines the upper limit of entropy generation in the earth system.

It is commonly assumed that processes in the earth system also exhaust this entropy limit, i.e., the maximum possible entropy is generated. This difference simultaneously sets an upper limit for the possible power that can be performed in the earth system, since this power is dissipated in the earth system.

If in the future more power from technology is to be converted in the earth system, then correspondingly more dissipated power must be released. For this purpose, the surface temperature of the earth must then be increased in order to radiate more heat. This is a particularly relevant aspect of all ideas to collect energy in the future using satellites via photovoltaics and then transmit it to the earth's surface.

Entropy balance

The entropy balance of the earth is balanced on average (Delgado-Bonal 2017):

$$\dot{S}_{earth} = \underbrace{\frac{4}{3}\cdot\frac{\dot{Q}_{sol}}{T_{sol}}}_{\substack{\text{incoming}\\ \text{from the sun}}} + \underbrace{\frac{4}{3}\cdot\frac{\dot{Q}_{earth}}{T_{earth}}}_{\substack{\text{emission}\\ \text{of earth}}} + \underbrace{\sum_i \dot{S}_i}_{\substack{\text{generated}\\ \text{in earth system}}} = 0$$

Here, the temperature T_{sol} is the temperature at which the sun emits its radiation, and T_{earth} is the temperature at which the earth emits its energy. The heat flux incident on the earth from the sun $\dot{Q}_{sol}$ is equal in magnitude to the heat flux radiated by the earth to its surroundings $\dot{Q}_{earth}$. The sum describes all processes in the earth system that generate entropy.

With this we can calculate how much entropy can be generated in the earth system, taking into account the signs of the heat fluxes. Positive is what is supplied to the earth:

$$\sum_i \dot{s}_i = -\frac{4}{3}\cdot\frac{\dot{q}_{sol}}{T_{sol}} - \frac{4}{3}\cdot\frac{\dot{q}_{earth}}{T_{earth}} = -\frac{4}{3}\cdot\frac{240\,\frac{\text{W}}{\text{m}^2}}{5776\,\text{K}} + \frac{4}{3}\cdot\frac{240\,\frac{\text{W}}{\text{m}^2}}{255\,\text{K}}$$

$$\sum_i \dot{s}_i = -0.042\,\frac{\text{W}}{\text{K}\cdot\text{m}^2} + 941\,\frac{\text{W}}{\text{K}\cdot\text{m}^2} = 899\,\frac{\text{W}}{\text{K}\cdot\text{m}^2}$$

Since we do not account for effects from the scattering of radiation in the atmosphere here, our result is a good estimate but not an exact value (Kleidon 2016).

7.2 Processes in the Atmosphere

The spatial and temporal variation of solar incoming radiation drives processes in the atmosphere: energy differences generate wind. The global variation in supplied heat and radiated heat between the poles and the equator requires large-scale structures to transport large amounts of heat over these great distances. These large structures begin first on a small and local scale. From many small processes, the large flows in the atmosphere and in the oceans emerge.

Dry Atmosphere

First we consider dry air. The air movement is driven by solar radiation, which is absorbed by the ground and increases its temperature. The air directly at the ground is in direct thermal contact with the ground and is heated. This is the portion of the solar incoming radiation that is transferred to the atmosphere as sensible heat – on a global average 24 W m^{-2}. The air heated in this way expands as its temperature increases and its density decreases. Therefore, it detaches from the ground and rises. The surrounding cooler and therefore denser air flows in. This rising lighter air expands due to the decreasing pressure with altitude: with dry air and without mixing with other air, no heat is added or released after the air has detached from the ground. This is an adiabatic and therefore isentropic expansion (Linow 2022). With the reduction of pressure, the temperature of the air also decreases. In the adiabatic case and for the quite good assumption that air is an ideal gas, the following applies

$$T_2 = T_1 \cdot \left(\frac{p_2}{p_1}\right)^{\frac{\kappa-1}{\kappa}} = \left(\frac{\rho_2}{\rho_1}\right)^{\kappa-1} \text{ and } \rho_2 = \rho_1 \cdot \left(\frac{p_2}{p_1}\right)^{\frac{1}{\kappa}} = \left(\frac{T_2}{T_1}\right)^{\frac{1}{\kappa-1}} \tag{7.1}$$

Here, κ the isentropic exponent. For air at atmospheric conditions, $\kappa = 1.40$. For dry and cloud-free atmosphere, this is a good description.

Moist Atmosphere – Air and Water

Everything changes when water evaporates or condenses: solar incoming radiation leads to the evaporation of water from open bodies of water, from moist ground, and from plants. On average, this amounts to 88 W m^{-2}. The evaporation rate, i.e., the mass flow of water transitioning from liquid to gaseous, depends on several factors:

- For open bodies of water the atmosphere is in direct contact with water, and the evaporation rate then depends on the temperature of the water, the temperature of the air, and the relative humidity of the air (Linow 2022).
- Wind can increase the evaporation rate, as it leads to a faster exchange of air at the interface with the water. Through waves and especially heavy seas with spray, the interface between water and air is greatly enlarged.
- Over land surfaces the evaporation rate also depends on the temperature and the relative humidity of the atmosphere, as well as additionally on the availability of water from the ground. Soil chemistry influences how strongly water is bound. Soil structure influences how quickly water can rise to the surface.
- Vegetation strongly influences evaporation: on the one hand, this occurs through the available surface area and the structure of the leaves, through which evaporation essentially takes place. On the other hand, this occurs through the physiological state of the plant, i.e., whether it is currently willing to release water or actively prevents this, and finally through the interactions between vegetation and soil.

Moist and warm air also rises and initially cools adiabatically as pressure decreases. If the relative humidity of the air exceeds $\varphi = 100\ \%$, condensation of the moisture occurs. Depending on the temperature of the air, droplets or ice crystals form, and thus clouds. The latent heat of evaporation of the moisture is released to the extent that the water condenses again. Overall, the temperature of the moist air with condensation decreases more slowly than that of dry air. If the air contains only little moisture, fair-weather clouds form at the upper end of convection and the air does not rise further. If the air contains a great deal of moisture and thus a great deal of latent heat, it can rise very high and forms, e.g., thunderclouds that reach up to the tropopause. At this point, the water cycle of the earth begins.

In Figure 7.4 the differences between the behavior of dry air and moist air are illustrated for a simple case: when dry air rises adiabatically, i.e., without absorbing heat, expansion occurs. The pressure p decreases and the temperature T also decreases accordingly. If this air subsequently descends again to the initial altitude z_1, the initial conditions are also reversibly restored. In the example in Figure 7.4 this change of state of the air is forced by the geography of a mountain.

For moist air, a different temperature change results from the latent heat released when water vapor condenses and rains out: during condensation, heat is released, which is transferred to the air. At the same time, the amount of water in the air decreases due to the rain. When the pressure subsequently rises back to the initial value, the temperature rises and the water evaporates. Nevertheless, the air has changed irreversibly, as there is a lower amount of water, a higher temperature, and a lower relative humidity. This can be observed well during Foehn weather conditions in the Alps.

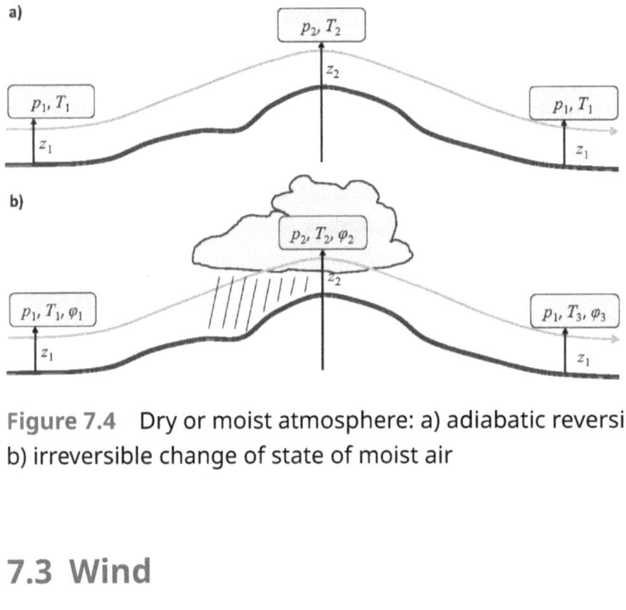

Figure 7.4 Dry or moist atmosphere: a) adiabatic reversible change of state of dry air, b) irreversible change of state of moist air

7.3 Wind

The heat flux emitted by the earth into space through radiation is approximately constant over the entire earth's surface – namely 240 W m^{-2}. This is evident because the temperature of the stratosphere, from which the largest proportion of this emission originates, remains approximately constant from pole to equator. The heat flux locally radiated into the earth over the course of a year, however, depends on the distance from the equator. Significantly more heat is absorbed near the equator than at the poles (Figure 7.3). This heat must be transported to the poles. This is accomplished via large-scale air and ocean currents and in particular via the transport of moist air.

Currents are a form of mechanical power. For these currents to occur, the heat flux of the sun must first be converted into mechanical power. From a temperature gradient between the equator and the pole, a heat flux arises, and this generates, as in a heat engine, mechanical power. The mechanical power in the earth system is found

- In the kinetic energy of the wind,
- In the kinetic and potential energy of the waves on bodies of water,
- In the kinetic energy of ocean currents in the waters,
- In the potential energy of water vapor and the water droplets in the air, and as a consequence
- In the potential and kinetic energy of the waters on land.

Figure 7.5 shows the typical structure of air currents of the Holocene that develops from this. Three large cells form, in which air circulates. The Hadley cell at the equator is a distinctly stable structure that migrates somewhat northward or southward with the seasons. The polar cell is stabilized by the large ice masses and the low amount of heat supplied there even in summer. Between these cells, the less stable Ferrel cell mediates the transport of matter and heat. At the boundary regions of these cells, strong air currents are found at great altitude. These air currents stabilize the position of the cells. Due to the loss of snow and ice around the North Pole, this structure is currently changing, as more and more heat is being introduced there in summer.

Efficiency of wind generation

The heat from solar radiation is supplied to the atmosphere at approximately the mean temperature of the earth's surface of 288 K today. The atmosphere releases this heat via radiation back to the surroundings of the earth (outer space) at a temperature of 255 K. If we now assume that the atmosphere is a kind of heat engine that converts the solar heat flux into power, then we can use the Carnot efficiency to calculate the maximum possible conversion efficiency for this conversion:

$$\eta_{atm} = 1 - \frac{T_{out}}{T_{in}} \cong 1 - \frac{255\ \mathrm{K}}{288\ \mathrm{K}} = 0.115$$

This fraction of the solar heat flux incident on the earth's surface could be used at most in earth system to transport and distribute this heat within the system. The mechanical power generated in this way is dissipated again in earth system. On average it amounts to approximately

$$p_{\max} = \frac{P_{\max}}{O_{earth}} = \eta_{atm} \cdot \dot{q}_{in} = 0.115 \cdot 165\ \frac{\mathrm{W}}{\mathrm{m}^2} = 19.0\ \frac{\mathrm{W}}{\mathrm{m}^2}$$

The actual value is lower, since a portion of the solar radiation is not absorbed at the earth's surface (Kleidon 2016). The mechanical power in earth system determined in this way is not fully available for human use, e.g. for the generation of wind energy, since harvesting it completely would bring the atmospheric air currents to a standstill.

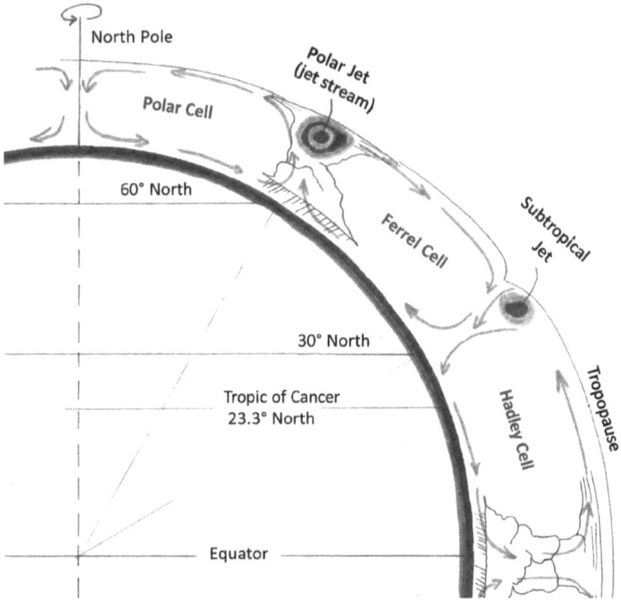

Figure 7.5 Global structure of the large air currents

7.4 Water Cycle

The atmospheric water cycle begins with the evaporation of water and the transport of water by wind. When the water rains out over the sea, the cycle is directly closed. Relevant for human use is rain over land, as this enables life there.

Potential energy Rain on land mostly falls on land above sea level. The rained-out water has a usable potential energy due to its difference in elevation relative to the sea. A power is then associated with this energy and the flowing water:

$$P = \dot{m} \cdot g \cdot \Delta z \tag{7.2}$$

Here Δz is the altitude above sea level at which rain falls. This potential energy causes the water to flow down to the sea. In the process, potential energy is converted into kinetic energy of the flowing water. Ultimately, the energy is dissipated by turbulence and friction, but also by the transport of sediment and the reshaping of river courses.

The total potential power deposited in this way on the land surface is determined by

$$P_{rain,land} = \int_{land} \dot{m}_A \cdot g \cdot \Delta z_A \cdot dA \tag{7.3}$$

Here $\dot{m}_A$ is the local mass flow of rainwater minus evaporation in the area element dA. This integral can in principle be determined numerically using local data. A simpler approach is, for example, to know an average rainfall amount for a region and to describe the geographical distribution using elevation distributions. A rough global estimate can be found in Table 7.2. This is based on the rather imprecise approximation

$$P_{rain,land} = \int_{land} \dot{m}_A \cdot g \cdot \Delta z_A \cdot dA \simeq \overline{\dot{m}} \cdot g \cdot \int_{land} \Delta z_A \cdot dA = \overline{\dot{m}} \cdot g \cdot \overline{\Delta z} \qquad (7.4)$$

In reality, rainfall is very unevenly distributed over the earth's surface and only a part of it is usable:

- A considerable portion of the water is needed by plants. For this, plants must allow the water to evaporate from their leaves so that it can flow through them and, for example, transport nutrients.
- A considerable portion of the water is needed in agriculture. This water also evaporates from plants or the soil.
- A portion of the water evaporates directly from bodies of water. This is particularly relevant for water surfaces in dry, warm climate zones, i.e., reservoirs and irrigation canals in the desert.
- Many bodies of water are too small to be used technically. Mill ponds in pre-industrial society served many purposes simultaneously. They are, for example, the mill pond for a mill that operates only a few weeks per year, or a fish farm. Such bodies of water are not sufficient for generating large amounts of electricity.
- Especially in mountainous areas, bodies of water are often difficult to access.
- A considerable portion of the water does not flow above ground, but as groundwater that is inaccessible for hydropower.

In Table 7.2 the global flows of the water cycle are summarized. Precipitation values sometimes vary considerably from year to year and across seasons at the regional level.

The energy for evaporation is calculated from the enthalpy of vaporization of water. At an ambient temperature of 15 °C, this is r = 2500 kJ kg^{-1}. The value for the released potential energy and potential power of precipitation over land given in Table 7.2 is roughly estimated using Equation 7.4. It was assumed that precipitation over the land surface is homogeneously distributed and that the land surface has a mean elevation of 800 m (Kleidon 2016). The difference between the supplied heat of vaporization and the potential energy drives the weather and the atmospheric transport of heat toward the poles of the earth.

Table 7.2 Water Cycle of the Earth (Source: Kleidon 2016), Mass Flow Rates $\dot{m}$, Potential Energy E_{pot} and Dissipated Power

	Process	$\dot{m}$ in kg a^{-1}	$\dot{m}$ in kg s^{-1}	E_{pot} in J a^{-1}	Power in W
a	Evaporation (sea)	$434 \cdot 10^{15}$	$13.8 \cdot 10^{9}$	$1090 \cdot 10^{21}$	$34.4 \cdot 10^{15}$
b	Precipitation (sea)	$398 \cdot 10^{15}$	$12.6 \cdot 10^{9}$	0	0
c	Transport (sea → land)	$36 \cdot 10^{15}$	$1.14 \cdot 10^{9}$	–	–
d	Evaporation (land)	$71 \cdot 10^{15}$	$2.25 \cdot 10^{9}$	$178 \cdot 10^{21}$	$5.62 \cdot 10^{15}$
e	Precipitation (land)	$107 \cdot 10^{15}$	$3.39 \cdot 10^{9}$	$-0.84 \cdot 10^{21}$	$-0.026 \cdot 10^{15}$
f	Inflow (land → sea)	$36 \cdot 10^{15}$	$1.14 \cdot 10^{9}$	–	–
	Sum a + d	$505 \cdot 10^{15}$	$16.0 \cdot 10^{9}$	$1262 \cdot 10^{21}$	$40.0 \cdot 10^{15}$

Potential for hydropower in Germany

In this example, the potential for hydropower in Germany is roughly estimated. The total area of Germany is 357 340 km^2 and the mean precipitation is 700 mm a^{-1}. The total area can be divided into classes:

1. Areas located below sea level: 2 % of the area or 7147 km^2
2. Areas with an elevation of 0 m to 100 m: 35.9 % or 128 285 km^2
3. Areas with an elevation of 100 m to 500 m: 46.1 % or 164 733 km^2
4. Areas with an elevation of 500 m to 1000 m: 15.1 % or 53 958 km^2
5. Areas located above 1000 m: 0.9 % or 3216 km^2

With the chosen strong simplification of the given data, areas up to 100 m elevation make no contribution. Areas below sea level may require energy for drainage.

The range of low mountain ranges up to 500 m provides a potential energy of $112.9 \cdot 10^{15}$ J per year. The higher low mountain ranges between 500 and 1000 m provide $184.9 \cdot 10^{15}$ J per year and the high mountains $22.04 \cdot 10^{15}$ J per year. Since the annual rainfall varies greatly locally and is significantly higher in the high mountains, these are lower values. From these figures, the mean potential power is 10.14 GW.

The installed rated capacity today is 4780 MW, approximately half of the potential. The actual generated power is lower. Many potential sites were formerly used for mills or small power plants and can still be recognized in the landscape. The large number of dams in German waterways has a strong influence on the ecosystems. Natural flowing waters with their very high fish populations have become rare.

7.5 Greenhouse Effect of the Earth

The earth's atmosphere influences the energy flow between the sun and the earth's surface, as Figure 7.2 shows based on the radiation fluxes and Figure 7.6 based on the spectra. As one result, the surface of the earth experiences much smaller temperature fluctuations between day and night or across the seasons compared to celestial bodies without an atmosphere, such as our moon. Secondly, the mean temperature at the surface of the earth is significantly higher than that of the moon. This is the result of the greenhouse effect of the earth's atmosphere. Only through this greenhouse effect is more complex or even abundant life on earth possible.

The spectral properties and the concentrations of individual gases determine the strength of the effect. These greenhouse gases have the property that they transmit radiation well in the range of short-wave solar incoming radiation, i.e., approximately between 400 nm and 1500 nm, while at the same time strongly absorbing long-wave radiation in the infrared spectral range:

- Oxygen and nitrogen and the noble gases such as argon do not absorb appreciably in either the visible or the infrared spectral range. They are not greenhouse gases.
- Water vapor has a large number of absorption bands in the infrared, between which there are wide regions without absorption. This makes water vapor an important greenhouse gas. The concentration of water vapor in the atmosphere depends on the temperature in the earth system. At higher temperatures, the atmosphere can absorb more water vapor, since the saturation pressure of water increases. Strong feedbacks exist between the weather, the climate, and the water vapor concentration. Unlike the other greenhouse gases, the residence time of water vapor in the atmosphere is very short and the exchange between bodies of water and the atmosphere occurs very rapidly. Water vapor thus directly follows changes in the climate. For these reasons, water vapor is counted as a greenhouse gas, but this will not be discussed further here.
- Carbon dioxide is, alongside water vapor and methane, the most important natural greenhouse gas and thus significantly influences the temperature of the earth's surface, even though it is present in the earth's atmosphere only as a trace element. Carbon dioxide is distributed nearly uniformly throughout the atmosphere. Spatial and seasonal variations are measurable but small.
- The absorption spectra of carbon dioxide, methane, nitrous oxide, and other greenhouse gases partially overlap. As a result, they mutually close the transmission windows of the other gases.
- Methane reacts chemically in the upper atmosphere with reactive radicals present there, such as OH. It is oxidized to carbon dioxide within a few years. For methane to exist in the atmosphere, there must be sources and a steady influx of methane into the atmosphere. The methane concentration of the atmosphere responds within a few years to changes in the influx

The current composition of the earth's atmosphere is listed in Table 17.3 in Section 17.3. Information on the greenhouse effect of important gases can be found in Table 17.4 in Section 17.3. The greenhouse effect of individual gases is determined essentially by their concentration and their residence time in the atmosphere, i.e., the time over which they act as a greenhouse gas. As a third decisive factor, their spectroscopic properties in the infrared determine their greenhouse effect. This is specifically the spectral range in which they strongly absorb infrared radiation. If they absorb primarily in spectral ranges where water and carbon dioxide do not absorb, their effect is stronger. Important greenhouse gases and their properties are listed in Table 17.4 in Section 17.3.

Solar Incoming Radiation

Figure 7.6 shows the solar incoming radiation, as it exists outside the earth's atmosphere. This is the emission spectrum of the sun (AM 0). For comparison, the emission of a black body emitter with a surface temperature of 5776 K is also shown. This spectrum can be used quite adequately as an approximation for the complicated real spectrum of the sun. The third spectrum shown illustrates how the heat at ground level is spectrally distributed (AM 1.5). This spectrum can be recorded at ground level under a normal clear atmosphere and without clouds. It typically corresponds to an angle of incidence of 48°. At this angle, the radiation of the sun must pass through 1.5 times the air mass that it would pass through at normal incidence. This condition is called AM 1.5 (1.5 times the atmospheric mass).

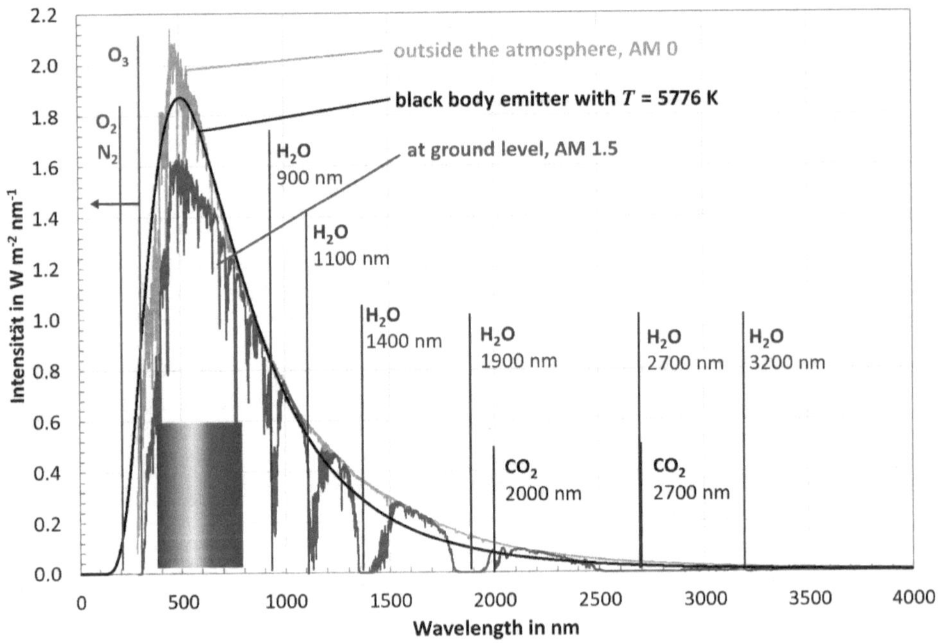

Figure 7.6 The solar spectrum outside the atmosphere and at ground level, as well as the spectrum of a black body radiator at 5776 K: relevant absorption bands are labeled.

For this typical spectrum at ground level:

- The maximum intensity is at 520 nm in the green spectral range.
- Towards shorter wavelengths, all spectra drop off rapidly. Solar radiation below 400 nm is absorbed by ozone (O_3) in the stratosphere and below 200 nm by oxygen and nitrogen. This is what makes life on the earth's surface possible, because UV-C radiation destroys chemical bonds, including the bonds in genetic material (DNA). Since this portion cannot reach the ground, plants and animals are able to thrive there.
- In the infrared, strong absorption bands are visible at ground level: only a portion of the radiation reaches the ground. Part of it is absorbed by water vapor and carbon dioxide in the atmosphere. Since the absorption by water vapor depends on the current local concentration, these are average values.

The ground and the upper atmosphere emit heat depending on their temperature. This radiation is emitted in the far infrared with a maximum at approximately 10,000 nm.

Earth's Climate – a Retrospective

Figure 7.7 shows the temporal trends of carbon dioxide and mean surface temperature over the last 66 million years, i.e., since the last major mass extinction. This was caused by the impact of an asteroid and ended the age of the dinosaurs. The data show how closely CO_2 and temperature are correlated. CO_2 is, by virtue of its properties, the dominant greenhouse gas in the earth's atmosphere. For this reason, the greenhouse effect of all other greenhouse gases is expressed relative to CO_2.

Earth's climate has been significantly influenced for about 3 million years, and due to the current position of the continents, by the variation of solar irradiance in the northern part of the Northern Hemisphere, particularly at around 65° North. The total solar irradiance far to the north, and in particular the share of irradiance in summer, varies strongly due to the Milanković cycles. These cycles were first described by Milutin Milanković and are based on a superposition of several effects (Ruddiman 2006):

1. The earth's axis does not remain fixed in space; rather, its orientation itself rotates once every 26,000 years: the point of the earth's orbit closest to the sun shifts from northern summer to northern winter and vice versa. This causes the heat radiated in summer to vary, particularly near the poles, over this period.
2. The tilt of the earth's axis changes periodically between 22.1° and 24.5°. This period lasts approximately 41,000 years and leads to significant fluctuations in solar irradiance near the poles.
3. The earth's orbit varies in shape between nearly circular and slightly elliptical with a cycle of approximately 100,000 years. As a result, solar irradiance varies along the orbit between a minimum of 2% when the orbit is nearly circular and a maximum of 23% in the elliptical case. Together with the rotation of the earth's axis, this leads to a sensible variation in the heat radiated in summer near the poles.

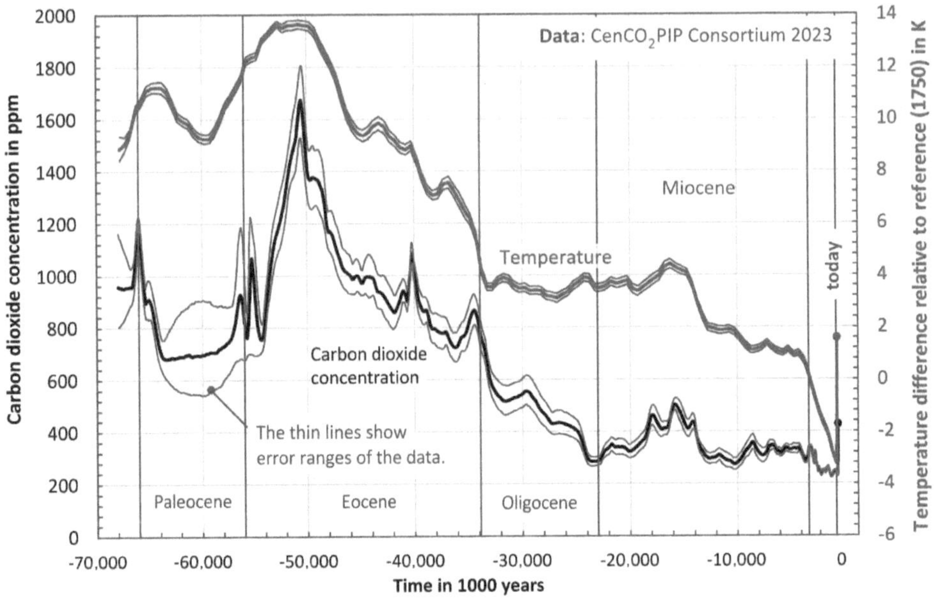

Figure 7.7 Trend of atmospheric CO_2concentration and mean temperature since the end of the dinosaurs 66 million years ago (data source: CenCO$_2$PIP Consortium 2023)

Starting from this driver, ice on land can temporarily survive the summer when solar irradiance there is particularly low: glaciers form, which exert further influence on the climate through their high reflectivity, as more solar radiation is reflected. An ice age begins. Conversely, when solar irradiance reaches a critical value, these glaciers will melt in summer. The cause of the strong rapid climate fluctuations of the last 2.7 million years can thus be well described.

Measurements in ice cores show that the concentration of carbon dioxide drops to approximately 220 ppm at the glacial maximum and reaches approximately 280 ppm at the climate optimum (Figure 7.8). This is currently the natural range of variation of CO_2concentration in the earth's atmosphere and the atmospheric driver of the dynamic climate of the last 2.7 million years.

With the beginning of agriculture approximately 8,000 years ago, a deviation of the temporal development of the concentration of carbon dioxide and also methane from the expected natural concentrations is observed: the temperature of the earth has been remarkably constant for about 10,000 years – in the geological epoch of the Holocene. This is despite the fact that solar irradiance at 65° North has decreased significantly and the formation of large glaciers there would actually have been expected long ago. These changes can be explained by the influence that the spreading of agriculture and the associated emissions of CO_2 and CH_4 have on the climate (Ruddiman et al. 2020).

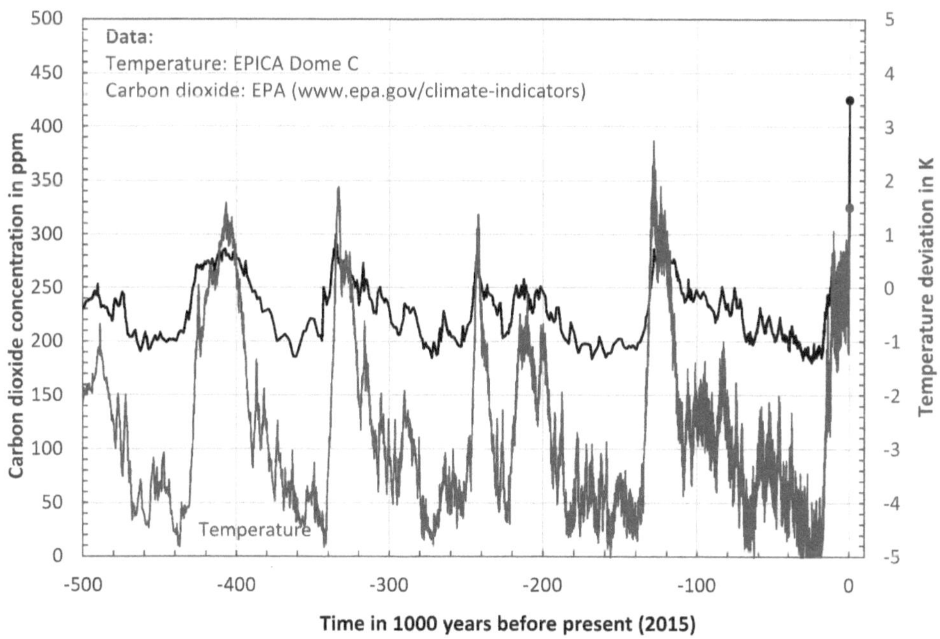

Figure 7.8 Course of the CO_2concentration in the atmosphere and the mean temperature over the last 500,000 years (data source carbon dioxide: EPA)

Radiative Forcing

The effect of greenhouse gases is described by a measure that quantifies the change in the radiation balance at the earth's surface due to the concentration of these gases in the atmosphere. The deviation from a reference value is used as the measure. The reference value is the conditions before the beginning of industrialization, i.e., around 1750. The measure is called radiative forcing according to (IPCC 2018):

> *"Radiative forcing is a measure of the influence a factor has in altering the balance of incoming and outgoing energy in the Earth-atmosphere system and is an index of the importance of the factor as a potential climate change mechanism. (...) radiative forcing values are for changes relative to preindustrial conditions defined at 1750 and are expressed in Watts per square meter (W/m²)."*

For CO_2 the radiative forcing as a function of concentration is calculated approximately with

$$\Delta F_{CO_2}(t) = 5.35\ \mathrm{W\ m^{-2}} \cdot \ln\left(\frac{r(t)}{r_0}\right) \tag{7.5}$$

In this equation, $r(t)/r_0$ is the ratio of the current concentration $r(t)$ of CO_2, relative to the measured concentration of CO_2 in the year 1750, defined as r_0 = 280 ppm. Other greenhouse gases are assigned their own radiative forcings (see Chapter 17).

In the steady-state case, the atmosphere is in equilibrium with the radiative forcings. If the radiative forcings change, the atmosphere or climate changes accordingly, starting from the change in mean temperature toward

$$\Delta T_{atm} = \lambda \cdot \Delta F \tag{7.6}$$

The value of climate sensitivity commonly used today is $\lambda = 0.8\ \mathrm{K\ W^{-1}\ m^2}$. A significantly higher value is currently being discussed (Hansen et al. 2023). From this value and with these equations, the expected mean change in atmospheric temperature can be determined for an emitted quantity of carbon dioxide via the change in concentration in the Earth's atmosphere. This is also possible for another determined value of a radiative forcing.

This equation contains the assumption that the relationship between radiative forcing and its effect on the Earth's temperature is linear. This assumption is only valid for a certain range around the stationary reference state. With very strong radiative forcings, feedbacks and nonlinear effects occur. Clouds, glaciers, snow and ice, as well as aerosols in the atmosphere can act as negatively acting radiative forcings. They lower the temperature of the atmosphere.

Global Warming Potential (GWP)

Dealing with radiative forcings is cumbersome and unnecessarily complicated for many considerations. Simpler is the use of the GWP. Here, the effect of a greenhouse gas over a specific time period is compared with that of CO_2 and expressed as a factor. Table 17.4 in Section 17.3 gives this value for relevant gases as GWP 20 (over 20 years) and GWP 100 (over 100 years). In principle, GWP values must be accompanied by the time period for which they apply, since some greenhouse gases disappear from the atmosphere much faster than CO_2 (e.g. CH_4) and others much more slowly (e.g. CF_4).

The actual determination of GWPs is a complex process in which the mutual spectral interactions between the gases must also be taken into account. Therefore, these values are always subject to a certain degree of error even for the most important current greenhouse gases (CO_2, CH_4 and N_2O) (Etminan et al. 2016). Shine et al. discuss GWP as a metric and provide a comparison with other possible indicators (Shine et al. 2005).

I want to know more about climate and the greenhouse effect!

There is a great deal of literature on the subject:

- Easy to access, factually sound and visually appealing is this small book: *Nelles, D./Serrer, C.:* Kleine Gase – große Wirkung. Der Klimawandel. Selbstverlag. 2018. *www.klimawandel-buch-de*
- Similar in nature is this book: *Lesch, H./Scorza-Lesch, C./Theis-Bröhl, K.:* Den Klimawandel verstehen. Ein Sketchnote-Buch. Springer, Berlin 2021

- Concise and relevant is the following book: *Rahmstorf, S./Schellnhuber, H.-J.:* Der Klimawandel. Diagnose, Prognose, Therapie. C. H. Beck, München 2019
- The comprehensive starting point is of course the current assessment report of the IPCC, Working Group I, "Physical Science Basis" (*www.ipcc.ch*)

Another level is offered by specialist articles on individual aspects or journals such as Nature Climate Change. Current topics are discussed there or relevant data are made accessible. Of particular importance at present are these publications:

- *Hansen, J. E./Sato, M./Simons, L./Nazarenko, L. S./Sangha, I./Kharecha, P./Zachos, J. C./Schuckmann, K. von/Loeb, N. G./Osman, M. B./Jin, Q./Tselioudis, G./Jeong, E./Lacis, A./Ruedy, R./Russell, G./Cao, J./Li, J.:* Global warming in the pipeline. In: Oxford Open Climate Change 3, 2023, kgad008. *https://doi.org/10.1093/oxfclm/kgad008*
- *Romanello, M./Walawender, M./Hsu, S.-C./Moskeland, A./Palmeiro-Silva, Y./Scamman, D./Ali, Z./Ameli, N./Angelova, D./Ayeb-Karlsson, S./Basart, S./Beagley, J./Beggs, P. J./Blanco-Villafuerte, L./Cai, W./Callaghan, M./Campbell-Lendrum, D./Chambers, J. D./Chicmana-Zapata, V./Chu, L. et al.:* The 2024 report of the Lancet Countdown on health and climate change. Facing record-breaking threats from delayed action. In: The Lancet 404, 2024, pp. 1847–1896

The question of when we could have known that we were playing with the climate is answered by the first specialist article to estimate this quantitatively. It dates from 1896:

Arrhenius, S.: On the influence of carbonic acid in the air upon the temperature of the ground. In: Philosophical Magazine and Journal of Science 5, 1896, pp. 237–276. *https://www.tandfonline.com/doi/pdf/10.1080/14786449608620846*

7.6 Carbon Dioxide in the Atmosphere

The atmospheric concentration of carbon dioxide and all other greenhouse gases changes over time. In the natural climate regime, the concentration of the greenhouse gases carbon dioxide, methane, nitrous oxide, and water is relevant. The concentrations of these four are closely linked to each other and to the external drivers from the Milanković cycles (Ruddiman 2020).

Rapid removal Carbon dioxide is released into the atmosphere by a number of natural processes and removed from the atmosphere again by processes with different half-lives. Natural processes that increase the atmospheric concentration of carbon dioxide include volcanism, fire, and oxidation of other carbon containing substances, as well as processes in soils and vegetation. Carbon dioxide can disappear from the atmosphere through transition into other reservoirs, in particular bodies of water and vegetation, as well as through weathering processes of rock.

In Table 7.3 a compilation of the essential processes and their typical half-lives that have changed the concentration of carbon dioxide so far can be found (Archer/Brovkin 2008). If a single, not too large pulse of carbon dioxide is added to the atmosphere, this mass of carbon dioxide will be removed from the atmosphere over time and the previous state will be restored. The temporal behavior over short time periods can be described by a simple function (Joos et al. 1996):

$$\frac{m_{CO_2}(t)}{m_{CO_2}(t_0)} = 0.26 \cdot \exp\left(\frac{-t}{3.4}\right) + 0.24 \cdot \exp\left(\frac{-t}{21}\right) + 0.18 \cdot \exp\left(\frac{-t}{70}\right) + 0.14 \cdot \exp\left(\frac{-t}{420}\right) + 0.18 \quad (7.7)$$

Here the time is entered in years. The assumption is that the earth system is otherwise in equilibrium and that the pulse does not fundamentally disturb this equilibrium. The function describes the fraction of the original mass that is still present in the atmosphere after a time t. This function only represents the fast processes such as mixing with the ocean. Long-term weathering, which binds carbon dioxide in rock with a half-life of approximately 100,000 years, is not represented here. Rather, these are the 18% that remain as a constant in the equation after a long time. The temporal progression of the function is illustrated in Figure 7.9.

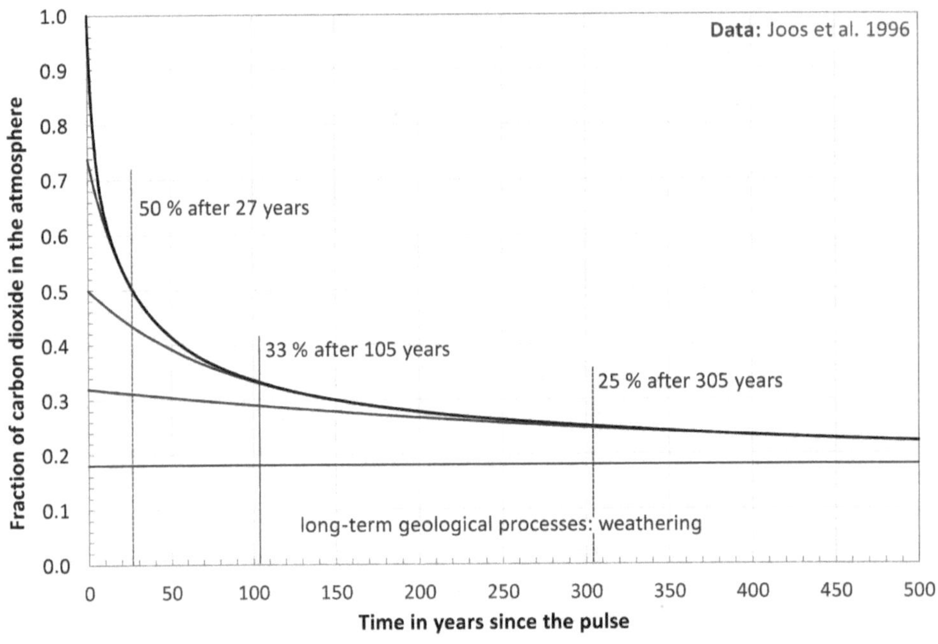

Figure 7.9 Residence time of a pulse of CO_2 in the atmosphere according to Equation 7.7: the CO_2 transitions into other reservoirs, in particular the ocean (explanation in the text).

Equation 7.7 is a strong simplification. The actual emission does not occur as a simple pulse but takes place continuously. The concentration of carbon dioxide in the atmosphere depends on the magnitude of the pulse as well as on atmospheric interactions.

Some of the processes are currently changing: the uptake in soils and forests has almost come to a standstill globally as of 2023. The transition into the oceans slows down with the formation of very cold deep water, e.g. in the context of the Atlantic Meridional Overturning Circulation (AMOC). This is the northern part of the Gulf Stream.

Tipping points Changed greenhouse gas concentrations lead to systemic effects in the earth system: other processes change or systems transition into a different state. They can do this linearly or non-linearly, or they follow a hysteresis behavior (Figure 7.10). Of particular importance are tipping points, at which an entirely new state of the system is established when the driver exceeds a certain value. It is these complex interactions between greenhouse gases and other systems that have caused the dramatic changes in the Earth's climate over the last 3 million years.

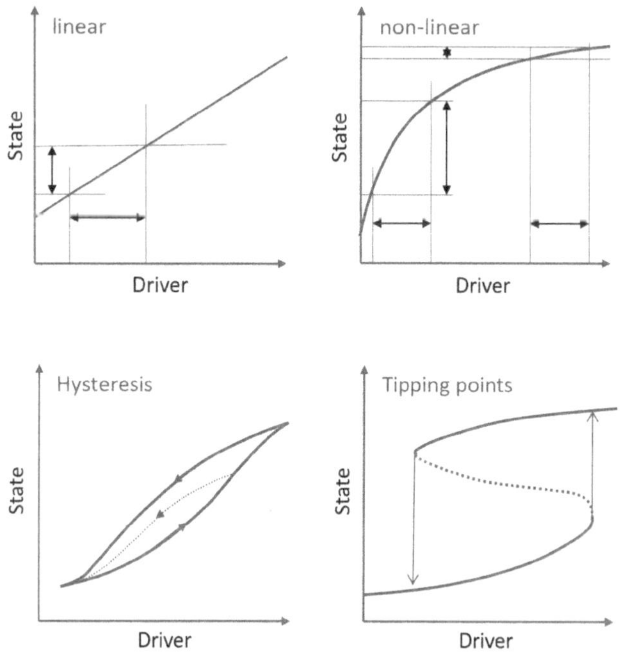

Figure 7.10 Possible responses of earth systems to the change of a driver, e.g. the radiative forcing of the earth

Such tipping points can lead to the release of considerable additional greenhouse gases as a result of the change of state, massively amplifying the change.

Geological removal Weathering by geological processes can be divided into a reaction with $CaCO_3$, which can bind a considerable proportion of the carbon dioxide within a few tens of thousands of years. In addition, there are slowly occurring reactions with silicate rocks. Both processes can be accelerated technically (see Chapter 14).

Table 7.3 Reservoirs and Processes That Influence the Atmospheric Concentration of CO_2 (Source: Archer/Brovkin 2008)

Reservoir	Mass	Time period	Process
Atmosphere	$700 \cdot 10^{12}$ kg		
Ocean	$38{,}000 \cdot 10^{12}$ kg	Centuries	Natural exchange via water surface – distribution via ocean currents – sink and source regions long-term reaction to pH value changes
Land		Millennia	Erosion of limestone on land, binding of CO_2 in sediment on land or in the ocean
Biosphere on land	$500 \cdot 10^{12}$ kg	Decades	Reaction of vegetation to climate
Soil	$15\,000 \cdot 10^{12}$ kg	Centuries	Reaction of soil to vegetation
Permafrost	$400 \cdot 10^{12}$ kg	Centuries	Reaction to climate: thawing, then decomposition of peat
Crude oil	$200 \cdot 10^{12}$ kg	Decades	Human extraction and combustion
Natural gas	$200 \cdot 10^{12}$ kg	Decades	Human extraction and combustion
Coal	$2000 \cdot 10^{12}$ kg	Decades	Human extraction and combustion
		> 1 000 000 a	Formation of coal from peat deposits – if corresponding large land areas can develop their geological dynamics undisturbed
Methane hydrates	$> 1000 \cdot 10^{12}$ kg	Presumably centuries	Response to ocean temperature and sea level

Where does the heat come from during global warming?

In this task we ask ourselves: What amount of heat is needed to warm the Earth's atmosphere and the parts of the oceans in direct contact with the atmosphere by 0.5 K (this is the difference between 1.5 °C and 2.0 °C global warming in the IPCC climate targets)? The working hypothesis is that this amount of heat comes from the combustion of fossil energy carriers.

What is the heat required to raise the average temperature of the atmosphere by 0.5 K?

Using the total mass of the atmosphere already determined previously, the following heat is required for this:

$$Q_{atm} = m_{atm} \cdot c_{p,air} \cdot \Delta T = 5.218 \cdot 10^{18}\ \text{kg} \cdot 1.004\ \frac{\text{kJ}}{\text{kg} \cdot \text{K}} \cdot 0.5\ \text{K} = 2.619 \cdot 10^{21}\ \text{J}$$

What is the heat required to raise the temperature of the upper 500 m of the oceans by 0.5 K?

First we determine the mass of this layer, assuming that approximately 70 % of the Earth's surface is covered by oceans:

$$m_{OO} = \Delta z \cdot O_{earth} \cdot 0.7 \cdot \rho_{water} = 500\ \text{m} \cdot 509.5 \cdot 10^{12}\ \text{m}^2 \cdot 0.7 \cdot 1000\ \frac{\text{kg}}{\text{m}^3} = 178.3 \cdot 10^{18}\ \text{kg}$$

From this follows the required heat:

$$Q_{OO} = m_{OO} \cdot c_{p,water} \cdot \Delta T = 178.3 \cdot 10^{18}\ \text{kg} \cdot 4.190\ \frac{\text{kJ}}{\text{kg} \cdot \text{K}} \cdot 0.5\ \text{K} = 373 \cdot 10^{21}\ \text{J}$$

This figure corresponds well with a detailed analysis that determined a heat uptake in the oceans of approximately 300 · 10^{21} J for the period from 1980 to the present (Cheng et al. 2017).

How much coal would have to be burned to generate this heat?

The result is a mass of

$$m_{coal} = \frac{Q_{atm} + Q_{OO}}{H_{I,coal}} = \frac{2.619 \cdot 10^{21}\ \text{J} + 373 \cdot 10^{21}\ \text{J}}{25\ \frac{\text{MJ}}{\text{kg}}} = 15.0 \cdot 10^{15}\ \text{kg}$$

How does the CO_2 content in the atmosphere change as a result of burning this coal?

The mass of CO_2 from this coal amounts to, if we assume that this coal consists only of carbon,

$$m_{CO_2} = m_{coal} \cdot \frac{M_{CO_2}}{M_C} = 15.0 \cdot 10^{15}\ \text{kg} \cdot \frac{44.01\ \frac{\text{kg}}{\text{kmol}}}{12.01\ \frac{\text{kg}}{\text{kmol}}} = 55.0 \cdot 10^{15}\ \text{kg}$$

The CO_2 content of the atmosphere is given in volume fractions. In order to be able to calculate this, we determine the standard volumes from the masses of the atmosphere and the additional CO_2 respectively (we treat the atmosphere as if it were entirely present under standard conditions).

The standard volume of the air of the atmosphere without this additional carbon dioxide is

$$V_{N,atm} = \frac{m_{atm} \cdot R_{air} \cdot T_N}{p_N} = \frac{5.218 \cdot 10^{18}\ \text{kg} \cdot 287.2\ \frac{\text{J}}{\text{kg} \cdot \text{K}} \cdot 273.15\ \text{K}}{101\ 325\ \text{Pa}} = 4.040 \cdot 10^{18}\ \text{Nm}^3$$

The standard volume of the added CO_2 is

$$V_{N,CO_2} = \frac{m_{CO_2} \cdot R_{CO_2} \cdot T_N}{p_N} = \frac{55.0 \cdot 10^{15}\ \text{kg} \cdot 188.9\ \frac{\text{J}}{\text{kg} \cdot \text{K}} \cdot 273.15\ \text{K}}{101\ 325\ \text{Pa}} = 28.0 \cdot 10^{15}\ \text{Nm}^3$$

The volume fraction r of this added CO_2 would therefore be

$$r_{CO_2} = \frac{V_{N,CO_2}}{V_{N,atm} + V_{N,CO_2}} = \frac{28.0 \cdot 10^{15}\ \text{Nm}^3}{4.04 \cdot 10^{18}\ \text{Nm}^3 + 28.0 \cdot 10^{15}\ \text{Nm}^3} = 0.00688 = 6880\ \text{ppm}$$

This number is only valid for the case that all the CO_2 remains in the atmosphere. However, a considerable proportion migrates into other media.

What increase in CO_2 in the atmosphere is required for the temperature rise?

The difference between +1.5 °C and +2.0 °C corresponds to a difference in the CO_2 content of the atmosphere of 49.3 ppm, if all other influences are neglected. This is calculated in each case from the radiative forcing required to warm the mean temperature of the Earth's atmosphere to 1.5 and to 2.0 °C:

$$\Delta F_{CO_2}(2.0\ \mathrm{K}) = \frac{\Delta T}{\lambda} = \frac{2\ \mathrm{K}}{0.8\frac{\mathrm{m^2 \cdot K}}{\mathrm{W}}} = 2.5\ \frac{\mathrm{W}}{\mathrm{m^2}}$$

and

$$\Delta F_{CO_2}(1.5\ \mathrm{K}) = \frac{1.5\ \mathrm{K}}{0.8\frac{\mathrm{m^2 \cdot K}}{\mathrm{W}}} = 1.88\ \frac{\mathrm{W}}{\mathrm{m^2}}$$

The corresponding CO_2concentrations in the atmosphere are

$$r_{CO_2}(1.5\ \mathrm{K}) = r_0 \cdot \exp\left(\frac{\Delta F_{CO_2}(\Delta T)}{5.35\ \mathrm{W\ m^{-2}}}\right) = 280\ \mathrm{ppm} \cdot \exp\left(\frac{1.88\ \frac{\mathrm{W}}{\mathrm{m^2}}}{5.35\ \frac{\mathrm{W}}{\mathrm{m^2}}}\right) = 398\ \mathrm{ppm}$$

and

$$r_{CO_2}(2.0\ \mathrm{K}) = r_0 \cdot \exp\left(\frac{\Delta F_{CO_2}(\Delta T)}{5.35\ \mathrm{W\ m^{-2}}}\right) = 280\ \mathrm{ppm} \cdot \exp\left(\frac{2.50\ \frac{\mathrm{W}}{\mathrm{m^2}}}{5.35\ \frac{\mathrm{W}}{\mathrm{m^2}}}\right) = 447\ \mathrm{ppm}$$

The difference between the two temperatures would therefore mean an addition of CO_2 corresponding to a volume fraction of 49 ppm.

This simple estimate shows that it is not the waste heat from combustion or the dissipation of the work generated in the earth system that causes global warming. The heat required to warm the oceans and atmosphere is two orders of magnitude higher than could be released from the combustion of fossil fuels.

The heat that brings about this effect is a portion of the sun's radiation that can no longer leave the earth system as easily due to the rising concentration of greenhouse gases. Only at a higher temperature of the atmosphere and the ground does the radiative flux return to equilibrium.

7.7 Humans in the Earth System

Pleistocene The intensification of the use of the earth system by us humans can be traced along the geological eras. All the essential developments leading to modern humans took place during a period of strong climatic fluctuations. This was the Pleistocene, referring to the last 2 million years. The climate fluctuated particularly strongly between ice ages and warm periods due to the Milanković cycles and the special position of the continents (Figure 7.8). Already during this period, approximately since 50,000 years ago, the effects of human economic activity on the earth system can be recognized, particularly in the decline of megafauna and its massive extinction at the end of the last ice age on almost all continents (Bergmann et al. 2023). Megafauna refers to all animal species over 45 kg in weight.

Holocene The end of the last ice age marks the beginning of the Holocene, that is, the period in which modern humans developed agriculture, cities, governments, and all the other elements of our modern civilization. The absence of mammoths, forest elephants, bison, horses, or rhinoceroses created an ecological niche that enabled the first agriculture at the beginning of the Holocene, since only through the absence of large herbivores could cultivated areas remain free from regular destructive visits. In the Holocene, agriculture and the associated additional emissions of the greenhouse gases carbon dioxide and methane led to a first human influence on the climate. Through a complex interaction between land use, emissions, and the atmosphere, the nearly constant climate of the last 10,000 years was made possible (Ruddiman 2020). This nearly constant climate was presumably the prerequisite for our further technological development.

Anthropocene By now, the impact of human economic activity on the earth system is so great that it is argued that we humans are now the dominant geological force in the earth system and that this should be designated as the Anthropocene. Here, the determination of what exactly constitutes the beginning of the Anthropocene or by what it is precisely characterized has not been concluded, as there are many meaningful and possible points in time for a beginning (Ellis 2016; Waters et al. 2015).

Food Supply and Agriculture

Humans require a very regular and consistent supply of high-quality food. The minimum required energy intake, i.e., the basal metabolic rate, is approximately 100 W. Humans can perform much higher outputs in the short term and also sustain over 100 W in the long term given sufficient fitness and adequate food supply. Humans have a thermal efficiency of at most approximately 0.3. This means that at most approximately 30% of the energy supplied through food can be delivered as work. To perform work, additional food beyond the basal rate must be consumed. Work here encompasses all physical human activities as well as, to a certain extent, mental activities.

The adequate supply of food for all people now requires considerable shares of the land area of the earth. In Table 7.4 the existing land areas of the earth (Ricardian Land), their assignment to the major potential biomes (ecosystems) and the share of their human use in the year 2000 are shown. The missing share contains settlements and infrastructure as well as Antarctica (Ramankutty et al. 2008). To what extent land areas not yet used can also be developed for human use without the associated damage outweighing the benefits is at least extremely questionable. Essential negative aspects of a further expansion of the use of land areas in earth system are:

- **Destruction of the hydrological system of tropical rainforests:** The Amazon and Congo regions are self-irrigating structures. Once the forest as a water reservoir has been removed, desert can form there. The amount of carbon released by the clearing would result in a considerable additional greenhouse effect.

- **Destruction of the retreat areas of essential species such as insects and birds for agriculture:** If these retreat areas are absent or fall below critical proportions of the landscape, the populations of pollinating insects, for example, will collapse. Pollination can in principle also be carried out by humans, but this would be an extremely laborious and time-critical task that would tie up large parts of the population.
- **Destruction of ecosystem services such as coastal protection, flood protection and water storage:** If these services are absent, they must be replaced by technology at corresponding energetic and financial costs, or the previous use of the areas thus supplied must be abandoned.
- **Salinization of soils in dry regions such as grassland, shrubland, savanna or desert, through unsuitable irrigation:** Salinized soils are no longer available for fertile vegetation over geological time periods.

Table 7.4 Land Areas and Their Use in the Year 2000 (Source: Ramankutty et al. 2008)

Potential biome (ecosystem)	Area in km^2	Share of land area	Arable land in km^2	Share of arable land in biome	Pasture land in km^2	Share of pasture land in biome
Total land area of the earth	149 430 000	100.0 %				
Tropical evergreen forests	16 770 000	11.2 %	1 810 000	10.8 %	1 480 000	8.8 %
Tropical deciduous forests	5 860 000	3.9 %	1 580 000	27.0 %	1 430 000	24.4 %
Temperate coniferous forests	3 610 000	2.4 %	720 000	19.9 %	370 000	10.2 %
Temperate evergreen forests	1 130 000	0.8 %	270 000	23.9 %	230 000	20.4 %
Temperate deciduous forests	4 840 000	3.2 %	1 460 000	30.2 %	820 000	16.9 %
Boreal evergreen forests	5 980 000	4.0 %	90 000	1.5 %	100 000	1.7 %

Potential biome (ecosystem)	Area in km²	Share of land area	Arable land in km²	Share of arable land in biome	Pasture land in km²	Share of pasture land in biome
Boreal deciduous forests	2 220 000	1.5 %	40 000	1.8 %	50 000	2.3 %
Mixed forests	14 960 000	10.0 %	1 160 000	7.8 %	710 000	4.7 %
Savanna	19 180 000	12.8 %	3 020 000	15.7 %	6 490 000	33.8 %
Grassland (prairie, steppe)	14 290 000	9.6 %	2 740 000	19.2 %	7 250 000	50.7 %
Dense shrubland	5 990 000	4.0 %	1 070 000	17.9 %	1 870 000	31.2 %
Open shrubland	11 940 000	8.0 %	870 000	7.3 %	5 150 000	43.1 %
Desert	15 240 000	10.2 %	130 000	0.9 %	1 200 000	7.9 %
Tundra	7 010 000	4.7 %	40 000	0.6 %	920 000	13.1 %
Polar desert, rock, ice	1 210 000	0.8 %	0	0.0 %	20 000	1.7 %
Total	130 230 000	87.2 %	15 000 000	11.5 %	28 090 000	21.6 %

Technosphere

Human uses of the earth system now extend far beyond the procurement of food. The technosphere defines the totality of things created by humans. This includes not only current things and artifacts, but also parts of the earth that have been fundamentally influenced or altered by humans. Table 7.5 gives a rough estimate of essential components, with the data being substantiated in Zalasiewicz et al. (Zalasiewicz et al. 2011). Table 7.5 shows the earth's surface covered by certain structures, the mean layer height of the structure, the mean density of the solid components, as well as the mass determined from these and the proportion of this structure in the total mass of the technosphere. Obviously included are all above-ground structures such as buildings, roads, airports, harbors, quarries, mines, railways, landfills, canals, dams and levees. This also includes the underground portions of structures such as foundations, water supply and sewage systems, energy supply, underground railways and subways, as well as covered landfills. Not included in Table 7.5 is the technosphere of the atmosphere, i.e., for example, the amount of carbon dioxide in the atmosphere. At ambient pressure, this amount is approximately 1 m thick.

Many components of the technosphere must be maintained and renewed, with the usual energy costs. Further components such as landfills and waste storage facilities must be secured. Degraded ecological functions of damaged or destroyed ecosystems must be replaced and compensated by ever more technology. The Anthropocene demands considerable care for our technology.

Table 7.5 Estimation of the Mass of the Technosphere (Source: Zalasiewicz et al. 2011)

Structure	Area in km^2	Thickness in m	Density in $kg\ m^{-3}$	Mass in Mt $= 10^9$ kg	Share in %
Urban areas	3 700 000	2.00	Approx. 1500	11 100	36.9
Rural settlement	4 200 000	1.00	Approx. 1500	6300	20.9
Rural roads	500 000	0.50	Approx. 1500	380	1.3
Railway	30 000	0.50	Approx. 1500	20	0.1
Pastureland	33 500 000	0.10	Approx. 1500	5030	16.7
Arable land	16 700 000	0.15	Approx. 1500	3760	12.5
Planted forests, forestry and plantations	2 700 000	0.10	ca. 1000	270	0.9
Reservoirs	200 000	1.00	ca. 1000	200 000	0.7
Land and erosion	5 300 000	0.10	ca. 1500	800	2.7
Trawler-fished seabed	15 000 000	0.10	ca. 1500	2250	7.5
Total	81 830 000			30 110	

References

Archer, D./Brovkin, V.: The millenial atmospheric lifetime of anthropogenic CO_2. In: Climatic Change 90, 2008, pp. 283–297

Bergmann, J./Pedersen, R./Lundgren, E. J./Lemoine, R. T./Monsarrat, S./Pearce, E. A./Schierup, M. H./ Svenning, J.-C.: Worldwide late Pleistocenen and early Holocene population declines in extant megafauna are associated with Homo sapiens expansion rather than climate change. In: Nature Communications 14, 2023, p. 7679

CenCO$_2$PIP Consortium – The Cenozoic CO$_2$ Proxy Integration Project Consortium: Toward a Cenozoic history of atmospheric CO_2. In: Science 382, 2023, eadi5177

Cheng, L./Trenberth, K. E./Fasullo, J./Boyer, T./Abraham, J./Zhu, J.: Improved estimates of ocean heat content from 1960 to 2015. In: Science Advances 3, 2017, pp. e1601545

Czersky, H.: Blue machine. How the ocean shapes our world. Torva, London 2023

Delgado-Bonal, A.: Entropy of radiation: the unseen side of light. In: Nature Scientific Reports, 7, 2017, p. 1642

Ellis, E.: Involve social scientists in defining the Anthropocene. In: Nature 540, 2016, pp. 192–193

EPA – United States Environmental Protection Agency: Climate Change Indicators in the United States. *https://www.epa.gov/climate-indicators*

Etminan, M./Myhre, G./Highwood, E. J./Shine, K. P.: Radiative forcing of carbon dioxide, methane, and nitrous oxide. A significant revision of the methane radiative forcing. In: Geophysical Research Letters 43, 2016

IPCC: Global Warming of 1.5 °C. An IPCC Special Report on the impacts of global warming of 1.5 °C above pre-industrial levels and related global greenhouse gas emission pathways, in the context of strengthening the global response to the threat of climate change, sustainable development, and efforts to eradicate poverty. 2018. *https://www.ipcc.ch/sr15*

Joos, F./Bruno, M./Fink, R./Siegenthaler, U./Stocker, T./Le Quéré, C./Sarmiento, J. L.: An efficient and accurate representation of complex oceanic and biospheric models of anthropogenic carbon uptake. In: Tellus 48, 1996, pp. 397–417

Kleidon, A.: Thermodynamic Foundations of the Earth System. Cambridge University Press 2016

Linow, S.: Angewandte technische Thermodynamik. Carl Hanser Verlag, München 2022

Ljing, C./Trenberth, K. E./Fasullo, J./Boyer, T./Abraham, J./Jiang Zhu: Improved estimates of ocean heat content from 1960 to 2015. In: Science Advances 3, 2017, e1601545

McGlade, C./Ekins, P.: The geographical distribution of fossil fuels unused when limiting global warming to 2 °C. In: Nature 517, 2015, pp. 187–190

Ramankutty, N./Evan, A. T./Monfreda, C./Foley, J. A.: Farming the planet: 1. Geographic distribution of global agricultural lands in the year 2000. In: Global Biogeochem. Cycles 22, 2008

Richerson, P. J./Boyd, R./Bettinger, R. L.: Was agriculture impossible during the Pleistocene but mandatory during the Holocene? A climate change hypothesis. In: American Antiquity 66, 2001, pp. 387–411

Ruddiman, W. F.: Plows, plagues & petroleum. How humans took control of climate. Princeton University Press 2006

Ruddiman, W. F./He, F./Vavrus, S. J./Kuzbach, J. E.: The early anthropogenic hypothesis. A review. In: Quarternary Science Review 240, 2020, 106386

Shine, K. P./Fuglestvedt, J. S./Hailemariam, K./Stuber, N.: Alternatives to the global warming potential for comparing climate impacts of emissions of greenhouse gases. In: Climatic Change 6, 2005, pp. 281–302

Stephens, G. L./Juilin, L./Wild, M./Clayson, C. M./Loeb, N./Kato, S./L'Ecuyer, T./Stackhouse, P. W./Lebsock, M./Andrews, T.: An update on Earth's energy balance in light of the latest global observations. In: Nature Geoscience 5, 2012, pp. 691–696

Wallace, J. M./Hobbs, P. V.: Atmospheric science. An introductory survey. Academic Press, Amsterdam 2006

Waters, C. N./Zalasiewicz, J./Summerhayes, C./Barnosky, A. D./Poirier, C./Gałuszka, A./Cearreta, A./Edgeworth, M./Ellis, E. C./Ellis, M./Jeandel, C./Leinfelder, R./McNeill, J. R./Richter, D. D./Steffen, W./Syvitski, J./Vidas, D./Wagreich, M./Williams, M./Zhisheng, A./Grinevald, J./Odada, E./Oreskes, N./Wolfe, A. P.: The Anthropocene is functionally and stratigraphically distinct from the Holocene. In: Science Vol. 351, Issue 6269, 2015, aad2622

Zalasiewicz, J./Williams, M./Waters, C. N./Barnosky, A. D./Palmesino, J./Rönnskog, A.-S./Edgeworth, M./Neal, C./Cearreta, A./Ellis, E. C./Grinevald, J./Haff, P./Ivar do Sul, J./Jeandel, C./Leinfelder, R./McNeill, J. R./Odada, E./Oreskes, N./Price, S. J./Revkin, A./Steffen, W./Summerhayes, C./Vidas, D./Wing, S./Wolfe, A. P.: Scale and diversity of the physical technosphere. A geological perspective. In: The Anthropocene Review 4, 2011, pp. 9–22

Part II – Implementation

Part I ends with the introduction of the idea of the Anthropocene. There, the Anthropocene was named as a description of the current state of the earth: we humans have for some time been the most powerful geological force in the earth system. Exactly since when is an important question, which we will not investigate here.

More pressing for us is the question of how we want to live in the future and how we can supply ourselves with sufficient energy within the framework our planet offers us. Part II addresses this question on the basis of the foundations from Part I of this book. We begin on a fully planetary scale, as this is first and foremost the relevant perspective for exploring boundary conditions and limits: what is actually possible in the near future? The subsequent chapters then examine various energy technologies and their very concrete contributions. In Chapter 14 we examine possibilities and inputs required to reduce the concentration of greenhouse gases in the atmosphere or even to engage in geoengineering.

The Anthropocene is associated with great responsibility for us and our actions. We can ignore this responsibility or try to reject the legacy – to our own detriment and that of all living beings on our planet. That would be the path into the Hellocene. Alternatively, we can accept this responsibility and together shape a good future within the scope of our possibilities. This then encompasses responsibility for all living things, for societal flourishing, and for the reverence associated with the task. This is the Anthropocene in its full and fundamental meaning, in which care is at the core of our actions.

The focus of this book is on the essential aspect of future energy supply for human societies. Chapter 8 develops the framework within which a future energy system can serve as a long-term foundation for flourishing in the Anthropocene. The pressing questions of justice – for whom is the energy system? for what purpose does the system serve? – can only be touched upon here. Answers to these questions are, however, an essential component of a flourishing future.

The following chapters look at concrete possibilities and contributions of heat (Chapter 9), of hydrogen and its derivatives (Chapter 10), of wind and water (Chapter 11), and of photovoltaics (Chapter 12). The particular challenges of storing and transporting electricity are the subject of Chapter 13. Chapter 14 deals with the question of how we can succeed in keeping our earth's climate within a range that is adequate for us, and what input we must expend for this. Chapter 15 summarizes our findings and offers ideas for our further course of action and design.

Who is meant by "we"?

The short answer is: all of us, that is, everything that lives. The very unfortunate separation between "human" and "environment" or between "mind" and "body" I try to avoid, as the terms cannot be meaningfully separated. We cannot break these down into parts that function independently of one another. Humans exist within the living earth system or they do not exist at all. Mind and consciousness are embodied in complex living beings.

The question of what is particularly valuable is a difficult one that always demands our full attention. Every individual human being has their own inviolable dignity, and every other conscious living being (i.e., vertebrates, cephalopods, many insects) likewise. Other claims recede behind these.

What does this have to do with me?

Why should planetary matters concern me? After all, I live here in this city and hold this passport. And I certainly cannot change such planetary things anyway ...

At its core, this context is about self-efficacy. What do I need to know and be able to do in order to help shape the future? How do I recognize ideas and people with whom it may be possible together to shape our good future?

8
Energy Transition – the Task

This chapter develops an evaluation matrix for energy systems. This matrix then serves as the perspective from which we examine individual technologies in the subsequent chapters. To this end, we must occasionally adopt a planetary perspective here in order to capture relevant framework conditions. In the following chapters, we can then return to specific conditions.

This chapter begins by describing the current state of our earth and identifying the guardrails within which we have good opportunities to shape an appropriate future and living conditions for ourselves. We then take a closer look at this goal of future adequate living conditions. After that, relevant methods are discussed before the evaluation criteria for energy systems are finally motivated.

8.1 State and Trajectories

Where do we stand today? How are we doing as inhabitants of a lonely small planet on which nearly all of us will live and die (current state), and in which direction can our planet develop with us? The topics relevant to an energy system are planetary boundaries (the guardrails). Within these boundaries, we need in particular a climate that is tolerable for us and a great deal of life around us (biodiversity) as the basis of health and supply. Added to this, for us humans, are raw materials and resources for our technology (see Chapter 5). First, we take a brief look back. How did we arrive at our current state?

Our current energy system

The affordable provision of sufficient energy at any time and in any place is the central foundation of our current social structure. Everything else follows from this (Fizaine/Court 2016; Smil 2017). All other aspects, particularly economic ones, depend on this prerequisite (Hall/Klitgard 2018). For this reason, the evaluation of technical concepts and solutions intended to ensure the long-term provision of sufficient and affordable energy is of great importance.

For a deeper understanding of this statement, our energy history is helpful. This refers to the development of the technical provision of energy and its repercussions on human societies. Table 8.1 provides an overview of relevant steps. It is noteworthy that the third and fourth industrial revolutions are each not based on technology that provides energy itself or does so in a more energy-efficient manner. The last row in Table 8.1 is to be understood as an outlook on those technologies that enable a further provision of affordable energy when the provision of fossil energy carriers is no longer energetically favorable, socially desired, or financially viable. This can be understood as a kind of *backstop technology* in the sense of Solow: backstops are those alternatives that are applied when today's (fossil) technologies become relatively too expensive or are exhausted.

Table 8.1 Key Technical Development Steps or Innovations in the Provision of Cheaper Energy: Shown Are 11 Steps, Designated As Industrial Revolutions (IR), Their Beginning, and Their Energetic Benefit for Human Societies.

No.	IR	Point in time	Technology	Energetic benefit
1	–	Approx. 2 million years ago	Fire	Better utilization of food, lower input for digestion, possibilities for greater brain performance: optimizes social developments, language, and hunting behavior
2	–	More than 400,000 years ago	Hunting weapons	New game, better supply of food and other resources: optimizes social developments, language, and hunting behavior
3	–	Well more than 50,000 years ago	Sewing: clothing from hides, tents, and boats	Expansion of the habitat into all climate zones: new game enables new cultures
4	–	Approx. 10,000 years ago	Agriculture and clothing from fibers	Better and more regular utilization of soils: enables higher population density
5	–	Approx. 8,000 years ago	Animal husbandry: cattle and sheep	Use of marginal areas (steppe, mountains) for the production of food

No.	IR	Point in time	Technology	Energetic benefit
6	–	Approx. 4200 years ago	Horse and wheel	Enable transport, the driving of machines, and rapid warfare
7	–	Approx. 4000 years ago	Water wheels	Enable the driving of machines, irrigation for agriculture, and the milling of grain
8	–	Approx. 2000 years ago	Wind wheels	Enable the driving of machines, the milling of grain, the irrigation of agricultural land, and the drainage of marshland
9	First	1713 to 1850	Coal and steam engine	Enable the driving of machines, transport and the railway as well as mining, steel, artificial fertilizers, and the generation of cold
10	Second	1879 to 1920	Electrification	Enables the driving of machines, transport as well as light, electric heat, and vacuum
–	Third	1980	Computers	A possible indirect energetic benefit is expected to arise from the improved calculability and predictability it enables: efficiency.
–	Fourth	2020	Digitalization and AI	A possible indirect energetic benefit is expected to arise from the improved calculability and predictability it enables: efficiency.
11	–	Now, later, or never!?	Replacement for fossil energy carriers	Preservation of the possibility of civilization

8.1.1 Planetary Boundaries

The planetary boundaries are a model that spans the framework within which human civilization is safely possible in the long term (Richardson et al. 2023). Table 8.2 provides an overview of the processes in the earth system relevant to these boundaries, the maximum tolerable stress or impairment of these processes by humans over the longer term, as well as the current status. Six of nine boundaries are currently exceeded. The definition of many further boundaries would be possible, but these nine are sufficient to cover all essential aspects. It is a planetary model, but each of these boundaries, the named processes, and their actual transgression has very concrete local effects everywhere and also right here.

Table 8.2 Overview of the Planetary Boundaries, Each As a Change in Planetary Processes, the Control Variables Used for Quantification, the Pre-Industrial Reference, the Identified Boundary, and the Current Value of the Variable (for Detailed Explanation and Justification See Richardson et al. 2023)

Process	Variable	Reference	Boundary	Current
Earth climate	Atmospheric CO_2concentration	280 ppm	≤ 350 ppm	425 ppm
	Anthropogenic radiative forcing	0 W m^{-2}	≤ 1.0 W m^{-2}	+2.91 W m^{-2}
Change in biosphere integrity	Genetic diversity and extinction rate	1 E/MSY	≤ 10 E/MSY	> 100 E/MSY
	Functional diversity and share of NPP appropriation by humans	1.9 %	≤ 10 %	30 %
Stratospheric ozone depletion	Ozone concentration	290 DU	> 276 DU	261 DU, increasing
Ocean acidification	Carbonate ion concentration	3.44 Ω_{arag}	> 2.75 Ω_{arag}	2.8 Ω_{arag}, decreasing
Bio- and geochemical substance flows	Phosphorus mass flow freshwater → ocean	0	≤ 11 TG a^{-1}	23 TG a^{-1}, increasing
	Nitrogen mass flow	0	≤ 62 TG a^{-1}	190 TG a^{-1}, increasing
Land-system change	Share of living forest in the potential area for forest	100 %	Tropical: ≥ 85 % temperate: ≥ 50 % boreal: ≥ 85 %	Tropical: 60 % temperate: 30 % boreal: 60 %
Freshwater change	Disturbance of surface water and groundwater	9.4 %	≤ 10.2 %	50 %
	Disturbance of plant-available water	9.8 %	≤ 11.1 %	50 %

Process	Variable	Reference	Boundary	Current
Atmospheric aerosol loading	Interhemispheric difference	0.03 AOD	≤0.1 AOD	0.076 AOD
Novel substances	Share of synthetic substances released without adequate review	None	None	Exceeded

E/MSY: Extinctions per Million Species-Years
NPP: Net Primary Production
DU: The Dobson Unit is a measure of absorption by ozone.
AOD: The Aerosol Optical Depth is a measure of the reduction of solar incoming radiation by aerosols.

These planetary boundaries set a clear framework for any future energy system, both at the planetary and at the very local level.

Climate overheating The goal is to avoid an irreversible and massive deterioration of the foundations of life in the earth system. Since the planetary boundary has been substantially exceeded, this means the immediate end of all greenhouse gas emissions:

- The use of fossil energy carriers – regardless of the application – should end immediately.
- The use of biomass and agriculture are affected, as nitrogen and phosphorus fertilization must be immediately reduced and restructured. Nitrogen appears in two boundaries, since the conversion of nitrogen fertilizer into nitrous oxide is a major greenhouse gas source.
- The use of hydropower in warm regions and on rivers with high transport of plant material is affected, since reservoirs emit methane and contribute to global warming through their dark surface.

Biosphere Ecosystems and species are the building blocks of the living earth. They perform a great many functions that cannot be replicated technically. Adhering to this boundary requires that large contiguous areas not be used intensively or industrially. Light and caring use by indigenous peoples, on the other hand, is explicitly desired here: indigenous forms of use often contribute to biodiversity.

From an energy perspective, this is of great importance, since areas cannot be sacrificed arbitrarily for extraction or industrial agriculture. Biomass and land area are only very limitedly available.

Stratospheric ozone Sufficient ozone in the stratosphere is a prerequisite for life on land, as ozone absorbs DNA-destroying components of sunlight. The Montreal Protocol and its amendments regulate this. The transition to non-ozone-depleting refrigerants that are not greenhouse gases is currently underway. From an energy perspective, this is not a particular hurdle or requirement. It only requires some care in the design of systems.

Acidification of the oceans The decreasing pH of the oceans is a direct consequence of the rising CO_2concentration in the atmosphere. It is foreseeable that this could lead to the destruction of food chains and thus of life in the oceans. From a human perspective, this concerns the loss of fisheries and thus hunger (for measures see climate heating).

Bio- and geochemical substance flows Phosphorus and reactive nitrogen are the most relevant substances causing massive changes in ecosystems on land and in the sea: ecosystems become impoverished or collapse. Less industrial fertilization requires a preserving agriculture that tends, maintains, and systematically improves its soils. Agricultural yields will decline, so that significantly less biomass will be available for energy use.

Land system change Sufficient integrity of as many ecosystems as possible requires that they exist in large contiguous areas. This is necessary so that the larger animals essential for biodiversity and productivity can live in the ecosystems in sufficient numbers. From an energy perspective, this is of great importance, since areas cannot be sacrificed arbitrarily for extraction or industrial agriculture. Biomass and land area are only very limitedly available.

Freshwater change From an energy perspective, several processes are of importance:

- The use of water bodies to dispose of waste heat from fossil power plants: this causes rivers to overheat and be severely damaged. This ends with the end of fossil fuels (see preceding discussion).
- Dams destroy the migration routes of fish and thus destroy the ecosystems of the river, the nutrient transport from the sea to land (via the fish), and ecosystems in the ocean. Dams prevent sediment transport and thus destroy river deltas. This sets a limit on the use of hydropower.
- The irrigation of land areas reduces runoff, sediment transport, and seasonal fluctuations in rivers, thereby destroying their ecosystems. The salinization of soils destroys agricultural land. Taking this boundary into account reduces the availability of biomass.
- Wastewater from industrial processes poisons water bodies. Adequate treatment of wastewater or processes without such toxic wastewater are somewhat more expensive and energy-intensive. They affect the cost of energy.

Atmospheric aerosol loading Industrial emissions of fine particles can be minimized at large sources with little effort (see Chapter 14).

Difficult are diffuse sources such as biomass fires in households, tire abrasion, and open-cast mining. Lack of vegetation on fallow fields and the burning of fields are further important sources. Some of this could be reduced through greater care and appropriate economic practices.

Added to this is the 'Faustian bargain' of sulfur emissions from shipping: the reduction of the previously massive sulfur emissions in recent years is contributing to the current rapid global warming (Hansen et al. 2025).

Novel substances The adverse effects arise from a multitude of processes and impacts, including among others the following:

- Direct poisoning of humans, animals, and ecosystems: illness and premature death burden living conditions in a wide variety of ways. They contribute to the destruction of ecosystems and societies. Illness and death reduce physical and mental performance. They reduce productivity.
- The hormonal effects of many substances additionally affect reproductive capacity and impair the future of human societies and ecosystems.
- The alteration of genetic material for all living beings in the earth system affects them in manifold ways.
- Added to this are the unknown effects of known and new substances (trade secrets).

The decision to use only known and harmless substances is often associated with only a small additional input. The input for the treatment of emissions is usually low compared to the social or ecological damage they cause. It is unclear whether forgoing new, unknown, or toxic substances requires any relevant technical or energy input. Presumably, things work very well without them too.

Summary

From these boundaries, the following key conclusions follow:

- The end of fossil energy carriers is urgent. It cannot be postponed.
- Biomass as an energy carrier or raw material is actually barely available. We need the land and soils first for food and for our living world.
- Care for our world requires changes in actions and processes. For this we will need resources and energy.
- We will have to clean up after ourselves, for example by removing greenhouse gases from the atmosphere (see Chapter 14)

8.1.2 Climate Overheating

The international scientific consensus on the state of knowledge about our industrial global warming is published every five years by the International Panel on Climate Change (IPCC). The process is roughly as follows: scientists compile a report, which is then edited together with government representatives in an intensive process to reach a consensus. Fundamentally, the process at the IPCC is characterized by great restraint. In particular, there is an understanding that statements should err on the safe side and be as far as possible from dramatic descriptions (*"erring on the side of the least drama"*, see Brysse et al. 2013). This means that the IPCC's predictions are to be classified as extremely conservative, since the earth system has continuously followed the IPCC's worst-case scenario for many years (Wallace-Wells 2019) or is currently exceeding it (Hansen et al. 2025). A first departure from this restraint was shown by the IPCC with the 1.5 °C report (Masson-Delmotte et al. 2017), which presented in an accessible form the horrors of a global warming of +1.5 °C above the baseline temperature compared to the much more difficult to bear consequences of a global warming of +2.0 °C.

The +1.5 °C were exceeded for the first time in 2024, and exceeding +2.0 °C can realistically no longer be prevented. Therefore, this observation by William Gibson concerns us all:

> *"the future is already here – it's just not very evenly distributed"*

The consequences of the collapse of our climate, which has been stable for many millennia, are clearly visible. Their impact on people's lives is very unevenly distributed regionally. The ongoing temperature rise is itself highly dynamic. At the same time, many processes in the earth system are sluggish and respond with a certain delay to changes in their boundary conditions (Hansen et al. 2013). This means that many changes have been set in motion today that will only fully unfold in the coming years. There is no new stable state and no certainty that the weather will now remain as it is. Instead, we are currently observing an increase in variations in our region. Individual years and individual events are becoming more extreme, while the average shifts with high dynamics toward a point that we humans have determined and will determine through our current and future greenhouse gas emissions. Due to this strong dynamics with an accelerating rate of change and the non-linear onset of systemic feedbacks, this development is very difficult for us humans to grasp. In fact, only linear changes over short periods of time are easily understandable or imaginable for us humans.

What Is Happening?

An optimistic assessment of the space of possibilities for human development in the context of future global warming is offered by Steffen et al. and Wallace-Wells (Steffen et al. 2018; Wallace-Wells 2019). Optimistic here means that a decision space is seen which, with rapid massive action, sees the possibility of continuing to shape civilization and a good life in a very different and greatly changed world. In this context, Wallace-Wells (Wallace-Wells 2019) provides an easily accessible overview of many essential and current sources on the expected and currently foreseeable horrors of further global warming (known-knowns, see Fig. 2.4):

1. Direct heat death occurs for humans when the wet-bulb temperature of the atmosphere exceeds 35 °C for a few hours. These conditions are occurring locally and temporarily from now on in India and Arabia.
2. Famine, caused for example by (i) dryness and drought, by (ii) reduction of the nutrient content of plants due to the rise in atmospheric CO_2content, by (iii) die-off of coral reefs and ocean fish, by (iv) extreme weather, by (v) reduction of the yield of many crops at elevated temperatures, and by (vi) loss of low-lying areas due to sea level rise
3. Drowning due to flooding caused by extreme weather events poses a danger. In the long term, events due to sea level rise (melting of glaciers including Greenland and Antarctica) are added.
4. Wildfires and forest fires spreading from forests, scrubland or savannas into cities with the consequences: (i) burning, (ii) loss of crops, (iii) loss of home and homeland, (iv) far-reaching, often global health consequences due to the massive emissions
5. Extreme weather events such as severe storms, prolonged droughts or extreme rainfall with these consequences: (i) death from the event itself, (ii) loss of property, crops and home, (iii) health consequences, (iv) trauma and long-term mental illness, (v) destruction of infrastructure
6. Loss of usable water in many regions due to: (i) higher evaporation rate during heat periods, (ii) stronger long-term weather cycles, (iii) shifting of desert zones into currently densely populated regions (Mediterranean area), (iv) poisoning by wildfires or damaged infrastructure, (v) disappearance of glaciers as seasonal water reservoirs
7. Loss of the oceans as food sources due to: (i) acidification (carbonic acid from atmospheric CO_2 dissolves in the sea), (ii) suffocation of animals (less O_2 can remain dissolved in warmer water), (iii) direct overheating
8. Stronger winds and more wildfires cause more air pollution. Aerosols and dust have a variety of short-term and long-term health consequences.
9. The higher concentration of carbon dioxide reduces the cognitive abilities of all people.
10. Tropical infectious diseases are spreading as the habitats of their vectors (mosquitoes, ticks) expand.
11. Violent death, injury and trauma. Violence between people increases with ambient temperature. Wars over water and resources arise. States are destabilized by climate refugees.
12. Permanent shrinkage of the economy (growth at a negative rate, see chapter 5), e.g. due to (i) more expensive resources, (ii) frequent severe destruction (extreme weather events, flooded regions), (iii) disappearing markets, (iv) war, (v) more expensive food, and (vi) reduced human productivity

13. Increase in the number of traumatized people and mental illnesses, and thus a greater burden on society through care and nursing
14. Systemic interactions between these aspects

These horrors are part of the changes in the earth system in the range of warming of +1.5 °C to +3.5 °C. This is the range that will be maintained in the best case. Not included in the list are the known-unknowns and the unknown-unknowns. Such horrors are mostly hidden in less easily understandable scientific articles. As an example, Hansen describes on the basis of geological data the steady-state earth system at +1 °C (Hansen et al. 2016).

Today's emissions of industrially produced greenhouse gases thus have fast-acting and short-term effects on the earth system. This is the most commonly chosen time frame up to the year 2100. The enumeration presented above also refers to this near future (Masson-Delmotte et al. 2017). In fact, these effects extend much further into the future. Many consequences, such as the equilibration of sea level, will still require several hundred years (Archer 2009). Thus, every human society will have to deal with the consequences of industrial global warming for much longer than the period of fossil fuel use will last. For this reason too, the estimates of the unwritten-off costs that fossil carbon actually generates over the long future are very high and range from €10,000 to €750,000 per tonne of carbon (van Renssen 2019).

The key scientific question in 2019 was whether there is still a safe range of further CO_2emissions, i.e., a range in which the emission of CO_2 correlates approximately linearly with the rise in atmospheric temperature (this would be the behavior described in Chapter 7). All IPCC statements on a CO_2budget and the +1.5 °C and +2.0 °C climate targets depend on this assumption. Evidence is mounting that this may not be the case, and that we have already reached the range of nonlinear responses of the earth system (Hansen et al. 2023). In this case, a far more pessimistic view of the possibilities of human civilization in the earth system might be appropriate (Bendell 2018).

The energy costs of the necessary adaptation to the ongoing climate change increase very significantly in a nonlinear manner with total greenhouse gas emissions, particularly when livable conditions in the earth system are to be maintained for as many people as possible at the same time (Kotz et al. 2024).

Climate Adaptation

The ongoing collapse of our climate system, which has been stable for many millennia, will in any case be emotionally difficult in addition to the technical and financial costs. Global warming will cause many people to lose their homes, as they sink beneath the sea or dry out. People who are able to remain in one place will experience the loss of their familiar surroundings. This too can be emotionally difficult.

The change in weather and vegetation alters the familiar home of all people today. This process can be observed everywhere on earth, and its pace will increase significantly in the coming years. This change can occur gradually or in rather catastrophic surges. It is a process that will only slow down when the earth system reaches a new steady state. This can take centuries. This sets a framework for the ongoing task of climate adaptation:

- In principle, the space for human civilization, for decisions, and for the ability to shape the future independently becomes dramatically smaller with increasing concentrations of greenhouse gases and temperature, until it disappears at a currently unknown threshold: adaptation is only possible to a limited extent.
- All communities will have to continuously adapt their infrastructure. No technical solution and no technical approach can be regarded as permanent.
- All communities will have to repeatedly restructure and adapt their public services, depending on the changing available resources. In addition to water, this concerns the possibilities of food production.
- All communities will have to continuously adapt their rules, their procedures, and their values. Many traditions make little sense in a constantly changing world.
- All communities will have to invest more effort in actively understanding the changes in their environment. In particular, the constant readiness to adapt and change as a core competency will become essential for the survival of communities. The memories of elderly people from other temperature eras will be worth less. In return, their experience in dealing with change and adaptation counts all the more.
- Every individual will have to expend much more effort on changes than was necessary in a stationary world. Regular periods for reorientation, for grieving over losses, and for realignment are necessary. For this, our society urgently needs appropriate rituals.

These aspects themselves require energy and resources. They demand resilience.

Action Is the Cheaper Option

Not starting right now and acting with maximum speed, great seriousness, and globally relevant measures is grossly negligent. This is the international scientific consensus (Masson-Delmotte et al. 2017). There are boundary conditions here that follow from the characteristics and energy costs of the possible renewable technologies, but also from the requirement to achieve this transition with the lowest possible carbon dioxide emissions. The financial costs of a transition are substantially lower than any delay or inaction (Kotz et al. 2024).

Only very short-term, purely financial planning (e.g., the currently common requirement in industry that an investment yields returns in less than two years) arrives at a different conclusion. This is what the term *discounting the future* or *writing off the future* means. Every form of postponement, of waiting for greater certainty and for decisive events, is in itself already a decision against action. All the facts summarized here have been known for approximately 50 years, and since that time we have been living in the age of political waiting and postponement, i.e., the targeted prevention of sensible actions (DeLay 2024). Out of consideration for the particular interests of the fossil fuel industry, the entire world population has been and continues to be exposed to the horrors of a far too hot earth in the coming millennia.

Super-wicked problems

Already in Chapter 2 it was shown that the challenges of an energy transition, approached with the urgency required, sufficient seriousness, and evidence-based decision-making foundations, are not linked to simple solutions but belong to the wicked problems. The availability of resources imposes additional constraints. This makes the energy transition, the steering of global warming, and the stewardship of the earth system within tolerable limits for humanity into problems for which a separate class has been named: the super-wicked problems. These are characterized by four additional aspects (Levin et al. 2012).

(1) Time is running out

The snail's pace of normal politics, where problems disappear through procrastination, generational change, or shifting priorities and boundary conditions, is completely unacceptable. These problems must be addressed now. The necessary measures must be taken today. There is no more time to wait another 50 years for research results. But there is also no longer any need to, because the knowledge is there.

In the usual course of political development, groups that have lost out in a distributive debate on a social issue and do not see their claims or expectations adequately considered can regroup and restart the debate in a few years. This luxury does not exist for the topics of energy, climate, and biodiversity. In addition, there can be no soft compromise on these three core problems, where some are allowed to pollute more and others get more climate protection. This situation is unusual for typical political mechanisms.

The longer we wait on these problems, ignore or deny them, limit ourselves to further research, or negotiate without results, the more urgent they become. Any position that defers action is already part of the problem. It thereby destroys possible, less extreme options for action.

(2) The originators of the problem are themselves searching for a solution

Everyone who tries to be part of the solution also contributes to the problem. In particular, we as well-educated people from wealthy countries require a great deal of energy. We cannot avoid using the energy- and resource-intensive infrastructure of our region. A boundary between the guilty and the innocent is hardly possible to draw. Everyone is part of the problem. Any boundary between people who cause more or less of the problem is artificial in the social middle. The borderline cases are simple (billionaires with a 100-meter private yacht, private jet, and weekly intercontinental flights versus subsistence farmers), but the middle is complex. This situation makes it difficult for each individual, because measures always affect nearly the entire society. It is not a negotiation process between different groups in a society, but rather there are negotiation processes within each individual: How much sufficiency am I willing to endure? How much energy do I want to use? What price can I pay for energy?

(3) There is no central authority

The three core problems circled here require, in the best case, a strong global authority that can plan, define, implement, and monitor the execution of measures (SDG 17). It must be an authority capable of prosecuting violations of the law and enforcing the law. International politics is fragmented, weak, and easily blocked entirely. There are no means to announce more than shared good intentions. Few regional alliances or major states are systematically addressing these problem areas. Especially in international politics, priorities and spaces of possibility are constantly shifting.

(4) As a result of these first three aspects, political responses discount the future

> *"Partly as a result, policy responses discount the future irrationally." (Levin et al. 2012)*

The necessity of a rapid, sweeping, and effective policy that tackles the three problem areas of energy, climate, and biodiversity within planetary boundaries is scientifically undisputed. Extensive, reliable, and robust information is clearly available to everyone at all times. Nevertheless, this information is ignored or undervalued in politics by most actors: the decision-making horizon of political decisions is short-term, and measures are supposed to give one's own voters the feeling of positive change before the next election. All approaches to tackling the super-wicked problem areas bring long-term benefits or are the basic prerequisite for securing one's own state and society in the long term, but they entail short-term inconvenience and costs.

How long will fossil energy carriers last (and do we want to burn them all)?

Based on the data from McGlade and Ekins, the following reserves and resources of fossil energy carriers exist (McGlade/Ekins 2015):

- Approximately 17 ZJ of lignite and 60 ZJ of hard coal,
- Approximately $7.10 \cdot 10^{12}$ barrels (bbl) crude oil and thus approximately 43 ZJ
- Approximately $680 \cdot 10^{12}$ Nm^3 natural gas and thus approximately 24.4 ZJ

What mass and volume of hard coal and lignite do these values correspond to?

Hard coal has a density of approximately 800 kg m^{-3} and a heating value of 25 MJ kg^{-1}. To a good approximation, hard coal consists exclusively of carbon and lignite to 50% (mass fraction). Lignite has a density of approximately 1000 kg m^{-3} and a heating value of 9 MJ kg^{-1}. Darmstadt, with its 160,000 inhabitants, extends over an area of 112 km^2.

Then

$$m_{coal} = \frac{Q_{coal}}{H_{I,coal}} = \frac{60 \cdot 10^{21}\ \text{J}}{25\ \frac{\text{MJ}}{\text{kg}}} = 2.4 \cdot 10^{15}\ \text{kg}$$

and

$$m_{lignite} = \frac{Q_{lignite}}{H_{I,lignite}} = \frac{17 \cdot 10^{21}\ \text{J}}{9\ \frac{\text{MJ}}{\text{kg}}} = 1.9 \cdot 10^{15}\ \text{kg}$$

Thus the volume of lignite is $1.9 \cdot 10^{12}$ m^3 or 1900 km^3 and for hard coal $3.0 \cdot 10^{12}$ m^3 or 3000 km^3. A heaped pyramid made of hard coal, in which the height equals half the side length, would have an edge length of

$$b = (6 \cdot V)^{\frac{1}{3}} = (6 \cdot 3000 \text{ km}^3)^{\frac{1}{3}} = 26 \text{ km}$$

It would cover an area of almost 700 km^2 and be three times higher than Mt. Everest.

What are the static reserves of coal, natural gas, and crude oil?

The total current global energy demand is 650 EJ and roughly 30 % each is covered by the three primary energy carriers. The static reserves follow from the respective annual consumption of 195 EJ each:

- For coal it is approximately 385 years (that is 40 years less than in the last edition of this book).
- For crude oil it is approximately 210 years (that is about 30 years less than in the last edition).
- For natural gas approximately 115 years are given (that is about 20 years less than in the last edition).

What are the dynamic reserves of the three energy carriers?

For a further growth rate of global energy demand of 2 % per year we can use formula 6.25 and obtain (where we additionally subtract 10 years since publication of the paper)

$$\Delta t_{coal} = \frac{1}{z} \cdot \ln\left(\frac{Q_{coal}}{\dot{Q}_{coal}} \cdot z + 1\right) = \frac{1}{0.02} \cdot \ln\left(\frac{77 \cdot 10^{21} \text{ J}}{195 \cdot 10^{18} \text{ J}} \cdot 0.02 + 1\right) = 99 \text{ a}$$

$$\Delta t_{crude} = \frac{1}{z} \cdot \ln\left(\frac{Q_{crude}}{\dot{Q}_{crude}} \cdot z + 1\right) = \frac{1}{0.02} \cdot \ln\left(\frac{43 \cdot 10^{21} \text{ J}}{195 \cdot 10^{18} \text{ J}} \cdot 0.02 + 1\right) = 74 \text{ a}$$

$$\Delta t_{NG} = \frac{1}{z} \cdot \ln\left(\frac{Q_{NG}}{\dot{Q}_{NG}} \cdot z + 1\right) = \frac{1}{0.02} \cdot \ln\left(\frac{24.4 \cdot 10^{21} \text{ J}}{195 \cdot 10^{18} \text{ J}} \cdot 0.02 + 1\right) = 53 \text{ a}$$

Here the growth rate $z = 0.02$ and $x(t_0) = 180$ EJ is the current consumption for each of the three energy carriers. This yields

- For coal approximately 90 years,
- For crude oil approximately 65 years and
- For natural gas approximately 45 years.

These time periods are significantly shorter than the static reserves.

What is the volume of carbon dioxide released when the coal is burned?

Here we assume that the mass of hard coal is pure carbon and 50 % of the mass of lignite is carbon. The mass of carbon is thus $m_{C,K} = 3.34 \cdot 10^{15}$ kg. The mass of carbon dioxide is

$$m_{CO_2} = m_C \cdot \frac{M_{CO_2}}{M_C} = 3.34 \cdot 10^{15} \text{ kg} \cdot \frac{44.01 \frac{\text{kg}}{\text{kmol}}}{12.01 \frac{\text{kg}}{\text{kmol}}} = 12.24 \cdot 10^{15} \text{ kg}$$

Then the volume of CO_2 at standard conditions is

$$V_{N,CO_2} = \frac{m_{CO_2} \cdot R_{CO_2} \cdot T_N}{p_N} = \frac{12.24 \cdot 10^{15}\ \text{kg} \cdot 188.9\ \frac{\text{J}}{\text{kg}\cdot\text{K}} \cdot 273.15\ \text{K}}{101\ 325\ \text{Pa}} = 6.23 \cdot 10^{15}\ \text{Nm}^3$$

We calculate the standard volume for crude oil and natural gas accordingly. For crude oil we additionally need the mass fraction of carbon in crude oil. This is 0.85. The volume is then approximately $1.43 \cdot 10^{15}\ \text{Nm}^3$. Assuming that natural gas consists only of methane, the volume of CO_2 equals the volume of CH_4 or $0.680 \cdot 10^{15}\ \text{Nm}^3$. In total this amounts to $8.34 \cdot 10^{15}\ \text{Nm}^3$.

How does the carbon dioxide content of the atmosphere change?

We have already calculated the standard volume of the atmosphere in Chapter 7. It amounts to $3.986 \cdot 10^{18}\ \text{m}_N{}^3$. If we introduce the CO_2volume just determined into the atmosphere and assume that – roughly as before – about 1/3 of the carbon dioxide remains in the atmosphere, then the future concentration is approximately

$$r_{CO_2} = r(2024) + \frac{1}{3} \cdot \frac{V_{N,CO_2}}{V_{N,atm}} = 425 \cdot 10^{-6} + \frac{1}{3} \cdot \frac{8.34 \cdot 10^{15}\ \text{Nm}^3}{3.986 \cdot 10^{18}\ \text{Nm}^3} = 1122\ \text{ppm}$$

From this follows a radiative forcing (see Chapter 7) of

$$\Delta F = 5.35\ \frac{\text{W}}{\text{m}^2} \cdot \ln\frac{r}{r_0} = 5.35\ \frac{\text{W}}{\text{m}^2} \cdot \ln\frac{1122\ \text{ppm}}{280\ \text{ppm}} = 7.43\ \frac{\text{W}}{\text{m}^2}$$

This would lead to a change in mean temperature (relative to the reference year 1750) of

$$\Delta T = \lambda \cdot \Delta F = 0.8\ \frac{\text{K} \cdot \text{m}^2}{\text{W}} \cdot 7.43\ \frac{\text{W}}{\text{m}^2} = 5.94\ \text{K}$$

This is a temperature increase that no one who can experience it would want, as it would likely mean the end of humanity as a species.

8.1.3 Why the Energy Transition Now?

The energy provided by fossil fuels is currently very inexpensive on the market and – if we disregard the ethical problems of its origin – reliable. So why should we strive for an energy transition now, toward the more costly renewable energy sources with lower security of supply? We will examine this question in the following.

Finite Resources

Figure 8.1 summarizes the findings from Chapter 5 and Chapter 6. To ensure a constant supply, finite resources such as fossil fuels and minerals require ever greater input into extraction and processing:

- Ever more capital is required for extraction and provision, since the easily accessible and higher-concentration deposits were extracted first.
- The energy required for extraction increases as reserves are increasingly extracted from remote, deep, and poor deposits.

- The capital and material tied up in extraction increases as the processes for separating raw material from waste rock become more complex and the facilities therefore larger.
- Technological developments may be able to slow this process, but cannot reverse it.
- As a result, the required price for the provided energy carrier and the extracted material increases over time, whereby the cost increase can be externalized to investors (fracking) or states and their taxpayers (subsidies). These costs do not need to be visible on the market.

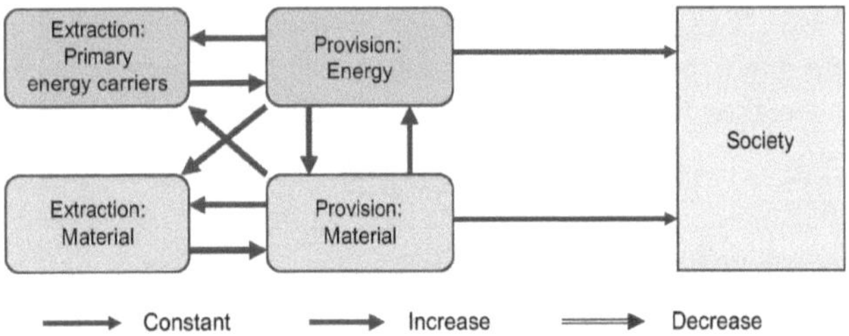

Figure 8.1 Finite resources and the constant supply of society with material and energy lead to an increase in consumption within the supply system.

If the reserves or inputs for individual energy carriers reach a critical point, this system can enter a crisis. If essential materials are no longer available in sufficient quantities, substitute technologies must be employed that generate additional costs or enable lesser utility. If individual primary energy carriers can no longer be provided in sufficient quantities and at a price still acceptable to the market – both are always closely linked – it can very quickly become impossible to continue guaranteeing a supply of that energy carrier.

Resources and Geopolitics

A special role is played here by the two energy carriers diesel and kerosene, as they are not only essential for the infrastructure of Figure 8.1 today, but also represent the current foundation of military capabilities. These two energy carriers are the indispensable prerequisite for the projection of influence or power, as they move naval and ground forces at high speed and make an air force possible in the first place.

Should either the confidence in the relative inexhaustibility of these essential energy carriers dry up, or the price become no longer bearable for society in some regions of the world, regional or geopolitical crises may occur. These crises can arise because states can no longer maintain their order or because distribution conflicts over remaining reserves emerge. This situation is illustrated by Figure 8.2. This can lead to considerable disruptions or the long-term collapse of the supply of other resources as well. Since the easily accessible reserves of fossil energy carriers are foreseeable to be exhausted, such disruptions with considerable consequences for the supply of all states and regions may occur.

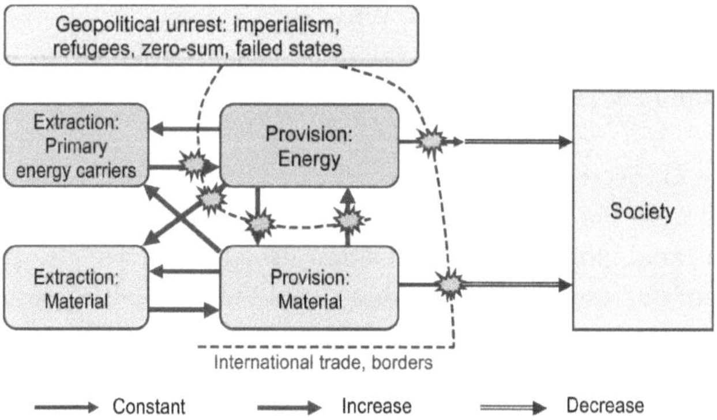

Figure 8.2 Geopolitical effects disrupt the provision of energy and material.

A functioning regional energy system based on renewable energy carriers offers the advantage that it is only marginally disrupted by such events. Renewable energy carriers also require material and energy for their operation. They therefore cannot be completely decoupled from global developments, but they help greatly in preserving stability. However, the infrastructure of the energy transition must already be sufficiently developed before the onset of massive disruptions in the supply of fossil energy carriers or other important resources, so that it can seamlessly take over or continue its function. Subsequent construction in a crisis situation and within a few weeks as a "war program" is then not possible.

Energy Transition and Local Economy

The background to this discussion is the question of whether individual states should simply start their energy transition now or wait for a joint global effort (the "the Chinese" argument). One argument against starting now is that the necessary additional costs of an energy transition would disadvantage one's own economic location, and that therefore only a possibly achievable future, yet-to-be-negotiated international agreement – formulated with particular regard to the interests of today's established energy-intensive industries, the large fossil fuel industry, and the entrenched geopolitical power structures, and ultimately preventing individual states from paying more for energy than others – would be indispensable for further economic growth. Only this would be the right path to allow the states of the international community to jointly and evenly integrate renewable energy technologies to a certain degree with a long-term perspective. This or something similar is argued in particular by self-proclaimed advocates of the "free market", who also fight against any form of subsidy for renewable energies, yet have no problem with the direct (Coady et al. 2015) and indirect subsidies (externalization of the costs of a climate catastrophe) for fossil energy carriers. The result of this line of argument would be the complete abandonment of any societal shaping of the future and remaining in the status quo until a fully developed crisis has destroyed these power structures. Whether a reconstruction of a similar civilization would then be possible or would fail due to a lack of resources is difficult to assess today (Diamond 2011).

Finite Resources and Climate Catastrophe

A different perspective arises from the viewpoint of industrial global warming (Section 8.1.2). This also makes a rapid global energy transition necessary. If the effects of global warming are incorporated into the relationships from Figure 8.2, the result is Figure 8.3, because industrial global warming will itself become a driver of geopolitical instability through its sensible consequences: famines and refugee flows can lead to unrest, the collapse of states, and war. In terms of societal consequences, this is fundamentally comparable to those of the obvious drying up of cheap fossil raw materials. At the same time, all societies require additional resources for their adaptation to climate change and the structural and technical transformation needed for it.

For this situation too, it is very advantageous if the energy transition has been advanced far enough to ensure the regional supply of renewable energy. The alternative would be to believe that the supply of raw materials would magically remain secured forever and without consequences.

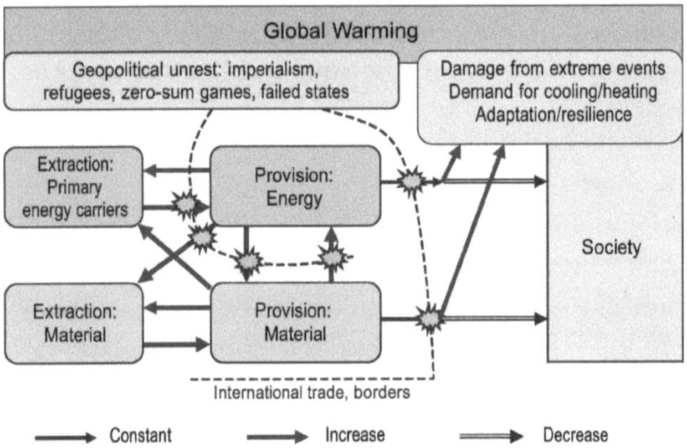

Figure 8.3 Adapting to climate warming requires additional energy and resources in an increasingly difficult world.

8.2 Goals – Anthropocene Instead of Hellocene

Our energy system is in a difficult transitional state: it may currently provide a great deal of energy at low cost, but under conditions that will not be sustainable for long. Neither are fossil energy carriers available for a sufficiently long time, nor will the consequences of the devastation caused by their use for us humans on our living earth remain socially bearable for much longer under a "business as usual" scenario (Kotz et al. 2024; Hansen et al. 2025).

In this context, our energy system is initially just one aspect of the fundamental question of how we avoid the Hellocene. Other topics intervene, demand consideration, or capture our attention. The supply of energy is the fundamental prerequisite of every human society. More complex societies and civilizations in particular are limited by this, and therefore their energy system is the focus. Conversely, this also limits the complexity that a (future) energy system can have (Tainter 1990). From a human perspective, we need a very concrete framework of values and goals against which we can weigh all decisions in terms of how we succeed in living well in the long term. This – and not the question of what technology we want – is the anchor point of careful decisions. This already includes the decision to transform our energy system so that it benefits us in the long term (within planetary boundaries). There is no such thing as no decision; there is only the decision to change nothing (currently often framed as "pragmatic and measured") – and this then with the foreseeable consequences for us (DeLay 2024).

Sustainable Development

One formulation of a society designed for long-term duration that incorporates fundamental aspects of justice into the future is that of sustainable development. This is formulated as a goal in many nastions' sustainability strategies. Whereby the source, the Sustainable Development Goals of the UN (SDGs, UN 2015), is legally binding for us. This principle thus finds its way, alongside many other competing goals, into the actions of governments and other organizations. Within the framework of this book, a few core theses suffice for further specification:

1. Sustainable development is not a defined end state, but describes a process. An end state in the thermodynamic sense is a system in equilibrium. As is argued elsewhere, this can only be a dead system. The living earth is a stationary flow system; therefore, the concept and definition must fit accordingly.
2. The process of sustainable development is characterized by wicked problems. These arise directly from the finiteness of the system, thus from planetary boundaries and finite resources.
3. Sustainable development is understood as an evolutionary process within a finite system and is subject to constant change. The process of sustainable development is fundamentally only conceivable within the finite earth system. The process is ex-

pressly not characterized by long-term linear or exponential changes. The process depends on the dynamically changing state of the living earth and the feedbacks within this system.

4. Our sustainable development takes place within the earth system or not at all.
5. It is not a lifestyle choice of individual people. It is the foundation of human life and must be understood as a long-term core task as part of state responsibility and public welfare provision (Grunwald 2010).
6. At its core, it contains demands for justice at all levels: that is, between people, regions, classes, and generations.

Sustainable Development

The term sustainable development is used, when it concerns long-term oriented action within an earth system that is finite in every respect. This term is associated with intergenerational equity (DIN ISO 26000):

> *"Development that meets the needs of the present without risking that future generations will be unable to meet their own needs.*
>
> *NOTE Sustainable development combines the goals of high quality of life, health, and prosperity with social justice and maintains the earth's capacity to support life in all its diversity. These social, economic, and environmental goals are interdependent and mutually reinforcing. Sustainable development can be seen as a way of expressing the overarching expectations of society as a whole."*

The concept of sustainability originates from forestry: there it was understood as a form of management that maintains a forest at all times with the highest possible yield and for a variety of competing uses (Carlowitz 1713): In Hans Carl von Carlowitz's time, forests were in poor condition due to the many different intensive uses and often yielded low and declining outputs. Finding the right measure for balancing different claims and forms of use, and thus maintaining the value of the forest for the owner in the long term, was his problem. His client and owner was the territorial prince, who wanted to secure and improve living conditions and the economy in his domain over the long term. His problem is also ours, if we understand ourselves as owners of our world and wish to secure our human actions within planetary boundaries over the long term (Richardson et al. 2023).

By now the term *sustainable* is used in an inflationary and often distorting manner, so that currently sustainable development – with reference to the corresponding goals of the United Nations – is clearer here (UN 2015).

Sustainable Development Goals and energy

The Sustainable Development Goals (SDGs) of the United Nations (UN 2015) serve as a good orientation for what sustainable development can be. These encompass 17 target areas of a large global agenda, whereby these goals are specifically to be achieved by 2030 and cannot simply be addressed long after the end of the terms of office of today's politicians. They are therefore urgent, comprehensive, and formulated in a sufficiently concrete and measurable way. At the same time, these 17 major goals are already obviously contradictory among themselves. This characterizes the wickedness of the underlying problems. Their value is only increased by this, as it becomes clear that these goals can only be achieved through genuine effort and through a weighing of goods and values. Furthermore, the SDGs can be summarized in three groups (UN 2015).

Keeping our planet viable

The first group is central to long-term life on earth. The goals formulated here are therefore non-negotiable. They cover the planetary boundaries.

These goals are already discussed in 8.1.1.

Justice and adequate living conditions for all

As a second group, social aspects are summarized. These follow basic human needs. Here too, the first goals (SDG 2) are non-negotiable.

It is about justice, it is about sharing, it is about dignity for all of us. Implementing this is a major challenge – especially under the conditions of climate change. A future regenerative energy system will no more than our fossil system enable a billionaire lifestyle for all of us. Our energy system, together with global climate disruptions, will dictate how and where we can live (Vince 2023).

A long-term economy

As a third group, dependent on the previous categories, economic goals are formulated. In these goals, unlike in the previous ones, a belief in growth is still contained. Thinking about these goals with limited resources while at the same time enabling justice and reduced inequality presents a particular challenge. A future energy system will ultimately set the framework within which we can shape this. It is therefore worthwhile to implement the transformation quickly and carefully.

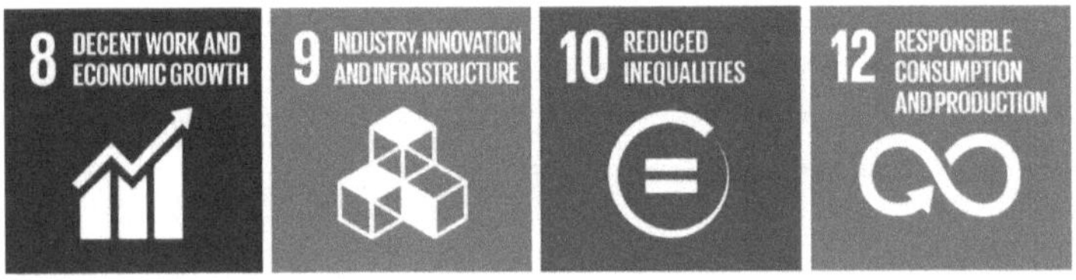

Prerequisites for the implementation of these SDGs

The goals of the SDGs do not include rapid adaptation to regional climate change. The SDGs implicitly assume a climatically static or only slowly changing world. Explicitly, a very high degree of growth and technology optimism is found in the statements on the four economic SDGs. If temporally and spatially unlimited economic growth were possible, the social goals addressed could in principle be achieved without paying greater attention to the central theme of equality of opportunity (SDG 10). In a finite and limited world, this is, as part of justice, one of the core issues.

In summary, it becomes clear that the SDGs are not purely technical topics, but that technology and technological development must often make a relevant contribution. Regardless of this, there is a concrete need for energy and resources to implement the individual goals. All goals, when thought of concretely and locally, directly become wicked problems.

Decent living conditions for all

The goals of the SDGs are initially not formulated in energetic or material terms. However, these goals can be directly translated into an energy demand and also into required resources. A relevant applied possibility is to formulate real living conditions under which a life in dignity and with social participation is possible (decent living standards). From these conditions, it would then be possible, if needed, to determine what monetary value this corresponds to (Kikstra et al. 2021, see also Table 8.3):

Nutrition Food in sufficient quantity and with appropriate nutritional content must be produced, prepared, and stored. This means that the energy demand of a long-term harmless agriculture, for clean cooking stoves, and for food refrigeration is part of the calorific value of adequate nutrition.

Housing (shelter) This is more than a tarpaulin. It encompasses a durable and permanent structure, sufficient space, heating and, where applicable, cooling, as well as appropriate clothing. Biomass can only be considered here as a fuel if it can be used without critical health effects and with high efficiency (Brown 2022).

Health The basic prerequisite is access to clean drinking water. A sanitation facility whose wastewater is separated from daily life, hot water for personal hygiene, and access to a public health system are elementary.

Transport There should be a publicly accessible infrastructure that enables (motorized) transport with which all basic needs – in the city and in the countryside – can be met. This includes buses and railways, roads and vehicles, and the energy for their propulsion. Motorized individual transport is not included.

Society An adequate life includes education, communication, and access to information that enables participation in society.

The energy requirements vary with the climate and other framework conditions. They amount to approximately 17 GJ per year per person and globally for 2040 to approximately 160 EJ per year (Kikstra et al. 2021). This is significantly below the current global energy throughput of approximately 650 EJ. This energy demand can be well represented in the long term under strict climate protection scenarios and would offer additional room for prosperity for all.

Table 8.3 Individual Human Needs and the Hierarchy of Needs According to Maslow

Need level	Concrete needs
Basic physical needs	Food, air, water and excretion
Protection and safety	Physical integrity, protection from the elements, shelter, warmth (and cold), protection from trauma
Affection	Family, group belonging, communication and religion
Recognition	Respect, participation, and self efficacy
Self-actualization	Creativity, expression, fashion and decoration
More recognition	Status, influence and power

8.3 Shaping Transformation

The successful design and implementation of transformation processes is an art in itself and is not the subject of this chapter. This section instead examines central concepts that are discussed individually or in combination as approaches.

8.3.1 Paradigms of Sustainable Development

The material design of sustainable development within planetary boundaries is based on three intertwined fundamental paradigms of consistency, sufficiency and efficiency (Heyen et al. 2013). From an engineering perspective, these paradigms each have specific meanings.

Efficiency
Efficiency encompasses the following core question: How can I manufacture or operate a product in such a way that I require as few resources as possible?

✓ **Efficiency**

We understand efficiency always as our human benefit divided by our human input. We often also understand it as my personal private benefit divided by my personal input. This makes it clear that efficiency always allows only a very one-sided picture. All expenditures that are not directly and immediately relevant to the person asking the question are systematically excluded. This includes all expenditures that can be externalized: to other regions, to other people, to future generations, into the living earth system.

These expenditures, thus hidden from view, have the unpleasant property of becoming visible again and – e.g. as legacy burdens – then demanding our time, attention and resources. In doing so, we tend again toward "efficient solutions" that extend, enlarge and shift the problem (Schlaudt 2024).

Efficiency can be regarded as the driving force behind the development of mechanical engineering: Newcomen's first steam engine from 1713 was economically viable only at a coal mine, since it worked directly at its energy source there and avoided the comparatively high costs of animal feed. The steam engine had to become more efficient in order to be used at other locations and for other tasks at all. For this, the mechanics had to be understood. Machine elements had to be developed and manufactured reliably. Materials had to be understood and processes developed in order to produce them in sufficient quality. Thermodynamics developed alongside the steam engine and from the question of how such a machine could be designed with the maximum possible technical efficiency. With the scientific answering of these questions, our current understanding of the role and responsibility of engineers emerged. From the rather small efficiency of Newcomen's steam engine of approx. 0.3%, the efficiency of current steam power plants of 45% was developed. This current efficiency is close to the thermodynamically possible maximum.

270 years were needed for this efficiency increase by a factor of 100 and toward the technical maximum. This period is the typical development timeframe for technology (Smil 2017). With the success of the technical development of the steam engine, innovations became possible in many further segments of technology, allowing considerable further efficiency improvements in processes and plants there (Smil 2017). The pursuit of maximum efficiency is a core element of engineering sciences and part of the intrinsic motivation of engineers.

At the same time, efficiency is also part of good economic management. For von Carlowitz too, efficiency in the use of forests was a relevant concept. Maximizing the efficiency of a technical process is certainly a problem in the sense of VDI 2221-1, but usually not a wicked problem in the true sense. The wicked aspect develops only from what is not taken into account: many of the current wicked problems are the result of efficient economic management. Because fossil fuels are brought to market efficiently, at high speed and at minimal cost, they are quickly depleted and are currently rapidly leading to climate catastrophe. Efficiency alone cannot solve the problems it has created.

Consistency

Consistency means acting within delimited and mutually compatible material flows:

- How can I ensure that I reuse as many material resources as possible over and over again, so that ultimately no new ones need to be extracted?
- How can I design products so that they consist of as pure or easily separable materials as possible?
- How can I ensure that organic substances remain usable as organic substances?
- How can I ensure that I do not poison or pollute the living world? How do I ensure that only those substances enter our world that definitely belong there and do not produce unwanted changes?

The requirement for consistent, i.e. delimited, material flows is directly reflected in the planetary boundaries. Living organisms, ecosystems and humans can only cope with a certain amount of toxic, mutagenic, and hormonally active substances. Beyond that, we living beings become ill, can no longer produce offspring, or die. At least regionally, we have exceeded these limits. That is the most pressing reason for consistency. And as we now know, everything comes back to us, no matter where we bury it, dump it, or have it incinerated (Schlaudt 2024).

The second reason is closer to the topic of this book: mixing generates entropy and separation requires valuable work. It is often not energetically worthwhile to extract individual substances from mixtures. In some cases, it is economically impossible to separate mixtures again. This applies to many alloys and to all of microelectronics: there, individual devices and components are essentially mixtures at almost the atomic level.

Only through carefully separated and guided material flows can it succeed in reusing valuable resources over and over again. This careful design and action in cycles is clearly distinct from efficiency. We often determine early on which of the two concepts should take priority in the engineering search for solutions. Consistent product design thinks from the perspective of use and end of use:

- Products should have the longest possible service life and be designed to be sufficiently robust for this purpose.
- They should be designed to be as completely repairable as possible in order to maximize their service life.
- Updates and upgrades for new functions should be possible at any time.
- The product should either be able to be put to a new use as a whole, or this should be done with its components as far as possible at the end of its service life.
- Products are designed to be easily disassembled for this purpose. Disassembly is carried out non-destructively.
- At the end of complete disassembly, there are components made of clearly identified material that can be reused materially without problems and without losses (European Commission 2015).

These technical requirements only appear simple at first glance: serious consistency demands a different approach to material selection than efficiency. It requires a different product design. It necessitates new business models, which currently often only become viable through regulatory requirements. Consistently applied, this leads to comprehensive standardization of many components and technical materials, i.e., a reduction in material diversity. New trade-off processes emerge, as the goals of efficiency and consistency are contradictory.

Sufficiency

Sufficiency is a societal question that cannot be resolved privately (Grunwald 2010). Its core question is: What do we need to be able to lead a good life?

The socially necessary reduction to a much smaller quantity of products that are actually needed for the continued existence of society, for material well-being, and for a good life, and that thus reflect the goals of the SDGs, requires new patterns of thinking and different expectations in society. Sufficiency leads directly to wicked problems, in which societal and technical levels are deeply intertwined.

A future within planetary boundaries will be a sufficient future. This makes it clear that any technology designed for a long-term future will become sufficient (Bihouix 2020; SRU 2024). With regard to consistency and sufficiency, it will be necessary to classify applications of energy in order to supply only the affordable ones and, first and foremost, the necessary ones.

Efficiency is quickly exhausted

We often hear that efficiency will save us. Let us therefore take a closer look at this.Figure 8.4 illustrates the idea using the example of global coal consumption. This is approximated by a simple growth curve with z = 3 %. For three time periods, an increase in the efficiency of all technology by 10 % over a period of 30 years each is then assumed. This results in three additional growth curves. The model roughly represents the increase in the efficiency of electricity generation from coal. The curves together approximate the actual coal consumption surprisingly well. All four curves are ultimately identical, but merely shifted in time. The effect occurs later, but then with equal intensity. Efficiency can therefore make no contribution to shaping human activity within planetary boundaries. The effect of efficiency is a different one:

> *"It is wholly a confusion of ideas to suppose that the economical use of fuels is equivalent to a diminished consumption. The very contrary is the truth."*

This insight is over 150 years old and comes from William Jevons in the year 1865. The additional benefit of more efficient technology is not left unused; instead, we continually fill it with something new (Fix 2024).

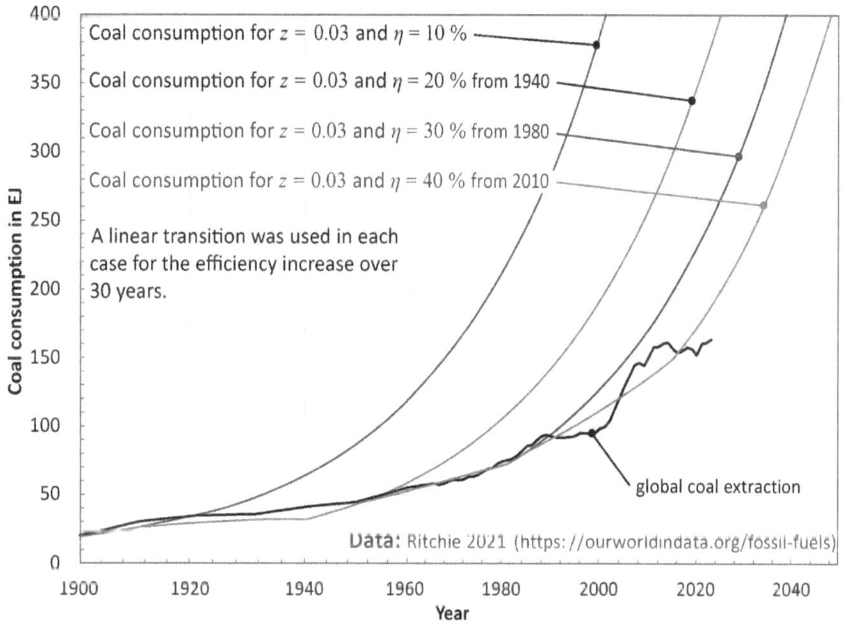

Figure 8.4 Efficiency gains shift growth somewhat into the future (data source: Ritchie 2021).

Classification of applications

A basis is needed with which we can decide which needs are to be supplied in the future, and at what level of intensity, in an energy system that cannot satisfy all needs. It is therefore proposed here to assign the planned use of certain forms of energy to one of four classes:

(0) Obstructive and energy-policy-endangering uses

These involve the use of energy or resources for purposes that actively hinder decarbonization. They are uses that waste resources or energy in an irresponsible manner, thereby endangering the energy supply of society, the economic foundations of society, or its essential basic functions.

(1) Non-essential use

These are uses that are not necessary for the decarbonization of society and are not essential for health, safety, or the functioning of society. Society functions very well without this form of use. These uses can readily be dispensed with even when there are no viable alternatives for them.

(2) Substitutable uses

The objective of the use is essential for society, but alternative technologies exist with which this objective can be achieved more quickly, more energy-efficiently, more economically, and with fewer impacts. Such substituting technologies must exhibit a sufficient technology readiness level to be deployed now at the required scale.

(3) Essential uses

These are applications that are necessary for the functioning of society and that cannot be substituted by any other technology available today. This includes central aspects of health and safety infrastructure as well as essential processes in the basic materials industry.

The concept is applied with regard to hydrogen as an energy carrier in Chapter 10. Questions of availability, EROI, consistency, and sufficiency are incorporated here.

Measuring sustainable development

How can the sustainability of a product, a process, or a measure be assessed?

Comparative assessment – VDI 4605

A simple comparative assessment is offered by the method of VDI 4605. This VDI guideline has been designed primarily for the assessment of infrastructure for the provision of energy, but the method can also be applied to other products. The assessment system used refers specifically to the SDGs. This then gives rise to the individual aspects of the assessment matrix that are to be compared.

Methodologically, it is a simple comparison between several solutions to a technical problem or between various similar products. Points are awarded relative to a benchmark product if the product under consideration is better (+1) or worse (–1). The respective totals of points achieved are then compared.

The method can readily be extended in the spirit of the usual assessment method for technical solutions according to VDI 2221-1, by additionally weighting individual aspects differently where this can be meaningfully justified, and by using measured or calculable values for comparison where applicable. The method provides a good initial orientation; however, it is not designed to capture large problems, wicked problems, or problems that involve complex interactions.

Systemic consideration

Systems that develop sustainably within their boundaries are characterized, among other things, by the fact that mass flows across the system boundaries balance out over the long term and assume a net value of zero. Otherwise, there would be an outflow or accumulation of the flowing quantity within the system (Neumann/Churchill 2011).

If a process is to run over the long term, its conversion rates must be sustainably maintainable. It is not significant whether this is a technical, a natural, or a social process. Of fundamental importance here is that not only the system in which the process takes place, but also its environment connected to it via material flows, must sustain this process, since effects are supplied and removed. Effects arise through energy, substances, social resources, or entropy. The environment must not, in the long term, lose its processes that are part of the system's needs. These requirements of the system on processes of the environment include the absorption of environmental impacts and in particular harmful effects in a long-term healthy and viable environment. Thus the system is always only a subsystem of the environment.

These framework conditions can be described by a rate equation (Neumann/Churchill 2011), where for the rates of each individual resource the following applies:

$$r_{prod}(t) + r_{reg}(t) - r_{cons}(t) - r_{depl}(t) - r_{abs}(t) \geq 0$$

At any point in time, strictly speaking, the rates for generation r_{prod} and regeneration r_{reg} of each resource must be equal to or greater than the rates for consumption r_{cons}, irreversible extraction r_{depl} and the required input for the absorption of by-products r_{abs}. For systems whose temporal development can be surveyed, it is possible to allow short-term variations and to maintain a balance over the longer term; however, at no time may irreversible damage occur in the environment, such as through the loss of relevant processes, ecosystems, or species.

$$\int_t \left(r_{prod}(t) + r_{reg}(t) - r_{cons}(t) - r_{depl}(t) - r_{abs}(t)\right) \cdot dt \geq 0$$

This is a mathematical formulation of the requirement for sustainable development within planetary boundaries. Here, for the temporal change of all quantities x that can flow across the system boundaries, the following applies in general

$$\frac{dx}{dt} = \dot{x}_{in}(t) + \dot{x}_{out}(t) + \sum_i r_i(t)$$

In this equation, any deviation from 0 is an indication of non-sustainable processes.

8.3.2 Resilient Systems

Energy systems should reliably provide the required energy at all times and under as many circumstances as possible. At the same time, we see that there are events that overwhelm and damage a system (Linow 2020). Energy systems should be designed for two different situations:

Hardened Systems should be designed for expected and foreseeable events. The primary concern here is to harden the infrastructure against impacts that might otherwise lead to a failure. In addition, disturbances are to be avoided or minimized. The system remains functional and integral. In particular, a possible failure affects the dependent systems only in the short term and leaves no relevant damage.

In a world with strong dynamic changes, this is no simple undertaking, since the impacts, their magnitude, and their direction change. In wealthy societies, the functioning of the energy infrastructure is taken for granted: there is someone who is responsible and accountable for it. In addition, energy systems there have typically been built out with redundancy and reliability.

Hardened systems are fundamentally brittle, i.e., they cannot withstand arbitrarily strong events. Instead, they tend to break irreversibly when a critical threshold is exceeded. This is impressively demonstrated by dams, which hold up without problems up to a certain water level. However, if their resistance is exceeded, they can fail with catastrophic consequences.

Resilient A second form of preparation is enabling the system to actively cope with unforeseeable damaging and disruptive events by adapting itself. It is about dealing with events that significantly impair and damage the system. The associated failure of the energy supply is spatially and temporally massive and important social functions are impaired. A resilient system can absorb this; a non-resilient system depends on outside help. If this help does not come, the system can subsequently be irreversibly damaged or destroyed. Resilience is a combination of different properties that together ensure that a single accident or extreme event remains a crisis and does not become a prolonged catastrophe (Thierney 2014).

In wealthy societies, the resilience of the energy system is initially outsourced to the grid operator. This operator ensures that the grid functions at all times and is quickly restored after massive events. The current energy transition, which also entails a massive restructuring of the grids and generation, offers on the one hand the opportunity to adapt the system to new and changing hazards. On the other hand, it often seems nearly impossible to carry out changes (new construction of power lines or power plants). This jeopardizes the function and resilience of the system.

Enabling Robustness and resilience are a lived practice that develops well before an event occurs (Sendai Framework 2015):

- This includes the capability to fundamentally anticipate a disaster and its possible impacts and to be appropriately prepared for its occurrence, whether through inherently sufficiently robust or through flexible and adaptive systems. This includes energetic and financial costs to assess the risk and to design the systems appropriately.

- Adequate and obvious precautionary measures are necessary. In practice, this means appropriately securing particularly vulnerable elements and placing them outside particularly endangered zones. This includes additional energetic and financial costs.
- In the event of a disaster, people should be comprehensively and appropriately informed, those affected should be enabled to evacuate in time, and options for basic protection should be provided. This presupposes a sufficient degree of responsible, i.e., fact-based and forward-looking, governance.

Essential for achieving resilience are the community's options for action after the disaster. This presupposes properties that were developed beforehand:

- This includes the ability to quickly resume essential services and basic public welfare activities after a disaster has subsided. This encompasses the supply of drinking water, protection from the elements, food, the maintenance of order, and necessary communication. If the required technical functions are fundamentally simple, robust, and operable by laypeople, they can often be restored directly on-site with local resources. This is a fundamental technical requirement for all components of essential public welfare.
- There must be the possibility of providing rapid initial medical assistance to those affected on-site and of carrying out at least a minimal level of healthcare. For this purpose, basic means and basic knowledge for this care must be available on-site. The care includes psychological and spiritual support.
- There must also be the possibility for the affected community on-site to shape the reconstruction and necessary reorganization itself.
- A framework for mourning and remembrance must be created, as well as a framework for understanding the disaster.

Since the process of mourning, reorientation, and new beginnings requires time, traumatic events must not follow one another too quickly, as otherwise every capacity and possibility for resilience is exhausted. The community disintegrates under an overly intense onslaught of individual crises. What remains is lawlessness, violence, and failed states.

Resilience is a major demand. For this, risks must be known and the necessary governance must exist to prepare and design settlements appropriately. Large cities are not very resilient against larger, less frequently occurring events, as these are mostly financially written off and modern cities are planned according to other boundary conditions such as profit maximization and efficiency. Structuring cities in such a way that they do not become deadly traps for their inhabitants during extreme weather events, natural disasters, prolonged power outages, or other major disasters is extremely difficult. Making large cities resilient against possible future events will tie up resources, capital, and energy over long periods of time.

Rural settlements have, due to population pressure or negligent planning in both poor and wealthy regions alike, encroached into heavily threatened or unsafe locations where they are frequently exposed to disasters. Such regions pose a particular danger to the states in which they are located, as it is here that the available resources for governance are first exhausted. Such regions are therefore often the starting point for a local or large-scale collapse of state order.

Resilient infrastructure is characterized by the fact that the failure of individual elements does not cause the collapse of the system, but rather individual elements continue to function. Resilience almost always comes at the expense of efficiency, as it means that fully functional elements exist at all levels and that the failure of the connection to a central control unit does not represent a fundamental problem. Resilience requires more resources or a focus on what is necessary in robust form. Resilient infrastructure means local infrastructure, so that required technology can be procured quickly and without complications. This also means that global supply chains with only one or very distant suppliers are fundamentally not resilient. Local and small-scale infrastructure is generally somewhat less efficient.

8.3.3 Digitalization

What contribution does digitalization make in addressing the problem areas of energy, climate, and resource scarcity? What resources are necessary for this?

What It Is Not About

The development, use, maintenance, and end-of-life treatment of technical systems uses digital systems in many processes. Dealing with digital tools has become an indispensable part of engineering education. Engineers significantly develop digital tools and often use them as soon as they emerge.

Fundamental digital tools are computer-aided engineering (CAE) as well as the use of programmable logic controllers for machines and systems (PLC). CAE encompasses design software (CAD), but also uses many numerical methods for calculating the properties of components or systems, such as finite element calculations (FEM) or the simulation and prediction of the behavior of fluids and their interaction with components or systems (computational fluid dynamics, CFD). In professional life, the use of software for enterprise resource planning (ERP) is part of everyday work, as is the development and use of an impressive variety of other digital tools. These are therefore an integral part of technology, of technical solutions, and of the professional world of engineers. At the same time, our civilization and technology are fundamentally also possible and conceivable without digital elements (but not without steel, cf. Smil 2016).

The Promise
These established tools and their use are, however, not the real core of the promises of a future digitalization. This is attributed — currently, for example, as "artificial intelligence" — a significantly more far-reaching shaping power. The range of claims about what digitalization will enable is very large (Bridle 2020; Selke 2023).

Solutions
Industry 4.0 (whereby this hype has already passed), big data, and artificial intelligence (AI) are supposed to transport us into a new future in which we leave all earthly problems behind. The goal here is supposed to be the singularity, i.e., the moment at which our technology begins to develop further without us. Until then, digitalization is supposed to unlock our true potential through total networking and disclosure of all data, it is supposed to fix all the errors (*bugs*) present in us and our society, and it is supposed to enable the digital world-improvement entrepreneurs to achieve the monopoly position that allows them to advance their disruptive human-improvement program (Bridle 2020). The term *bugs* comes from programming language and refers to errors in program code that can be removed by simple testing and searching, i.e., by solving tasks. The great challenges of our future are thereby regarded, like all the other *bugs* of the earth system or the technosphere, as purely technically solvable: no social negotiation process is to be invoked, as only the digital applications (or today *apps*) — avoiding all state interference and avoiding non-digitally controlled social processes — could solve this for us. This credo is at the center of the digital agenda and is a purely technically conceived *solution* (Nachtwey/Seidl 2017).

> *"The internet is making the world better, not just by giving us better gadgets and more information, but by reshaping society, root and branch. We now have the technology to solve problems that have plagued humanity for centuries, making old institutions and old rules obsolete and replacing them with computation" (Slee 2016, cited in Nachtwey/Seidl 2017).*

We may now safely replace *internet* with *AI*. From this perspective, all problems become mere tasks that can be solved for us by *apps* and an exponentially more resource-intensive digital technology, if only we place sufficient faith in it. We document this faith through total trust in digital things and digitally mediated content in all areas of our lives.

Digital Technology As a System
One of the hallmarks of our modern technology is that its mode of operation remains hidden from us. Unlike with steam locomotives, we can no longer directly perceive how something works or where forces are acting. This condition is particularly pronounced with digital technology. Complex calculations take place in geometrically simply designed housings with smooth surfaces and glowing screens. But we remain in the dark (Bridle 2020).

Like every other major technology, the elements of digital technology also require a number of fundamental prerequisites. Since these are hidden and are overlooked from the perspective of solutionism, it seems necessary to address some important aspects here in the sense of a life cycle assessment (see Chapter 4). Fundamental elements of digital infrastructure include:

- Data centers, i.e., buildings in which data is stored and processed in servers. Such data centers today have rated capacities in the gigawatt range. In data centers, the supply of electricity must be particularly uniform; even disturbances in a single half-cycle can mean data loss. For this reason, data centers additionally have their own generators to ensure the electrical supply.
- The infrastructure of global data connections also counts among these. These are fiber optic lines for longer distances and further technologies for the connection to end customers, as well as intermediate stations that route the data in the right direction.
- Part of the digital infrastructure are the transmitter and receiver units for data that is transmitted by means of electromagnetic radiation.
- In addition, end devices designed as human-machine interfaces are included, or end devices of the so-called Internet of Things (IoT), in which all artifacts, i.e., all elements of human technology, are intended to exchange data with one another (the hype around IoT has already passed as well).
- Programs that perform computational operations and that must be programmed beforehand are part of this.
- Furthermore, the infrastructure for the provision of electricity is included — specifically everywhere where digital technology is to be used, since digital technology cannot function without electricity.
- The infrastructure for the manufacture of all physical devices of digitalization is included, from the extraction of resources to disposal.
- The infrastructure for monitoring all users is an essential part. The mirroring of all data at a second commercial and third intelligence-agency level is not a basic prerequisite for the functioning of digital networked technology, but it is an element of today's manifestation and is relevant for resource and energy demand (Bridle 2020).

All these elements must be manufactured and operated. Unlike in other industries, the service life of capital goods is exceptionally short. Noteworthy for us is the number of essential, i.e., non-replaceable chemical elements and compounds that are required and from which digital technology is composed. This number of essential elements exceeds that of any other technology and encompasses the complete periodic table. Table 8.4 presents a few of these elements and provides some of the required industrial processes for their production in each case. It is intended to illustrate the required interdependencies by way of example.

Figure 8.5 illustrates a central, yet not publicly discussed question. Some essential metals of digitalization will foreseeably no longer be available in sufficient quantities unless technology succeeds very quickly in recycling digital equipment completely. And even then it is unclear whether the reserves and resources of these critical elements are sufficient to enable the future growth of digital technology. It is therefore unclear whether digital technology can continue to accompany us at the costs we are accustomed to today.

Already Table 8.4 gives a first impression of the network of dependencies. Digital technology does not stand alone. It is extremely dependent on a complete, global, and very complex technological and economic system. This network of fundamental, non-digital technologies provides all the required chemical elements, compounds, components, processes, plants, buildings, and products in the required purity, form, and quantity. The basis for this is the global, at all times orderly and very reliable supply of high-quality electrical energy. Many essential processes (and not only data centers) are dependent on an absolutely uninterrupted and continuous supply of electricity.

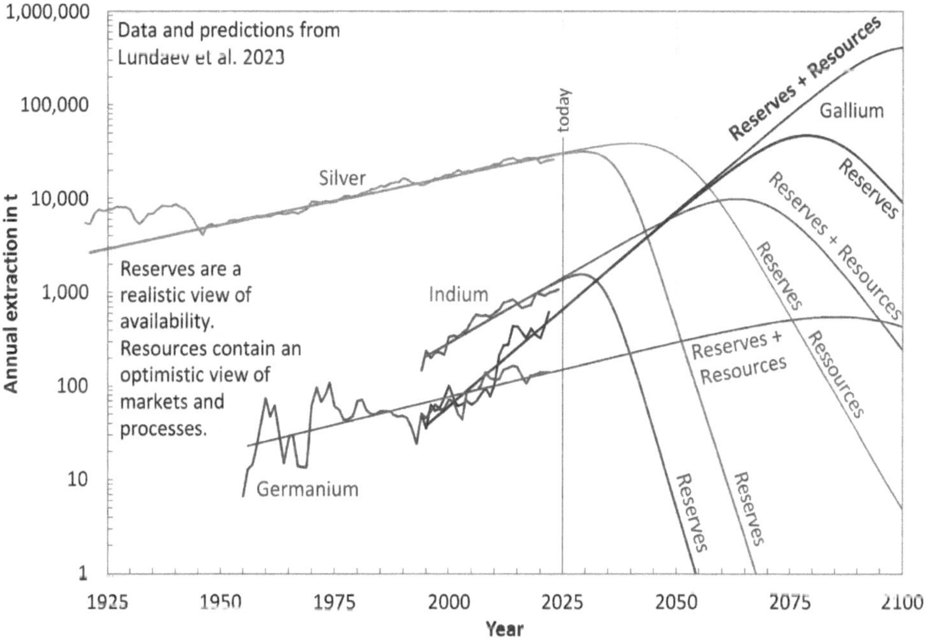

Figure 8.5 Critical metals of digitalization and their future availability (data and forecasts from Lundaev et al. 2023)

Table 8.4 Some Processes and Materials Without Which Today's Digital Technology Would Not Exist: Each Individual Element Is a Conditio Sine Qua Non. Only a Few Elements of the Process Chains Are Listed.

Material	Process	Application	Energy carrier
High-purity silicon	Electric arc furnace/Siemens process/Czochralski process	Base material for semiconductor circuits starting material for quartz glass	electricity chemical energy
Iron, steel	Blast furnace/steel-works/rolling mill ...	Housings, racks buildings, fences power pylons and power lines	coal natural gas electricity
Display glass, 'Gorilla Glass" This is sodium-free glass with an untouched surface.	Overflow downdraw process	Displays	Electricity
Dispersion-hardened platinum	Platinum metallurgy	Furnace for overflow downdraw	Electricity chemical energy
Ultra-pure fused silica	Fused silica furnaces	Wafer stepper optics/optical fibers	Electricity chemical energy
Ultra-pure fused silica, doped fused silica	Fiber drawing furnace	Optical fibers for data transmission	Electricity
Copper	Various hot processes	Electrical conductors for supply	Natural gas Electricity chemical energy
Gold	Precious metal metallurgy	Electrical contacts bond wires	Electricity chemical energy
Tantalum	Metallurgy for refractory metals	Small high-performance capacitors	Natural gas chemical energy
Plastics	Chemical process engineering	housings insulators	Natural gas, crude oil chemical energy
Fluorine, hydrofluoric acid	Dissolution of precious metals and refractory metals etching of fused silica	Processes for the production of fused silica, tantalum	Natural gas chemical energy

At the same time, this information shows that a complete energetic assessment of the digital technosphere must take into account at least all direct inputs for operation, but also all inputs for the manufacture and disposal of all required precursor materials, plants, buildings, and devices. This also includes, on a proportional basis, the required infrastructure for the provision of the necessary energy itself.

Studies that determine the current energy consumption of digital technology, can, if only due to the data situation and the very rapid technological development, only map parts of the overall system. Figure 8.6 shows an approximation (Castro et al. 2024). Here, a value of 0.00002 J per year and per byte is used as the specific electricity demand. This is a lower bound. The upper bound is set in the publication at 0.00005 J per year and per byte. It becomes clear that the electricity demand does not increase exponentially, but can be well described by a third-degree polynomial. How difficult it is to obtain reliable data here is discussed by (Chen 2025). This also illustrates the influence that significantly different system boundaries have on results.

At the same time Figure 8.6 illustrates that further growth in the energy demand of digital technology raises fundamental questions: it is unclear how the supply of electricity can be ensured in the future. The question arises of how the then very scarce electricity should be distributed between essential needs (in hospitals) and digital technology. It appears neither possible nor desirable to continue enabling the current growth in the energy and material demand of digital technology.

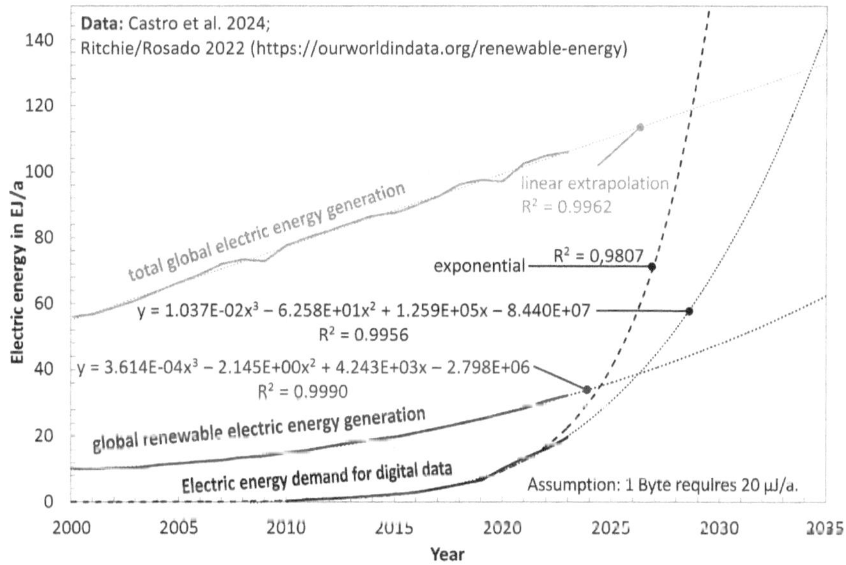

Figure 8.6 Electricity demand for digital technology and generated electricity, data and extrapolations (data sources: Castro et al. 2024; Ritchie/Rosado 2022)

Digitalization and Efficiency

Many of the devices of digitalization are more energy-efficient than their predecessor generations. This form of efficiency is easy to measure and directly recognizable as operating costs. The efficiency of devices is defined as energy consumption per computing operation or per bit transported. Nevertheless, digital technology today is not efficient: requirements focus on the speed and reliability of data transmission and response time. At the same time, a feature of modern software is its hasty programming (it is itself full of *bugs*) and its impressive demand for computing operations and storage space for simple tasks (Bridle 2020). This forces a high unnecessary, additional consumption of resources and energy across the entire digital technosphere.

The promise of digitalization in engineering is the increase of efficiency. There are digital tools that keep this promise. To what extent this applies to the currently hyped technologies, however, can only be determined through a systemic analysis in which the inputs and costs are fully assessed. In many cases, activities are not digitalized but relocated – often to clickworkers in the global south.

Digitalization and Consistency

Aspects of the reuse of components or materials are mostly not taken into account in the design of digital technology. Usage periods are short, since the ever-increasing demand for storage and computing capacity quickly makes devices obsolete. Digital technology therefore consumes resources.

Digital technology today is dissipative, i.e., it rapidly and thoroughly disperses the elements and connections required for it in the earth system. This fundamentally distinguishes it from a great many other sectors, each of which creates accessible material repositories in civilization. It is therefore unclear how long digital technology will remain materially producible in the future.

Digitalization and Resilience

Digital tools enable monitoring, predictions and assessments. They can warn communities quickly and accurately about approaching dangers. If data are available, digital tools can be used after a disaster in the distribution and allocation of resources for relief. In this sense, digital tools are essential elements for warning and preparation and thus support resilience.

Digital devices are an impressive source of misinformation and lies, conspiracy narratives and bullshit. Digital technology thus massively prevents the required community and the required engagement with the facts of the world. It destroys resilience – often quite deliberately as a business model.

Digital tools presuppose an uninterrupted, stable and sufficient supply of electricity as well as the integrity of digital communication channels, the necessary sensors and the elements of the digital system. Electronic infrastructure must be kept permanently dry and within a narrow temperature range. If elements fail due to a disaster, the electrical infrastructure must first be restored. Only then can the digital infrastructure be put back into operation. If learning-capable systems are affected, these may first need to be retrained. Only then are digital tools operational again. Digital tools thus tie up, during the critical period immediately after a disaster, the resources that are urgently needed for essential actions and services, without themselves contributing to direct supply.

As Thierney explains, it represents a major challenge for organizations, even in the technological domain, to engage appropriately with relevant risks or disasters (Thierney 2014). The balancing process between the required resilient infrastructure and the degree of digitalization presents itself as a wicked problem, since conflicting goals collide here.

Digital Tools

In the currently developing future energy supply, controlling consumption as a function of generation will be an important element. Digital tools that combine forecasting and control are needed for this:

Smart Grid One possibility for better matching the generation and consumption of electricity is hidden behind the concept of the Smart Grid. By networking all consumers and generators and the intensive communication between all integrated components that this makes possible, generation and consumption can become more plannable. More important is the second aspect, in which consumption is to be regulated and thus adapted to the generated electrical power – whether through price signals or otherwise depends on further parameters. This can, in the best case, significantly reduce the need for storage. Only this goal makes the additional inputs required for it necessary or tolerable from a societal perspective.

Modeling Good models, robust scenarios and detailed simulations are urgently needed for managing and solving wicked problems. Since every attempt and every action counts when managing wicked problems, evidence-based decision-making is necessary. One of the possibilities for preparing decisions is to simulate or compare things.

Mapping supply chains Shared databases across supply chains allow a large number of coordinations and alignments to be organized jointly. Compliance with regulatory requirements is just one of many possibilities: beyond that, key performance indicators can be aggregated across the life cycle. It becomes possible to select variants or to optimize a supply chain with regard to socially determined goals.

8.4 Evaluating Options

The transition of the energy supply from today's readily available fossil energy carriers to the use of long-term usable regenerative energy flows of the earth system places special demands on provision and use. These requirements were known and understood in all societies before the beginning of the first industrial revolution. They were a fixed component of the organization of society and life. Through the widespread and inexpensive use of fossil fuels and electricity, today's demand-oriented provision of energy has developed from a supply-oriented use of currently available energy. This current state is to be maintained – at least this is implicitly contained in most plans or scenarios for the future of energy supply. This gives rise to additional requirements for regenerative energy systems that must be taken into account. Whether these requirements can or should be realistically implemented is one of the core questions of this book.

Energy density Fossil energy carriers allow a very high concentration of power at one location. Few, large power plants are needed to provide a lot of electricity. Storage facilities and tanks can be small and thus allow long-haul aircraft and ocean-going ships for intercontinental supply chains.

The power density in the earth's energy flows is limited and mostly quite low (see Chapter 7). As a result, regenerative forms of energy must be harvested over large areas and with large installations. A centralized infrastructure hidden either in old industrial sites or in remote locations must become a decentralized technology distributed across the land and visible everywhere. This leads to considerable acceptance problems.

Fluctuation The energy flows of the earth system fluctuate strongly: The daily and annual cycle of the sun determines how much solar incoming radiation is maximally available. Clouds cast shadows. Precipitation and wind vary with the weather. Larger weather influences such as dry periods, storms, calms, inversion weather conditions, gusts, thunderstorms, etc. affect the energy that can be generated. Thus the power that can be harvested at a location varies. This leads overall to yields that vary locally and temporally for each of the important energy flows. These fluctuations can be smoothed out on a large scale via a sufficiently powerful transmission network. They place additional demands on generation, use and the required storage (see Chapter 13). The use of regenerative energy sources is associated with additional energetic costs for a society when a purely demand-oriented supply remains the goal.

Location of generation Many renewable energies cannot be harvested well everywhere. There are preferred locations where particularly high yields are possible. Solar incoming radiation varies significantly with distance from the equator and the local climate. In addition, the efficiency of photovoltaics and solar thermal energy depends on the ambient temperature. Wind blows primarily over the sea or over mountain ridges. Usable ocean currents or tides are only available at a few favorable locations.

Generation and consumption are therefore often spatially far apart. While fossil primary energy carriers could previously be transported relatively easily, electricity must be transported over sometimes very long distances to meet demand where it exists today. The alternative of large consumers relocating to regions with high supply security is conceivable, but involves technical effort and significant social changes.

Storage Fossil fuels can be stored exceptionally well. With appropriate integration of storage facilities into the transport and processing chain, fossil energy carriers can be stored with exceptionally low effort and losses such that seasonal fluctuations in demand are easily absorbed. This is the current state of affairs.

Electrical energy can only be stored with greater effort or with high losses. How this is done and what losses then occur is discussed in Chapter 10 and Chapter 13.

Demand vs. generation Demand-oriented generation is only limitedly possible with many renewable energy technologies, and intermediate storage is associated with considerable effort and relevant losses (see preceding discussion). The demand-oriented provision of electricity is therefore a challenge. One possible solution would be a paradigm shift toward supply-oriented use: electricity is used when it can be generated.

Instead, almost only approaches are discussed that aim to maintain the previous demand oriented model. To what extent this can succeed depends on societal acceptance of the associated additional infrastructure and costs, on the feasibility of technical solutions, the socially tolerable price for electricity, and on the resources required for the technical solutions.

Transformation The fundamental restructuring of an energy supply, such as we are now experiencing as the energy transition, takes many years. It cannot be started only when the old infrastructure has ended its function (Smil 2017). The most important options for action are:

- Expansion of the supply network to distribute available power over larger areas and thus improve overall security of supply
- Relocation of industrial consumers to the vicinity of secure generation
- Transformation of usage so that consumption can vary with generation – i.e., from demand-driven to generation-driven
- No energy transition and thus the end of our civilization with the exhaustion of fossil resources or through extreme climate change – depending on which strikes first (Kotz et al. 2024).

Assessment of Future Possibilities

An assessment of the suitability of individual technical solutions with regard to the transformation of our energy system must take into account a large number of their properties. This section provides an overview of how individual options or several different approaches can be systematically evaluated: the aim of such assessments is to obtain guidance on which method is the right one for a specific purpose, a region, or a concrete application.

A useful, simple, and fairly comprehensive list of requirements for possible future technologies for the provision of energy is reproduced in Table 8.5. This evaluation matrix is continuously evolving. It was inspired by a blog that addresses similar topics to this book (*https://dothemath.ucsd.edu*). The list uses a simple classification based on the traffic light scheme. If a technology is to play a major role in the future, it must perform well in all aspects. Often the lowest rating determines the overall assessment:

- **High/Central:** These are all technologies that can play a major role in the future.
- **Medium/Contribution:** These are technologies that can make a contribution or whose use is associated with certain societal restrictions.
- **Low/Niche:** These are technologies that could only be used with major restrictions or that are suitable for special applications in niches.

This allows a fairly good assessment of the state of the art and the discussion. Such a simple approach is sufficient for many discussions and applications. Table 8.5 is taken up again as a summary of the book in Chapter 15.

Table 8.5 Evaluation Criteria for Energy Technologies

Aspect	High or central + + +	Medium or contribution + +	Low or niche +
Potential: what power can be generated with the method in the long term (centuries)?	A large amount of power is available (sufficient to meet today's demand or more). more than 600 EJ/a	A relevant amount of power is available that makes a contribution to supply, but this technology alone is not sufficient. between 600 EJ/a and 60 EJ/a	No high power is available. The contributions are small or limited. less than 60 EJ/a
Heat: can the technology provide heat for heating?	Yes	Via electricity as electric heat or via heat pumps	no
Heat: can the technology provide heat for industrial processes?	Yes	Via electricity as electric heat/for low-temperature processes, if applicable also via heat pumps	no
Electricity: can the technology provide electricity?	Yes	Generally yes, but high input or low efficiency	No (or it makes no sense)
Transport: can the technology provide high-quality fuel?	yes These are liquid fuels with good calorific value. For ships, solid fuels are acceptable.	Yes, but via electricity	No (or it makes no sense)
Difficulties – current technical difficulties	Low technical challenges in application	High technical challenges or complex and failure prone technology	Very high or unresolved technical challenges

Aspect	High or central + + +	Medium or contribution + +	Low or niche +
Technology readiness level	At least TRL 7 The technology can be purchased and deployed today.	TRL 4 to TRL 6 The technology can with high probability be purchased in the foreseeable future.	Less than TRL 4 The technology may possibly become available at some point in the future.
Costs: what does the energy cost compared to today's prices (net excluding taxes and levies)?	Costs comparable to today's expectations Heat: less than 0.06 €/kWh Electricity: less than 0.12 €/kWh	Costs somewhat higher, but manageable Heat: 0.06 €/kWh to 12 €/kWh Electricity: 12 €/kWh to 0.25 €/kWh	Costs significantly higher and difficult to bear Heat: above 0.12 €/kWh Electricity: above 0.25 € kWh
Social costs: who bears the actual long-term costs of impacts?	Low externalization of costs These are included in the price or clearly covered by taxes and levies.	Externalized costs are socially known and accepted. They are clearly lower than the energy costs. within planetary boundaries	High or unknown long-term perpetual burdens neocolonial transfer of impacts exceeding planetary boundaries
Acceptance: is the technology socially accepted?	yes There is no or negligible opposition.	Relevant resistance A social negotiation process is necessary (not In my backyard, NIMBY).	no not enforceable in society (build absolutely nothing anywhere near anything, BANANA)
Risk: are the (long-term) risks socially acceptable?	Low and acceptable risks All risks are known.	Risks do not pose a threat to society, its continued existence, or the health of large parts of it.	Risks threaten society as a whole, its continued existence, or the health of larger parts of it.
Backyard: can the technology be used by anyone in their backyard?	Yes	No, but the technology can be implemented near population centers	no far away from people

Aspect	High or central + + +	Medium or contribution + +	Low or niche +
Complexity: can the technology be built, maintained, repaired, or modified on-site using common means?	Yes	Partially (spare parts are required)	no only one supplier globally vulnerable to disruptions in global goods transfer extremely delicate processes patents
Efficiency	> 50 %	10 % to 50 %	< 10 %
Fluctuations: does the technology always generate power or only sometimes?	Always, independent of weather or time of day capacity factor above 0.8	Dependent on external factors, but plannable capacity factor 0.8 to 0.2	Not plannable capacity factor less than 0.2
Energy Returned on Energy invested (EROI)	> 15	7 to 15	Less than 7
Storage: is the energy provided storable?	yes The cycle loss is below 10 %.	With effort The cycle loss is between 10% and 50%.	no The cycle loss exceeds 50%.
Resources: are the required resources available for the technology?	Yes	Basically yes, but they bind a very large amount of material with implications for availability	Not suitable for global implementation
Recycling: can the bound resources be easily reused?	yes recycling losses below 1%	With losses recycling losses between 1% and 10%	no recycling losses above 10%
Planetary boundaries: all planetary boundaries can be maintained.	yes with margin	barely The budget is essentially filled by this technology alone (30% to 100%).	no

Outlook – sharpening the imagination

Shaping the future first requires of us the ability to imagine its possibilities. Today we can observe a reckless lack of imagination in public discourse: the loudly proclaimed visions of the future are a mixture of unfounded technological optimism combined with the denial of our rapidly changing framework conditions of available energy, looming consequences of climate change, resource scarcity, and fading vitality – a continuation of today's conditions that denies any recognition. Scientifically grounded analyses of possible or already firmly booked and increasingly occurring horrors are dismissed by many actors as wrong, inappropriate, or unhelpful. This lack of imagination also pervades current literary fiction, where emptiness mostly prevails. Ghosh analyses this condition with great clarity for us (Ghosh 2017). Examples that can open our imagination in a clear way are impressively rare. Therefore, I provide some ideas here.

In science fiction, Kim Stanley Robinson with his very optimistic but entirely possible visions of the future, and Cixin Liu, are the well-known exceptions. The *Science in the Capital* trilogy by Kim Stanley Robinson was a realistic yet quite hopeful look at the political possibilities of the USA – possibilities that are currently being systematically destroyed by present-day actors, making these novels appear sentimental today. His *The Ministry for the Future* also looks optimistically into the realm of the possible. Cixin Liu thinks through human behavior and possible technological developments with great sharpness and clear consequence. His science fiction has no room for silly omnipotence fantasies. He lets dust, blood, love, death, hope, meaninglessness, and meaningfulness arise directly from our current conceptions of the future. Insanely funny is the idea of his digital apocalypse in *Curse 5.0*.

There are many stories about the end of the world. Few capture our most important causes. Some are at the same time wonderfully hopeful, like Kate Sawyer's *The Stranding* or Megan Hunter's *The End We Start From*. Ned Beauman, on the other hand, lets us in *Venomous Lumpsucker* participate in the absurdity of certificates as part of a technofix – here extinction certificates for species. In the background of Omar El Akkad's *American War* we can catch a glimpse of what a realistic future might look like, with a society that, despite its civil war in the foreground, has managed to contain many major problems for the time being. The four *Southern Reach* novels by Jeff VanderMeer can be read as attempts to grasp the enormity of our situation. Perhaps Old Jim in *Absolution* comes closest to our situation?

Literature for broadening horizons

Beauman, N.: Venomous Lumpsucker. Hodder And Stoughton 2023

El Akkad, O.: American War. Alfred A Knopf 2017

Ghosh, A.: The Great Derangement. Climate Change and the Unthinkable. The University of Chicaco Press 2017

Hunter, M.: The End We Start From. Ingram 2017.

Liu, C.: The Wandering Earth. Head of Zeus 2018

Robinson, K. S.: The Ministry fr the Future. Orbit 2021

Sawyer, K.: The Stranding. Hodder & Stoughton, London 2021

Vandermeer, J.: Southern Reach. Annihilation, Authority, Acceptance

Vandermeer, J.: Absolution. 4th Estate, London 2024

Literature

Archer, D.: The Long Thaw. How humans are changing the next 100,000 years of earth's climate. Princeton University Press 2009

Bendell, J.: Deep adaption. A map for navigating climate tragedy. IFLAS Occasional Paper. 2018. *http://www.iflas.info*

Bihouix, P.: The age of low tech. Towards a technologically sustainable civilization. Bristol University Press 2020

Bridle, J.: New Dark Age. Technologie und das Ende der Zukunft. C. H. Beck, Munchen 2020

Brown, A.: Just enough. Lessons from Japan for sustainable living, architecture, and design. Ingram 2022

Brysse, K./Oreskes, N./O'Reilly, J./Oppenheimer, M.: Climate change prediction. Erring on the side of least drama? In: Global Environmental Change 23, 2013, pp. 327–337

Carlowitz, H. C. von: Sylvicultura oeconomica Oder Haußwirthliche Nachricht und Naturmäßige Anweisung Zur Wilden Baum-Zucht. Braun, Leipzig 1713

Castro, V./Georgiou, M./Jackson, T./Hodgkinson, I. R./Jackson, L./Lockwood, S.: Digital data demand and renewable energy limits. Forecasting the impacts on global electricity supply and sustainability. In: Energy Policy 195, 2024, 114404

Chen, S.: AI's energy problem. In: Nature 639, 2025, pp. 23–25

Coady, D./Parry, I./Sears, L./Baoping, S.: How Large Are Global Energy Subsidies? IMF working papers. International Monetary Fund 2015. *https://www.imf.org/external/pubs/ft/wp/2015/wp15105.pdf*

DeLay, T.: Future of denial. The ideologies of climate change. Verso, London 2024

Diamond, J.: Kollaps. Warum Gesellschaften überleben oder untergehen. Fischer 2011

DIN ISO 26000: Leitfaden zur gesellschaftlichen Verantwortung von Organisationen

Europäische Kommission: Den Kreislauf schließen – Ein Aktionsplan der EU für die Kreislaufwirtschaft. COM (2015) 614 final. Europäische Kommission, Brüssel.*https://ec.europa.eu/transparency/regdoc/rep/1/2015/DE/1-2015-614-DE-F1-1.PDF*

Fix, B.: A tour of the Jevons paradox. How energy efficiency backfires. 2024. *https://economicsfromthetopdown.com*

Fizaine, F./Court, V.: Energy expenditure, economic growth, and the minimum EROI of society. In: Energy Policy 95, 2016, pp. 172–186

Grunwald, A.: Wider die Privatisierung der Nachhaltigkeit. Warum ökologisch korrekter Konsum die Umwelt nicht retten kann. In: Gaia 19, 2010, pp. 178–182

Hall, C. A. S./Klitgard, K.: Energy and the wealth of nations. Springer, Cham 2018

Hansen, J./Kharecha, P./Sato, M.: Assessing „Dangerous Climate Change." Required Reduction of Carbon Emissions to Protect Young People, Future Generations and Nature. In: PLOS ONE 8, 2013

Hansen, J. E./Kharecha, P./Sato, M./Tselioudis, G./Kelly, J./Bauer, S. E./Ruedy, R./Jeong, E./Jin, Q./Rignot, E./Velicogna, I./Schoeberl, M. E./Schuckmann, K. von/Amponsem, J./Cao, J./Keskinen, A./Li, J./ Pokela, A.: Global Warming Has Accelerated. Are the United Nations and the Public Well-Informed? In: Environment: Science and Policy for Sustainable Development 67, 2025, pp. 6–44

Hansen, J./Sato, M./Hearty, P./Ruedy, R./Kelley, M./Masson-Delmotte, V./Russell, G./Tselioudis, G./Cao, J./Rignot, E./Velicogna, I./Tormey, B./Donovan, B./Kandiano, E./Schuckmann, K. von/Kharecha, P./ Legrande, A. N./Bauer, M./Lo, K.-W.: Ice melt, sea level rise and superstorms. Evidence from paleoclimate data, climate modeling, and modern observations that 2 °C global warming could be dangerous. In: Atmos. Chem. Phys. 16, 2016, pp. 3761–3812

Hansen, J. E./Sato, M./Simons, L./Nazarenko, L. S./Sangha, I./Kharecha, P./Zachos, J. C./Schuckmann, K. von/Loeb, N. G./Osman, M. B./Jin, Q./Tselioudis, G./Jeong, E./Lacis, A./Ruedy, R./Russell, G./Cao, J./Li, J.: Global warming in the pipeline. In: Oxford Open Climate Change 3, 2023, kgad008

Heyen, D. A/Fischer, C./Barth, R./Brunn, C./Grießhammer, R./Keimeyer, F./Wolff, F.: Mehr als nur weniger. Suffizienz: Notwendigkeit und Optionen politischer Gestaltung. Öko Institut Working Paper 3/2013

Kikstra, J. S./Mastrucci, A./Min, J./Riahi, K./Rao, N. D.: Decent living gaps and energy needs around the world. In: Environmental Research Letters 14, 2021, 095006

Kotz, M./Levermann, A./Wenz, L.: The economic commitment of climate change. In: Nature 628, 2024, pp. 551–557

Levin, L./Cashore, L./Bernstein, S./Auld, G.: Overcoming the tragedy of super wicked problems. Constraining our future selves to ameliorate global climate change. In: Policy sciences, 45, 2012, pp. 123–152

Linow, S.: Creating Resilience, Minimizing Vulnerability of Communities. In: *Leal Filho, W./Azul, A./ Brandli, L./Lange Salvia, A./Wall, T. (Hrsg.):* Affordable and Clean Energy. Encyclopedia of the UN Sustainable Development Goals. Springer, Cham 2020

Lundaev, V./Solomen, A. A./Le, T./Lohrmann, A./Breyer, C.: Review of critical materials for the energy transition. An analysis of global resources and production databases and the state of material circularity. In: Minerals Engineering 203, 2023, 108282

Masson-Delmotte, V./Zhai, P/Pörtner, H.-O./Roberts, D./Skea, J./Shukla, P. R./Pirani, A./Moufouma-Okia, W./Péan, C./Pidcock, R./Connors, S./Matthews, J. B. R./Chen, Y./Zhou, X./Gomis, M. I. /Lonnoy, E./Maycock, T./Tignor, M./Waterfield, T. (Hrsg.): IPCC Special Report 2018. Global Warming of 1.5 °C. An IPCC Special Report on the impacts of global warming of 1.5°C above pre-industrial levels and related global greenhouse gas emission pathways, in the context of strengthening the global response to the threat of climate change, sustainable development, and efforts to eradicate poverty. Cambridge University Press, Cambridge 2017

McGlade, C./Ekins, P.: The geographical distribution of fossil fuels unused when limiting global warming to 2 °C. In: Nature 517, 2015, pp. 187–190

Nachtwey, O./Seidl, T.: Die Ethik der Solution und der Geist des digitalen Kapitalismus. Institut für Sozialforschung an der Johann Wolfgang Goethe-Universität 2017. *https://www.ifs.uni-frankfurt.de*

Neumann, M./Churchill, S. W.: A General Process Model of Sustainability. In: Ind. Eng. Chem. Res. 50, 2011, pp. 8901–8904

Renssen, S. van: Looking past the horizon of 2100. In: Nature Climate Change 9, 2019, pp. 349–351

Richardson, K./Steffen, W./Lucht, W./Bendtsen, J./Cornell, S. E./Donges, J. F./Drüke, M./Fetzer, I./Bala, G./Bloh, W. von/Feulner, G./Fiedler, S./Gerten, D./Gleeson, T./Hofmann, T./Huiskamp, W./Kummu, M./Mohan, C./Nogués-Bravo, D./Petri, S./Porkka, M./Rahmstorf, S./Schaphoff, S./Thonicke, K./Tobian, A./Virkki, V./Wang-Erlandsson, L./Weber, L./Rockström, J.: Earth beyond six of nine planetary boundaries. In: Science Advances 9, 2023

Ritchie, H./Rosado, P.: Fossil fuels. October 2022. *https://ourworldindata.org/fossil-fuels*

Schlaudt, O.: Zugemüllt Eine müllphilosophische Deutschlandreise. C. H. Beck, München 2024

Selke, S.: Technik als Trost. Verheißungen künstlicher Intelligenz. transcript Verlag, Bielefeld 2023

Sendai Framework for Disaster Risk Reduction. 2015. *https://www.unisdr.org/we/coordinate/sendai-framework*

Slee, T.: What's Yours is Mine. Against the Sharing Economy. OR Books, New York 2016

Smil, V.: Still the iron age. Iron and steel in the modern world. Butterworth-Heinemann 2016

Smil, V.: Energy and civilization. A history. The MIT Press 2017

SRU – Sachverständigenrat für Umweltfragen: Suffizienz als „Strategie des Genug." Eine Einladung zur Diskussion. 2024. *https://www.umweltrat.de*

Steffen, W./Rockström, J./Richardson, K./Lenton, T. M./Folkea, C./Liverman, D./Summerhayes, C. P./Barnosky, A. D./Cornell, S. E./Crucifix, M./Donges, J. F./Fetzer, I./Lade, S. J./Scheffer, M./Winkelmann, R./Schellnhuber, H. J.: Trajectories of the Earth System in the Anthropocene. In: PNAS 115, 2018. pp. 8252 –8259

Tainter, J.: The collapse of complex societies. Cambridge University Press 1990

Thierney, K.: The Social Roots of Risk. Producing Disasters, Promoting Resilience. Stanford University Press 2014

UN – United Nations: Transforming our World. The 2030 Agenda for Sustainable Development. 2015. *https://sdgs.un.org*

VDI 2221-1: Entwicklung technischer Produkte und Systeme – Modell der Produktentwicklung

VDI 4605: Nachhaltigkeitsbewertung

Vince, G.: Das nomadische Jahrhundert. Wie die Klima Migration unsere Welt verändern wird. Piper 2023

Wallace-Wells, D.: The Uninhabitable Earth. Life after warming. Random House 2019

9 Providing Heat

The technical use of heat is deeply connected to what constitutes human culture and civilization. The targeted transfer of heat is indispensable for the basic materials industry and for a great many technical processes in the life cycle of nearly all goods. For technical processes, a large number of heat sources and variants exist, depending on the required temperature, the necessary process atmosphere, and further requirements of the individual processes (Pfeifer et al. 2018).

This chapter first explains important terms, then looks back at fossil fuels and atomic processes, and finally discusses the potential of organic fuels as well as geological heat sources and heat pumps for a future energy supply. Regeneratively produced hydrogen as a substitute for fossil fuels is the subject of Chapter 10.

Application of heat in engineering

Basic materials industry

Heat is required by many basic engineering processes that form the basis of the industrial revolution and that also represent the foundation of our technical civilization today (Smil 2017). In particular in the production of basic materials the targeted use of heat is necessary, as some important examples illustrate:

- Iron is produced from iron ore in a blast furnace. In this process, high-quality coal serves as a reducing agent that removes the oxygen from the iron oxide and at the same time provides the heat required to produce liquid cast iron at over 1500 °C (Smil 2016).
- Steel is either obtained from cast iron from a blast furnace, in which case heat is supplied indirectly from outside via oxygen, or remelted from steel scrap in an electric arc furnace (Smil 2016).
- Aluminum is produced by first oxidizing purified bauxite in a rotary kiln with the addition of heat to form aluminum oxide (Al_2O_3). The aluminum oxide is then reduced in electric furnaces at approximately 950 °C. Carbon serves as the reducing agent in this process.

- Nearly all other metals require reduction processes and thermal processes for their processing from the extracted ore (Wiberg 2009).
- For Portland cement, limestone together with silicon dioxide and aluminum oxide is fired in a rotary kiln at 1450 °C. In this process, the limestone releases the carbon contained in it as carbon dioxide ($CaCO_3 \rightarrow CaO + CO_2$). These rotary kilns are today fired with fossil fuels.
- Silicon is produced in electric arc furnaces from quartz sand (SiO_2): the sand is reduced at very high temperature. The added charcoal provides the carbon that reacts as a reducing agent with the silicon dioxide ($SIO_2 + C \rightarrow Si + CO_2$). The heat is required for this chemical reaction and because the product is liquid silicon.
- Crude oil is fractionated in refineries, i.e. separated into various products according to boiling points. For this process, the crude oil is heated by means of steam to at least the condensation temperature of the heaviest fraction that is to be separated.
- The production of explosives, fertilizers and other nitrogen-containing compounds via the Haber-Bosch process is carried out today using natural gas, but can in principle also be performed with coal (see box on artificial fertilizers in Chapter 9.4 and Chapter 10).
- Plastics are produced from natural gas or volatile components of crude oil (naphtha), with the necessary processes requiring heat.

Propulsion

The industrial revolution begins with the first steam engines, i.e., the first machines that convert heat into work. These first steam engines use heat from burning coal (England): the heat is used in a boiler to vaporize water and to generate cyclic or rotary motion with the steam – as so-called shaft work. In the later developed internal combustion engines, gasoline, diesel or fuel gases are used to generate heat in a cylinder, which causes the engine to perform a cyclic motion through gas expansion. In gas turbines, kerosene or natural gas is burned to directly drive a shaft (Linow 2022). Each of these machine types required specific fuels, and over time fuel and engine co-evolved. Such combustion engines, together with directly electrically driven motors, are the essential foundation for the provision of propulsion.

Food and space heating

Heat is essential for human life. The use of fire is a fundamental step on the path to modern humans and our extinct close relatives. Fire enables cooking. Cooking breaks down food and makes it much easier to digest. As a result, cooking apes require less energy for digestion and can use this energy, e.g., for more intensive brain activity (Smil 2017).

Only heat for warming tents or other living spaces makes the settlement of less suitable climate zones possible. With the technology package of fire for heat, sewing for clothing, and means of transport, the settlement of the entire accessible world by our ancestors becomes understandable and feasible. The fireplace or the stove still represents the center of the home for many people today.

Cooking If the essential energetic activity of a household is the preparation of food and thus cooking, this is still mostly done today with biomass: these are households in a rural, essentially self-sufficient society. The directly available energy there is organic material, for the provision of which a considerable portion of time, energy, or income often has to be expended (Smil 2017). Only in technologized societies do other forms of providing heat or energy for cooking become possible:

- wood, dung, straw etc. require good stoves so that the health effects from emissions remain within limits. Such stoves, some with very high energy efficiency, were developed in many cultures (Brown 2012).
- Coal as a fuel requires stoves that can effectively discharge the flue gases. Good low-sulfur hard coal makes for a very pleasant fire.
- Town gas and coke oven gas offer greater convenience than the fuels mentioned above, since the flame ignites easily and can be used immediately. They offer high output combined with high flexibility. However, both present a high hazard from the fuel (carbon monoxide) as well as a risk of explosion in the event of misuse.
- Natural gas and liquefied petroleum gas are similarly quick and convenient to obtain as town gas and coke oven gas. At the same time, the direct health hazard is much lower than with town gas or coke oven gas.
- Electricity eliminates any effort for supplying air and removing flue gases. Here, immediately available heat is present without fire.
- Microwaves allow food to be warmed and the complete dispensing with cooking in the narrower sense.
- Delivery services relieve the individual entirely from food preparation and having their own stove. At the same time, energy demand increases.

Indoor climate Space heating and increasingly cooling often require a considerable share of the total energy demand of households. The possibility of using energy grows with the available income of the household. If more energy is available for tempering living space, this can be used:

- The actively temperature-controlled living area is growing larger. Larger portions of activities can be moved indoors and become independent of weather conditions.
- It is possible to forgo adapting one's clothing and actions to the local climate, the season, or the time of day.
- Activities that without air conditioning carry a health risk, such as hard physical labor in the midday heat or personal hygiene in severe frost, can be carried out at any time when temperature control is available.

Permanently heated living space serves as one way of displaying one's own energetic wealth – as so-called *conspicuous consumption*. Wealth can thus be demonstrated through an unusually energy-intensive temperature control of the living space, such as exceptionally high indoor temperatures in cold weather or very low indoor temperatures in hot weather. In this sense, heating and cooling against open windows can also be understood as a display of energetic wealth.

9.1 Use of Fuels

Today's technically relevant fuels are either organic or they are fossil fuels formed from organic substances through geological processes. The actual source of the energy is the radiation of the sun, which is captured and stored by plants. In the future, hydrogen from renewable sources may take over some of the roles of fossil fuels (see Chapter 10).

Fuels react with air. They combust. In the chemical reactions that take place, considerable amounts of thermal energy can be released. The heat released during complete combustion in air depends only on the converted mass of the fuel and a substance-specific property:

heating value The heating value H_I (inferior heating value or, in older usage, lower heating value H_U) indicates the total specific amount of heat released in a chemical reaction and thus the maximum usable heat, whereby the condensation of the water vapor produced during combustion is explicitly excluded from the usable energy.

calorific value The calorific value H_S (superior heating value or, in older usage, upper heating value H_O, sometimes also referred to as the gross calorific value) is the standard enthalpy of combustion ΔH of the reaction. The value includes the heat released during a conversion under standard conditions (1 bar and 25 °C) including the condensation of the water produced during combustion.

The heating value is always lower than the calorific value. If the fuel contains no hydrogen, both values are equal. The heating value is used for all common combustion processes in which the flue gases do not condense. This applies to gas turbines, internal combustion engines, gas boilers, industrial furnaces, and most combustion systems. The calorific value is only used for condensing boilers as well as for food and feed. Heating values and calorific values are normally stated with reference to the usual form of delivery:

- For gaseous fuels, they are preferably stated relative to the volume of the gas at standard conditions (index N).
- For liquid fuels, the data are preferably given for the liquid volume.
- For solid fuels, the data refer to the mass.

Data for many fuels can be found in Table 17.6 to Table 17.8 (see Section 17.5).

Thus the heat generated by a mass or a volume of fuel, without condensation, is

$$Q = m \cdot H_I = V_N \cdot H_{I,N} = V_N \cdot \rho_N \cdot H_I \tag{9.1}$$

With condensation of the water produced, the heat is

$$Q = m \cdot H_S = V_N \cdot H_{S,N} = V_N \cdot \rho_N \cdot H_S \tag{9.2}$$

Temperature of combustion

The usable temperature, at which heat from the chemical reaction of combustion can be transferred to the process, depends on the properties of the combustion and the fuel (Pfeifer 2018):

- The adiabatic combustion temperature is the maximum temperature that the reaction products of combustion with air can reach when the heat released by the reaction only increases the temperature of the flue gases. Theoretical values for many technical fuels are approximately 2200 °C (when fuel and air are supplied at standard conditions).
- In the combustion zone, heat is released primarily by radiation to the surroundings. Water vapor and carbon dioxide radiate mainly in the infrared spectral range, soot particles also in visible light (glow of a candle flame). Overall, no adiabatic condition exists and the actual temperature is significantly lower.
- If pure oxygen is used as the oxidizing agent (oxyfuel), the adiabatic combustion temperature for many fuels reaches approximately 3000 °C. Important areas of application are the cutting of steel with acetylene and oxygen and the production of quartz glass.
- If air and fuel are preheated, the combustion reaches a correspondingly higher temperature. Preheating of fuel and air is typically carried out using the hot flue gases. This is one of the classic and very effective methods for increasing energy efficiency in thermal processes.

The maximum temperature that can be achieved by combustion is also limited by the fact that at high temperatures, increasing amounts of latent heat are generated: the reaction products exist as radicals (e.g. H, O, OH, CH), whereby these radicals bind energy. The heat of reaction is only released when the flue gas cools down, through the combination of the unstable radicals into stable molecules such as H_2, O_2, H_2O and CO_2.

Particularly high temperatures are achieved when the material is directly exposed to the flame or the radiation of the flame. This is utilized in blast furnaces, glass melting furnaces, rotary kilns, etc. In many processes, the material for furnace walls, turbine blades, etc. defines the maximum temperature at which heat can be provided.

Oxygen demand and air demand

For hydrocarbons, i.e. substances composed only of atoms of carbon (C), hydrogen (H) and oxygen (O), the following applies to the oxygen demand, i.e. the number of oxygen molecules required per fuel molecule, for stoichiometric combustion

$$\mathrm{C}_m\mathrm{H}_n\mathrm{O}_o + \left(m + \frac{n}{4} - \frac{o}{2}\right) \cdot \mathrm{O}_2 = m \cdot \mathrm{CO}_2 + \frac{n}{2} \cdot \mathrm{H}_2\mathrm{O}$$

Stoichiometric combustion means that exactly sufficient oxygen is present for the complete chemical conversion of the fuel to flue gas, i.e. all molecules of the fuel can react consuming all the oxygen. The symbols m, n and o in this formula denote the number of atoms in the fuel molecule. The minimum oxygen demand for complete combustion is for a mixture of fuels correspondingly

$$O_{2,\min} = \sum_i \left(m_i + \frac{n_i}{4} - \frac{o_i}{2}\right) \cdot y_{C_mH_nO_o,i} - y_{O_2,B}$$

It is thus composed of the oxygen demands of the individual fuels according to their coefficients and concentrations, minus any oxygen already contained in the fuel. The mole fractions y_i of the components describe the composition here:

$$y_i = \frac{n_i}{\sum_i n_i} = \frac{n_i}{n}$$

With the composition of air (21 % volume fraction oxygen and 79 % inert gases: 78 % nitrogen and 1 % argon), the minimum air demand for stoichiometric combustion is

$$L_{\min} = \frac{1}{0.21} \cdot O_{2,\min}$$

The air-fuel ratio then describes the type of combustion reaction:

$$\lambda = \frac{L}{L_{\min}}$$

Here L is the actual air fraction of the combustion. An air-fuel ratio of $\lambda = 1$ corresponds to stoichiometric combustion. $\lambda < 1$ is rich combustion and unburned fuel remains. $\lambda > 1$ is lean and oxygen remains. The minimum oxygen requirement is a ratio and describes how many moles of oxygen are needed to convert one mole of fuel, or how many moles of air are needed to convert one mole of fuel.

These relationships are well applicable to simple fuels for which the chemical formula is known. However, especially for mixtures or biological fuels, an elemental analysis of the fuel is preferable: in that case the coefficients m, n and o can also become real numbers (more on this in Linow 2022).

Combustion efficiency

The combustion efficiency serves to describe the efficiency of a plant in which heat is released by combustion and then transferred to a product. The combustion efficiency indicates the fraction of heat that is no longer present in the flue gas but has left the system via the product or via wall losses of the furnace. The content presented here also refers to the formulas and relationships prescribed in regulations (Industrial Emissions Directive). The balance of heat flows in a stationary thermal apparatus is thus (Linow 2022)

$$\dot{Q}_{load,in} - \dot{Q}_{load,out} + \dot{Q}_{fuel} + \dot{Q}_{air} - \dot{Q}_{fluegas} - \dot{Q}_{loss} = 0$$

For the mass flows through the thermal apparatus the following applies

$$\dot{m}_{load,in} - \dot{m}_{load,out} + \dot{m}_{fuel} + \dot{m}_{air} - \dot{m}_{fluegas} = 0$$

The combustion efficiency indicates what fraction of the chemical energy of the fuel is not released via the flue gas. For the case where the supplied enthalpy of the mass flows of air and fuel can be neglected (not preheated), it is

$$\eta_F = \frac{\dot{Q}_{combustion} - \dot{Q}_{fluegas}}{\dot{Q}_{combustion}} = 1 - \frac{\dot{Q}_{fluegas}}{\dot{Q}_{combustion}} = \frac{\Delta\dot{Q}_{load} + \dot{Q}_{loss}}{\dot{Q}_{combustion}}$$

The so-called Siegert formula determines the flue gas loss in %, which is directly related to the combustion efficiency:

$$q_A = 100\,\% \cdot (1 - \eta_F) = (\vartheta_{fluegas} - \vartheta_{air}) \cdot \left(\frac{A_2}{21 - \mathrm{O_2}} + B\right) = (\vartheta_{fluegas} - \vartheta_{air}) \cdot \left(\frac{A_1}{\mathrm{CO_2}} + B\right)$$

In this equation O_2 is the oxygen content of the dry flue gas in % and CO_2 is the carbon dioxide content of the dry flue gas in %. This formula is easy to use, since there are inexpensive measuring instruments that reliably measure the oxygen or carbon dioxide content of the flue gas as well as its temperature. For the Siegert formula to be applied, the water content of the flue gas must therefore be taken into account accordingly. This is done by the common measuring instruments used by, e.g., chimney sweeps. In Table 9.1 coefficients for important fuels are summarized.

Table 9.1 Coefficients of the Siegert Formula for Common Fuels

Fuel	A_1	A_2	B
Coke oven gas	0.29	0.60	0.11
Town gas	0.35	0.63	0.011
Natural gas L, natural gas H	0.37	0.66	0.009
Propane, liquefied petroleum gas	0.42	0.63	0.008
Fuel oil	0.50	0.68	0.007

In order to better estimate the technical effort involved in cleaning flue gases from power plants, and thus to prepare for Chapter 14, the calculations in the two following examples represent the energetic benefit associated with the emission of 1000 m^3 of flue gas.

Flue gas and power plant output for hard coal combustion

In the VDI guidelines for the description and selection of flue gas cleaning systems, many specifications refer to 1000 m^3 flue gas. Here, the flue gas is meant at standard conditions, dry and for coal as fuel with 6% oxygen content. This follows from the applicable laws, here the European Industrial Emissions Directive (IED) 2010/75/EC and the Federal Immission Control Act (BImSchG) or, for power plants, the 13th BImSchV and the TA-Luft. The question is therefore: what energy quantity corresponds to 1000 m^3 of flue gas for high-quality hard coal (i.e., pure carbon) as fuel? This is a surprisingly complex task, which is therefore carried out in individual steps. We need the figure for the consideration of the efforts involved in flue gas cleaning in relation to the CED.

Step 1: The chemical reaction

We assume for simplicity that the coal consists only of carbon. The only chemical reaction of combustion taking place is then

$$C + O_2 = CO_2$$

Thus the combustion air contains exactly as many molecules as the flue gas, and the volumes of combustion air and flue gas are also equal at standard conditions.

Step 2: The reference oxygen

For the flue gas to contain the specified 6% reference oxygen, more air must be added than is required for stoichiometric combustion. This additional portion of the total supplied combustion air V_L does not react with the coal. The non-reacting fraction $V_{R,air}$ depends on the value of the reference oxygen fraction and can be calculated, since the following applies with the oxygen concentration in fresh air of 21%:

$$V_{R,air} = \frac{0.06}{0.21} \cdot V_R = \frac{0.06}{0.21} \cdot 1000 \text{ Nm}^3 = 285.7 \text{ Nm}^3$$

Thus $V_{air} = V_R - V_{R,air} = 1000 \text{ Nm}^3 - 285.7 \text{ Nm}^3 = 714.3 \text{ Nm}^3$ of air reacts with the coal.

Step 3: Amount of substance of carbon

Only the oxygen from the reacting air reacts with the coal, thus

$$V_{air,O_2} = V_{air} \cdot 0.21 = 714.5 \text{ Nm}^3 \cdot 0.21 = 150.0 \text{ Nm}^3$$

To determine the amount of coal from this, we use the amount of substance, since the reaction equation is an equation of amounts of substance. As shown in part 1 of the problem, each oxygen molecule will react with one atom of carbon. So if we have the amount of substance of O_2, we also know the amount of substance of C. Using the equation of state of the ideal gas in molar form and the universal gas constant $R = 8314 \text{ J kmol}^{-1} \text{ K}^{-1}$, we obtain

$$n_C = n_{O_2} = \frac{p_N \cdot V_{air,O_2}}{R \cdot T_N} = \frac{101\,325 \text{ Pa} \cdot 150.0 \text{ Nm}^3}{8314 \frac{\text{J}}{\text{kmol} \cdot \text{K}} \cdot 273.15 \text{ K}} = 6.693 \text{ kmol}$$

Amount of substance and mass can be converted into each other via the molar mass. Thus 1000 m_N^3 of flue gas corresponds to a mass of

$$m_C = n_C \cdot M_C = 6.693 \text{ kmol} \cdot 12.01 \frac{\text{kg}}{\text{kmol}} = 80.38 \text{ kg}$$

Step 4: Heat and work

We determine the heat released by the combustion using the lower heating value. This depends on the type of coal and for very pure coal such as anthracite is approximately $30 \text{ MJ kg}^{-1} < H_I < 32.5 \text{ MJ kg}^{-1}$. Thus the heat corresponding to the flue gas volume is approximately

$$Q = m_C \cdot H_{I,C} = 80.38 \text{ kg} \cdot 32.5 \frac{\text{MJ}}{\text{kg}} = 2612 \text{ MJ}$$

Typical plant efficiencies for coal-fired power plants are between 30% and 40%. This means that approximately 900 MJ of electrical energy is generated, or 1000 m^3 of flue gas corresponds to approximately 900 MJ of electricity provided.

Flue gas and power plant output with natural gas firing

In this task as well, the electricity corresponding to a flue gas volume of 1000 $m_N{}^3$ is determined. The solution approach is similar to the previous example, however the reference oxygen for power plants with natural gas firing, such as gas turbines, is set at 3%.

Step 1: The chemical reaction

We assume for simplicity that natural gas H is used as fuel and that it consists only of methane. The chemical reaction is then

$$CH_4 + 2 \cdot \left(O_2 + \frac{0.79}{0.21} \cdot N_2\right) = CO_2 + 2 \cdot H_2O + 7.5 \cdot N_2$$

where all reactions take place in the gas phase. Methane, air, and also the flue gas are ideal gases under standard conditions. Thus the combustion air contains a total of 10.5 molecules for one reacting molecule of methane.

Step 2: The reference oxygen

For the flue gas to contain the specified 3% reference oxygen, more air must be added than is required for stoichiometric combustion. This additional portion of the total supplied combustion air V_L does not react with the natural gas. The non-reacting fraction $V_{B,O2}$ depends on the level of the reference oxygen and can be calculated, since with an oxygen concentration in fresh air of 21% the following applies

$$V_{R,air} = \frac{0.06}{0.21} \cdot V_R = \frac{0.03}{0.21} \cdot 1000\ \mathrm{Nm}^3 = 142.9\ \mathrm{Nm}^3$$

Thus $V_{R,stoich} = V_R - V_{R,air} = 1000\ m^3 - 142.9\ m^3 = 857.1\ m^3$ of flue gas has been produced from the stoichiometric combustion of methane with air.

Step 3: The supplied volume of methane

As determined above, with stoichiometric mixing we obtain from 1 m^3 of methane 10.5 m^3 flue gas, each at standard conditions. Thus the volume of methane is

$$V_{CH_4} = \frac{V_{R,stoich}}{10.5} = \frac{851.7\ \mathrm{Nm}^3}{10.5} = 81.11\ \mathrm{Nm}^3$$

Step 4: Heat and work

The heat released by combustion is determined using the lower heating value. For natural gas H this is approximately 36 MJ m^{-3}. Thus the heat is

$$Q = V_{CH_4} \cdot H_{I,CH_4} = 81.11\ \mathrm{Nm}^3 \cdot 36\ \frac{\mathrm{MJ}}{\mathrm{kg}} = 2920\ \mathrm{MJ}$$

Typical for pure gas turbine power plants is an efficiency of 40%, while CCGT power plants achieve a plant efficiency of 60%. Thus, approximately 1769 MJ of electrical work is generated in a CCGT power plant, or 1000 m^3 of flue gas corresponds to approximately 1760 MJ of electricity.

9.2 Fossil Fuels

Covering fossil fuels serves as a reference point, since technology and our expectations of technology have developed alongside them. For this purpose, we examine important fields of application.

Fuels

The processes for converting chemical energy from fuel into heat, the subsequent conversion of heat into work, and the optimal or maximum conversion efficiency achievable in the conversion of heat into work are the subject of thermodynamics (Linow, 2022): The thermodynamic efficiency is then precisely the fraction of heat that can be converted into work, and this fraction is also limited for the Carnot efficiency by the achievable temperatures. Today's fossil fuels were developed together with their associated engines:

- Methane in the form of natural gas or biogas is mostly used in gas turbines. Methane can also be used in gas engines, diesel engines, or fuel cells. In principle, it is possible to fire boilers for steam turbines with natural gas.
- Gasoline is used exclusively in Otto engines. For other applications it is too dangerous, as it contains many highly volatile components.
- Diesel is used in diesel engines. Many other fuels such as heavy fuel oil, kerosene, or natural gas are also suitable for use in diesel engines.
- Kerosene is used primarily in aviation. It is comparable to gasoline or light diesel, but without the explosion hazard typical of gasoline. Kerosene can be used in Otto engines or in gas turbines.
- Coal, lignite, peat, and to a lesser extent crude oil, natural gas, or wood, as well as nuclear fuels and heat from solar thermal power plants are used to generate the superheated steam for steam turbines.

Why Liquid Fuels for Transport?

Fossil liquid fuels (gasoline, kerosene, diesel) set the expectations for the transport sector:

- They all have a high energy density of more than 25 MJ kg^{-1} or more than 25 MJ l^{-1}: This means that only fairly small tanks are required.
- The conversion of the energy carrier into kinetic energy takes place in the engine without residues: no combustion residues and no ash remain in the vehicle. Only exhaust gases are produced.

- Storage in the vehicle is relatively safe: the risk of fire or explosion is considered manageable. When used properly, they are not excessively toxic. No pressure vessels are required as tanks. They do not cause electrical fields or voltages that could pose a hazard.
- The transfer into the vehicle is technically simple. It takes place quickly and relatively safely.
- None of these fuels produces exhaust gas with a high proportion of toxic substances. Toxic substances in the exhaust gas are understood here as substances that are so toxic or released in such quantities that special effort is required for their treatment or control. The typical pollutants are either not sufficiently toxic (carbon dioxide) or do not occur in such quantities (carbon monoxide, nitrogen oxides, sulfur dioxide) that immediate fatal consequences are to be expected. The harmful health effects of the emissions act in the long term and are massive, but are currently socially accepted (Fuller et al. 2022; Huang et al. 2020).

These are currently the most cost-effective liquid fuels. They are cost-effective as long as the devastation of countries, ecosystems, and the Earth's climate associated with their use is not taken into account. Until now, the low overall efficiency (well to wheel) has therefore also been tolerable.

This creates a first hurdle for every new technology intended to replace these fossil fuels. Every new technology starts small and is therefore inevitably more expensive. It must enable a different benefit that the existing technology does not offer, if the transition is to occur through market forces alone. Fundamental difficulties can be identified in all previous attempts to replace fossil fuels in the transport sector.

Basic Materials Industry

Fossil fuels are used or preferred in almost all processes of the basic materials industry used or preferred:

- Many processes only became possible with the favorable availability of fossil primary energy carriers and have since developed over a long period of time, also with a view to optimal use of these fuels.
- In most cases, fossil energy carriers are the most cost-effective way to carry out the process. They are cost-effective again as long as the devastation of countries, ecosystems, and the Earth's climate associated with their use is not taken into account.
- Other energy carriers or raw materials are not available in sufficient quantity or quality and at comparable prices.

In many sectors, efforts are being made to further develop processes so that they can operate entirely without fossil energy carriers (more on this in Chapter 10).

Energy supply for Estonia

Estonia is in trouble with the European Commission, because Estonia has the highest per capita CO_2emission in the EU and is not meeting its climate targets. Help Estonia!

The only energy raw material of Estonia is oil shale, which is currently mined in open-cast mining: oil shale is either burned directly or processed into gas and oil products. 70 % of the extracted oil shale is used for the generation of electricity. 90 % of Estonia's electricity and 70 % of its CO_2emissions come from oil shale. Per year, approximately $11.0 \cdot 10^9$ kg of oil shale are currently (2018) mined and a total (all sources) of 8.8 TWh of electricity is generated. The power plants used for this have a combined rated capacity of 2380 MW. For historical reasons, they are located directly on the border with Russia.

Oil shale is a mixture of limestone and organic material. The organic material typically has a molar ratio of 1.5 carbon atoms to every 1 hydrogen atom. The heating value is 15 MJ kg^{-1}. On average, 40 % of the extracted mass is organic (i.e., C and H atoms). The ash has a density of 2000 kg m^{-3} on the spoil dump.

Estonia has a land area of 42 388 km^2. Of this, 50 % is forested. It is located at 58° north latitude. The usable growth of wood in northern European forests is on average 7.7 m^3 per hectare and year. Wood is often used as fuel in the form of pellets. The heating value of wood pellets is 17 MJ kg^{-1} with a density of the wood pellets of 650 kg m^{-3}.

Figure 9.1 Oil shale processing in Kohtla-Järve, Estonia (the ash dump on the left)

What are the masses of carbon and hydrogen burned in one year?

From the given data, the following can be written for the proportion of combustible atoms:

$$m_C + m_H = m_{shale} \cdot 40\ \% = 11.0 \cdot 10^9\ \text{kg} \cdot 0.40 = 4.4 \cdot 10^9\ \text{kg}$$

From this, an amount of substance can be determined – the amount of substance of particles consisting of 1.5 C and 1 H:

$$n_{1.5\,C+1\,H} = \frac{m_C + m_H}{1.5 \cdot M_C + 1 \cdot M_H} = \frac{4.4 \cdot 10^9\ \text{kg}}{1.5 \cdot 12.01\ \frac{\text{kg}}{\text{kmol}} + 1 \cdot 1.008\ \frac{\text{kg}}{\text{kmol}}} = 231.3 \cdot 10^6\ \text{kmol}$$

From this value, the amount of substance of the two atoms then follows: each of the particles consisting of 1.5 C and 1 H contains 1.5 C atoms and 1 H atom:

$$n_C = 1.5 \cdot n_{1.5\,C+1\,H} = 1.5 \cdot 231.3 \cdot 10^6 \text{ kmol} = 347.0 \cdot 10^6 \text{ kmol}$$

$$n_H = 1 \cdot n_{1.5\,C+1\,H} = 231.3 \cdot 10^6 \text{ kmol}$$

This also gives the masses of the two atoms as

$$m_C = n_C \cdot M_C = 347.0 \cdot 10^6 \text{ kmol} \cdot 12.01 \frac{\text{kg}}{\text{kmol}} = 4.170 \cdot 10^9 \text{ kg}$$

$$m_H = n_H \cdot M_H = 231.3 \cdot 10^6 \text{ kmol} \cdot 1.008 \frac{\text{kg}}{\text{kmol}} = 0.233 \cdot 10^9 \text{ kg}$$

If you find this approach hard to digest or even outlandish – the verification shows that it is correct (and I have not found any other way to determine the masses from the sparse data).

What is the annual CO $_2$ emission of Estonia?

This follows from the given data as

$$m_{CO_2} = \frac{m_{C,shale}}{70\,\%} \cdot \frac{M_{CO_2}}{M_C} = \frac{4.170 \cdot 10^9 \text{ kg}}{0.70} \cdot \frac{44.01 \frac{\text{kg}}{\text{kmol}}}{12.01 \frac{\text{kg}}{\text{kmol}}} = 21.8 \cdot 10^9 \text{ kg}$$

What is the specific CO $_2$ emission of electricity from oil shale?

The specific CO_2 emission is

$$\frac{m_{CO_2,shale}}{W_{el,shale}} = \frac{m_{CO_2,shale}}{90\,\% \cdot W_{el}} = \frac{21.8 \cdot 10^9 \text{ kg}}{0.90 \cdot 8.8 \cdot 10^9 \text{ kWh}} = 2.75 \frac{\text{kg}}{\text{kWh}}$$

Even German lignite has a lower specific CO_2emission.

What efficiency does the generation of electricity from oil shale in Estonia have?

The heat obtained from burning the oil shale is

$$Q_{zu,shale} = 70\,\% \cdot m_{shale} \cdot H_{I,shale} = 0.70 \cdot 11.0 \cdot 10^9 \text{ kg} \cdot 15 \cdot 10^6 \frac{\text{J}}{\text{kg}} = 115.5 \cdot 10^{15} \text{ J}$$

The electricity generated from oil shale amounts to

$$W_{el,shale} = 90\,\% \cdot W_{el,Est} = 0.90 \cdot 8.8 \cdot 10^{12} \text{ Wh} = 28.5 \cdot 10^{15} \text{ J}$$

This gives the efficiency of the power plants from Soviet times as

$$\eta_{shale} = \frac{W_{el,shale}}{Q_{in,shale}} = \frac{W_{el,shale}}{Q_{in,shale}} = \frac{28.5 \cdot 10^{15} \text{ J}}{115.5 \cdot 10^{15} \text{ J}} = 0.247$$

What is the utilization rate of the oil shale power plants?

The utilization rate follows from the available rated capacity as

$$\psi_{shale} = \frac{W_{el,shale}}{P_{N,shale} \cdot \Delta t} = \frac{28.5 \cdot 10^{15} \text{ J}}{2380 \text{ MW} \cdot 365 \cdot 24 \cdot 3600 \text{ s}} = 0.380$$

The power plants are therefore not really operating at full capacity.

Can Estonia replace the demand for oil shale with wood from its own forests?

From the forest area and the possible yield of the forests, a heat output of

$$Q_{wood} = \frac{\Delta m}{\Delta t}\Big|_{wood} \cdot H_{I,wood} \cdot 50\ \% \cdot A_{Est} = \frac{\Delta V}{\Delta t}\Big|_{wood} \cdot \rho_{wood} \cdot H_{I,wood} \cdot 50\ \% \cdot A_{Est}$$

$$Q_{wood} = 7.7\ \frac{\text{m}^3}{\text{ha}} \cdot 650\ \frac{\text{kg}}{\text{m}^3} \cdot 17\ \frac{\text{MJ}}{\text{kg}} \cdot 0.5 \cdot 42\ 390\ \text{km}^2 = 180.3 \cdot 10^{15}\ \text{J}$$

This is approximately 1.5 times the heat currently generated with oil shale. Under the reasonable assumption that the power plants can achieve significantly better efficiencies, demand can be reduced by a further approximately 30%. In principle it is therefore possible, but it is unclear whether sufficient wood is actually available given current uses:

- Wood is needed today in rural areas for heating.
- Timber and export as wood pellets are important economic uses.
- Nature conservation restricts the use of all forests. In addition, many forests stand on poor soils, so it is unclear whether the yield is really achievable in the long term.

This example illustrates how a picture can emerge from a few quickly gathered figures. In the case of Estonia, it becomes clear that the transition of the energy supply will be difficult, since the alternatives each have recognizable disadvantages. However, continued supply with oil shale is also difficult. The material lies only a few meters below the earth's surface, but the usable layer is not very thick and mining leaves behind a devastated landscape that cannot be used further.

9.3 Photosynthesis and Plant Growth

Photosynthesis is a complex chemical reaction in which the chlorophyll in plant leaves, together with some other organic molecules, undergoes a cyclic process: energy from incoming radiation is absorbed by these molecules. In doing so, they change and mediate a chemical reaction. In this mediated chemical reaction, carbon dioxide from the air is broken down to build glucose. Chlorophyll and the other molecules involved act as catalysts. Only photons with the correct energy can mediate the necessary changes in the catalysts. The chemical process that takes place is complex and involves many partial reactions. A whole series of photons of suitable energy must be absorbed for the cycle. The result of photosynthesis is a molecule with higher bound energy than that of the starting materials. Heat is bound as chemical energy, which can then be released and made usable in the plant's metabolism, as food for animals, or through combustion.

Photosynthesis is the process in the earth system in which the incoming heat of the sun is converted into chemical energy usable by almost all living organisms. First, plants use the process and the nutrients thus produced for their own growth. The plant material formed in this way constitutes the basis of life for all animals and fungi that cannot themselves convert energy from solar radiation. A portion of the dead plant material is not immediately consumed but is deposited. Over geological time periods, these deposits have formed the reserves of fossil energy carriers. Thus, the photosynthesis of long-extinct plants is the current foundation of our civilization.

Photosynthesis as a process is not very efficient in converting solar radiation into usable energy. This is partly due to the chemical processes involved and partly due to the fact that the most intense portion of solar radiation, green light, is not used in photosynthesis: this light is scattered by leaves. This is why they appear green.

In order for plants to use sunlight as effectively as possible, many other prerequisites are necessary for the plants. Table 9.2 lists essential factors. In principle, even one of these factors can impair the harvest. All conditions must be favorable for plants to yield high output. This relationship is Liebig's law or the law of the minimum (Mitscherlich 1909).

Light, water and all essential nutrients for plants cannot be replaced by other substances, technology or capital. This is where weak sustainability fails. Each individual one of these substances cannot be replaced and must simultaneously be present in the soil in a chemical compound accessible to the plant in sufficient quantity (i.e., not too little and not too much). The soil must have sufficient layer thickness, structure and biodiversity, which is often only ensured through geological, chemical and biological processes in a complex interplay.

Table 9.2 Essential Factors for the Growth of Plants

Factor	Effects
Intensity of light	The plant or its cell organelles in the leaves, the chloroplasts, in which photosynthesis takes place, require sufficient intensity for the chemical reaction to proceed. Different types of plants can cope well with different intensities, but high yields require direct solar irradiation. Almost all plants can cope with maximum solar incoming radiation.
Duration of light	The plant or its leaves must maintain their metabolism day and night, whereby the intensity of the metabolism is influenced by the ambient temperature or the temperature of the leaf. Over the course of the day, the energetic benefit from photosynthesis must at least offset the energetic inputs of the leaf for its metabolism. This does not apply to every single day, but is necessary on average. The leaf and the plant require a positive EROI.
	For this reason, many plants cease their activity in winter, since photosynthesis is then not worthwhile on a net basis far to the north or south: they shed their leaves. Long nights and in particular long nights at high temperatures significantly reduce yield or can prevent plant growth entirely, since metabolism then proceeds intensively and all nutrients produced during the day are consumed.

Factor	Effects
Water	Plants require a steady and sufficient supply of water. Drought phases often have long-lasting effects, since the plant must first regenerate, as roots, leaves or branches abandoned during the drought must first be newly formed. Too much water can lead to the death of plants, as the roots or other flooded parts of the plant suffocate or are mechanically damaged.
Nitrogen	An adequate supply of nitrogen bound in a form suitable for plants is essential for their growth. Nitrogen can be present in the soil in a form accessible to the plant. Some plants can virtually fix nitrogen from the air themselves, as they live in symbiosis with nitrogen-fixing fungi. Nitrogen can be added via fertilizers.
Phosphorus, potassium and magnesium	An adequate supply of the elements phosphorus, potassium and magnesium is essential for plant growth. These elements can be present in the soil in a suitable form. They are often added via fertilizers. An excessively high concentration in the soil leads to salinization, i.e., the plant can no longer absorb water from the soil and dies of thirst.
Further essential nutrients	The essential nutrients for plants include sodium, potassium, magnesium, calcium, aluminum, carbon, silicon, nitrogen, phosphorus, oxygen, sulfur, selenium, fluorine, chlorine, bromine, iodine, iron and molybdenum. All of these elements must be present in the soil in sufficient quantity and in a chemically usable form for plants so that the plant can absorb them for its growth. An excessively high concentration often has a toxic effect.
Salt	Many plants have a tolerance range for salts in the soil that they can tolerate without restrictions. Many essential nutrients are present in the soil as soluble salts, but they must not be present in too high a concentration, as the plant would then no longer be able to extract water from the soil. Different species tolerate different levels of salt concentration.
Carbon dioxide	Carbon dioxide is one of the raw materials of photosynthesis, which converts available carbon dioxide into sugar using light. Leaves exchange gases with their surroundings through small openings that can be opened more or less widely by the plant depending on ambient conditions. These openings allow gases from the atmosphere into the leaf and simultaneously allow water vapor out of the leaf. Plants are forced to find a balance between the supply of CO_2 and the loss of water. At very low CO_2 content in the atmosphere, such as 220 ppm at the maximum of the last ice age, this strongly limits the locations where plants can grow, as drought stress sets in very quickly.
	An oversupply of carbon dioxide also has effects on plants. Although sufficient CO_2 is then available for photosynthesis even in dry locations, since the leaf openings for gas exchange hardly need to be opened. However, the content of valuable fats and proteins in the plant material decreases significantly with increasing CO_2content of the atmosphere.

Factor	Effects
Oxygen	Plants require an adequate supply of oxygen to their roots. For this, the soil must be sufficiently loose and not too wet. Different plants have very different requirements in this regard.
Predators	Protection against animals that want to eat the plant and especially its tender leaves is necessary. Otherwise the plant must laboriously regenerate its leaves. At the end of the last ice age, the megafauna disappeared, i.e., all the large and hungry herbivores such as mammoths, rhinoceroses, elephants, horses, aurochs or bison. These animals shaped the landscape through their feeding behavior. Only through their extinction did agriculture become possible. Today the relevant predators are domestic animals (such as goats and cows) or insects (such as locusts). Species introduced from other regions through international supply chains are of particular relevance.
Diseases	Protection against fungi or bacteria is necessary. Fungi and bacteria can attack and destroy living plants. Particular relevance is gained by species introduced through international supply chains from other regions.
Weather	Suitable weather for plants provides sufficient sunlight and sufficient water. Temperatures are neither too low nor too high. Frost destroys leaves and organs that are not designed for it. Excessively high temperatures lead to high metabolism without benefit. Conversely, drought, flooding, and extreme weather events such as storms and thunderstorms represent threatening risks. In Central Europe with its severe winter storms, it makes sense for trees to reduce their wind resistance during the storm season by shedding their leaves. This minimizes storm damage to their crowns.
Tempera-ture	All major grain varieties are today used in many regions at the upper end of their possible temperature range: many species show a significant reduction in yield at higher ambient temperatures, as self-consumption through their own metabolism increases with temperature. Many plants have quite narrow temperature windows for important developmental steps such as germination or flowering. These steps either proceed optimally only within this range or can only take place within this range. A great many plants cannot cope with low temperatures or frost at all: they are not frost-hardy and die when exposed to frost. Other plants are fundamentally frost-hardy, but their leaves, flowers, or fruits can die from frost. For tropical plants, the minimum tolerable temperature can be significantly above the frost threshold. These plants depend on the protection of a rainforest in order to survive.

The use of plants for human purposes leads to a trade-off process on our intensively used planet:

1. Areas used for the production of food are not simultaneously available for the cultivation of energy crops.
2. Areas occupied and mostly sealed for industrial facilities, cities, or infrastructure are no longer available for the cultivation of plants.
3. The resource land can only be occupied once.
4. In addition, the resource soil is limited. As a result, rapid changes in use between sealing and agriculture can hardly be carried out.
5. Areas used by us are not available for natural ecosystems and wilderness.

The efficiency of photosynthesis along its energy chain and for the optimal case is represented Table 9.3. Often the efficiency is significantly lower, as individual conditions in reality deviate considerably from the optimum. In the temperate latitudes, photosynthesis does not take place in many plants for many months. Therefore, of the 178,000 TW of available solar radiation power, ultimately only about 40–140 TW are converted and made usable through photosynthesis (Kleidon 2016).

Table 9.3 The Energy Conversion Chain of Photosynthesis in the Optimal Case (Source: Kleidon 2016)

Share	Process
100 %	Of 100 % of sunlight falling on a leaf, ...
53 %	... 53 % of this radiation is present as light usable by the plant, i.e., in the spectral range between 400 nm and 700 nm.
37 %	Of this light incident on the leaf surface, a maximum of 70 % reaches the chloroplasts, i.e., the cell organelles in which photosynthesis takes place.
9 %	Of the chemical energy thus absorbed, a maximum of 32 % is used to form glucose as sugar. The remainder is waste heat, i.e., dissipation in the chemical processes.
5 %	Of the glucose formed in the leaf, the leaf itself requires 40 % under good site conditions. The other 60 % of the chemical energy is available to the plant. With these maximum 5 %, roots and flowers, fruits and stalks or branches are formed. Energy is also required to ward off predators or combat diseases.

Soil as a limited resource is currently being massively eroded by forms of agriculture that are not gentle on the soil. Continents are losing substrate and soil at a very high rate — much faster than would occur through geological processes. As a consequence, agriculture can only be practiced for a limited time: usable soils are being systematically reduced in area and depth (Wilkinson/McElroy 2006). In some fertile regions, the underlying bedrock will be reached in the foreseeable future.

9.4 Organic Fuels

Organic fuels were the only fuels in most regions until well into the 19th century. Coal was systematically developed as a fuel in England and Scotland after the country had been deforested to its most remote areas (Smil 2017). Organic fuels were also replaced by fossil fuels in other regions because they were no longer available in sufficient quantities. Although methods for the long-term management of forests as a basis for the supply of fuel and building materials were developed — the beginning of sustainability as a concept and as an economic practice (von Carlowitz 1713) — this was not sufficient from the early 19th century onwards to meet industrial demand.

A second reason for the replacement of biological by fossil fuels lies in the partly superior technical properties of fossil fuels:

- Wood, wood pellets and charcoal are fundamentally comparable to hard coal, however high-quality hard coals have a higher mechanical load-bearing capacity, thus allowing larger blast furnaces for steel production than would be possible with charcoal.
- Electrodes for the reduction of metals or silicon can in principle also be produced from organic precursor material, however today's electrodes are manufactured from synthetic precursor material, since only in this way can the required large dimensions and homogeneity be achieved.
- Anaerobically obtained gas from biomass can be a substitute for natural gas through drying and purification, but often does not reach the calorific value.
- From organic material, liquid fuels can be obtained in a wide variety of ways that have similar properties to gasoline, diesel or kerosene from fossil crude oil.

Calorific value Usually the calorific value of organic fuels is somewhat lower, but still comparable to the calorific value of fossil fuels (see Table 17.8 in Section 17.5).

The Potential of Organic Energy Carriers

In political discussions, organic fuels are assigned an important role for the future. The classic organic fuels are wood, straw, dung and charcoal. In the future, additional fuels produced from biomaterial are to be used:

- Plant oils as precursors for kerosene, gasoline or diesel
- Ethanol or methanol from plant material as a substitute for gasoline
- Wood pellets, i.e. granules mechanically formed from wood
- Biogas e.g. from anaerobic conversion of plant residues, dung or biowaste
- Further substances such as dimethylether (DME), which are also synthetically produced from organic precursors, as a substitute for diesel
- Biomaterial is also intended to be used as a raw material for the production of plastics, pharmaceuticals and many other products that are currently manufactured from fossil materials.

In order to assess whether these concepts are viable, it must be clarified whether sufficient starting material is available for large-scale use. Since fossil fuels were developed at a time when the demand for fuels was considerably lower than today, and since sufficient organic material was not available at that time, it is clear that organic fuels are nowhere near sufficient in supply.

EROI of Organic Energy Carriers

For use as an energy carrier, it is crucial whether the EROI is sufficiently high. For this purpose, a complete life cycle assessment is necessary. In the calculation for mechanized agriculture, the inputs for soil cultivation, for maintenance and harvesting, for machinery, for transport, for technical processes of processing as well as for extraction of fertilizers and their production are included. Organic materials are widely dispersed across agricultural land and must be collected with an expenditure of energy, while fossil energy carriers are mostly present in highly concentrated form. The EROI is therefore often extremely unfavorable and does not justify many uses.

Table 17.9 in Section 17.5 provides an overview of achievable yields of typical energy crops and the nitrogen requirements to be met for them. In Table 17.9, the proportion N is the ratio of the energy bound in nitrogen (CED) to the energy yield. These figures are relevant both with regard to the availability of energy from biomass and with regard to their energetic assessment.

Data on the EROI of fuels from biomass can be found in the literature. As a reference value and for an overview of the inputs, the data on food from Table 9.4 are used here. The values for the EROI become significantly more favorable for cultivation with draft animals when the work of the draft animals is excluded. This is permissible when the draft animals provide for themselves, i.e. graze primarily on uncultivated land and no dedicated resources need to be deployed for them.

Table 9.4 Typical EROI Values for Food (Source: Pimentel 2009)

Product	Method	Yield in kg ha^{-1}	EROI
Wheat	Manual cultivation in Kenya	1780	3.3
Wheat	Industrial cultivation in the USA	2900	2.6
Corn	Manual cultivation in India or Indonesia with draft oxen	1200	1.1
Corn	Industrial cultivation in the USA	9400	4.1
Rice	Manual cultivation in the Himalayas with draft oxen	1830	0.8
Rice	Industrial cultivation in the USA	7616	1.4
Soybeans	Industrial cultivation in the USA	2600	3.7

For the production of liquid fuels from biomass, there is a reliable database in which a large number of technical variants have been compiled comparatively. Table 17.15 in Section 17.9 summarizes the key data. In the conversion of biomass, further use of the waste as fertilizer in agriculture is always already included. Emissions from changes in land use are not included, however. These are particularly critical in the case of palm oil. The rather high greenhouse gas emissions from fuels derived from biomass are attributable to the emissions of N_2O associated with agriculture as well as to cultivation on former peatland areas with their massive CO_2 emissions (JRC 2014).

Availability of Organic Energy Carriers

To assess what contribution fuels from biological material can make to the energy supply, the following aspects of availability are relevant:

- There is always a competing use with the production of food on the land or the use of organic material as a precursor for other products (furniture, household items, houses).
- Furthermore, there is often a competing use with other objectives such as the protection of biosystems or biodiversity.
- The total available area for cultivation as well as available soil and required soil quality are each limiting factors.

Even if the EROI would allow intensive use, there is simply not enough available (Moriarty 2022; Smith et al. 2021).

Synthetic fertilizer

The condition for high yields in agriculture is that all necessary prerequisites for the plants' rapid and undisturbed growth are available at all times (Chapter 7). In industrial agriculture, this translates into water, pesticides, and the provision of the required nutrients via fertilizers:

- Potassium is often mined, processed and sold in various forms. Potassium is found in special salt deposits and can be suitably extracted there. The cumulative energy input for the provision of potassium is therefore quite low. Values of around 3 MJ kg^{-1}.
- Phosphorus is only available in limited economically relevant deposits. Reserves exist in some countries of the world such as China, the USA and Russia, with larger resources essentially only in Morocco. Phosphorus therefore belongs to the critical resources. The energy input appears to be comparable to that of potassium.
- Nitrogen fertilizer is produced using the Haber-Bosch process (see also Chapter 10): fossil fuels are used to convert nitrogen from the air into suitable fertilizers. The specific energy input when using natural gas as synthesis gas is typically 27 to 30 MJ kg^{-1} NH_3. If coal is used to produce the required hydrogen, the value is approximately 50 MJ kg^{-1}. Currently, the average energy input for ammonia synthesis is 35 MJ kg^{-1} or, based on pure nitrogen, approximately 43 MJ kg^{-1} (Smil

2017). Since ammonia as a gas cannot be used for fertilization, it is subsequently converted after ammonia synthesis into nitric acid HNO_3 or most commonly urea CH_4N_2O. In the case of urea, more nitrogen is bound relative to the mass of the fertilizer. The specific energy for urea is typically 55 MJ kg^{-1} based on nitrogen.

Fertilizers are applied in modern agriculture using energy and other resources. Energy-relevant aspects include:

- Extraction of raw materials, processing or production as well as transport to the user
- Monitoring of agricultural areas, sampling and analysis where necessary
- Application of fertilizers using machinery: the diesel fuel required is particularly significant here.
- Additional measures that become necessary due to unintended effects of fertilizers: fertilizers that leave the agricultural area have effects either in bodies of water (eutrophication) or in the atmosphere (N_2O as a greenhouse gas,).

Figure 9.2 illustrates how affordable energy promotes the production of food or energy crops. At the same time, relevant effects of poor practice and the resulting environmental impacts are shown (Andrews-Speed et al. 2015). Conversely, Figure 9.2 thus illustrates what effects increasingly scarce fossil energy carriers or significantly increased requirements for reducing the environmental impacts of agriculture will have on the supply of food or energy crops.

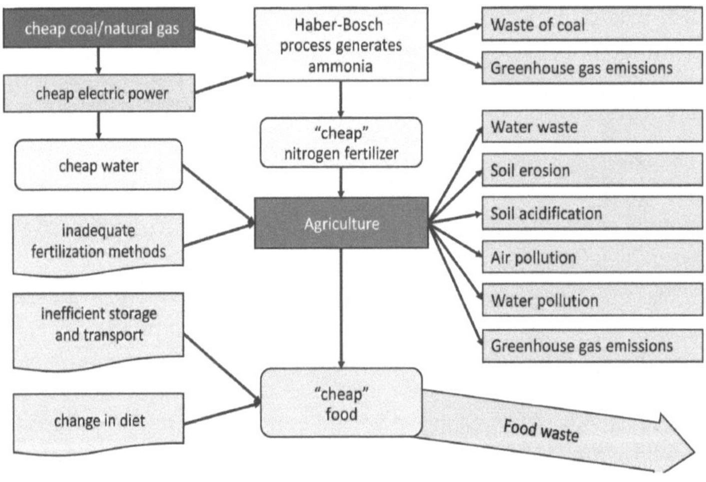

Figure 9.2 Relationship between affordable energy, affordable fertilizers, affordable food or energy crops and relevant environmental impacts (based on Andrews-Speed et al. 2015)

Supplying the Studstrup power plant in Aarhus

This example illustrates how a power plant that was originally fired by fossil fuels was converted to a renewable energy source from biomass. The example considers the old coal-fired power plant Studstrup near Aarhus in Denmark, which was converted to wood pellets by its operator Dong Energy in 2016:

- The rated electrical power is 360 MW and additionally up to 515 MW of waste heat can be supplied as district heating, since Aarhus is heated almost entirely by district heating. The utilization factor of the power plant is approximately ψ = 40 % (operation in winter only).
- Approximately 800,000 t of wood pellets are consumed per year. Wood pellets have a heating value of approximately 17 MJ kg^{-1} and their density is 650 kg m^{-3}.
- The usable growth of wood in northern Europe (birch and spruce) is 7.7 m^3 per ha and year.
- The power plant is supplied with pellets from Estonia and Latvia (Figure 9.3). The sea route from Riga to Aarhus is approximately 1150 km. The average distance from the loading station to the port in the Baltic states is 200 km (railway, standard gauge) and the average distance from the forest to the loading station is 25 km (narrow-gauge rail). **NOTE:** In practice, wood pellets in the Baltic states are today primarily transported by truck. The assumption thus already anticipates elements of a transport transition.

Figure 9.3 Pellets ready for loading in the Baltic states

What is the energy content of the annual quantity of wood pellets used, the annual electrical energy output, and the efficiency of electrical energy generation?

The chemical energy contained in the pellets is

$$Q_{pellets,a} = H_{I,pellets} \cdot m_{pellets,a} = 17\ \frac{\text{MJ}}{\text{kg}} \cdot 800 \cdot 10^6\ \text{kg} = 13.6 \cdot 10^{15}\ \text{J}$$

In one year the power plant generates an amount of electricity of

$$W_{el,a} = P_{N,el} \cdot \psi \cdot \Delta t = 360\ \text{MW} \cdot 0.40 \cdot (365 \cdot 24 \cdot 3600\ \text{s}) = 4.541 \cdot 10^{15}\ \text{J}$$

The overall efficiency is then

$$\eta = \frac{W_{el,a}}{Q_{pellets,a}} = \frac{4.541 \cdot 10^{15}\ \mathrm{J}}{13.6 \cdot 10^{15}\ \mathrm{J}} = 0.334$$

Taking into account the full use of waste heat as district heating (very cold winter day), the power plant achieves a maximum of approximately 80 % utilization of the supplied energy.

What is the energy content of the usable annual increment from one hectare of forest in the Baltic states?

The area yield is

$$q_{growth} = \frac{Q}{A} = \frac{V_{growth,a}}{A} \cdot \rho_{pellets} \cdot H_{I,pellets}$$

$$q_{growth} = 7.7\ \frac{\mathrm{m^3}}{\mathrm{a}} \cdot 650\ \frac{\mathrm{kg}}{\mathrm{m^2}} \cdot 17\ \frac{\mathrm{MJ}}{\mathrm{kg}} = 85\,090\ \frac{\mathrm{MJ}}{\mathrm{ha}} = 8.509\ \frac{\mathrm{MJ}}{\mathrm{m^2}}$$

What is the efficiency of converting sunlight into biomass at 55° north latitude?

A simple way to determine the actual mean solar irradiation is to read the global horizontal irradiation from *https://globalsolaratlas.info*. For the Baltic states this is slightly above 1000 kWh m^{-2} a^{-1}.

From this, the efficiency of converting solar irradiation into biomass follows as:

$$\eta_{photo} = \frac{\dot{q}_{growth}}{\dot{q}_{sun}} = \frac{8.509\ \frac{\mathrm{MJ}}{\mathrm{m^3 \cdot a}}}{1000\ \frac{\mathrm{kWh}}{\mathrm{m^3 \cdot a}}} = \frac{8.509\ \frac{\mathrm{MJ}}{\mathrm{m^3 \cdot a}}}{3600\ \frac{\mathrm{MJ}}{\mathrm{m^3 \cdot a}}} = 0.00236$$

What is therefore the energy-related land requirement of the power plant?

$$A = \frac{Q_{pellets,a}}{q_{growth,a}} = \frac{13.6 \cdot 10^{15}\ \mathrm{J}}{8.509\ \frac{MJ}{\mathrm{m^2}}} = 1.598 \cdot 10^9\ \mathrm{m^2} = 1598\ \mathrm{km^2}$$

What is the additional energy input per year for transporting the pellets from the forest to the power plant?

The energy input consists of the average transport inputs required to transport the necessary mass of wood pellets to the power plant. The factor of 2 for the outward and return journey is certainly assumed to be too large, since energy consumption when empty is always lower:

$$W_{narrow\text{-}gauge\ rail} \cong e_{narrow\text{-}gauge\ rail} \cdot 2 \cdot l_{narrow\text{-}gauge\ rail} \cdot m_{pellets}$$

$$W_{narrow\text{-}gauge\ rail} \cong 650\ \frac{\mathrm{J}}{\mathrm{km \cdot kg}} \cdot 25\ \mathrm{km} \cdot 800 \cdot 10^6\ \mathrm{kg} = 26.0 \cdot 10^{12}\ \mathrm{J}$$

$$W_{rail} \cong e_{rail} \cdot 2 \cdot l_{rail} \cdot m_{pellets} = 250\ \frac{\mathrm{J}}{\mathrm{km \cdot kg}} \cdot 2 \cdot 200\ \mathrm{km} \cdot 800 \cdot 10^6\ \mathrm{kg} = 80.0 \cdot 10^{12}\ \mathrm{J}$$

$$W_{ship} \cong e_{ship} \cdot 2 \cdot l_{ship} \cdot m_{pellets} = 160\ \frac{\mathrm{J}}{\mathrm{km \cdot kg}} \cdot 2 \cdot 1150\ \mathrm{km} \cdot 800 \cdot 10^6\ \mathrm{kg} = 294.0 \cdot 10^{12}\ \mathrm{J}$$

This results in a total additional requirement of approximately 400 · 10^{12} J.

If the means of transport are heated not with diesel but with pellets: what is the additional demand for cultivated area?

Steam turbines in large ships have approximately 33% efficiency for converting heat into work, diesel engines typically achieve 35%, and modern steam locomotives achieve 12%.

The ship can be operated with a steam turbine, so the following applies

$$Q_{ship,a} = \frac{W_{ship}}{\eta_{turbine}} = 836.40 \text{ TJ}$$

The railways, each calculated as a steam engine, have

$$Q_{rail,a} = \frac{W_{narrow\text{-}gauge\ rail} + W_{rail}}{\eta_{steam\ loco}} = 883.30 \text{ TJ}$$

From this, the additional forest area for pellets is determined:

$$A_{transport} = \frac{Q_{ship,a} + Q_{rail,a}}{\frac{Q}{A}} = 202.1 \text{ km}^2$$

This example illustrates the rather low energy density resulting from the use of organic fuels, and the consequent need for additional infrastructure for collecting and transporting the fuels. In principle, this power plant could also be supplied with wood from Denmark. However, in that case no further power plant in Denmark could be supplied from Danish forests, since Jutland has a total area of 29,000 km^2, of which only a few percent is productive forest.

9.5 Nuclear Fission

A great fascination emanates from the hidden and dangerous possibilities of nuclear energy. Due to the complexity of the required processes, the development of military applications and fundamentally civilian-usable technology proceeded in parallel from the very beginning. In all civilian variants of nuclear technology, usable heat is first generated. This heat is then almost always used in a steam turbine to generate electricity.

In nuclear fission very heavy and therefore unstable atoms are induced to decay at a much higher rate than would occur naturally. In nuclear fission in reactors, uranium (specifically the isotope ^{235}U) or plutonium (specifically ^{239}Pu) splits into lighter elements that together contain less energy than the decayed atom, as Figure 9.4 illustrated. The energy difference of the atomic nuclei is released during the decay process as particles or radiation. Through technical control of the decay reaction, the rate at which individual nuclei decay can be influenced:

- In a bomb, after ignition the reaction proceeds as fast as possible. The aim is to allow as many of the uranium or plutonium atoms present in the bomb to decay before the bomb overheats and explodes due to the reaction released — depending on the purpose of the bomb, either the highest possible explosive force or the highest possible neutron yield is set.
- In a nuclear reactor a continuous stream of decay reactions is established, which as closely as possible meets the power demand of the connected power plant. As many of the fissile atoms contained in the nuclear fuel as possible should decay, and as

far as possible all the energy released should be transferred within the reactor to an easily usable heat carrier such as water.

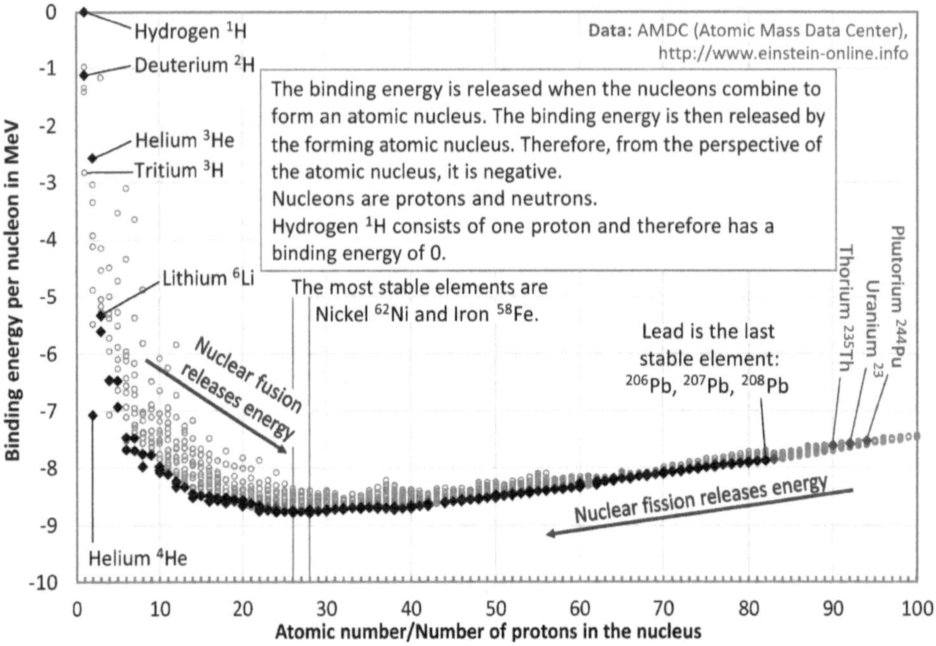

Figure 9.4 Where does the energy in nuclear reactions come from? Through the nuclear reaction, atomic nuclei with overall higher binding energy are formed. This difference in binding energy is released during the nuclear reaction (data source: AMDC, Einstein-Online).

Electricity The releasable heat relative to the mass of the nuclear fuel is many orders of magnitude greater than heating values. For ^{235}U this value is $76 \cdot 10^{12}$ J/kg. In the nuclear reactor, the decay products of the ongoing nuclear reaction are captured and decelerated. This releases heat. In civilian reactors, this heat is used to generate steam and then to generate electricity via a steam turbine. Due to the technical requirements arising from the required plant safety, the efficiency achievable in this way is not as high as with other heat sources. Reactors and the steam cycle are typically subcritical, so that an efficiency of 33% is common. This efficiency will not be different for new reactor concepts that are being considered or planned.

disposal The treatment of radioactive residual materials presents a technical and societal challenge. The spent fuel rods, the reactor, and other components contain a large mass of radioactive elements with in some cases long half-lives. So far there is no concept for how these materials can be safely disposed of in a final repository. There are currently "permanent" interim storage facilities that are intensively guarded and maintained.

Due to the unresolved disposal issue, it is difficult to conclusively determine a CED for this technology today. This also means that a conclusive determination of the EROI is not possible. In principle, the financial expenditures for the disposal of power plants and radioactive waste can be assumed to be low in planning through appropriate depreciation (Narbel et al. 2014).

The energy and material expenditures of interim or final storage must be borne entirely by future generations. The resources required for this must be provided by them to guard our waste, without any benefit to themselves.

Future use The future role of nuclear fission is unclear (IEA 2019). This can be attributed to a few central aspects:

- Electricity from nuclear energy is comparatively expensive. This is due to the impressive investment costs and the resulting depreciation required. The operating costs of a nuclear power plant are also quite high due to the special safety requirements. The perpetual costs of final disposal are not priced in here (Präger et al. 2024).
- Environmental impacts along the utilization chain of uranium and plutonium make its use undesirable for many societies.
- At the current consumption of 60,000 t/a of uranium, the known reserves are expected to last approximately 130 years. These are all known or more probably suspected deposits up to extraction costs of approximately 250 €/kg. The supply of uranium already requires significant investments in the medium term to develop these deposits as reserves (NEA 2023).

Thus, uranium-based nuclear technology is not available as a significant long-term option for energy supply (Hirschhausen 2023). In particular, it cannot replace other renewable forms of energy.

9.6 Nuclear Fusion

In nuclear fusion very light atoms are caused to combine into heavier atoms. In principle, a great deal of energy is released in the process, as Figure 9.4 shows. Nuclear fusion occurs naturally in the cores of stars, where the high pressure and high temperature allow the repulsive forces that otherwise prevent atoms from merging to be overcome.

Technically, nuclear fusion has been achieved for many decades by briefly exposing the atoms to be fused to sufficient pressure and temperature inside an atomic bomb explosion to enable a fusion reaction (hydrogen bomb). The art here is to generate an explosion that is sufficiently powerful and at the same time acts sufficiently symmetrically inward. This method is fundamentally not suitable for civilian use in earth system.

Fusion Power Plant?

Currently, two technical routes are being pursued with which a power plant using nuclear fusion could one day be built: either with a plasma of sufficient temperature or by igniting small fuel pellets with lasers. That it is possible under these conditions to carry out a fusion reaction in principle is considered certain:

- First short-term experiments with plasma, in which significantly more energy is to be released through fusion than is absorbed by the plasma from the heating, are already planned, starting approximately 2039 at the ITER experiment (Dittmar 2019; Kleimaier et al. 2024).
- The National Ignition Facility has reported several events in which more energy was released through fusion than was delivered to the pellets by means of lasers.

This part has a TRL 3. The next steps are to increase the energy release by approximately a factor of 1000 so that electricity can ultimately be fed into the grid. In laser-driven fusion, the cycle count must be increased by at least a factor of 1,000,000 (from 1 per day to 10 per second) to achieve a continuous sufficient energy conversion for a power plant. In plasma fusion, the duration of constant reactions must accordingly be extended by approximately a factor of 1,000,000 (from approximately 5 s to approximately 1 year). Concepts still need to be developed for rapidly and safely removing the reaction products from the reactor without affecting the plasma or the laser beams in the process.

Due to the relatively low requirements of pressure and temperature, only the fusion of deuterium ^{2}H with tritium ^{3}H is feasible. The reaction taking place would then be $^{2}H + ^{3}H \rightarrow ^{4}He + n + 17.6$ MeV. The helium atom formed carries 3.5 MeV of kinetic energy and the neutron 14.1 MeV. Tritium is radioactive with a half-life of 12.3 years. Therefore, the tritium in hydrogen bombs must be replaced regularly. Currently there are approximately 50 kg of tritium worldwide. A fusion reactor will in the future require approximately this mass for one year of operation (Abdou et al. 2021).

Currently, this is the technological status for the relevant components of a power plant:

- It is unclear from what material the reactor wall can be made. This is the wall that encloses the space in which the fusion is to take place. All known materials do not meet all the required specifications.
- The requirements for all materials traversed by the neutrons from the fusion are high: these neutrons must be slowed down but must not be captured: many metals capture neutrons and then form (often) radioactive isotopes. This causes these metals to change their properties and may need to be replaced. Concepts are lacking for how this can be done during operation. Neither humans nor robots can quickly

reach damaged areas or enter the reactor due to the initially high radioactivity. Reactors must decay for approximately 100 to 1000 years after operation before they can be dismantled or repaired, hence TRL 2.

- Outside the reactor wall, a steam generator and a tritium breeder are to be located. The neutrons are to react in the breeder with lithium ^{6}Li to form helium and tritium — ideally all of them. For this purpose, the reactor would need to be surrounded by many hundreds of tons of lithium. At this position, the steam generator must also be located simultaneously: this consists of pipes through which water flows at high pressure and high temperature. The lithium should have the lowest possible temperature so that the breeding reaction achieves a high efficiency. One idea is to surround the reactor with a tank of molten salt at high temperature made of LiF-BeF_2, in which the breeding reaction is then to take place. Since this has not yet been tested, TRL 2 results.
- The tritium must be extracted from the lithium safely and completely. It must neither escape by diffusion nor must lithium or tritium come into contact with air. Only concepts exist for this (TRL 2).
- For the lasers in laser-driven fusion, a further increase in efficiency by a factor of 100 is still to be achieved, so that it becomes possible to generate the electricity required to operate the lasers using the fusion power plant. Whether this is possible would need to be demonstrated by further technical development. There is no clear roadmap (TRL 2).

So far, no experiments are possible because there is no sufficiently intense source of neutrons with 14.1 MeV. It is important to use such neutrons, since the reactions change with the energy of the neutrons, i.e., previous experience cannot simply be transferred.

The scale and complexity of current projects show that fusion will require impressive quantities of valuable, expensive, and rare elements. Many of the materials used will be unusable for at least a longer period due to radioactivity. All components exposed to neutron radiation can only be used for a few months. After that, they must be replaced. It is to be feared that the EROI of the first generations of power plants (should the technical problems prove solvable after all) will be far below 1.

While optimism about the feasibility of fusion was quite justified at the beginning of research, the massive intensive research projects of the last 70 years have ultimately identified ever more aspects that make the realization of fusion probably impossible (Kleimaier et al. 2024; Englert/Kopp 2024).

EROI of future fusion power plants

If it does work out – if one miracle or another occurs – what benefit would these power plants then provide? A first indication is given by Gaio et al. for plasma fusion (Gaio et al. 2022): at a generator output of approximately 1 GW, approximately 500 MW are required internally. This puts the EROI at 2. The planned power plant will only ever run for a few minutes at a time, so the energy must be stored intermediately in order to generate a kind of reliable baseload. This reduces the EROI.

If the impressively high material input of critical and rare metals, the complex manufacturing of components, and the disposal and guarding of the highly radioactive reactors are taken into account, the EROI decreases further significantly. Currently, for plasma fusion, an EROI = 1 ± 1 is therefore to be expected under optimistic assumptions. So far, no experience with real facilities indicates that this value could be significantly higher for laser-driven fusion.

9.7 Solar Thermal Energy

Solar thermal energy is one way to provide process heat without combustion. Here, the thermal radiation of the sun, visible and infrared light, is utilized.

9.7.1 Collector Without Concentrator

The simplest form of technology for utilizing solar thermal radiation are collectors, which do not concentrate the radiation of the sun (Figure 9.5). For solar thermal collectors without concentration of the incoming radiation, the useful power, i.e., the heat flow provided (Quaschning 2023)

$$\dot{Q}_{no\text{-}conc} = I_{inc} \cdot A_{abs} \cdot \tau \cdot \alpha - A_{abs} \cdot k_{sys} \cdot (T_{abs} - T_0) \tag{9.3}$$

with the area of the collector A_{abs}, the heat losses related to the area due to thermal conduction k_{ges}, the temperature of the surroundings T_0, the temperature of the absorber T_{abs}, the fraction of radiation α, which is absorbed by the absorber, and the fraction of radiation τ, which is transmitted through the glass cover of the collector onto the absorber.

The incident intensity of the radiation is

$$I_{inc} = I_0 \cdot \cos(\theta) + I_{diffuse} \tag{9.4}$$

This intensity is composed of a fraction of the direct radiation of the sun at ground level I_0 at an angle θ and a diffuse fraction of scattered radiation $I_{diffuse}$ from the blue sky or from clouds from all directions. Heat losses k_{tot} arise primarily from heat conduction from the absorber to the surroundings. The radiation incident on the module is first reduced by the cover, through which only the transmitted fraction τ passes. Of this radiation, only the fraction α is actually absorbed by the absorber. Glass has approximately 4% reflection losses per surface. In addition, there may be losses in transmission due to impurities in the glass (iron) as well as dust, dirt and scratches on the surface. Good absorbers achieve α well above 95%. Thus the efficiency of converting the incident intensity into usable heat, i.e. heat that can be extracted from the absorber, is

$$\eta = \alpha \cdot \tau - k_{inc} \cdot \frac{T_{abs} - T_0}{I_{inc}} \tag{9.5}$$

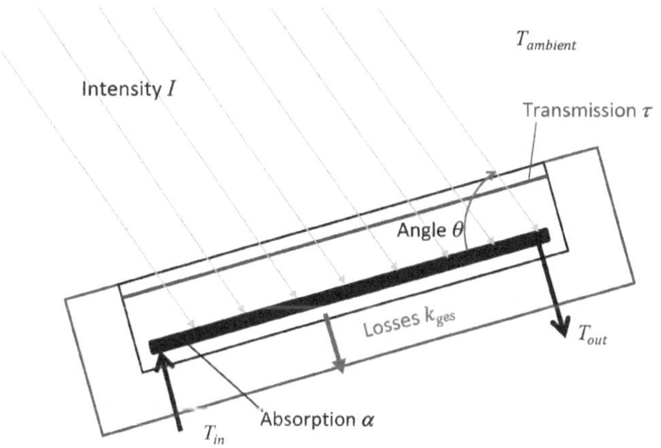

Figure 9.5 Principle of a solar thermal module without concentration

What is typically observed is an efficiency that is better described by a quadratic function, since convective influences and radiation contribute as nonlinear effects to the heat losses. The parameters of the equations always depend on the actual collector. The transmission of the cover can decrease significantly over time due to adhering dust, dirt, scratches and aging effects.

The typical application of this design is the generation of domestic hot water for household use or also as support for space heating. The achievable temperature of the water does not meet the usual requirements of industrial processes.

9.7.2 Solar Thermal with Concentration

The typical structure of a solar thermal module with concentration of solar radiation is shown in cross-section in Figure 9.6. The absorber is a pipe through which a medium flows that absorbs the heat. This is often a thermal oil. Around the absorber, a glass pipe is arranged. This pipe reduces the convective heat loss from the absorber to the surroundings. The absorbed radiation now consists of a small direct fraction of radiation that strikes the absorber directly,

$$\dot{Q}_{conc,dir} = I_{inc} \cdot A_{abs} \cdot \tau \cdot \alpha \tag{9.6}$$

and a larger fraction directed onto the absorber by the concentrator:

$$\dot{Q}_{conc,refl} = I_{inc} \cdot (A_{conc} - A_{abs}) \cdot \rho \cdot \gamma \cdot \tau \cdot \alpha = I_{inc} \cdot A_{conc} \cdot z \cdot \rho \cdot \gamma \cdot \tau \cdot \alpha \tag{9.7}$$

Here ρ is the reflectivity of the reflector and γ is a quality factor of the reflector (fraction of the reflected radiation that is directed onto the absorber). In addition, there is a geometric factor z, which describes the fraction of the radiation that falls onto the concentrator via the reflector:

$$z = \frac{A_{conc} - A_{abs}}{A_{conc}} \tag{9.8}$$

All areas here are projections onto a plane perpendicular to the direction of incidence of the light: in operation, the concentrator is oriented so that it is exactly perpendicular to the direction of propagation.

Thus the usable heat quantity is

$$\begin{aligned} \dot{Q}_{conc} &= \dot{Q}_{conc,dir} + \dot{Q}_{conc,refl} - \dot{Q}_{loss} \\ \dot{Q}_{conc} &= I_{ges} \cdot (A_{abs} + A_{conc} \cdot z \cdot \rho \cdot \gamma) \cdot \tau \cdot \alpha - \pi \cdot A_{abs} \cdot k_{sys} \cdot (T_{Abs} - T_0) \end{aligned} \tag{9.9}$$

The efficiency of utilizing the incident radiation for absorbers with a circular cross-section is

$$\eta = \left(\frac{1}{\kappa} + \left(1 - \frac{1}{\kappa}\right) \cdot \rho \cdot \gamma\right) \cdot \alpha \cdot \tau - \frac{\pi}{\kappa} \cdot k_{sys} \cdot \frac{T_{abs} - T_o}{I_{inc}} \tag{9.10}$$

Here a concentrator ratio is used:

$$\kappa = \frac{A_{conc}}{A_{abs}} \tag{9.11}$$

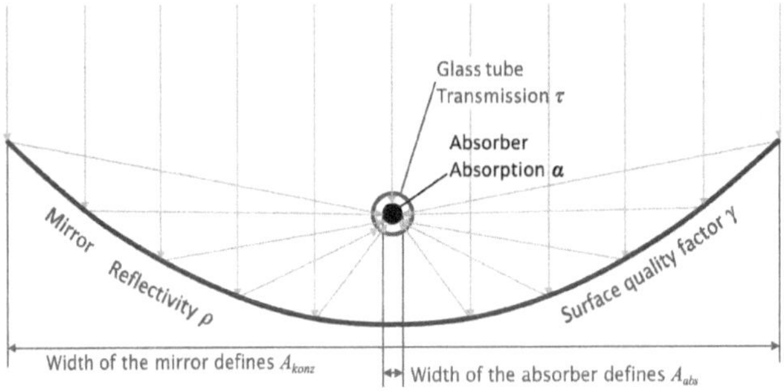

Figure 9.6 Principle of a solar thermal module with concentrator (explanation in the text)

Design Types Two variants are technically relevant:

1. A parabolic trough concentrator is the one-dimensional variant, as shown in cross-section in Figure 9.6 and achieves a temperature of approximately 400 °C at favorable locations. The technical implementation is overall quite simple, since all parabolic troughs must be tracked toward the sun equally and in only one direction.
2. A heliostat concentrator consists of heliostat fields, in which a large number of individually sun-tracked mirrors concentrate the radiation onto a common absorber. At the absorber, a temperature above 1000 °C is thereby possible with a sufficient number of these so-called heliostats. The technical effort is considerable, since each heliostat must be moved individually in two directions.

Parabolic troughs are suitable for generating low-temperature process heat or for steam power plants. Heliostat power plants could in principle provide process heat at the melting temperature of many metals; however, no suitable materials are available for this use for the absorber, which would then have to serve as the melting vessel.

Maximum Concentration The maximum useful concentration for a one-dimensional parabolic trough concentrator, as shown in cross-section in Figure 9.6, is

$$\varkappa_{1D} = \frac{1}{\sin(\phi)} = \frac{1}{\sin(0.533\,°)} \leq 107 \tag{9.12}$$

For the concentration of radiation onto a point, the maximum useful concentration is

$$\varkappa_{0D} = \frac{1}{\sin^2(\phi)} = \frac{1}{\sin^2(0.533\,°)} \leq 11\,600 \tag{9.13}$$

In this equation, ϕ is the acceptance angle, which is typically the angle corresponding to the diameter of the sun in the sky (with ϕ = 32 arc minutes = 0.533°). Actually implemented systems exhibit lower concentrator ratios due to typical manufacturing tolerances. All concentrator modules must track the sun so that the sun's radiation is always imaged onto the absorber. The required tracking accuracy depends on the concentration ratio. The actually usable intensity therefore depends only indirectly on the position of the sun above the horizon, as long as the individual concentrator modules do not shade each other.

Solar thermal power plant in Ashalim

In Ashalim, Israel, there is a solar thermal power plant, which was built at the time by GE. The researched parameters of the power plant are:

- Rated power: 110 MW
- Projected annual electrical yield: 320 GWh
- Area of the mirror field: 3.15 km²
- The absorber is mounted on a tower 240 m high. The heliostat field consists of 55,000 units. Each heliostat has two mirrors, each with an area of 2.3 m · 3.3 m.

- Liquid salt is used as the heat transfer medium in the tower and as an intermediate heat storage. The steam turbine uses a supercritical steam cycle.
- The annual solar irradiation in Ashalim averages 2100 kWh m^{-2} (solargis.com) and the maximum irradiation in Ashalim is 1090 W m^{-2} (30° northern latitude, i.e., the sun is nearly at the zenith in summer). The site receives approximately 100 mm of precipitation per year. It is a desert.

What is the efficiency of the entire system?

What is meant here is the efficiency of converting the radiant power incident on the site into electricity. This efficiency depends on the angle of incidence of the sun and is maximum at perpendicular irradiation:

$$\dot{Q}_{\max} = I_{\max} \cdot A = 1090\ \frac{\mathrm{W}}{\mathrm{m}^2} \cdot 3.15 \cdot 10^6\ \mathrm{m}^2 = 3.434 \cdot 10^9\ \mathrm{W}$$

Then the efficiency is

$$\eta_{area} = \frac{P_N}{\dot{Q}_{\max}} = \frac{110 \cdot 10^6\ \mathrm{W}}{3.434 \cdot 10^9\ \mathrm{W}} = 0.0320$$

A comparable value follows from the mean values.

What is the utilization rate of the system?

The utilization rate is derived from the rated power and the projected annual yield:

$$\psi = \frac{\overline{P}}{P_N} = \frac{W_{el,a}}{P_N \cdot 1\ \mathrm{a}} = \frac{320 \cdot 10^9\ \mathrm{Wh}}{110 \cdot 10^6\ \mathrm{W} \cdot 365 \cdot 24\ \mathrm{h}} = 0.332$$

This value is quite close to the technically possible maximum.

What rated power and annual yield would a typical photovoltaic system have at this location?

If a_{PV} = 80% of the plot area is covered with PV modules and the efficiency of the modules is estimated at η_{PV} = 20%, then the annual yield follows from the incoming radiation:

$$W_{el,PV} = \dot{q}_{sol} \cdot A \cdot \eta_{PV} \cdot \alpha_{PV} = 2100\ \frac{\mathrm{kWh}}{\mathrm{m}^2} \cdot 3.15 \cdot 10^6\ \mathrm{m}^2 \cdot 0.80 \cdot 0.20 = 3.81 \cdot 10^{15}\ \mathrm{J}$$

The rated power from the maximum intensity and the area is

$$P_{N,PV} = I_{sol} \cdot A \cdot \eta_{PV} \cdot \alpha_{PV} = 1090\ \frac{\mathrm{W}}{\mathrm{m}^2} \cdot 3.15 \cdot 10^6\ \mathrm{m}^2 \cdot 0.80 \cdot 0.20 = 549 \cdot 10^6\ \mathrm{W}$$

Is photovoltaics or solar thermal energy to be preferred at this location?

This depends on a number of arguments:

- Due to dust storms, a high input for cleaning and a gradual reduction in efficiency for both system types is to be expected.
- Solar thermal energy offers the potential to store heat and thus generate electricity when the sun is no longer shining. The system could produce according to demand.
- Solar thermal energy loses nearly no efficiency at very high daytime temperatures (and it gets very hot there in summer), particularly when wet cooling towers are used. The efficiency of photovoltaics drops significantly with temperature, especially with silicon-based technology.

- The heliostats must be individually and very precisely tracked to follow the sun so that they each reflect the sun onto the absorber on the tower. They are therefore complex mechanical components requiring high maintenance, while photovoltaic modules require little maintenance apart from cleaning.

9.8 Refrigeration Machine and Heat Pump

Refrigeration Machine and heat pump use essentially identical technology at their core. Both transport a heat flow from a system at low temperature to a system at high temperature and use mechanical power to do so. They utilize a reverse thermodynamic cycle. The simplest and most common variant is the reverse Clausius-Rankine reference cycle, in which a suitable refrigerant flows in a circuit (Linow 2022).

Cold side The refrigerant absorbs the heat flow $\dot{Q}_{in}$ in a heat exchanger at low pressure. It absorbs latent heat (evaporation of the refrigerant) and some sensible heat (temperature increase), thereby increasing its enthalpy.

Compressor The gaseous refrigerant is mechanically compressed, further increasing its enthalpy. The power P_{mech} is used for this purpose. The compressor drive is typically electric.

Warm Side In a second heat exchanger, which simultaneously acts as a condenser, the refrigerant releases the heat flow $\dot{Q}_{out}$ here the latent heat is released through condensation of the refrigerant. In addition, the refrigerant also releases sensible heat. In the process and without other losses,

$$\dot{Q}_{out} = \dot{Q}_{in} + P_{mech} \tag{9.14}$$

Throttle In a fourth process step, the refrigerant is expanded. Typically, a simple throttle is used here, so that the liquid refrigerant transitions isenthalpically into the wet steam region and reaches a much lower temperature through the phase transition.

More complex processes are used depending on the application. In principle, several refrigeration machines can be arranged in series. In that case, the condenser of the first refrigeration machine serves as the evaporator of the second. However, most air conditioning systems and heat pumps operate according to this simple functional principle.

Efficiency

Energy Efficiency Ratio The key performance indicator of a refrigeration machine is the Energy Efficiency Ratio (EER) ε. This is the ratio of the benefit (here the "cooling", i.e., the heat flow absorbed on the cold side) to the input of the power applied at the compressor:

$$\varepsilon = \frac{\dot{Q}_{in}}{P_{mech}} \tag{9.15}$$

The Energy Efficiency Ratio evaluates how much power is required to produce the actual benefit, which is the removal of a heat flow from a system.

COP In heat pumps, the benefit is the heat flow provided on the warm side. This is described by the COP (Coefficient of Performance or heating performance factor):

$$\text{COP} = \frac{\dot{Q}_{out}}{P_{mech}} = \frac{\dot{Q}_{in} + P_{mech}}{P_{mech}} = 1 + \varepsilon \tag{9.16}$$

The thermodynamic reference process refers to the refrigerant as the system. Only power and heat flows within the refrigerant are considered. Technically relevant manufacturer specifications always refer to the power consumption of the entire unit including all auxiliary units. For an energy-related assessment, this is the relevant specification. The COP refers to a specific set of boundary conditions. It changes when these vary.

Temperature Spread A key influencing parameter on the COP is the temperature spread, i.e., the temperature of the cold side T_{in} and that of the warm side T_{out} as well as the difference ΔT.

The temperature difference ΔT can be influenced via the required heat exchangers: the heat exchangers determine what temperature difference between the incoming and outgoing heat flows and the refrigerant is needed to transfer the heat flow. Good liquid → liquid heat exchangers require approximately 2 K. For air → liquid and liquid → air, approximately 10 K are required.

Refrigerant The selection of the refrigerant influences the efficiency. Many suitable refrigerants exist, which differ in temperature range, safety, and ozone and climate impact:

- Ammonia (R717) often represents the optimum when temperatures below the freezing point of water may occur. However, ammonia is flammable and toxic.
- Many compounds of carbon, fluorine, or chlorine are generally suitable, non-toxic, and poorly flammable. However, many of them are ozone-depleting and are therefore prohibited under the Montreal Protocol. Many others are extreme greenhouse gases, so their use is coming to an end. Almost all of these substances are also persistent chemicals with long-term and in some cases unforeseeable environmental effects, and they therefore also fall under the EU F-gas regulation.
- Hydrocarbons such as propane (R290) and butane (R600) are very good refrigerants, non-toxic, without ozone reaction and with only a minor greenhouse effect. Therefore, they will predictably be used in many areas in the foreseeable future.
- Carbon dioxide (R744) is a very suitable refrigerant at higher pressure – particularly in the supercritical range.

- Water (R718) only has a sufficient density at higher temperatures. It then often represents the ideal refrigerant.

Heat Pumps

Heat pumps are particularly suitable for providing space heating. This applies especially when the temperature spread between the supplied heat $\dot{Q}_{in}$ to the heat pump and the flow temperature of the heating system are not too high. Figure 9.7 provides an orientation as to which COP values can be expected for which temperature ranges. Real heat pumps with their refrigerants will always cover only a portion of the range.

If the temperature of the supplied heat varies, the COP is not sufficient as an efficiency indicator. Particularly when ambient air is used as a heat source, additional key figures are helpful.

SCOP This term (Seasonal COP) denotes an efficiency averaged over a year. Calculation methods exist with which the SCOP can be estimated (VDI 4650-1).

SEER The Seasonal Energy Efficiency Ratio (SEER) indicates the actual ratio of heat provided by a heat pump to its electrical consumption. The SCOP is, where applicable, a manufacturer's specification and the SEER the actual result on site. Both can differ from each other.

Real heat pumps are limited in their heating capacity by their electrical connection and their compressor output. They can efficiently provide only a supply flow temperature in the range of approximately 45 °C to 55 °C. In addition, the heating capacity decreases with decreasing temperature T_{in} depending on the design.

For these reasons and in order to maintain large living areas at high indoor temperatures, there is a clear requirement for insulating buildings. However, the economic considerations depend on the historically large discrepancy between the costs of electricity and natural gas.

Geothermal energy offers the advantage for the use of heat pumps that heat can be supplied at an approximately constant and rather higher temperature. This reduces the electrical energy required for the compressor and the COP is significantly higher than for heat pumps that use ambient air (Goebel 2024).

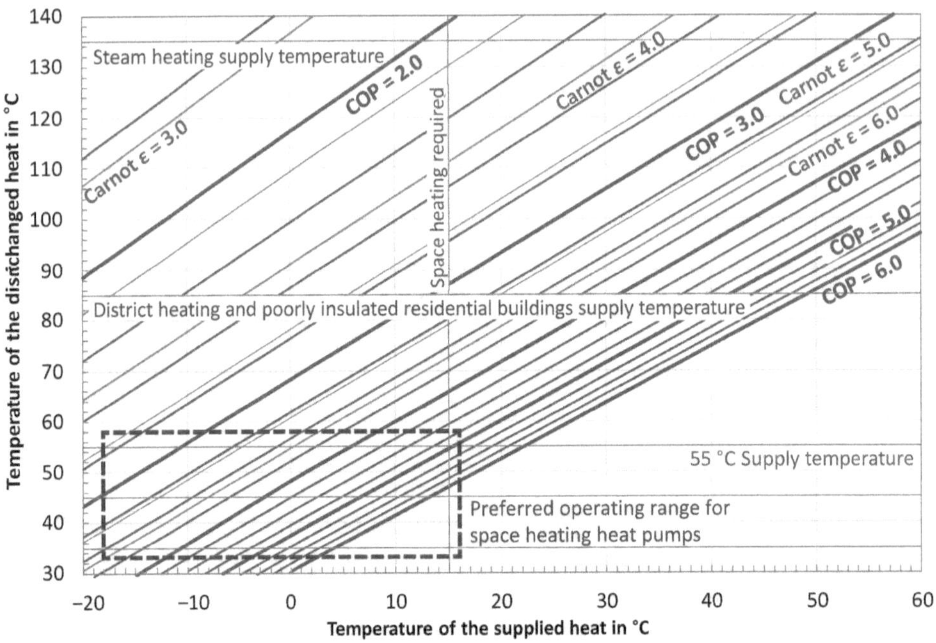

Figure 9.7 Typical operating range of heat pumps: ideal coefficient of performance and real COP for the transport of heat from the supplied temperature to the delivered temperature

Air Conditioning

Refrigeration machines have by now become almost indispensable in warm regions of the world. Air conditioning of living spaces, shopping centers and industrial facilities is spreading, particularly in cities. It represents a prerequisite for certain activities and ways of life. At the same time, this creates a high demand for a reliable supply of electrical energy.

Estimating the COP

The ideal efficiency describes the minimum mechanical power required to generate a heat flow from a lower temperature T_{in} to a higher temperature T_{out} :

$$\varepsilon_{Carnot} = \frac{T_{out}}{\Delta T} = \frac{T_{out}}{T_{out} - T_{in}}$$

For real heat pumps, the COP depends depends on the refrigerant used and the technical details. The refrigerant determines, for example, the temperature range over which the heat pump operates efficiently. Real heat pumps can be described by a quality factor. This indicates how closely a heat pump approaches the Carnot limit:

$$\eta_{HP} = \frac{\mathrm{COP}}{\varepsilon_{Carnot}}$$

Currently, this value is typically 0.6. Using this value, Figure 9.7 was generated. For real heat pumps, the lines of constant COP are not linear, and real heat pumps can only ever realize a portion of the diagram.

For a known quality factor and a given COP, the achievable temperature of the delivered heat can be determined from the temperature of the utilized heat:

$$T_{out} = T_{in} \cdot \frac{\varepsilon_{Carnot,eq}}{\varepsilon_{Carnot,eq} - 1} = T_{in} \cdot \frac{\frac{\mathrm{COP}}{\eta_{HP}}}{\frac{\mathrm{COP}}{\eta_{HP}} - 1}$$

9.9 Geothermal Energy and Ground Heat

Geothermal energy encompasses all methods in which heat from the subsurface is utilized or the ground is used as a heat storage medium. To extract the heat, the sensible heat present in the subsurface is used to heat a fluid (Quaschning 2023). The fluid is usually water. This heat exists as a temperature difference between the ground or rock and the surroundings, i.e., the usable range can extend from the mean temperature of the groundwater (which in winter is significantly above the air temperature) to liquid volcanic magma. Depending on the depth from which the heat is extracted, the requirements and technical possibilities differ:

Near-surface Geothermal Energy This encompasses all methods down to a depth of 400 m. The typical maximum depth is 150 m, since beyond that mining law often applies. The heat then has a temperature of around 10 °C in temperate regions and a heat pump (Section 9.8) is used to achieve a flow temperature sufficient for building heating. The effort for individual boreholes is low. Probes are not permitted everywhere, as they are often installed in aquifers.

The extractable heat from closed probes depends on the composition and structure of the subsurface. Typical values are around 50 W/m, so that a 100 m borehole yields approximately 5 kW. This extraction capacity results from the requirement not to cool down the subsurface. Associated with this is also the recommended minimum spacing between probes. This is 6 m or more, i.e., one probe requires a minimum footprint of approximately 36 m^2.

Medium-depth geothermal energy The second depth range is commonly understood as depths between 400 m and 1000 m. In this range, higher temperatures are present. This is not yet sufficient to be used directly as a flow temperature for space heating. Depending on the subsurface, boreholes can still be drilled with relatively little effort using well-drilling machines.

Deep Geothermal Energy Here the boundary is more of a temperature boundary (from 60 °C), since at that point the heat can be used directly as a supply temperature for space heating in energy-efficient modern buildings. In volcanically favored regions, heat at significantly higher temperatures can also be extracted. This can then also be used for steam power plants.

Heat Flux The mean heat flux from the interior of the earth to the surface is far too low to be used meaningfully. Near-surface geothermal energy therefore utilizes heat that flows downward from the earth's surface. For this reason too, combined variants that cool buildings in summer and transport heat into the subsurface are beneficial: the subsurface around the probes then acts as a seasonal heat storage.

With deep geothermal energy, heat in the subsurface will be extracted over time, reducing the temperature there. In central Europe there are no near-surface areas of high temperature, so geothermal energy here is used primarily for space heating. The potential for this is high (Quaschning 2023).

9.10 Electric Heating

Heat can be generated from electricity. For this, the required electricity must first be generated simultaneously. The efficiency of converting electricity into heat is not fundamentally limited; losses arise only through heat losses to the surroundings. This results in a very high efficiency in the process. Technically, heat transfer to the material is often the actual challenge; however, almost all technical processes can also be represented or implemented using purely electrical means (Pfeifer et al. 2018).

If process heat is to be generated non-fossil, then for higher temperatures only organic fuels and reducing agents as well as electricity remain. Other sources often cannot reliably provide the required temperatures. In the temperature range above space heating, the use of high-temperature heat pumps can be advantageous. In particular, water as a refrigerant offers a comparatively high COP from around 70 °C. With heat pumps, more heat is always provided than with the direct conversion of electricity into heat.

References

Abdou, M./Riva, M./Ying, A./Day, C./Loarte, A./Baylor, L. R./Humrickhouse, P./Fuerst, T.F./Cho, S.: Physics and technology considerations for the deuterium–tritium fuel cycle and conditions for tritium fuel self sufficiency. Nuclear Fusion 61, 2021, 013001

AMDC – Atomic Mass Data Center. *https://amdc.impcas.ac.cn//*

Andrews-Speed, P./Bleischwitz, R./Boersma, T./Johnson, C./Kemp, G./VanDeveer, S. D.: Want, Waste or War? The global resource nexus and the struggle for land, energy, food, water and minerals. Routledge, Abingdon 2015

Bogdanov, D./Satymov, R./Breyer, C.: Impact of temperature dependent coefficient of performance of heat pump on heating systems in national and regional energy systems modelling. In: Applied Energy 371, 2024, 123647

Brown, A.: Just Enough. Lessons in Living Green from Traditional Japan. Tuttle, Tokyo 2012

Carlowitz, H. C. von: Sylvicultura oeconomica Oder Haußwirthliche Nachricht und Naturmäßige Anweisung Zur Wilden Baum-Zucht. Braun, Leipzig 1713

Dittmar, M.: Status and Prospects of the ITER Plasma Physics Experiment. Is it time to terminate the project? Part I. Report commissioned by the Alliance 90/The Greens parliamentary group in the German Bundestag. 10.07.2019

Einstein-Online. *https://www.einstein-online.info*

Englert, M./Kopp, A.: Übersichtsstudie Kernfusion für den Klimabeirat Hessen. Ökoinstitut, Darmstadt 2024 (DOI: 10.5281/zenodo.15307836). *https://klimabeirat-hessen.de*

Fuller, R./Landrigan, P. L./Balakrishnan, K./Bathan, G./Bose-O'Reilly, S./Brauer, M./Caravanos, J./ Chiles, T./Cohen, A./Corra, L./Cropper, M./Ferraro, G./Hanna, J./Hanrahan, D./Hu, H./Hunter, D./Janata, G./Kupka, R./Lanphear, B./Lichtveld, M./Martin, K./Mustapha, A./Sanchez-Triana, E./ Sandilya, K./Schaefli, L./Shaw, J./Seddon, J./Suk, W./Téllez-Rojo, M. M./Yan, C.: Pollution and health. A progress update. In: The Lancet Planetary Health 6, 2022, e535-e547

Gaio, E./Ferro, A./Lampasi, A./Maistrello, A./Dan, M./Falvo, M. C./Gasparini, F./Lunardon, F./Magnanimo, A./Manganelli, M./Minucci, S./Panella, S./Proietti Cosimi, M./Ratti, D./Barucca, L./Ciattaglia, S./ Franke, T./Federici, G./Piovan, R.: Status and challenges for the concept design development of the EU DEMO Plant Electrical System. In: Fusion Engineering and Design 177, 2022, 113052

Gas- und Wärme-Institut Essen e. V. (Hrsg.): GWI-Arbeitsblätter. Verbrennungswert – Gaseigenschaft – Berechnungen. Vulkan Verlag, Essen 2012

Goebel, F.: Machbarkeitsstudie zur Nutzung von Wärmepumpen auf dem Mathildenhügel in Darmstadt. Bachelorarbeit. Hochschule Darmstadt 2024. *https://doi.org/10.48444/h_docs-pub-440*

Hirschhausen, C. von: Atomenergie. Geschichte und Zukunft einer riskanten Technologie. C. H. Beck, München 2023

Huang, Y./Unger, N./Harper, K./Heyes, C.: Global climate and human health effects of the gasoline and diesel vehicle fleets. In: GeoHealth, Volume 4, Issue 3, 2020, e2019GH000240

IEA – International Energy Agency: Nuclear Power in a Clean Energy System. Paris 2019. *https://www.iea.org/reports/nuclear-power-in-a-clean-energy-system*

JRC – Joint Research Center (Hrsg.): Well-to-Tank Report Version 4.a. JEC Well-to-Wheels Analysis. 2014. *https://ec.europa.eu/jrc/en/publication/eur-scientific-and-technical-research-reports/well-tank-report-version-4a-jec-well-wheels-analysis*

Kleidon, A.: Thermodynamic Foundations of the Earth System. Cambridge University Press, 2016

Kleimaier, M./Köln, K./Lens, H./Wieschollek, F.: Kernfusion – kein Beitrag zum Klimaschutz bis 2045 erwartet. VDE Hintergrund. Verband der Elektrotechnik, Offenbach 2024

Kummu, M./Heino, M./Taka, M./Varis, O./Viviroli, D.: Climate change risks pushing one-third of global food production outside the safe climatic space. In: One Earth 4, 2021, pp. 720–729

Linow, S.: Angewandte technische Thermodynamik. Carl Hanser Verlag, München 2022

Marek, R./Nitsche, K.: Praxis der Wärmeübertragung. Grundlagen – Anwendungen – Übungsaufgaben. Carl Hanser Verlag, München 2015

Mitscherlich, E. A.: Das Gesetz des Minimums und das Gesetz des abnehmenden Bodenertrages. Landwirtsch. Jahrb. 1909

Moriarty, P.: Can bioenergy once again become a major global energy source? In: Encyclopedia 2, 2022, pp. 1357–1369

Narbel, P. A./Hansen, J. P./Lein, J. R.: Energy Technologies and Economics. Springer, Cham 2014

NEA – Nuclear Energy Agency: Uranium 2022. Resources, Production, and Demand. OECD Publishing, Paris 2023

Pfeifer, H./Nacke, B./Beneke, F.: Praxishandbuch Thermoprozesstechnik. 2 Bände. Vulkan Verlag, Essen 2018

Pimentel, D.: Energy Inputs in Food Crop Production in Developing and Developed Nations. In: Energies 2, 2009, S. 1–24

Präger, F./Breyer, C./Fell, H.-J./von Hirschhausen, C./Kemfert, C./Steigerwald, B./Traber, T./Wealer, B.: Evaluating nuclear power's suitability for climate change mitigation: technical risks, economic implications and incompatibility with renewable energy systems. In: Frontiers in Environmental Economics 3, 2024

Quaschning, V.: Regenerative Energiesysteme. Technologie – Berechnung – Klimaschutz. 12. Auflage. Carl Hanser Verlag, München 2023

Smil, V.: Energy and Civilization. A History. MIT Press, Cambridge 2017

Smil, V.: Harvesting the biosphere. What we have taken from nature. MIT Press, Cambridge 2013

Smil, V.: Still the Iron Age. Iron and Steel in the Modern World. Elsevier, Amsterdam 2016

Smith, J./Farmer, J./Smith, P./Nayak, D.: The role of soils in provision of energy. In: Philosophical Transactions B 376, 2021, 20200180

Wiberg, N.: Hollemann · Wiberg. Lehrbuch der Anorganischen Chemie. 102. Auflage. Walter de Gruyter, Berlin 2009

Wilkinson, B. H./McElroy, B. J.: The impact of humans on continental erosion and sedimentation. In: GSA Bulletin 119, 2006, pp. 140–156

VDI, 4650-1: Berechnung der Jahresarbeitszahl von Wärmepumpenanlagen – Elektrowärmepumpen zur Raumheizung und Trinkwassererwärmung

10 Hydrogen and Its Derivatives

Hydrogen plays a special role in the transformation of our energy system. Hydrogen is – like coal, crude oil and natural gas – an energy carrier. Unlike fossil energy carriers, hydrogen cannot simply be extracted but must first be produced. At the same time, hydrogen can take over some of the tasks that fossil energy carriers have had until now. This chapter examines which these are and what prerequisites need to be clarified.

Hydrogen is difficult to handle:

- H_2 has an exceptionally low density as a gas, so that storage vessels quickly become too large or very little bound energy is present as hydrogen in storage vessels (see Table 17.6 in Section 17.5).
- H_2 has an exceptionally low boiling point at –253 °C, so that liquefaction is extremely energy-intensive. The density and energy density of liquid hydrogen are also very low.
- The compression work for hydrogen is exceptionally high. It is much higher than that of natural gas. Therefore, high compression should actually be avoided.
- H_2 diffuses through many materials and is then lost.
- H_2 embrittles many metals, so that special care is required in the selection of materials and monitoring of pressure vessels.

For these reasons, we must also look at hydrogen derivatives. This term refers to energy carriers that can be synthetically produced from hydrogen. These include ammonia, methanol, synthetic hydrocarbons and e-fuels.

10.1 Production of Hydrogen

Hydrogen can be generated either from fossil energy carriers or directly by electricity. Production from biomass is also possible in principle, but is not relevant due to the lack of available biomass, the complex technology required and the energy demand (see

Chapter 9). Table 10.1 and Figure 10.1 provide an overview of the production methods, the associated labeling as a "color" and the typical specific carbon dioxide emissions of the production methods.

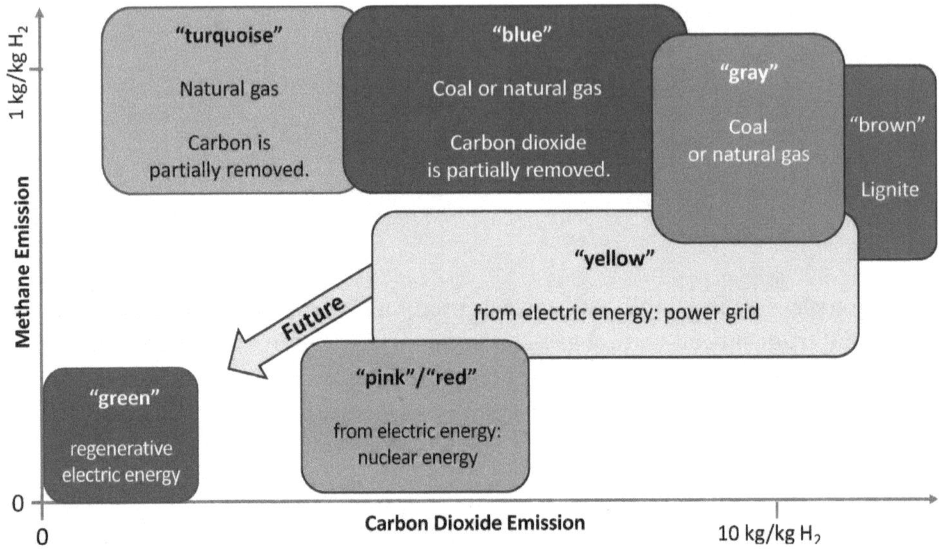

Figure 10.1 Production variants of hydrogen and their typical greenhouse gas emissions

Table 10.1 "Colors" of Hydrogen, the Meaning of the Colors and Typical Specific CO_2emissions from Production Without Upstream Chain (Based on Bareiß et al. 2019)

Color	Production	Specific emissions in kg $CO_{2,eq}$ per kg H_2
Brown	From lignite	Approx. 18 kg/kg
Black	From hard coal	Approx. 18 kg/kg
Gray	From natural gas or crude oil	Approx. 12 kg/kg
Blue	From natural gas with subsequent reduction of CO_2emissions by means of CCS	Currently, approximately 50% capture rate is achieved, i.e., approx. 10 kg/kg. At a long-term target capture rate of 90% for CO_2 approx. 3 kg/kg
Turquoise	From natural gas, whereby the contained carbon is partially separated as a solid	Currently, approximately 50% capture rate is achieved, i.e., approx. 10 kg/kg. At a long-term target capture rate of 90% for CO_2 approx. 3 kg/kg

Color	Production	Specific emissions in kg $CO_{2,eq}$ per kg H_2
Pink	Electrical, whereby the electricity originates from nuclear fission	Greenhouse gas emissions here are associated with the provision of nuclear fuel.
Yellow	Electrical, whereby the electricity originates from the current electricity mix	Currently 25 kg/kg to 28 kg/kg, with a decreasing trend for 2050, values significantly below 10 kg/kg are expected.
Green	Electrical, whereby the electricity was generated from renewable sources	Low depending on the manufacturing processes of the plants

10.1.1 Hydrogen from Fossil Energy Carriers

Fossil fuels can be converted into hydrogen and carbon dioxide by means of steam reforming into hydrogen and carbon dioxide. If natural gas is the fossil fuel, the first reaction is

$$CH_4 + H_2O \Rightarrow CO + 3 \cdot H_2$$

In this reaction, three moles of hydrogen are produced from one mole of methane.

If the reaction starts with coal, the relevant first reaction is

$$C + H_2O \Rightarrow CO + 1 \cdot H_2$$

In this reaction, one mole of hydrogen is produced from each mole of carbon. All other hydrocarbons (including biomass) yield correspondingly between 1 mol/mol and 3 mol/mol of hydrogen molecules per carbon atom in the fossil fuel.

These reactions require energy. This is provided by burning a portion of the fossil fuels, i.e., the actual yield is lower.

In a second step, the carbon monoxide is converted by a water-gas shift reaction used to produce somewhat more hydrogen.

$$CO + H_2O \rightleftharpoons CO_2 + H_2$$

For natural gas, the typical conversion efficiency based on the lower heating value of natural gas and of hydrogen is

$$\eta_{steam-ref} = \frac{m_{H_2} \cdot H_{I,H_2}}{m_{CH_4} \cdot H_{I,CH_4}} \cong 0.60 \text{ to } 0.70 \tag{10.1}$$

Purification The process gas produced from fossil fuel is purified of carbon dioxide, water, and other impurities, so that hydrogen of the desired purity is produced at the end. Depending on the required purity level of the hydrogen, these can be relatively simple processes or quite energy-intensive ones. In particular, the requirements for hydrogen that is to be transported over long distances in pipelines or that is to be liquefied are especially high here, in order to avoid corrosion (Cerbe/Lendt 2016).

Greenhouse gases Ostensibly, a large proportion of the carbon dioxide originating from fossil fuels is to be captured and geologically stored in the future during hydrogen production – so-called "blue" or "turquoise" hydrogen. To date, however, the technology required for this (Carbon Capture and Storage, CCS, see Chapter 14) has not been available with sufficient technological maturity despite decades of research.

A key motivation for the production of "blue" hydrogen may be to exploit fossil reservoirs more thoroughly. Carbon dioxide is already being injected into older oil or natural gas reservoirs today in order to extract more fossil fuels from them (enhanced oil recovery). This carbon dioxide frequently escapes again.

Bridge technology

Ostensibly, so-called "bridge technologies" would be needed to establish hydrogen in the market. This is about being able to continue using fossil raw materials, because these so-called "bridge technologies" are initially supposed to be grey, black, and brown, and possibly in the future blue or turquoise hydrogen. Brown and grey hydrogen are considerably more harmful to the climate in the overall balance than the direct use of fossil energy carriers - already due to the additional unavoidable energy losses. Investment in these technologies is therefore inadvisable.

Blue and turquoise hydrogen are essentially grey hydrogen, merely with the promise that a portion of the carbon from the fossil energy carriers will not enter the atmosphere. The extent to which producers of blue or turquoise hydrogen will keep their promise to at least not release a significant proportion of the carbon remains unclear at best (see Chapter 14).

The use of natural gas instead of coal is - purely in terms of the carbon dioxide released - less harmful to the climate than the use of coal. During the extraction and handling of natural gas, avoidable and in some cases also unavoidable emissions of methane as a potent greenhouse gas always occur. It is now becoming clear that natural gas therefore has a comparable or even higher greenhouse effect than coal. There is an example of this in Chapter 14 and the analysis by Howarth for LNG (Howarth 2024). Natural gas alone is not a "bridge technology" and must remain in the ground if we want to respect planetary boundaries.

With blue or turquoise hydrogen, methane emissions are not reduced. Therefore, this does not represent an improvement over brown or grey hydrogen. In summary, such "bridge technologies" are not sensible, but on the contrary detrimental to the transformation. Since, regardless of the chosen production variant, substantial investment in a hydrogen infrastructure is required in any case, this investment can be directed straight into the desired green technologies. Only in this way can a technology lock-in and the susceptibility of society to coercion be avoided.

10.1.2 Hydrogen from Electricity

The second process for the production of hydrogen uses pure water and electrolysis. The water is caused to split into hydrogen and oxygen in an electric field:

$$2 \cdot H_2O \Rightarrow O_2 + 2 \cdot H_2$$

The efficiency of real plants, based on the lower heating value of hydrogen, is

$$\eta_{ely} = \frac{H_{I,H_2}}{w_{el}} \cong 0.65 \text{ to } 0.70 \qquad (10.2)$$

With PEM electrolysis, hydrogen can already be provided at a pressure of 30 bar. This eliminates a relevant portion of the compression work. Current alkaline electrolysis plants produce hydrogen at lower pressure. At present, PEM electrolysis is well suited for fluctuating electricity supply, while alkaline electrolysis tends to remain dependent on a constant supply (Schmidt 2022).

10.2 Hydrogen Derivatives

Hydrogen can be further processed into other substances. Additional production processes are necessary for these derivatives, and each is particularly well suited for certain applications.

Ammonia The production of ammonia (NH_3) has been possible for about 100 years via the Haber-Bosch process (Smith et al. 2020). Ammonia itself is an intermediate product from which fertilizers or explosives are produced in the subsequent steps. The reaction between nitrogen and hydrogen takes place at high pressure and high temperature on an iron catalyst:

$$1 \cdot N_2 + 3 \cdot H_2 = 2 \cdot NH_3$$

Ammonia is relatively easy to store and transport, as it is liquid already at –35 °C or at less than 20 bar. However, ammonia is flammable, toxic, and smells quite strongly, while at the same time having no greenhouse effect. Ammonia can therefore be transported safely with appropriate safety precautions, and the following applications are being discussed:

- **Fuel for ships:** The combustion of ammonia in engines is complex if the emission of toxic nitrogen oxides is to be kept low, but this is achievable (TRL 7).
- **Transport option for hydrogen:** A disadvantage is the very low cycle efficiency $H_2 \rightarrow NH_3 \rightarrow H_2$. This is well below 50%. In addition, the conversion of ammonia into hydrogen has not yet been sufficiently mastered technically (TRL approx. 6).

Currently, it appears that ammonia and basic materials or products derived from it, such as fertilizers, will in the future be more likely to be imported into Central Europe, since these processes make sense where renewable electricity is available at particularly low cost (Galimova et al. 2023).

All other hydrogen derivatives are hydrocarbons.

Methane For the production of synthetic methane a carbon source is first required. This carbon can come as carbon dioxide from other hard-to-decarbonize industrial processes (cement production or reduction of metal ores) or directly from the atmosphere (as direct air capture, DAC, see Chapter 14). Through the Sabatier reaction at approximately 400 °C and using a nickel catalyst, methane can then be produced directly.

$$1 \cdot CO_2 + 4 \cdot H_2 \Rightarrow 1 \cdot CH_4 + 2 \cdot H_2O$$

This method is currently being further developed to achieve higher yields and energy efficiency, as the method is not yet very efficient.

Methanol Carbon dioxide and hydrogen are also required for the production of methanol. Via the reverse water-gas shift reaction, H_2 O and CO can then be produced from the CO_2 and the H_2.

$$CO_2 + H_2 \rightleftharpoons CO + H_2O$$

The water is condensed out and is lost to the process. In a second catalytic reaction, methanol is produced from the CO and H_2.

$$CO + 2 \cdot H_2 \Rightarrow CH_3OH$$

A total of 3 molecules of hydrogen per molecule of carbon dioxide are required to produce one molecule of methanol.

Methanol is liquid at ambient pressure and temperature, easy to store and transport, and it is a convenient fuel. For this reason, methanol is currently being discussed as a fuel for ships and trucks.

Hydrocarbons The production of synthetic hydrocarbons starts in the same way as that of methanol. Here too, carbon monoxide is first produced from captured carbon dioxide. The CO is then fed together with additional hydrogen into a reactor, where it forms longer-chain hydrocarbons at higher temperature and pressure via catalytic mediation.

Unlike the other processes, the result is a mixture (syncrude), which is then refined into different fractions in a manner comparable to crude oil. Diesel, kerosene, gasoline, naphtha, and other fractions can be separated out. The composition can be influenced by the ratio of CO to H_2 and further process parameters. This is the Fischer-Tropsch process, which was historically developed for the production of gasoline from coal and can also be used for the conversion of biomass or plastic waste into fuels and naphtha.

Naphtha is a mixture of short-chain hydrocarbons and is today an essential raw material for the synthesis of plastics and solvents. The individual basic materials or solvents are extracted from the naphtha or syncrude in refineries. Technically, it is no challenge to generate organic feedstocks in the future from syncrude in appropriately designed or further developed refineries (Vogt/Weckhuysen 2024). It makes sense to build these in combination with processes that reliably provide carbon from non-fossil sources (Fasihi et al. 2016).

Currently, approximately 9% of extracted crude oil and natural gas is further processed into basic materials and thus into organic substances. If the demand for plastics remains the same, this will in the future have to be met via hydrogen and hydrogen derivatives.

Conversion

What proportion of the energy is theoretically found in the products of the processes? First, we need the molar heating values. These are calculated from

$$H_{I,\mathrm{H_2},M} = H_{I,\mathrm{H_2}} \cdot M_{\mathrm{H_2}} = 120.0\ \frac{\mathrm{MJ}}{\mathrm{kg}} \cdot 2.016\ \frac{\mathrm{kg}}{\mathrm{kmol}} = 241.9\ \frac{\mathrm{MJ}}{\mathrm{kmol}}$$

$$H_{I,\mathrm{NH_3},M} = H_{I,\mathrm{NH_3}} \cdot M_{\mathrm{NH_3}} = 18.65\ \frac{\mathrm{MJ}}{\mathrm{kg}} \cdot 17.03\ \frac{\mathrm{kg}}{\mathrm{kmol}} = 317.6\ \frac{\mathrm{MJ}}{\mathrm{kmol}}$$

$$H_{I,\mathrm{CH_4},M} = H_{I,\mathrm{CH_4}} \cdot M_{\mathrm{CH_4}} = 49.94\ \frac{\mathrm{MJ}}{\mathrm{kg}} \cdot 16.04\ \frac{\mathrm{kg}}{\mathrm{kmol}} = 801.0\ \frac{\mathrm{MJ}}{\mathrm{kmol}}$$

$$H_{I,\mathrm{CH_3OH},M} = H_{I,\mathrm{CH_3OH}} \cdot M_{\mathrm{CH_3OH}} = 19.93\ \frac{\mathrm{MJ}}{\mathrm{kg}} \cdot 32.04\ \frac{\mathrm{kg}}{\mathrm{kmol}} = 638.6\ \frac{\mathrm{MJ}}{\mathrm{kmol}}$$

Using the reaction equations, the energy contents of reactants and products can then be related, and we thus calculate the theoretically maximum possible efficiency.

Ammonia Here the theoretically maximum possible efficiency is related to the heating value of the hydrogen used (from 3 mol H_2 2 mol $NH_3$are produced)

$$\eta_{HB} = \frac{2 \cdot H_{I,\mathrm{NH_3},M}}{3 \cdot H_{I,\mathrm{H_2},M}} = \frac{2 \cdot 317.6\ \frac{\mathrm{MJ}}{\mathrm{kmol}}}{3 \cdot 241.9\ \frac{\mathrm{MJ}}{\mathrm{kmol}}} = 0.875$$

Methane For methane the result is (from 4 mol H_2 1 mol $CH_4$is produced)

$$\eta_{Sab} = \frac{1 \cdot H_{I,\mathrm{CH_4},M}}{4 \cdot H_{I,\mathrm{H_2},M}} = \frac{1 \cdot 801.0\ \frac{\mathrm{MJ}}{\mathrm{kmol}}}{4 \cdot 241.9\ \frac{\mathrm{MJ}}{\mathrm{kmol}}} = 0.823$$

Methanol And for methanol (from 3 mol H_2 1 mol CH_3OH is produced) it is

$$\eta_{Meth} = \frac{1 \cdot H_{I,\mathrm{CH_3OH},M}}{3 \cdot H_{I,\mathrm{H_2},M}} = \frac{1 \cdot 638.6\ \frac{\mathrm{MJ}}{\mathrm{kmol}}}{3 \cdot 241.9\ \frac{\mathrm{MJ}}{\mathrm{kmol}}} = 0.880$$

The actual yields are always significantly lower.

E-fuels For e-fuels the calculation becomes difficult, as we would first need to determine the molar heating value of a typical mixture. Energetic yields of up to 60 % are achieved.

10.3 Hydrogen As an Energy Carrier

For Industrial Processes

There are in some cases considerable efforts to decarbonize industrial processes through electrification. However, this will foreseeably not succeed for all processes, and it is not certain whether this can be fully implemented for all electrifiable processes by 2045. Therefore, hydrogen will be needed in the long term in the basic materials industry as a fuel gas and reducing agent. The following areas of application may be cited as examples:

- Glass melting furnaces cannot be fully electrified.
- In cement production, electric heating is extremely difficult.
- The reduction of iron ore and other metal ores requires a reducing agent that can absorb the oxygen.
- In metallurgical processes, pure hydrogen may in some cases not be used.

Since the production of green hydrogen is associated with significant energy losses and massive investments, there is a clear cost driver to electrify as many processes as possible. For many industrial processes, electrical alternatives with a sufficient technology readiness level exist. Here, real technological possibilities and developments, which always depend on capital availability and sometimes on fortunate circumstances, determine whether processes are electrified in a timely manner.

Import or local production

Foreseeably, hydrogen will be available significantly more cheaply in other regions than here in central Europe. This development depends strongly on regional energy policy, the expansion of PV, wind and transmission networks (Fasihi/Breyer 2020). Based on this argument, there may be a relocation of companies in the basic materials industry. However, there are further aspects influencing decisions about where new plants are built, or whether existing plants can be successfully continued despite an unfavorable cost structure due to further aspects such as market access, demand, and strategic framework conditions.

Hydrogen produced locally on-site represents a comparably favorable alternative to the costly import when all aspects of the supply chain are taken into account (Galimova et al. 2023). For this, a sufficient generation of electricity and a reliable grid are needed here. A future circular economy, in which basic materials and materials are reused repeatedly, recycling loops are closed as far as technically feasible, and losses are deliberately minimized through design and use, requires all relevant metallurgical processes here in Europe. For strategic reasons alone, it makes little sense to ship old products and the raw materials contained in them to other regions, only to then buy back the (critical) raw materials contained in them at high cost. A seriously pursued circular economy creates massive pressure to retain and expand the basic materials industry in Europe. This can be achieved particularly well if the required energy carriers are produced here.

As a Fuel

Large parts of today's applications of fuels can be easily electrified. Due to the significantly higher systemic energy efficiency of electric drive systems (PV-to-wheel, wind-to-wheel) compared to systems with synthetic fuels, these electric drives will prevail economically.

Hydrogen itself is in most applications neither technically nor economically appropriate as a fuel: Due to its very low density and the impressive energy input required for compression or liquefaction, hydrogen is extremely difficult and loss-prone to transfer into vehicles (refueling). The required technology is expensive, sensitive, and maintenance-intensive. The energy losses are prohibitively expensive compared to other fuels.

Power Generation

Since hydrogen is first produced from electricity, it is energetically very unfavorable to use it later again for the generation of electricity. Hydrogen represents one of the options for stabilizing the power grid during times when renewable generation is insufficient. Suitable storage facilities and peak-load power plants are required for this purpose. Otherwise, the conversion of hydrogen to electricity appears economically unviable and technically unnecessary if sufficient renewable generation capacities are provided (Ram et al. 2022), sufficient grid capacity is available, smart load management is enabled, or – as is foreseeable – additional large battery storage systems become financially viable. Such storage options are discussed in Chapter 13.

Hydrogen can be a solution for special supply situations: Extreme peak loads or electricity during a dark doldrums period can then be generated at economically acceptable conditions via hydrogen-powered gas turbine power plants or combined heat and power units. A prerequisite for this is suitable hydrogen storage. If 10 days are to be bridged completely, approximately 2 % of the future annual electrical demand must be secured in this way. The demand to be determined depends critically on whether the entire usual demand is to be generated during such periods, or whether less relevant consumers can be switched off.

For District Heating and Combined Heat and Power Plants

Since hydrogen will foreseeably be far too expensive to regularly or reliably generate electricity in a plannable manner, the reliable generation of district heating base load via hydrogen is also ruled out. Many of the district heating networks can be supplied differently in the future: waste heat from data centers or other industrial sources, river heat or ambient heat in combination with large heat pumps are viable approaches and can provide base and medium loads cost-effectively. Hydrogen or storable waste (e.g., bulky waste, waste wood, etc.) can be used in special situations for peak loads.

Space Heating

Hydrogen is unsuitable for providing space heating or as a replacement for natural gas (Rosenow 2024; Irmisch et al. 2024; Scholand 2024):

- It is not foreseeable where the required quantity of hydrogen could come from at a price affordable for households.
- In particular, poorer households are not in a position to pay the foreseeably very high price resulting from strong demand and massive process losses.
- Hydrogen is indispensable in other applications. Here it is only one option, so these other applications take priority.
- The conversion costs for households are high. In particular, with a gradual transition (in which progressively more hydrogen would be added to the natural gas) households would face several conversions in quick succession.

Space heating with hydrogen instead of oil or natural gas?

Key aspects

Safety Hydrogen has its own combustion characteristics. Therefore, burners and, where applicable, safety devices must be adapted to the concentration of hydrogen. In particular, many natural gas fittings are not designed for use with hydrogen. When hydrogen is added to the natural gas network, the manufacturer's warranty therefore typically becomes void. The owner must then assess and declare CE conformity themselves. Otherwise, the system may not continue to be operated.

Hydrogen tends to leak more than natural gas. This makes the maintenance and upkeep of systems more important in order to maintain a comparable safety standard. It must be taken into account that the safety requirements for gas systems have tightened considerably since the end of the use of hydrogen-containing gases (many urban gas networks were initially operated with town gas, coke oven gas, or similar substances, i.e., mixtures containing hydrogen and carbon monoxide in varying concentrations as fuel). Therefore, operators and in particular end users must invest in the upgrading or new installation of their pipelines and networks so that these meet current requirements.

Seals, valves, meters, and other fittings of the natural gas network are not designed for hydrogen. They must be replaced.

Capacity The volumetric heating value of hydrogen is three times lower than that of natural gas. Therefore, the current natural gas network is not designed for the required capacity of hydrogen: increasing the pressure is impermissible in large parts of the network. Higher flow velocities require more powerful and more compressors in the network and are only partially feasible (Cerbe/Lendt 2016). Large parts of the natural gas network cannot simply be repurposed but must be rebuilt for the new fuel (Scholand 2024).

Efficiency Hydrogen is foreseeably one of the most inefficient options for space heating altogether. In Figure 10.2 this is illustrated for green hydrogen, whereby the actual systemic inputs for hydrogen are still considerably underestimated. This impressive difference will be reflected in prices on the market.

Regardless of this, it must be assumed that no society can afford this massive waste of energy without jeopardizing its existence. At the core is the argument of an unbearable, because far too low, EROI (see Chapter 4).

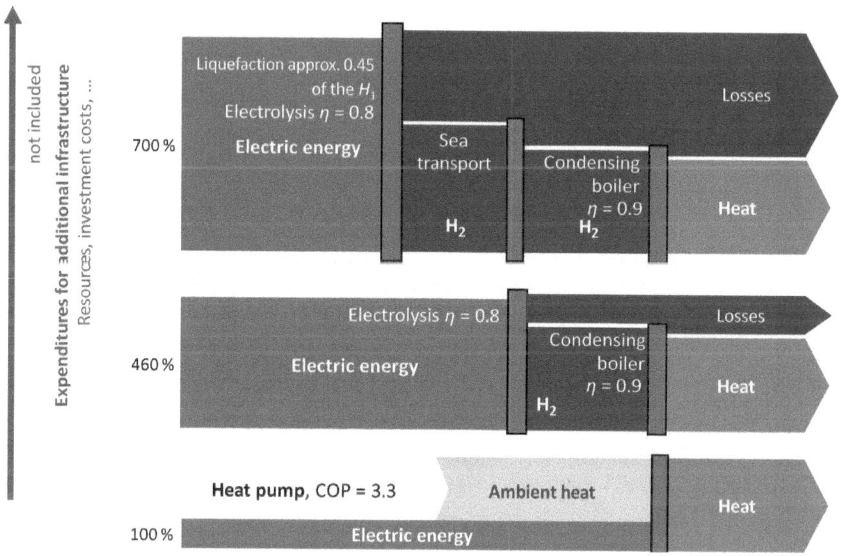

Figure 10.2 Electricity demand for the generation of space heating via imported cryo-hydrogen (top), feeding the hydrogen into the local low-pressure network (middle), and heat pump (bottom)

Blocking transformation

The idea of hydrogen as a possible replacement for natural gas, perhaps available at some point in the future, is currently preventing the urgently needed renovation of residential buildings: The impression arises that it is sufficient to wait for hydrogen and everything can remain as it is. Yet very many houses can already today be heated with heat pumps without significant restrictions. The urgently desirable, often necessary energy upgrading of residential buildings can otherwise be implemented within the framework of the usual maintenance cycles.

Ultimately, it is grossly negligent to let owners of owner-occupied residential property in particular fall into the trap of hydrogen hope: they are prevented from seeking and implementing now affordable and feasible options for adapting their residential property. The impression is also created that old boilers could be easily and cost-effectively converted to hydrogen. This depends heavily on what warranties the manufacturers of the boilers or conversion kits can then provide: the boilers are operated outside their originally intended use, which immediately voids all manufacturer liability. There is a great risk for owners of a massive loss in value of their future unheatable and thus uninhabitable residential building and of ruin. With this line of argument, the discussed blending of hydrogen into natural gas must also be classified as unsuitable and as blocking transformation.

10.4 E-Fuels

Four areas of use can be distinguished, for each of which specific criteria determine a possible use of e-fuels decide.

Passenger Cars E-fuels as hydrogen derivatives do not represent an economically viable technology as future fuels for broad individual mobility (Ueckert et al. 2021; Clausen et al. 2024). Starting from electricity and neglecting transmission losses, since these are similar for electrically powered passenger cars and for e-fuels, the typical plug-to-wheel efficiency of electric drives is approximately 0.80. The typical plug-to-wheel efficiency of e-fuels depends on the conversion efficiency of electricity into e-fuel of approx. 0.50, the losses in transport to filling stations or the Fischer-Tropsch-to-tank efficiency of 0.9, and the typical tank-to-wheel efficiency of passenger cars with internal combustion engines of 0.25 (or less). It is therefore approximately 0.11. Figure 10.3 illustrates the key contributions.

In summary, passenger cars running on e-fuel require 8 times the electricity of a comparable electrically powered passenger car. In addition, there is the impressive and yet-to-be-built industry for the production of e-fuels, which must be paid off through the price. Therefore, powering a passenger car with e-fuel should be approximately 10 to 15 times more expensive than an electric one. Therefore, e-fuels could be of significance in the prestige and luxury segment in the future.

Trucks Local commercial delivery traffic can be electrified very well. For trucks and buses on long-distance routes, e-fuels are currently being discussed as an option. On very long routes, hydrogen derivatives compete against overhead line and novel battery systems that are already being tested and enable considerably higher energy efficiency. Thus, electrified road systems such as overhead line charging on high-frequency routes sufficiently reduce battery weight and enable significantly higher efficiencies (Wietschel et al. 2017).

Aircraft The technological development of aircraft from today until around 2050 is well foreseeable. This forms the basis of the aviation industry's planning. By then, aircraft for very short routes (currently a few 10 minutes of flight time, in the medium term possibly city-hopping) will be able to be electrified.

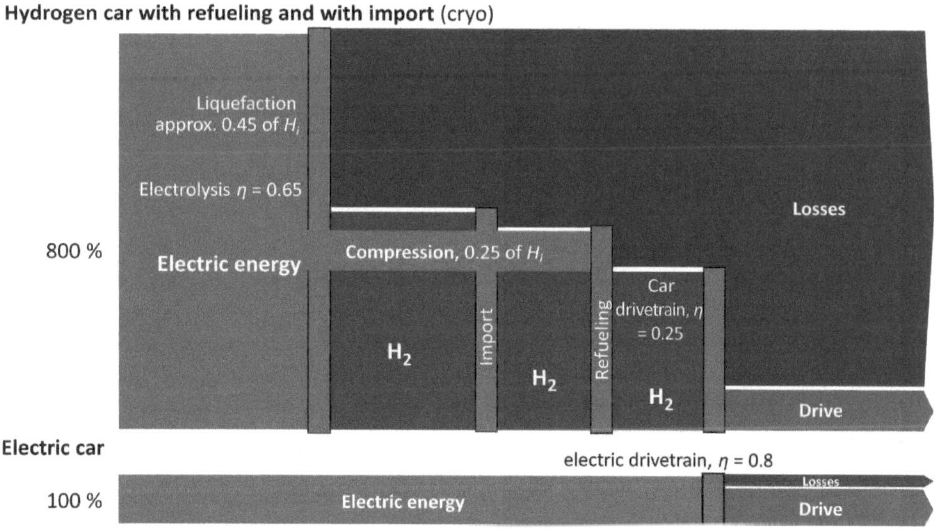

Figure 10.3 Comparison of the electricity required for the same propulsion benefit for imported hydrogen or electricity

Aircraft for longer routes will use (synthetic) kerosene at least until around 2050. Only for the period after 2050 is hydrogen being discussed as a fuel option for aviation. Even then, this will be a long-term transition, as the replacement of aircraft fleets takes place over decades.

Special Applications At the same time, there will be a clear future need for climate-neutral fuels for a few special applications: core areas of security and civil protection, in particular police, fire services, Federal Agency for Technical Relief, disaster relief and the military. Here, the supply of propulsion energy must be ensured in an appropriate form even when the electrical grid is not available.

None of these areas can be solved in any satisfactory way with hydrogen: First, the volume requirement for tanks in which gaseous hydrogen is stored even at high pressure is too high. Second, the energy input for liquefaction and in particular the losses from the tank during standstill periods due to boil-off are not acceptable. Boil-off losses from liquid hydrogen preclude its use especially in the area of disaster protection or preparedness, since hydrogen simply cannot be stored adequately for several days in this way. Pressure refueling is ruled out in the event of a power grid failure and cannot be accomplished with the stored hydrogen, since the latter is already consumed as compressor fuel for generating the required pressures.

This task is therefore foreseeably achievable, at least in part, only with liquid fuels that can be stored for a long time and easily transferred. These are e-fuels as hydrogen derivatives.

10.5 Essential Use

Hydrogen will remain scarce over a longer period of time, as the required infrastructure is only being built up. At the same time, there are a multitude of possibilities for using hydrogen. In view of strategic considerations, the decarbonization of central pillars of the economy and the management of a rapid but orderly development of a viable supply, it seems sensible to classify the possible uses. The aim of the classification is to be able to focus first on the essential uses of hydrogen.

Table 10.2 Classification of the Uses of Hydrogen and Hydrogen Derivatives According to Their Importance for the Economy and Society and Their Contribution to Decarbonization

Class	Applications
Class (3): essential	▪ Fuel for police, medicine, fire services, civil protection, military (insofar as this cannot or should not be electrified) ▪ Production of basic chemicals of hydrocarbon chemistry and ammonia ▪ Cement production (insofar as not electrifiable) ▪ Steel and iron production and processing (insofar as not electrifiable) ▪ Reduction processes in metallurgy and recycling processes (insofar as not electrifiable) ▪ Fuel for ocean-going ships ▪ Fuel for peak-load combined heat and power plants for electricity and district heating
Class (2): substitutable	▪ Space heating (heat pumps) ▪ Fuels for short-haul flights (electric or ground-based transport) ▪ Fuels for inland waterway shipping (foreseeable electrifiable) ▪ Fuels for trucks (foreseeable electrifiable)
Class (1): dispensable	▪ Fuel for historic vehicles ▪ Fuel for aircraft
Class (0): hazardous	▪ E-fuels for passenger cars ▪ Hydrogen addition to the natural gas network

Conversely, there is a risk that essential uses of hydrogen will be cannibalized or crowded out by other uses: while industrial applications are both price-critical and essential for the economic location, this does not apply to conspicuous consumption and luxury applications. In a free market, the conspicuous segment can therefore make essential uses impossible. This gives rise to the classifications for hydrogen and its derivatives as shown in Table 10.2.

10.6 Transport and Storage of Hydrogen

If the hydrogen is not produced at the point of use, it must be transported to the consumer in a suitable manner. Unlike other energy carriers, hydrogen is extremely problematic to transport. The reasons for this are:

- The very low density (Figure 10.4): the volumetric heating value under standard conditions is lower by a factor of three than that of methane. Pipelines and tanks must therefore be dimensioned correspondingly larger.
- There is a high energy input required for compression (compared, e.g., with natural gas, see the following discussion).
- The extremely low boiling point of –253 °C in combination with other material properties makes liquefaction extremely energy-intensive (see the following discussion).
- Liquid hydrogen (cryo) experiences high boiling losses due to heat transfer (owing to the extreme temperature difference between the tank and the surroundings) as well as due to ortho-para transitions. Sloshing during sea transport or in vehicles further increases the boiling loss (see the following discussion).
- The high diffusion rate through nearly all materials leads to noticeable losses, especially at high pressures.

Therefore, the transport of hydrogen is extremely expensive, energy-intensive, and, at least in the case of liquid hydrogen, additionally associated with (presumably economically unbearable) high losses.

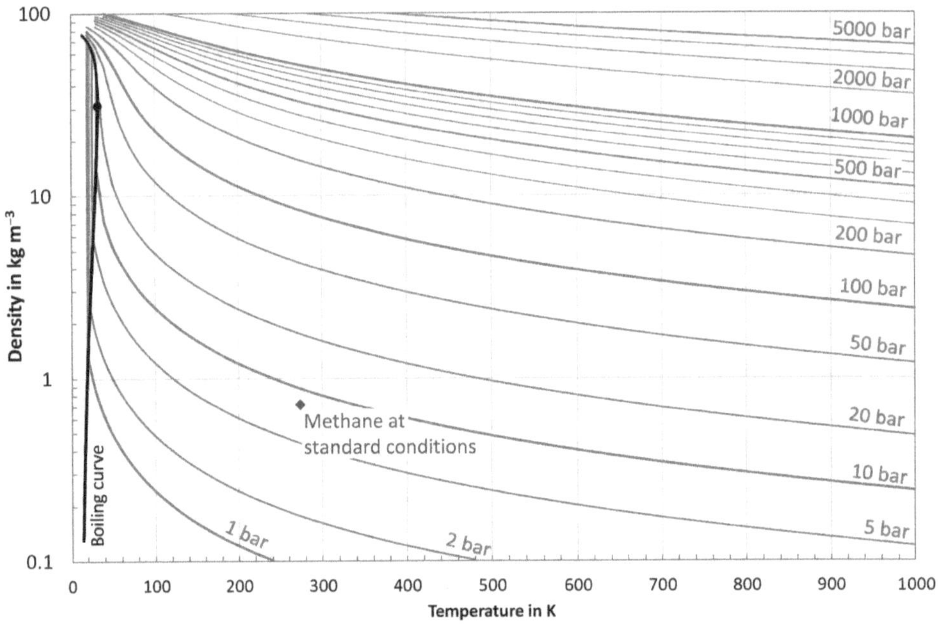

Figure 10.4 Density-temperature diagram for hydrogen

Pressure Tank Many substances are transported today in large pressure tanks: railway wagons and tank trucks make use of existing infrastructure and allow the cost-effective transport of large quantities of fuel and propellant. However, this is regularly not an option for hydrogen, since the mass of hydrogen that can be stored in conventional tanks is barely sufficient to bring the vehicle to its destination. Liquefied gas is transported, for example, in railway wagons of the ZAGS type. The permissible pressure is 17 bar. This allows 56 t of propane or $2.60 \cdot 10^{12}$ J to be transported. In such a wagon, 166 kg of hydrogen and thus $0.020 \cdot 10^{12}$ J can be accommodated: this is 1/100 of the energy.

For hydrogen high-pressure tanks in the range of 500 bar to 900 bar are being discussed. Such tanks require considerably more mass and, for their manufacture, considerably more embodied energy for a sufficiently pressure-resistant design, high manufacturing effort, close ongoing monitoring, and extreme caution in handling.

The hydrogen must be compressed to the required pressure at high energetic cost. Since such high-pressure tanks can only be safely designed with a rather small diameter, only a small storage volume can be realized. This solution appears to be economically viable only in exceptional cases and for small-scale needs.

Pipeline The transport of hydrogen by pipeline is an obvious approach. Existing pressure systems can continue to be used. Whether hydrogen pipelines can transport comparable amounts of energy as natural gas pipelines, or whether correspondingly more or larger diameters are needed, is under discussion. Due to the significantly lower viscosity of hydrogen, a higher flow velocity can be achieved with the same pumping power. Whether this velocity is desirable, however, must be clarified on a case-by-case basis (Cerbe/Lendt 2016).

Pipelines require a high initial investment. At the same time, their operating costs are quite low when compression is not taken into account. The compression power required is significantly higher for hydrogen than for all other gases (Galimova et al. 2023). Today, the required pumps are driven directly by the transported natural gas. Whether this will be sensible with the valuable hydrogen in the future, or whether pumps will be driven directly by electricity due to higher efficiency, remains to be decided.

It also remains unclear what portion of the existing network of natural gas pipelines can actually be used for pure hydrogen. In the low-pressure range, fewer problems are expected, but whether this is the case depends, due to higher leakage rates, not only on costs but also on current safety expectations and on the greenhouse effect of hydrogen (see Section 10.7).

In high-pressure pipelines, the embrittlement of steel poses a risk (Krieg 2012). Currently, the DVGW, based on random short-term investigations, assumes that this embrittlement does not affect existing pipelines to a critical extent. However, the existing network is intended to continue transporting fossil natural gas in the future, which could, if necessary and with sufficient availability, be enriched with hydrogen to 10% or, in the long term, even to 20%. This form of transition therefore does not constitute a contribution to climate neutrality by 2045.

Independently of this, a pure hydrogen network for Europe is currently being conceived (European Hydrogen Backbone, EHB). This network is intended to connect landing ports for cryo-hydrogen with consumers, and conceptually also transport pipelines that land green hydrogen from North Africa or gray/blue hydrogen from Saudi Arabia (Roth et al. 2024).

Compression Transports in pipelines or pressure tanks require the compression of hydrogen. It applies that the compression of hydrogen requires an extraordinarily large amount of energy compared to, e.g., natural gas. This can be estimated via the isentropic compression of the ideal gas, since hydrogen has an extraordinarily high specific heat capacity (Linow 2022; Roth et al. 2024). Figure 10.5 shows the specific compression work and Figure 10.6 relates this work to the energy content of the gas as the lower heating value. To compress hydrogen from 1 bar to 80 bar, approximately 20% additional hydrogen is required to operate the compressor. At 900 bar, it is approximately 100%.

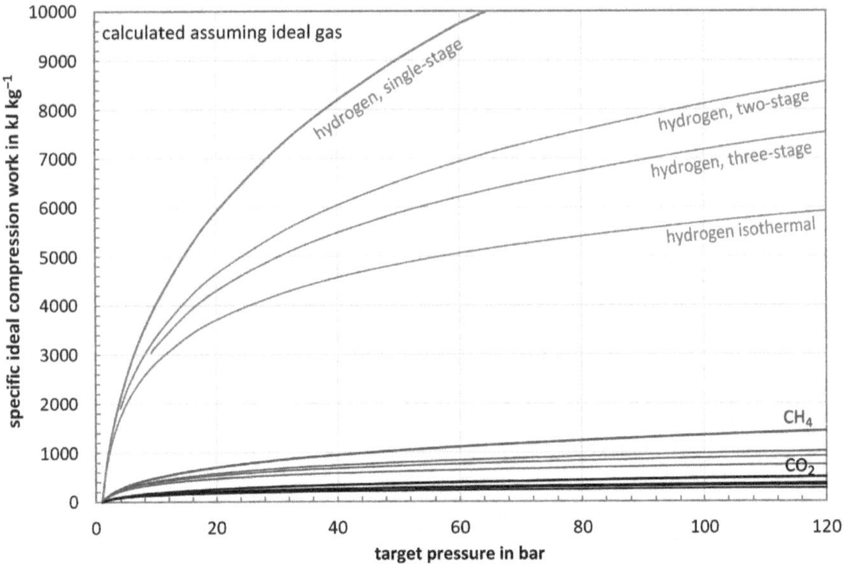

Figure 10.5 Specific ideal compression work for hydrogen, methane, and carbon dioxide

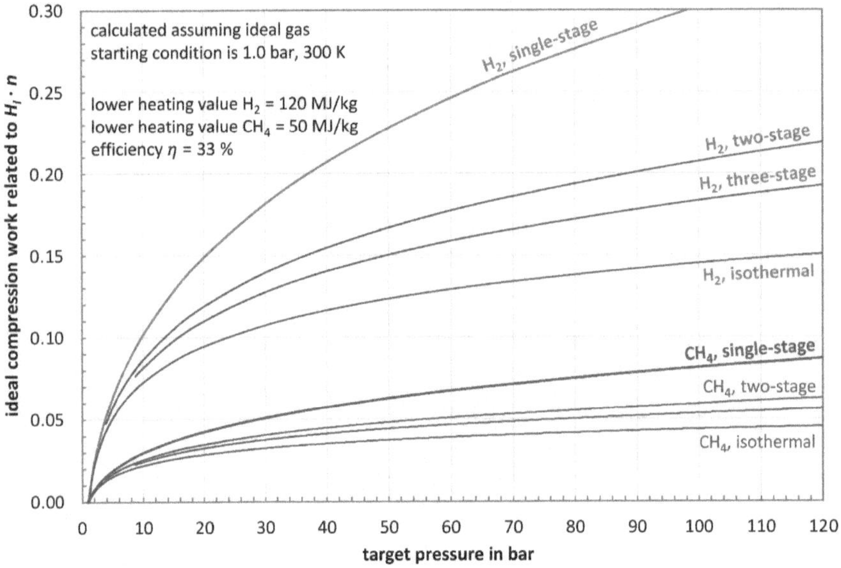

Figure 10.6 Specific compression work related to the lower heating value and the efficiency of a compressor driven by the fuel for hydrogen and methane

Special compressors are required for the compression of hydrogen. For admixtures, different compressors are again required from a certain change in physical properties, which on the one hand are adapted to the mixture and on the other hand can cope with the special requirements of hydrogen regarding materials and seals. A gradual transition from natural gas to hydrogen accordingly requires several recurring investments in suitable compressors. The higher the target pressure, the more energy and thus the higher the costs associated with compression. These costs must be borne by the consumer at the end of the pipeline.

Cryo Hydrogen cannot, unlike e.g. methane or nitrogen, be simply liquefied: hydrogen assumes a higher temperature above approximately 90 K when expanded through a valve (isenthalpic expansion) (reverse Joule-Thomson effect). This property necessitates that hydrogen must first be cooled by other refrigerants at least to the point where the actual liquefaction process can begin.

At low temperature and in equilibrium, hydrogen exists only as para-H_2. At room temperature, it is 25 % para-H_2 and 75 % ortho-H_2. The transition from ortho-H_2 to para-H_2 releases considerable heat and leads to rapid boil-off. Therefore, cold hydrogen must be catalytically converted to para-H_2 in order to liquefy it (Aziz 2021). After that, the actual liquefaction can take place. Current real liquefaction plants require approximately 50 MJ/kg of energy input. It is speculated that this value could be reduced to approximately 20 MJ/kg.

Liquid hydrogen can be transported, for example, by tanker ships. Here – as with LNG – boil-off losses occur. Stated numerical values for boil-off vary between approximately 2 % per day and 5 % per day (Aziz 2021). There are three processes that determine this:

- **Heat conduction through the tank insulation:** This can be reduced by using larger tanks and more elaborate insulation.
- Ongoing ortho-H_2- to para-H_2 transition with the associated heat release, if not all of the hydrogen has already been converted beforehand
- **Sloshing during ship movement:** Since hydrogen is light, this leads to greater movement of the ship and more sloshing than with other cargoes.

Unlike LNG, not all of the evaporated hydrogen can be used in the ship's engine here. Either the evaporated hydrogen is re-liquefied or released into the atmosphere.

During a sea voyage of 20 days, 33 % to 65 % of the originally loaded hydrogen can evaporate without re-liquefaction. The costs of the landed hydrogen increase accordingly due to these unavoidable losses.

Expansion Pipelines supply sub-networks that operate at lower pressure. When supplying natural gas, the natural gas must be tempered at these points where the gas is expanded (Joule-Thomson effect). Hydrogen exhibits a negative Joule-Thomson effect, so that the existing heating must then be switched to cooling. This would affect some 10,000 installations in the German network.

Transport with carrier substances In order to avoid the difficulties of transporting hydrogen in pure form, it is being discussed to transport hydrogen by means of chemical compounds or dissolved in other substances: either as a derivative, such as ammonia, methane or methanol (see preceding discussion), or dissolved in a liquid organic hydrogen carrier (Liquid Organic Hydrogen Carriers, LOHC).

LOHC is a type of oil in which a large amount of hydrogen can be dissolved using energy. The LOHC is transported in tanker ships and the hydrogen is extracted at the destination using heat.

Unavoidable energy losses occur during the production of the derivatives (see preceding discussion). Further considerable losses arise when the hydrogen is to be extracted again as a pure substance. Typically, significantly less than 50 % of the produced hydrogen can be utilized in this way when all involved processes are supplied with hydrogen as an energy carrier. The required infrastructure is impressively extensive and complex, making it a significant contributor to the cost structure. Therefore, this is considered today to be the most expensive transport option (Roth et al. 2024).

Fuel Hydrogen derivatives such as jet fuel and e-fuel, as well as the naphtha produced along the Fischer-Tropsch route as a feedstock for the chemical industry, can be transported very well. This appears to be a technically and economically viable transport option.

Electricity At the level of electricity, typically approximately 25 % of the originally generated electricity reaches its destination when the energy of the electricity is transported as hydrogen or a hydrogen derivative. Since, in addition, a rather impressive infrastructure must be depreciated and maintained, it can be assumed as a lower estimate that the costs of electricity at the feed-in point into the German grid are approximately a factor of 6 to 8 higher than the original generation costs. This electricity is then no longer competitive with regenerative electrical energy generated directly in Europe.

Geostrategic issues reduce the energetic benefit

Not included in the cost considerations so far are the expenditures for securing imports. The following factors are of importance in all import scenarios:

Sea routes The energy infrastructure on which Europe depends must be protected from access by malicious actors. Particularly in a world with massive disruptions caused by the climate crisis (and the associated migration movements, the possible collapse of states or regional orders, and regionally shifting centers of power), foreseeable high risks arise here for the transport infrastructure. Energy claims must be able to be enforced against other actors in order to ensure reliable delivery.

Seabed Submarine cables and pipelines on the seabed are vulnerable infrastructures that can be quickly sabotaged (Nordstream, Baltic Connector). Backup solutions are needed here. The question must be asked to what extent seabed pipelines represent a viable future transport option at all if they cannot be securely protected.

In principle, it is conceivable to enforce European claims globally through the deployment of an appropriate high-seas fleet and expeditionary corps. This generates impressive additional energetic costs that burden the import of hydrogen financially and energetically.

Local generation

In the case of direct generation of hydrogen in the region of the point of use, the following are eliminated

- Price risks, in particular since the price of hydrogen in the long term is determined by supply and not by costs,
- Climate impacts of the release of hydrogen from the transport chain and the associated risks,
- Risks of supply, and
- The in some cases massive losses of energy in the conversion chain.

Particularly for industrial applications, own generation can be a real option, since the costs of generation then become the determining factor. These depend on the one hand on the investment costs of the electrolysis plant and required intermediate storage. The second contribution, the cost of electricity, can be massively reduced in a market increasingly dominated by regeneratively generated electricity if electricity is drawn preferentially at times of high supply. Even today there are times when PV or wind must be curtailed to stabilize the grid. These times of very low or negative prices are attractive for the operation of electrolysis plants. Both the currently high price fluctuations and the times of low prices could decrease if, as is foreseeable, more battery storage and load management are integrated into the grid (see Chapter 13).

A further major advantage of a punctual start of hydrogen use and generation there is that this is already possible today. Many individual "no regret" decisions can then grow together appropriately and thus build a viable system. The risk of massive "stranded assets" is eliminated.

10.7 Greenhouse Gas

Hydrogen has a considerable greenhouse effect:

- Depending on the production process during generation, values up to the order of magnitude of 100 kg $CO_{2,eq}$ per kg of hydrogen arise from the emission of CO_2 and CH_4.
- Independently of this, the hydrogen itself produces a greenhouse effect: hydrogen is not a greenhouse gas. However, hydrogen interferes with atmospheric chemistry and prevents the breakdown of methane. Since methane is a potent greenhouse gas, hydrogen has an indirect greenhouse effect in this way. The Global Warming Potential (GWP) of hydrogen is therefore approximately 11.6 ± 2.8 as GWP 100. In the short term, the climate impact via methane is correspondingly higher and is closer to 35 for GWP 20 (Sand et al. 2023).

Even if only green (and possibly pink) hydrogen is used in the future, emissions cannot be avoided under any circumstances due to the following factors:

- Through unavoidable leakages in the pipeline system
- Through diffusion, in particular through pipelines, seals or tanks that are not diffusion-resistant
- Through venting during maintenance work on the pipeline network: Currently, the loss of imported natural gas in Germany through maintenance work via venting is approximately 1%, although this varies considerably from supplier to supplier.
- Through boil-off from cryo-tanks: Although optimistic and very optimistic figures can be found in the literature, current real values are 2% per day for very large stationary tanks and 5% per day for smaller tanks (Aziz 2021).

The release of hydrogen can be minimized by a number of measures, which in doing so shift the possibilities of a future hydrogen supply significantly towards local generation at the consumer's site:

- Minimization of network length through preferred generation at the point of use, instead of pipeline transport
- Use of suitable low-diffusion material for pipelines, i.e. dedicated new construction of the hydrogen network
- Avoidance of cryo-hydrogen: Re-liquefaction of the boil-off loss is realistically only sensible for very large tanks, i.e. for sea transport

Otherwise, the hydrogen emissions must be offset by suitable negative emissions (e.g. of carbon dioxide, see Chapter 14) in order to enable greenhouse gas neutrality. These are additional energy inputs that are included in the assessment of hydrogen.

References

Aziz, M.: Liquid hydrogen. A review on liquefaction, storage, transportation, and safety. In: Energies 14, 2021, p. 5917.

Bareiß, K./Rua, C. de la/Möckl, M./Hamacher, T.: Life cycle assessment of hydrogen from proton exchange membrane water electrolysis in future energy systems. In: Applied Energy 237, 2019, pp. 862–872

Cerbe, G./Lendt, B.: Grundlagen der Gastechnik: Gasbeschaffung – Gasverteilung – Gasverwendung. 8. Auflage. Carl Hanser Verlag, München 2016

Clausen, J./Altermatt, P. P./Beucker, S./Gerhards, C. /Linow, S.: Niedersachsen und die E-Fuels. Borderstep Institut 2024. *https://www.borderstep.de/wp-content/uploads/2024/05/Niedersachsens-Efuels-2024-erg.pdf*

Fasihi, M./Bogdanov, D./Breyer, C.: Techno-Economic Assessment of Power-to-Liquids (PtL). Fuels Production and Global Trading Based on Hybrid PV-Wind Power Plants. In: Energy Procedia 99, 2016, pp. 243–268

Fasihi, M./Breyer, C.: Baseload electricity and hydrogen supply based on hybrid PV-wind power plants. In: Journal of Cleaner Production 243, 2020, 118466

Galimova, T./Fasihi, M./Bogdanov, D./Breyer, C.: Feasibility of green ammonia trading via pipelines and shipping: Cases of Europe, North Africa, and South America. In: Journal of Cleaner Production 427, 2023, 139212

Galimova, T./Fasihi, M./Bogdanov, D./Breyer, C.: Impact of international transportation chains on cost of green e-hydrogen: Global cost of hydrogen and consequences for Germany and Finland. In: Applied Energy 347, 2023, 121369

Howarth, R. W.: The greenhouse gas footprint of liquefied natural gas (LNG) exported from the United States. In: Energy Science and Engineering 12, 2024, pp. 4843–4859

Irmisch, T./Peters, H./Hansen, W.: Achtung Kostenfalle. Risikoinvestition Wasserstoff in der kommunalen Wärmeplanung. 2024.*https://umweltinstitut.org/wp-content/uploads/2024/02/Infoblatt_Kostenfalle_Wasserstoff.pdf* und *https://umweltinstitut.org/energie-und-klima/wasserstoff*

Krieg, D.: Konzept und Kosten eines Pipelinesystems zur Versorgung des deutschen Straßenverkehrs mit Wasserstoff. Schriften des Forschungszentrums Jülich. Reihe Energie und Umwelt 144. Verlag des Forschungszentrums Jülich 2012. *https://juser.fz-juelich.de/record/136392*

Linow, S.: Angewandte technische Thermodynamik. Carl Hanser Verlag, München 2022

Ram, M./Bogdanov, D./Satymov, R./Lopez, G./Mensah, T. N. O./Sadovskaia, K./Breyer, C.: Accelerating the European renewable energy transition. LUT University and Greens European Free Alliance, Lappeenranta, Brussels 2022

Rosenow, J.: A meta-review of 54 studies on hydrogen heating. In: Cell Reports Sustainability 1, 2024, 100010

Roth, S./Abdelrahman, O./Hüllhorst, T./Sari, K./Griffith, H.: Bulk Transport Options for Green Molecules. Focus Area: Europe and MENA Region. 2024. *https://dii-desertenergy.org/publications*

Sand, M./Bieltvedt Skeie, R./Sandstad, M./Krishnan, S./Myhre, G./Bryant, H./Derwent, R./Hauglustaine, D./Paulot, F./Prather, M./Stevenson, D.: A multi-model assessment of the Global Warming Potential of hydrogen. In: Communications Earth & Environment 4, 2023, 203

Schmidt, T.: Wasserstofftechnik. Grundlagen, Systeme, Anwendungen, Wirtschaft. Carl Hanser Verlag, München 2022

Scholand, M.: Debunking the Hydrogen Hype. Why Europe should pursue electrification instead of Hydrogen for cooking and heating. 2024

Smith, C./Hill, A. K./Torrente-Murciano, L.: Current and future role of Haber-Bosch ammonia in a carbon-free energy landscape. In: Energy & Environmental Science 15, 2020, pp. 331–344

Ueckerdt, F./Bauer, C./Dirnaichner, A./Everall, J./Sacchi, R./Luderer, G.: Potential and risk of hydrogen based e-fuels in climate change mitigation. In: Nature Climate Change 11, 2021, pp. 384–393

Vogt, E. T. C./Weckhuysen, B. M.: The refinery of the future. In: Nature 629, 2024, pp. 295–306

Wietschel, M./Gnann, T./Kühn, A./Plötz, P./Moll, C./Speth, C./Buch, J./Boßmann, T./Stütz, S./Schellert, M./Rüdiger, D./Balz, W./Frik, H./Waßmuth, V./Paufler-Mann, D./Rödl, A./Schade, W./Mader, S.: Machbarkeitsstudie zur Ermittlung der Potentiale des Hybrid-Oberleitungs-Lkw. Studie im Rahmen der Wissenschaftlichen Beratung des BMVI zur Mobilitäts- und Kraftstoffstrategie. Auftraggeber: Bundesministerium für Verkehr und Digitale Infrastruktur (BMVI). 2017.*https://publica.fraunhofer.de/entities/publication/4e8883cf-5dab-4a8c-93be-d22cf04d212f/details*

11 Harnessing Flows

Mechanical power or electricity generated from a heat flow is the subject of Chapter 9 and Chapter 10. In this chapter, the focus is on processes and technical methods whose starting point is the material flows of the atmosphere or the water cycle. These material flows can be harvested to a certain degree. First, these material flows are used to provide mechanical power. This can then be used directly. Today it is mostly converted into electricity. Since mechanical work does not transport entropy, a theoretical efficiency of 1 can be achieved when converting work from one form to another. In practice, however, every form of energy conversion is associated with losses, which are typically caused by friction. This dissipation is lost to the usable work and generates entropy.

11.1 Hydropower

The maximum global potential for hydropower was already discussed in Chapter 7. Hydropower is mostly the conversion of energy present in water using a water wheel or a turbine into shaft energy (here energy that can be extracted from a rotating shaft). Usable energy in water can exist in three forms:

1. Due to a difference in height of the water relative to the surroundings as potential energy
2. Due to the kinetic energy of a flowing body of water or a current
3. Due to pressure that the water exhibits at a point of extraction relative to the surroundings

The utilization of mechanical power takes place either directly, e.g. in mills, sawmills, hammer mills or in the water machinery of mines (operation of pumps, elevators, cranes, hammer works), or indirectly via conversion to electricity in a generator.

To describe the energetic potentials of a flowing fluid, the Bernoulli equation for frictionless flows of incompressible fluids is a good starting point. In pressure representation it reads (Quaschning 2023)

$$\underbrace{p}_{\substack{hydrostatic\\pressure}} + \underbrace{\rho \cdot g \cdot \Delta z}_{height} + \underbrace{\rho \cdot \frac{c^2}{2}}_{\substack{flow-\\velocity}} = \text{const} \tag{11.1}$$

In the height representation the following applies

$$\underbrace{\frac{p}{\rho \cdot g}}_{\substack{hydrostatic\\pressure}} + \underbrace{g}_{height} + \underbrace{\frac{1}{g} \cdot \frac{c^2}{2}}_{\substack{flow-\\velocity}} = \text{const} \tag{11.2}$$

In this equation, p is the pressure, Δz the relative height and c the velocity. The Bernoulli equation describes the conservation of momentum in the fluid. Water can be very well regarded as an ideal liquid, i.e. as incompressible. Friction losses in flowing water can be neglected as a first approximation, since liquid water has a very low viscosity. The three terms of the equation correspond to the three forms of energy: pressure energy, potential energy and kinetic energy of flowing water. From this equation, the usable head of hydropower follows in general as

$$\Delta z_{use} = z_{in} - z_{out} + \frac{c_{in}^2 - c_{out}^2}{2 \cdot g} + \frac{p_{in} - p_{out}}{\rho_{fluid} \cdot g} \tag{11.3}$$

where the index *in* describes the water before and the index *out* describes the water after the hydropower plant or turbine. Here, the potential of pressure and velocity is thus represented as a head. This representation is helpful, since the pressure and velocity of flowing water result from a difference in height. The energy contained in the water and available for use is thus traced back to a potential energy. In real hydropower plants, one of the effects contained here is preferably exploited (Quaschning 2023): Overshot water wheels mostly use only the height term. Undershot water wheels primarily use the velocity. In Kaplan turbines and crossflow turbines, the flow velocity of the water is primarily used. Pelton turbines use the pressure difference.

In every case, the difference in height between the surface of the body of water far upstream and far downstream of the turbine is the source of the potential energy. This difference in height is converted in the conduit system near the turbine into usable kinetic power or usable pressure difference. For the precise design of turbines, the system boundary is relevant, which can be set differently depending on the question at hand. The maximum available power of the water follows from the usable head difference

$$P_{pot} = \dot{m} \cdot g \cdot \Delta z_{use} = \dot{V} \cdot \rho_{fluid} \cdot g \cdot \Delta z_{nutz} \tag{11.4}$$

The power actually generated additionally depends on the flow losses and the efficiency of the energy conversion in the plant. From an analysis of the pressure losses in real hydropower plants, the description of the actually achievable usable head is obtained as

$$\Delta z_{use} = z_{in} - z_{out} - \varsigma_{inlet} \cdot \frac{c_{inlet}^2}{2 \cdot g} - \varsigma_{pipe} \cdot \frac{c_{pipe}^2}{2 \cdot g} - \left(1 + \varsigma_{diff}\right) \cdot \frac{c_{diff}^2}{2 \cdot g} \tag{11.5}$$

This formula refers to the typical design of a power plant, as in Figure 11.1 shown. Important elements are taken into account with the head loss coefficient of the inlet structure and screen (1→2) ξ_{inlet}, the head loss coefficient of the pipelines between inlet and turbine (2→3) ξ_{pipe} and the head loss coefficient of the outlet or diffuser behind the turbine (4→5) ξ_{Diff}. Such head loss coefficients can be provided by the manufacturer.

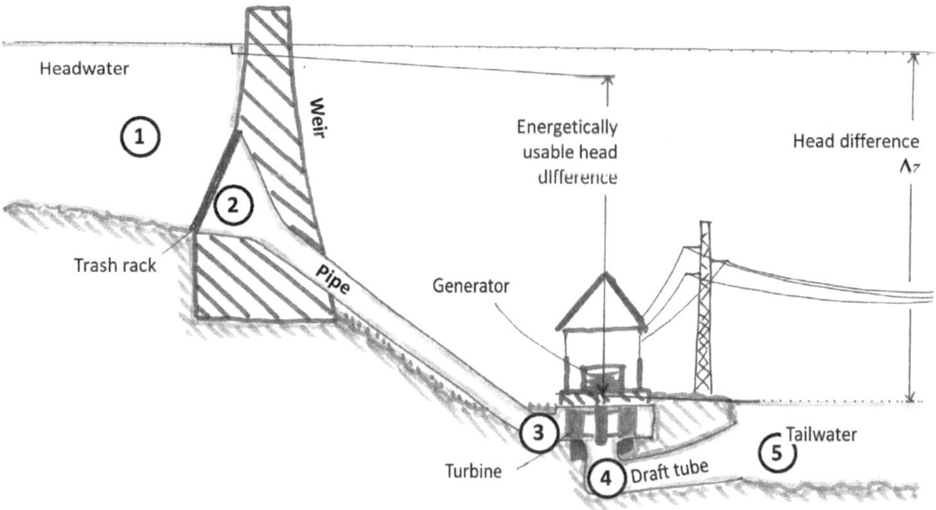

Figure 11.1 Layout of a hydropower plant: the numbers designate positions used in the text.

Regardless of this, the power actually available from the watercourse depends on the current flow rate. This varies with the seasons and with the weather. Current data and mean annual cycles for almost all potentially usable or used watercourses can be found on the websites of the responsible state authorities.

11.1.1 Technical Solutions

In order to be able to use hydropower, a flowing watercourse must be available for this purpose. This watercourse must be made accessible and developed in accordance with the intended use:

- A historically relevant option was river mills. For this purpose, a pontoon was anchored in the river on which a mill was mounted. River mills barely affect the watercourse. However, they can only make use of a very small proportion of the kinetic energy of the river. At the same time, they obstruct shipping, as they are preferably positioned in the area of the highest flow velocity in the middle of the river. Due to the high material input compared to the low possible yield, the EROI of this technology is extremely low.
- In mountainous areas with fast-flowing watercourses, small mills can be integrated directly into the current without requiring additional structures: the waterwheel then protrudes directly into the flow. Such mills only use a small proportion of the kinetic energy of the watercourse.
- With sufficient gradient of the watercourse, part of the stream or river is diverted and led along the slope via a channel that runs almost without gradient, until a sufficient gradient is reached opposite the watercourse in the valley floor to operate a mill. The water flows in the channel only when the mill is in operation. This method is particularly suitable when only a fraction of the power of the watercourse is needed, when many users are arranged in a small space but all require the water only intermittently, or when the watercourse would otherwise pose a danger to the mill during flooding. Such installations are characteristic of many low mountain range locations.
- With small watercourses, in flat terrain, or when precipitation varies greatly, the entire watercourse is dammed up by a few meters in order to then operate a mill with the impoundment pond thus created. Such mill ponds can also serve as energy storage when the mill is only used intermittently.
- So-called run-of-river power plants are used in particular in navigable rivers. In order to develop rivers for ships that are quite large relative to the watercourse, weirs with locks are installed at regular intervals. There, the water that is not needed for the locks can be channeled through a power plant. Out of consideration for shipping traffic and riparian residents, the water level in the watercourse may only fluctuate to a small extent. The dam is in this sense not a storage facility.
- Large reservoirs impound the water from a large catchment area. Such reservoirs are often designed to be large enough to compensate for seasonal fluctuations or multi-year dry periods. Such dams often have additional functions. They are intended, for example, to provide drinking water (Lake Mead on the Colorado), to reliably supply water for agriculture (Aswan Dam on the Nile), and to prevent flooding by the river (also the Aswan Dam on the Nile).

Hydropower in the Earth System

The variants mentioned at the beginning of Section 11.1.1 span the range of possible power plants from small installations that barely affect the watercourse to giant dams that strongly shape the great rivers of the earth and entire regions. The impacts of the power plant begin with the direct influence on the watercourse, but also have repercussions on the surrounding society and ecosystems. For the variants listed above, the impacts increase from top (low) to bottom (very high). These impacts are fundamentally greater the higher the dam and the larger the area of the reservoir, and the further the dam intervenes in the watercourse. Important aspects and key figures for their assessment are summarized in Table 11.1. Impacts must also be taken into account even if they do not affect one's own territory. The non-exhaustive list in Table 11.1 provides an overview of relevant issues.

Table 11.1 Impacts of Dams and Possible Indicators for Assessment

Aspect	Effects	Key figures
Mechanical power or electricity	The provision of mechanical power or electricity is intended.	Generated electricity, pumped water, mass of ground grain, etc.
Usable water	Evaporation from large open water surfaces	Change in volume flows
Flood protection	Prevention of recurring floods In wealthy countries, insurance is a good benchmark, as insurers are experienced in assessing real changes in risks. They assess according to established scientific criteria themselves.	Insurance costs of downstream residents
Dam failure	Devastating in the lower reaches	Insurance costs of downstream residents
Area for agriculture	Area changes due to the dam: ▪ Gains through irrigation ▪ Gains through potentially possible additional planting cycles (two harvests instead of one) ▪ Losses due to areas covered by the reservoir ▪ Losses due to erosion and subsidence in the river delta ▪ Losses due to development of agricultural areas made flood-safe by the dam	▪ Usable agricultural area ▪ Yields of the utilized areas ▪ Change in usable area

Aspect	Effects	Key figures
Soil for agriculture the change in soil should be accounted for separately from the pure area change.	The soil is flooded long-term by the reservoir. Sediment that would otherwise form new soil after flooding is deposited in the reservoir. Sediment in the reservoir is not soil, even if it consists of previously eroded soil.	■ Soil quality of the lost and newly developed areas ■ Sediment flows
Fertilization for agriculture	Soils regularly flooded by a river are fertilized free of charge by the river sediments deposited in this way. This fertilization must be replaced by technical fertilization. This results in additional costs for farmers. These additional costs must be covered by the yield and often lead to a change in the crops grown within the area of influence of the dam.	■ Costs for fertilizers ■ Cumulative energy demand for the fertilizers and the fertilization ■ Costs for treating the unwanted effects of fertilization In terms of energy, all inputs required for fertilization must be deducted from the power output of the power plant (as CED, see chapter 6).
Food: fish	Changes in the river regularly lead to a change in fish stocks: the loss of regularly flooded floodplains and other spawning grounds and the severing of migration routes significantly reduce biodiversity and fish stocks (thus, there is no longer any river fishing in Central European rivers, as there are no longer any significant fish stocks).	■ Balance of protein production for the population ■ Fish stocks and catch figures of the fishery along the entire river and in the delta This loss must be factored into the agricultural yield, as other sources of animal protein supply for the population are now required.

Aspect	Effects	Key figures
Energetic benefit and EROI	The energetic output must be balanced against the energetic input. In a complete system analysis, these aspects must also be taken into account: ▪ Protection of eroding coastlines ▪ Production of artificial fertilizer ▪ Decommissioning at end of life or if the dam does not function as hoped ▪ Military expenditures due to conflicts arising from water use	▪ CED ▪ EROI ▪ Net energy
Greenhouse gases	The accounting of greenhouse gas emissions must take into account the CO_2emissions from construction activities and cement production as well as the emission of methane from swamps and stagnant water surfaces, i.e. the reservoir.	Emissions of CO_2, methane, N_2O from fertilization and, where applicable, further greenhouse gases
Direct global warming	High absorption by open water surfaces compared to vegetation or soil	Altered radiation balance
Financial costs of the project	Direct costs are those for construction and maintenance as well as decommissioning. Indirect costs are costs for resettlement, additional construction and protective measures also along the lower reaches of the watercourse and at the coast.	▪ Full cost accounting ▪ Indirect costs ▪ Externalized costs ▪ Costs of all downstream users ▪ Cost-benefit analysis
Capital costs	Loans: direct and indirect costs of high national debt The indirect effect of national debt is often the long-term dependency on a creditor and on its strategic objectives in the region.	▪ Cost-benefit analysis ▪ Strategic assessment
Economy, food production	Loan costs, fertilizers and changes in ownership through irrigation all have impacts. Loans may necessitate long-term commitment to the production of export goods.	▪ Change in production ▪ Supply of the population ▪ Exports and imports

Aspect	Effects	Key figures
End of life	For large projects, it must be clarified in principle whether and how the dam can be decommissioned at any time.	Technology lock-in
Biodiversity	▪ Losses of ecosystems and foundations of the living world ▪ Loss of habitat, life forms, ecosystem services, and livelihoods	

Costs of Dams

On the topic of cost development and cost calculation of large projects there is extensive literature (Ansar et al. 2014, Flyvbjerg 2023). This shows how expectations and self-interest of capital-strong actors can lead to the cost-benefit analysis not being carried out with the required diligence and without taking into account all relevant risks. The benefit is often estimated with reference to the possible generated energy and agricultural yields, especially in such an environment, under decidedly optimistic assumptions that do not materialize in long-term operation. An assessment of current projects and possible alternatives using the example of the Mekong is given by Schmitt et al. (Schmitt et al. 2019). Here exists a strong conflict of interest between the generation of electricity on the one hand and the preservation of the watercourses as natural livelihoods and as suppliers of animal protein for their residents on the other (Grill et al. 2019).

Walchensee

The Walchensee power plant is a hydroelectric power plant that can also be used as an energy storage facility. It was built at the beginning of the electrification of Germany and still supplies electricity for the railway today. The power plant makes use of a natural favorable location, as the Walchensee is separated from the Kochelsee, which lies 200 m lower, only by a narrow ridge. In the power plant on the shore of the Kochelsee, four Francis turbines with a rated output of 18.0 MW each and four Pelton turbines with a rated output of 13.0 MW each are installed. The maximum volumetric flow rate at rated output is $\dot{V} = 84\ \mathrm{m^3\ s^{-1}}$. The Walchensee has an area of 16 km^2 and its water level can be lowered by a maximum of 6.6 m.

What is the efficiency of the power plant at rated output?

The theoretical power at the turbines, neglecting all losses, is

$$P_{th} = \dot{m} \cdot g \cdot \Delta z = \dot{V} \cdot \rho_{water} \cdot g \cdot \Delta z = 84\ \frac{\mathrm{m}^3}{\mathrm{s}} 1000\ \frac{\mathrm{kg}}{\mathrm{m}^3} \cdot 9.81\ \frac{\mathrm{m}}{\mathrm{s}^2} \cdot 200\ \mathrm{m} = 164.8\ \mathrm{MW}$$

The efficiency, as the ratio of the electrical rated output to the theoretical power, is therefore

$$\eta_{el} = \frac{P_{N,el}}{P_{mech,water}} = \frac{4 \cdot 18.0\ \text{MW} + 4 \cdot 13.0\ \text{MW}}{164.8\ \text{MW}} = 0.752$$

This is a rather low efficiency for a hydroelectric power plant. This efficiency could be attributed to the turbines and generators, which are no longer quite new. All of them are still originals from 1924.

What is the maximum operating duration at rated output and without inflow?

The maximum available volume in the Walchensee follows from the area and the permissible difference in water level in the reservoir:

$$V_{\max} = \Delta z_{Walchen} \cdot A_{Walchen} = 6.6\ \text{m} \cdot 16.0 \cdot 10^6\ \text{m}^2 = 105 \cdot 10^6\ \text{m}^3$$

At maximum outflow and without any inflow, the maximum operating duration is

$$\Delta t = \frac{V_{\max}}{\dot{V}_N} = \frac{105.6 \cdot 10^6\ \text{m}^3}{84\ \frac{\text{m}^3}{\text{s}}} = 1\ 257\ 000\ \text{s} = 349.2\ \text{h} = 14.6\ \text{d}$$

The Walchensee power plant has a significant influence on the Isar, which is partly diverted to fill the lake, as well as on the Loisach, which drains the Kochelsee: the Loisach has a mean natural flow of about 25 $m^3\ s^{-1}$, which can rise to 100 $m^3\ s^{-1}$ when the power plant is in operation.

11.1.2 Ocean Currents

A special variant of the use of hydropower is the use of ocean currents. Here we can distinguish the following variants:

1. True ocean currents such as the Gulf Stream flow with little temporal variation. Ocean currents can exhibit an impressive mass flow rate; however, they are extremely difficult to harvest, since the current mostly runs in the open sea, extends over a very large cross-sectional area, and thus has a rather low velocity.
2. Tidal currents result from ebb and flow. They therefore exhibit strong periodic fluctuations in direction and velocity. The timing and height of the tides can be predicted and planned very well. The current wind can slightly shift the timing of the tides and in particular cause them to run much higher or lower than normal.
3. Wave motions can in principle be utilized for the generation of electricity by a whole range of technical concepts.

All devices intended for use in the sea must be specially designed and manufactured for this purpose. The forces acting during storms and the impressive corrosion caused by saltwater limit possible concepts and materials. Both increase the necessary energy costs.

Should we use the Gulf Stream?

The Gulf Stream begins as the Florida current between Cuba and Florida and then flows across the North Atlantic (Figure 11.2). Depending on its position, it can have a volume flow rate of up to 150 Sv (Sv or Sverdrup is $10^6\ m^3\ s^{-1}$). The velocity in the Florida current at a volume flow rate of 30 Sv is up to 1.8 m s^{-1}. The seawater there has a density of approximately 1025 kg m^{-3}.

If we assume that the Gulf Stream flows as a homogeneous current with constant velocity, the power contained in it from the kinetic energy of the flow alone would be at most

$$P_{Florida\text{-}current} = \dot{V} \cdot \rho \cdot \frac{c}{2} = 30 \cdot 10^6\ \frac{\mathrm{m}^3}{\mathrm{s}} \cdot 1025\ \frac{\mathrm{kg}}{\mathrm{m}^3} \cdot \frac{\left(1.8\ \frac{\mathrm{m}}{\mathrm{s}}\right)^2}{2} = 49.8\ \mathrm{GW}$$

This power would be distributed over a cross-sectional area of

$$A_{Florida\text{-}current} = \frac{\dot{V}_{Florida\text{-}current}}{c} = \frac{30 \cdot 10^6\ \frac{\mathrm{m}^3}{\mathrm{s}}}{1.8\ \frac{\mathrm{m}}{\mathrm{s}}} = 16.7\ \mathrm{km}^2$$

In reality, the Gulf Stream occupies a larger cross-section and flows at a lower average velocity.

This power corresponds to approximately 75% of Germany's electricity demand, but is extremely difficult to harvest. In addition, only a fraction of the kinetic energy may be extracted, as otherwise the current would come to a halt, with unpleasant consequences for the global climate.

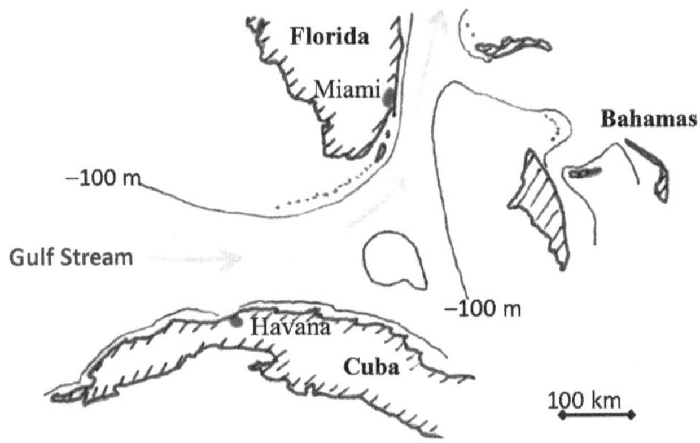

Figure 11.2 Course of the Florida current

11.1.3 Tidal Power Plants

Tidal currents with high kinetic energy form near coastlines. They are therefore much easier to harvest than ocean currents in the open sea. In a tidal power plant, the alternation between low tide and high tide is used to generate electricity. Ebb – the receding of the water – and flood – the rising of the water – are determined by the movements of the moon and sun and are therefore well and long-term plannable. The following factors are relevant for technical utilization:

- The usable tidal range – i.e., the difference in sea level between low tide and high tide – must be several meters. Projects typically start at around 6 m tidal range.
- A coastal configuration must be present that allows the straightforward construction of a dam to create a relevant height difference between the inside and outside. Deep bays are therefore particularly preferred.
- The risk to the dam from storms or extreme weather must not be too high.
- Further trade-off considerations are necessary, particularly with regard to shipping, coastal protection, fisheries, and ecological impacts.

Ultimately, there are therefore only a few suitable sites worldwide: France, Great Britain, Korea, and eastern Canada have suitable bays or coastal sections.

Key advantages of tidal power plants are that the producible power can be planned far in advance, since tidal calendars are available years ahead. The power is independent of clouds, and overall there is only a minor and at least short-term predictable influence of weather through storms.

The disadvantages arise from the rapid fluctuation of the producible power. In addition, there is no clear day-night cycle of generation, since the tides shift with the moon over the course of the month. This means that demand-driven generation is not possible, only generation-dependent utilization.

Swansea Seawall, a tidal power plant

In Swansea, Great Britain, a tidal power plant was planned: The plan envisaged separating an artificial lagoon of 11.5 km^2 area from the sea by a 9.5 km long dam. In Swansea, the tidal range is very high. It averages 6.2 m. Part of this height difference between the sea and the lagoon can be utilized by turbines. The lagoon was to be operated with a mean height difference (full minus empty) of 4.5 m. The plan Included:

- 16 turbines with a rated power of 20 MW each were to be installed.
- The projected annual electricity yield was 530 GWh.
- The maximum volume of water to flow through the turbines was to be 6000 $m^3\,s^{-1}$ flow.
- The turbines have a maximum efficiency of 91 %.
- The construction costs of the plant were estimated at 1470 M€.
- The generated electricity would later be purchased at an average of 0.06 € kWh^{-1}.
- At a minimum head difference of 1 m, the turbines could be switched on.

The following materials would be required for the construction of the plant:

1. 100,000 t of steel are required.
2. The volume of concrete can be estimated using the width of the turbine structure (500 m), the height (30 m), the depth (150 m), and an average fill fraction of 30 %. The fill fraction indicates what proportion of the cuboid is actually filled with concrete. Concrete has a density of 2200 kg m^{-3}. This should make the plant sufficiently heavy to prevent it from floating up.
3. The dam consists of a sand core, which is constructed using geotextiles (tubes) and secured on the outside with rocks.
4. The rock comes from Cornwall (100 km by sea and 20 km by road).
5. The average cross-sectional area of the dam is 1000 m^2 of sand, plus 700 m^2 for the boulder cover layer with an average density of 1500 kg m^{-3}.
6. On the dam crest there is a concrete roadway with a cross-sectional area of 5 m^2.
7. The investor states a service life of the plant of 120 years.

How long does the operator need to recoup the construction costs at the estimated electricity price?

If no capital or operating costs are taken into account, the result is

$$\Delta t_{payback} = \frac{C_{invest}}{W_{el} \cdot C_{el}} = \frac{1.47 \cdot 10^9\ €}{530\ \frac{\text{GWh}}{\text{a}} \cdot 0.06\ \frac{€}{\text{kWh}}} = 46\ \text{a}$$

The planning presumably assumes a significantly higher price for the generated electricity, or the power plant is to be financed from other sources as well.

At what head difference between the sea and the lagoon is the rated power reached?

From a head difference of

$$\Delta z = \frac{P_{pot}}{\dot{m} \cdot g} = \frac{P_{el}}{\eta \cdot \dot{V} \cdot \rho \cdot g} = \frac{16 \cdot 20\ \text{MW}}{0.91 \cdot 6000\ \frac{\text{m}^3}{\text{s}} \cdot 1027\ \frac{\text{kg}}{\text{m}^3} \cdot 9.81\ \frac{\text{m}}{\text{s}^2}} = 5.82\ \text{m}$$

the rated power is reached.

What power do the turbines deliver at a head difference of 1 m?

We cannot actually answer this question, since we do not know how the flow of seawater varies with the head. If we assume that it is constant, the minimum power is

$$P_{el} = \eta \cdot \dot{V} \cdot \rho \cdot g \cdot \Delta h = 0.91 \cdot 6000\ \frac{\text{m}^3}{\text{s}} \cdot 1027\ \frac{\text{kg}}{\text{m}^3} \cdot 9.81\ \frac{\text{m}}{\text{s}^2} \cdot 1.0\ \text{m} = 55\ \text{MW}$$

How quickly does the lagoon fill up at maximum flow?

At a constant maximum flow and a head difference of 4.5 m, the following time is required:

$$\Delta t = \frac{V_{lagoon}}{\dot{V}_{\max}} = \frac{A_{lagoon} \cdot \Delta h_{lagoon}}{\dot{V}_{\max}} = \frac{11.5 \cdot 10^6\ \text{m}^2 \cdot 4.5\ \text{m}}{6000\ \frac{\text{m}^3}{\text{s}}} = 8624\ \text{s} = 2.40\ \text{h}$$

What is the CED using the given data?

The energy input based on the data is

$$\text{CED}_{steel} = m_{steel} \cdot \text{ced}_{steel} = 2.77 \cdot 10^{15}\ \text{J}$$

Basalt is the gravel cover on the sand core. The sand is extracted directly on site. The following values result:

$$m_{basalt} = V_{basalt} \cdot \rho = l \cdot A \cdot \rho = 9.975 \cdot 10^9 \text{ kg}$$

$$\text{CED}_{basalt} = m_{basalt} \cdot \text{ced}_{basalt} = 2.39 \cdot 10^{15} \text{ J}$$

Transport by truck (the return journey is weighted at 0.5) gives

$$\text{CED}_{truck} = m_{basalt} \cdot 1.5 \cdot \Delta x_{truck} \cdot \text{ced}_{truck} = 0.212 \cdot 10^{15} \text{ J}$$

The transport of basalt by ship (the return journey is weighted at 0.5) results in

$$\text{CED}_{ship} = m_{basalt} \cdot 1.5 \cdot \Delta x_{ship} \cdot \text{ced}_{ship} = 0.539 \cdot 10^{15} \text{ J}$$

The mass of concrete is 1,589,500,000 kg.

Thus the CED is

$$\text{CED}_{concrete} = m_{concrete} \cdot \text{ced}_{concrete} = 1.04 \cdot 10^{15} \text{ J}$$

As the sum of these contributions, the energy input is at least $6.50 \cdot 10^{15}$ J. In addition to the data given, further energy costs must be taken into account, such as for the turbines and generators, the control technology, the high-voltage electrics, the construction of the sand fill and the geotextiles, as well as disposal.

What are the estimated EROI and the energy payback time?

Since no data on operation and maintenance are available, these are neglected here:

$$\text{EROI} = \frac{E_{year} \cdot \Delta t}{\text{CED}} = \frac{530 \, \frac{\text{GWh}}{\text{a}} \cdot 120 \text{ a}}{6.50 \cdot 10^{15} \text{ J}} = 35$$

The energy payback time would then be (no maintenance inputs)

$$\Delta t_{pb} = \frac{\text{CED}}{E_{year}} = \frac{6.50 \cdot 10^{15} \text{ J}}{530 \, \frac{\text{GWh}}{\text{a}}} = 3.4 \text{ a}$$

The actual EROI is lower, but lies within a good range. CED and EROI are to be understood here only as very rough first estimates that do not withstand deeper critical scrutiny, but can provide orientation. Far more critical for the planning and evaluation is the expected sea level rise over the lifetime of the facility. This will cause wave energy to increase significantly, thereby endangering the dam, so that a service life of 120 years appears illusory today. The project is currently no longer being pursued.

11.2 Wind Power

Wind turbines use the kinetic energy of the wind and convert it into mechanical energy or electricity. The wind wheel or wind turbine generates wave work in the process. The use of mechanical energy occurs either directly in mills and pumps or via conversion in a generator to electricity. Since this is an energy conversion chain in which mechanical power is converted into another form of power, the efficiency is in principle not limited by the second law of thermodynamics. A relevant limitation of the usability of the kinetic energy of the wind nevertheless exists.

11.2.1 Usable Energy

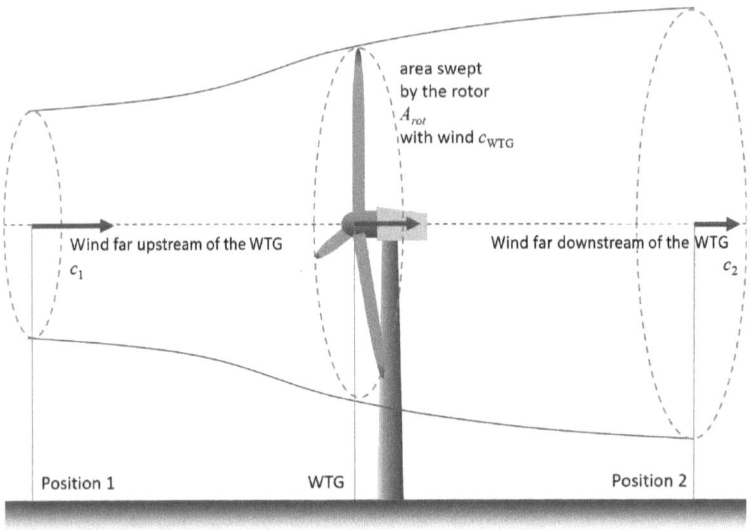

Figure 11.3 Simple model for wind power (see text)

For a simple configuration (Figure 11.3), in which irregularities in the terrain and the interaction of the airflow with the ground are neglected, the mass flow of the wind over an area A oriented perpendicular to the wind is in general

$$\dot{m} = \dot{V} \cdot \rho_{air}(p,T) \cong A_{rotor} \cdot \bar{c}_{wind} \cdot \rho_{air}(p,T) \tag{11.6}$$

with the mean velocity $\bar{c}$ of the wind and the mean density ρ of the air. In the simple model presented here, the density of the air within the rotor area is assumed to be constant. For this purpose, steady wind is assumed, blowing at constant velocity uniformly through the area spanned by the turbine. Real wind turbines must cope with temporally and spatially varying winds. The density of the air varies noticeably between bottom and top for a large rotor diameter.

From conservation of energy it follows that the power of the wind in front of the turbine must equal the power of the wind behind the wind turbine plus the extracted power of the wind power plant. This applies to the wind flowing through the rotor area:

$$P_{wind} = P_{kin} + P_{pressure} + P_{wpp} + P_{fluid} = \text{const} \tag{11.7}$$

The different forms of energy can be substituted here:

$$P_{wind} = \dot{m} \cdot \frac{\overline{c}^2}{2} + \dot{m} \cdot \frac{p}{\rho_{air}} + \dot{m} \cdot g \cdot h + P_{wpp} = \text{const} \tag{11.8}$$

This contains the Bernoulli equation (Equation 11.1). If the change in potential energy of the air is additionally neglected, then for positions far in front of or far behind the wind turbine

$$\dot{m} \cdot \frac{\overline{c}_1^2}{2} + \dot{m} \cdot \frac{p_1}{\rho_{air}} = \dot{m} \cdot \frac{\overline{c}_2^2}{2} + \dot{m} \cdot \frac{p_2}{\rho_{air}} + P_{wpp} \tag{11.9}$$

In this case, power is extracted only at a point downstream of position 1 but before position 2 from the wind turbine. Far away from the wind turbine, the pressure in front of and behind the wind turbine will be equal and it holds that

$$P_{wpp} = \dot{m} \cdot \frac{\overline{c}_1^2 - \overline{c}_2^2}{2} \tag{11.10}$$

The power contained in the wind that flows through a rotor area placed perpendicular to the wind is

$$P_{wind} = \dot{m} \cdot \frac{\overline{c}^2}{2} = \dot{V} \cdot \rho_{air} \cdot \frac{\overline{c}^2}{2} = A \cdot \overline{c} \cdot \rho_{air} \cdot \frac{\overline{c}^2}{2} = A \cdot \rho_{air} \cdot \frac{\overline{c}^3}{2} \tag{11.11}$$

The power of the wind increases with the third power of the wind velocity. It is therefore of great importance to find and exploit optimal sites for WPP. The velocity of the wind at the position of the rotor is taken as the mean value between the velocities far upstream and far downstream of the rotor:

$$\overline{c}_{rot} = \frac{\overline{c}_1 + \overline{c}_2}{2} \tag{11.12}$$

Thus the power that the wind power plant can extract, taking into account the mass flow at the position of the wind rotor, is

$$P_{wpp} = \dot{m} \cdot \frac{\overline{c}_1^2 - \overline{c}_2^2}{2} = A \cdot \rho \cdot c_{rotor} \cdot \frac{c_1^2 - c_2^2}{2} = A \cdot \rho \cdot \frac{\overline{c}_1 + \overline{c}_2}{2} \cdot \frac{\overline{c}_1^2 - \overline{c}_2^2}{2} \tag{11.13}$$

With these results we can now calculate the efficiency for extracting kinetic energy from the wind:

$$\eta_{Betz} = \frac{P_{wpp}}{P_1} = \frac{A \cdot \rho \cdot \frac{\overline{c}_1 + \overline{c}_2}{2} \cdot \frac{\overline{c}_1^2 - \overline{c}_2^2}{2}}{A \cdot \rho \cdot \frac{\overline{c}_1^3}{2}} = \frac{\overline{c}_1 + \overline{c}_2}{2 \cdot \overline{c}_1} \cdot \frac{\overline{c}_1^2 - \overline{c}_2^2}{\overline{c}_1^2} = \frac{1}{2} \cdot \left(1 + \frac{\overline{c}_2}{\overline{c}_1}\right) \cdot \left(1 - \frac{\overline{c}_2^2}{c_1^2}\right) \tag{11.14}$$

The maximum efficiency is the maximum of this function for η_{Betz}. This maximum value follows from the derivative of the power of the wind turbine with respect to the velocity c_2, since c_2 is itself a function of the extraction of power from the wind:

$$\frac{dP_{wpp}}{d\bar{c}_2} = \frac{d}{d\bar{c}_2}\left(A \cdot \rho \cdot \frac{\bar{c}_1 + \bar{c}_2}{2} \cdot \frac{\bar{c}_1^2 - \bar{c}_2^2}{2}\right) = \frac{d}{dc_2}\left(A \cdot \rho \cdot \left(\bar{c}_1^3 - \bar{c}_1 \cdot \bar{c}_2^2 + \bar{c}_1^2 \cdot \bar{c}_2 - \bar{c}_2^3\right)\right) \tag{11.15}$$

The area and the density are not dependent on c_2. Therefore the following must hold for the inner bracket:

$$\frac{d}{dc_2}\left(\bar{c}_1^3 - \bar{c}_1 \cdot \bar{c}_2^2 + \bar{c}_1^2 \cdot \bar{c}_2 - \bar{c}_2^3\right) = \left(2 \cdot \bar{c}_1 \cdot \bar{c}_2 + \bar{c}_1^2 - 3\bar{c}_2^2\right) = 0 \tag{11.16}$$

This is satisfied for

$$\bar{c}_2 = \frac{1}{3} \cdot \bar{c}_1 \tag{11.17}$$

Substituted into Equation 11.14 this gives $\eta_{Betz} = 0.59$. This is the theoretical upper limit for extracting power from the wind. A maximum of 59 % of the kinetic energy of the wind can be extracted.

In reality, the conditions around wind turbines are considerably more complex and an essential element is the generation of angular momentum in the wake of the wind rotor in addition to the reduction of wind velocity (Meyers/Meneveau 2011). Therefore, specific characteristics of the wind turbine enter into the calculation. Real conditions can be analyzed by experiment or by flow simulation (Quaschning 2023). The goal is to operate real wind turbines as close as possible to this ideal.

11.2.2 Optimal Placement of Wind Turbines

In wind farms individual wind turbines cannot be placed arbitrarily close together, as they interfere with each other and rob each other of wind. There are extensive analyses on this subject in which the various effects, such as mutual interference but also economic aspects, are evaluated (Meyers/Meneveau 2011; Satyov et al. 2022). As a rule of thumb for a flat geography, i.e., in flat terrain or offshore:

- For locations with mostly weak wind or where there is no pronounced main wind direction, the distance between the masts should be approximately 6 times the rotor diameter.
- For locations with good wind and a clear main wind direction, as is typical on the coast or offshore out at sea, the distance perpendicular to the main wind direction should be 6 times the rotor diameter and in the main wind direction 8 times.

These rules follow a balancing process in which, on the one hand, the wind turbines should be placed close together so that they can share common infrastructure, but on the other hand they should interfere with each other as little as possible and collectively be able to use as much wind as possible. These are rough estimates.

A more detailed analysis of a specific site may make other arrangements sensible. Inland and in hilly terrain, the wind turbines are placed on ridge crests, where wind is particularly easy to harvest or where the wind turbines reach particularly deep into the faster-blowing higher winds and thus achieve the highest EROI. Wind potential maps, on which such favorable sites can easily be identified, are available on the websites of the responsible state authorities.

Can Germany supply itself with electricity from onshore wind energy alone?

A typical wind turbine is a Siemens SWT 3.0-113. We use this design as it is optimized for operation inland and at moderate wind speeds. The rated power is P_N = 3.0 MW and the rotor diameter is d = 113 m. Since most wind turbines will presumably be located inland and in locations of moderate quality, we assume a capacity factor of ψ = 25 %. In principle, this capacity factor would need to be evaluated individually for each site. The area of Germany is 357 340 km^2.

What is the mean electrical power generated by the wind turbine under these assumptions?

This amounts to an annual average of

$$\overline{P}_{wpp} = \psi \cdot P_N = 0.25 \cdot 3.0\ \text{MW} = 0.75\ \text{MW}$$

What is the installation area of a single wind turbine in a wind farm?

The area requirement must be determined from the manufacturer's installation recommendations or the regional rules for the installation of wind turbines. In the closest arrangement (more typical for the North German Plain or offshore), the requirement is

$$A_{wpp} = 6 \cdot d \cdot 8 \cdot d = 48 \cdot d^2 = 48 \cdot (113\ \text{m})^2 = 612\ 900\ \text{m}^2$$

What is the typical area-specific power output of a wind farm?

In this example, the area-specific power output is

$$\overline{p} = \frac{\overline{P_{wpp}}}{A} = \frac{750\ 000\ \text{W}}{612\ 900\ \text{m}^2} = 1.224\ \frac{\text{W}}{\text{m}^2}$$

The literature gives approximately 1 W m^{-2} to 7 W m^{-2} for suitable sites, but not all sites in Germany are equally well suited.

What is the required area for generating 600 TWh per year?

600 TWh roughly corresponds to Germany's electricity demand. The number of installations required is therefore

$$N_{wpp} = \frac{W_{el,De}}{\overline{P_{wpp}} \cdot \Delta t} = \frac{600 \cdot 10^{12}\ \text{Wh}}{750\ 000\ \text{W} \cdot 365 \cdot 24\ \text{h}} = 91\ 320$$

This is approximately three times more than are currently installed in Germany. To install these, the following area is required:

$$A_{wind\text{-}park} = A_{wpp} \cdot N_{wpp} = 0.6129\ \mathrm{km}^2 \cdot 91\ 320 = 55\ 970\ \mathrm{km}^2$$

This corresponds to approximately 1/7 of Germany's land area.

This exercise illustrates the low energy density of renewable energy systems. If the wind turbines are placed at favorable sites, such as ridges or in regions with higher wind yields, the number can be significantly reduced, certainly by a factor of 2 to 3. Even with this number of wind turbines, however, the infrastructure for energy supply will be an important part of the landscape throughout Germany. This may be necessary with regard to future energy supply and yet be difficult to implement socially.

11.2.3 Properties of Wind

In order to design wind turbines safely and to adapt their rotors to the actual local conditions, the most reliable possible description of the wind conditions at the site is required. A key problem here is that it would be extremely costly to measure the wind at the later hub height with the required accuracy over long periods of time. Models are therefore needed with which the required information about the wind can be reliably represented. IEC 61400-1 contains these models and specifies their application for the design of wind turbines. Key parameters include the wind speed at height, the distribution of wind speed over time, and the rate at which the direction and strength of the wind can change. IEC 61400-1 provides two ways of describing the wind speed c as a function of height z above the ground.

Logarithmic A first option is a logarithmic function (Meyers/Meneveau 2011; IEC 61400-1):

$$c(z) = c_0 \cdot \left(\frac{\ln\left(\frac{z}{z_0}\right)}{\ln\left(\frac{z_0}{z_{rough}}\right)} + 1 \right) \tag{11.18}$$

Here c_0 denotes the wind speed at the reference height z_0. This reference height is often 10 m. However, it can be any other height for which sufficient data are available at this site. The parameter z_r is a measure of the roughness of the terrain. Open bodies of water follow the wind at the surface and are therefore extremely smooth, while cities form a particularly high-reaching boundary layer that disturbs and alters the free wind at height far upward. The function from Equation 11.18 represents the relative change in wind speed from the reference value with height. Table 11.2 lists values for the roughness of common terrain types.

Exponential A second description of wind speed with height is

$$c(z) = c_0 \cdot \left(\frac{z}{z_0}\right)^{\alpha} \tag{11.19}$$

Here the wind shear exponent α captures the behavior of the boundary layer, i.e., the roughness of the terrain. Values are given in Table 11.2 which yield a similar profile of wind speed with height as the roughness values. IEC 61400-1 specifies for common onshore cases and if no other values are available, $\alpha = 0.2$.

Both descriptions are shown in Figure 11.4 for some relevant values. The rougher the surface, the higher the hub of the wind turbine should be mounted, since with a rough surface the boundary layer extends much further into the atmosphere.

The temporal distribution of wind speed is described using statistical distributions, such as Rayleigh or Weibull distributions, where the parameters then depend on the location and its conditions. With these distributions, the capacity factor of installations can be well designed and optimized. In terms of risk assessment, it is often necessary to estimate rare extreme wind events from long-term data and climate projections (IEC 61400-1).

Table 11.2 Values for the Roughness z_{Rau} (Source: Quaschning 2023) and Corresponding Values for the Wind Shear Exponent α

Class	z_{rough}	α	Terrain type
1	0.0002 m	0.08	Open sea, large bodies of water
2	0.005 m	0.12	Flat land surface, tidal flats
3	0.03 m	0.15	Open flat terrain, steppe
4	0.1 m	0.17	Agricultural areas with low vegetation (winter)
5	0.25 m	0.21	Agricultural areas with tall vegetation (maize)
6	0.5 m	0.25	Parkland, savanna, individual trees or bushes
7	1 m	0.3	Regular structures such as forests (from medium canopy height), villages or suburbs
8	2 m	0.4	Cities with strong variation in building height

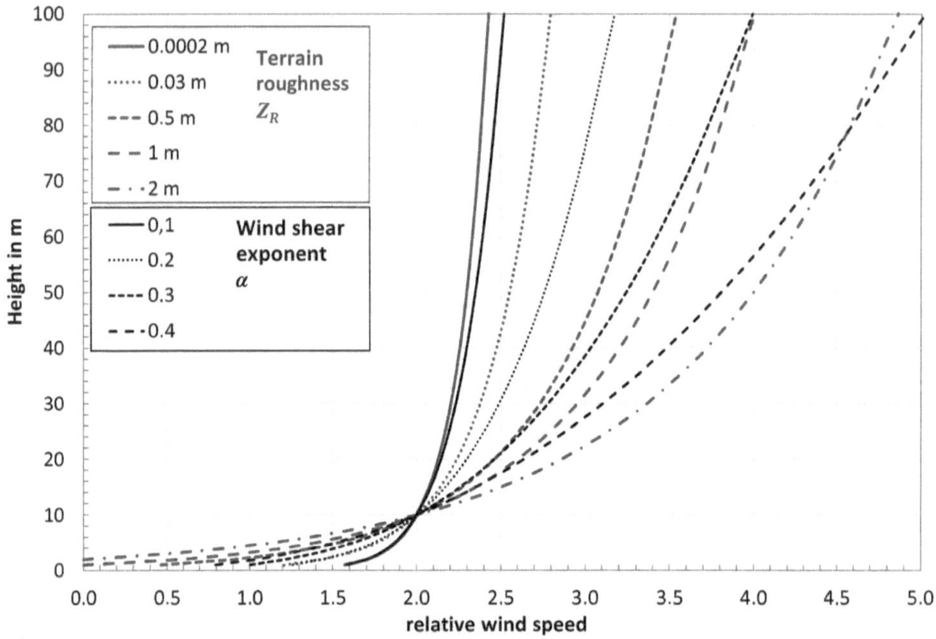

Figure 11.4 Variation of wind speed with height and as a function of the logarithmic and exponential description

Obtaining local wind data

A first data source is the global wind atlas (*https://globalwindatlas.info*). The data stored there give a very good impression of where favorable regions are located. The local resolution is not sufficient for planning at inland sites with undulating topography, hills, etc. A significantly higher resolution for Western Europe is provided by the New European Wind Atlas (*https://map.neweuropeanwindatlas.eu*). Data at very high resolution are available from national authorities or, in Germany, at the federal state level. In some cases, data material is available at the district level, allowing very precise information about specific sites.

Where to put the wind turbines?

Wind turbines have a number of impacts. Some of the impacts affect people and must be taken into account directly during planning. Other impacts influence public perception and public debate. Wind energy, together with photovoltaics, is the form of renewable energy that can achieve the widest spatial distribution, but which remains more visible as a large-scale technology than all other energy technologies:

- The tips of larger rotor blades sweep through the air at high speed, causing turbulence and noise generation. In small turbines, this tends to produce a constant rushing sound. In large turbines, the rotor blades pass close to the listener at certain time intervals, so that they can be heard individually.
- Large turbines produce a regular shadow flicker.
- Under suitable weather conditions, ice can form on rotor blades. This ice can break off in large chunks and be flung away. During freezing rain or fog, staying near the turbine can therefore be life-threatening.
- Wind turbines are preferably to be positioned on ridges, hills or mountain crests. Infrastructure is often absent there, so access roads must be built and electrical connections laid: this can represent a significant intervention in forests or ecosystems.
- Rotor blades can kill birds and bats. After domestic cats, cars, trucks, window panes and other infrastructure, wind turbines are also a cause of death for wildlife. Already due to the hub height of today's wind turbines, this affects only a few species and does not come close to the massive killing of birds and small animals by domestic cats. Loss et al. report for the USA 2,400,000,000 birds killed annually and 12,300,000,000 small animals killed annually by domestic cats (Loss et al. 2013).
- Wind turbines are preferably to be positioned on ridges, hills or mountain crests, as they yield the greatest output there and thus exhibit the highest EROI. They are then visible from a great distance. For example, in Darmstadt we can see the energy landscape of the Palatinate at a distance of 30 to 50 km in clear weather.
- Since the individual wind turbine has only a low rated power compared to fossil-fueled power plants, many wind turbines are needed to replace such a power plant. In order to generate relevant electrical power in the sense of an energy transition and to minimize transmission losses, wind turbines are needed at all favorable sites throughout the country. Wind energy is local and is potentially harvested everywhere, while fossil power plants are hidden from most people in old unattractive industrial areas (sacrifice zones). This makes NIMBY particularly pronounced in the case of wind energy.
- Large amounts of material are bound up in wind turbines. Foundations, masts, nacelles, generators, and electrical systems can be reused or refurbished at end of life, but today's rotor blades cannot. This is also reflected in the EROI of typical installations.
- Even large wind turbines, particularly onshore, exhibit rapid variation in the power they provide under many weather conditions. This is due to the typical large scale turbulence in the lower troposphere, which is also directly perceptible to people. A constant generation can therefore only be achieved through many widely distributed wind turbines and a correspondingly interconnected electrical grid.
- Wind turbines are disruptive or obstructive to airports, air traffic control, and other activities. As a result, many favorable sites are currently not being used.

Figure 11.5 Assembly of a 3-MW wind turbine in the Allgäu

Small or large wind turbines

Wind turbines with high hub heights and large rotor diameters are subject to strong criticism for the reasons mentioned in the preceding box. This does not apply to small installations with less than 50 kW rated power or mini-installations with rotor diameters in the range of meters. Criticism is particularly low for vertically operating small wind turbines based on the Savonius principle. Added to this is a widespread lack of understanding of the necessity to erect so many large installations: when wind turbines are compared with gas turbine power plants, wind turbines appear much larger, indeed disproportionately large.

Looking at the derivation of the efficiency, it becomes clear why wind turbines become large and have tall masts:

- The rated power depends directly on the swept area of the rotor. It therefore scales with the square of the rotor length. Larger wind turbines thus deliver significantly more power.
- Wind speed increases considerably with height, particularly in inland regions with high terrain roughness. Since the achievable power of a wind turbine increases cubically with wind speed, it is very important to reach positions with the highest possible wind speed, i.e., to erect the tallest possible masts.
- Small wind turbines bind, relative to rotor area, rather more energy (CED) than large wind turbines. However, since they do not reach into the high-yield upper wind layers, they generate far less power. They therefore have an exceptionally low EROI.
- The commonly used three-bladed rotors are very good in terms of their extraction efficiency. Values close to the technically achievable optimum are reached. Other designs, in particular those rotating about a vertical axis, exhibit considerably worse extraction efficiency.

- Particularly poor in every respect are the vertically operating Savonius rotors. Their extraction efficiency is at most 15% (Kaltschmitt et al. 2013). This makes them significantly worse than Dutch windmills from the 18th century. At the same time, they bind particularly large amounts of resources.

If, instead of relatively few, large, and widely visible wind turbines, the demand were to be met by small wind turbines, a significant supply of electricity in Germany would not be possible. At the same time, these small wind turbines might bind more energy than they could ever generate.

References

Ansar, A./Flyvbjerg, B./Budzier, A./Lunn, D.: Should we build more large dams? The actual cost of hydropower megaproject development. In: Energy Policy 69, 2014, pp. 43–56

Flyvbjerg, B.: How Big Things Get Done. Penguin 2023

Grill, G./Lehner, B./Thieme, M./Geenen, B./Tickner, D./Antonelli, F./Babu, S./Borrelli, P./Cheng, L./Crochetiere, H./Ehalt Macedo, H./Filgueiras, R./Goichot, M./Higgins, J./Hogan, Z./Lip, B./McClain, M. E./Meng, J./Mulligan, M./Nilsson, C./Zarfl, C.: Mapping the world's free-flowing rivers. In: Nature 569, 2019, pp. 215 221

IEC 61400-1: Wind energy generation systems – Part 1: Design requirements

Kaltschmitt, M./Streicher, W./Wiese, A. (Hrsg.): Erneuerbare Energien. Systemtechnik, Wirtschaftlichkeit, Umweltaspekte. Springer Vieweg, Berlin 2013

Loss, S. R./Will, T./Marra, P. P.: The impact of free-ranging domestic cats on wildlife of the United States. In: Nature Communications 4, 2013, 1396

Meyers, J./Meneveau, C.: Optimal turbine spacing in fully developed wind farm boundary layers. In: Wind Energy 15, 2011, pp. 305–317

Quaschning, V.: Regenerative Energiesysteme. Technologie – Berechnung – Klimaschutz. 12. Auflage. Hanser, München 2023

Satymov, R./Bogdanov, D./Breyer, C.: Global-local analysis of cost-optimal onshore wind turbine configurations considering wind classes and hub heights. In: Energy 256, 2022, 124629

Schmitt, R. J. P./Kittner, N./Kondolf, G. M./Kammen, D. M.: Deploy diverse renewables to safe tropical rivers. In: Nature 569, 2019, pp. 330–332

12

Direct Generation of Electricity

This chapter presents the relevant technical methods for the direct conversion of energy flows in the earth system into electrical energy. This is primarily photovoltaics, which converts sunlight into electricity.

12.1 Photovoltaics

Thermodynamic Limit of the Conversion of Light into Work

The same fundamental limits apply to the conversion of radiation into electricity as to any other form of conversion of heat into work, because radiation is a form of heat. In the ideal case, entropy does not increase. This is the reversible process with the theoretical maximum efficiency. The thermodynamic limit of the conversion of radiation (from the sun) into work, when the entropy is completely returned to the environment via radiation (Kleidon 2016), is

$$\eta_{max} = 1 - \frac{4}{3} \cdot G \cdot \frac{T_0}{T_{sun}} \tag{12.1}$$

In this equation, G is a geometry factor and T_0 is the temperature of the environment. For direct irradiation by the sun, the geometry factor is $G = 1$. This factor decreases when the radiation is not used in a directed manner or when it is incident diffusely. Thus, the thermodynamic limit of the conversion of solar radiation with $T_{sun} = 5776$ K and $T_0 = 288$ K under direct irradiation by the sun is approximately 93%. For diffuse incoming radiation (under clouds), this conversion efficiency is still approximately 79%. If, however, the entropy of the radiation is completely transferred to the environment via a heat flow that is not itself radiation, such as through cooling water or the evaporation of water, then the limit of conversion is

$$\eta_{max} = 1 - \frac{4}{3} \cdot \frac{T_0}{T_{sun}} \tag{12.2}$$

This is the usual case. The limit for directed radiation is 93% and that for diffuse radiation is 72%. This maximum conversion efficiency for the conversion of solar radiation into work is not even remotely observed in the natural processes of the earth system such as photosynthesis. Technical processes also do not reach this limit.

Operating Principle of Photovoltaics

Photovoltaics is the direct conversion of light into electricity. This takes place in photovoltaic cells. There, the photons of the incoming radiation directly separate electrons from their respective atoms and can thus generate a voltage between the two poles of the cell. When these two poles are externally connected electrically – through the electrical grid –, a current flows between them and the cell generates usable electricity. The light separates electrical charges in the cell, which cannot recombine directly but only via the connected electrical grid. The voltage that builds up between the poles is primarily a property of the specific structure and material of the semiconductor. It additionally depends on the intensity of the radiation.

The most important technical challenges in the design of photovoltaic cells are:

- As many as possible of the incoming photons should each separate one electron from an atom. The generated current depends directly on this.
- The charges separated by the photon must be spatially separated from each other as quickly as possible. Only then do they generate a voltage and only then does a current flow. If they recombine within the semiconductor, they produce no useful output.
- The voltage that builds up can be used as completely as possible for the generation of a usable electric current.

Modern PV cells and PV systems are designed with regard to these and several other requirements and achieve a very high efficiency of converting light into electricity.

Figure 12.1 shows the basic structure of a PV cell using the example of a silicon-based structure. In a central layer of a semiconductor, incident photons are absorbed and separate charges there. The entire further structure serves the purpose of generating as many such separated charges as possible and safely directing them from the semiconductor into a circuit. The absorbing layer can be quite thick, as in silicon solar cells, or very thin, as in thin-film solar cells. The inner layer, in which incident light is absorbed and charges are separated, is surrounded by additional functional layers (Quaschning 2023):

- All thin-film variants require a substrate for mechanical stability. This substrate is typically an architectural glass. The selection criteria for the type of glass are mechanical strength, cost, availability, electrical properties as an insulator, and optical properties.

- The semiconductor is doped layer by layer by incorporating foreign atoms. This doping then enables the spatial separation of the charges after they have first been electrically separated by the incident light. This allows a cell voltage to build up.
- Conductive layers are applied to conduct the current away. In silicon cells, aluminum is typically used as the back contact and conductive silver pastes on the front side. In thin-film cells, for example, molybdenum and indium tin oxide (ITO) are used.
- An anti-reflection coating on the sun-facing side is required to reflect as little light as possible. Reflection is particularly critical for silicon, and the defined application of the blue SiN anti-reflection coating is of great importance for the quality of the cell.

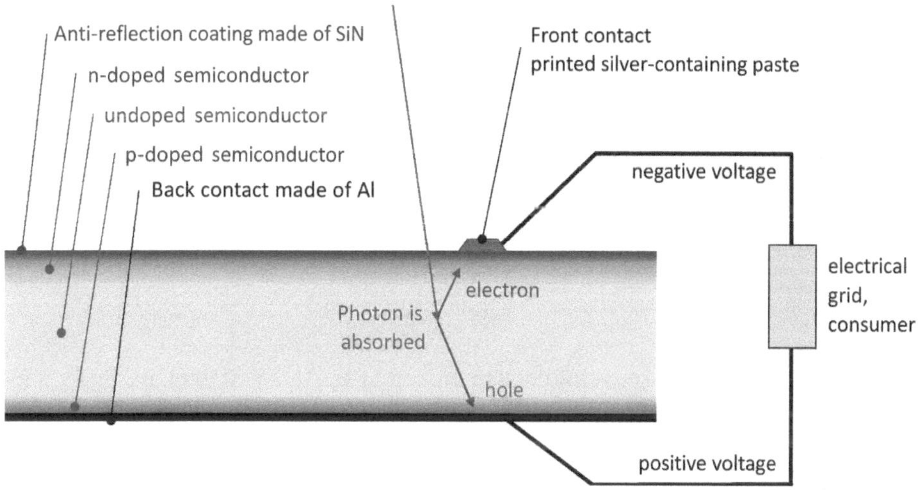

Figure 12.1 Typical structure of a PV cell (silicon as base material)

The cells are first manufactured individually. In a next step, all cells are tested and sorted by performance, so that cells with as similar characteristics as possible can be assembled into a module. In the module, all cells are contacted with each other and to the outside. They are mounted between glass panes as weather-resistantly as possible and enclosed in a frame for mechanical protection and mounting as a PV module. All properties and all production steps together define the performance of the individual module.

Efficiency of Photovoltaics

The actual conversion efficiency of a photovoltaic cell depends on the specific properties of the semiconductor and on the spectral distribution of the incident light. This behavior is described by the Shockley-Queisser model. For the AM 1.5 spectrum, which is important as a calculation and test case, the resulting theoretical efficiency as a function of the band edge of the semiconductor in Figure 12.2 is shown (Rühle 2016). AM 1.5 is the typical case in which the sun is approximately 42° above the horizon and the light passes through 1.5 times the air mass that it would pass through at normal incidence (Section 12.4).

The actually achievable efficiency of the cells depends on a few factors:

- The refractive index of the semiconductor material defines its reflectivity. This must be reduced, e.g. in the case of silicon, by additional anti-reflection coatings in order to increase the yield.
- Production-related defects in the semiconductor reduce the quality of the layer sequence and thus the yield of the processes. The better it is possible to produce the intended structure homogeneously and without defects, the higher the efficiency.
- Necessary functional elements such as conductor tracks reduce the proportion of the surface through which the light enters the semiconductor.
- Depending on the temperature, unavoidable leakage of charge carriers occurs in the cell. This effect reduces the efficiency with the temperature of the cell.

Figure 12.2 also shows how close cells available on the market today with TRL 9 operate to the thermodynamic limit.

photocurrent The current provided by a cell is

$$I_{cell} = I_{photo} - I_0 \cdot \left(\exp\left(\frac{e_0 \cdot U}{k_B \cdot T}\right) - 1\right) \tag{12.3}$$

where I_{photo} is the photocurrent induced in the cell by the incident light. Basically, a PV cell behaves like a diode. The saturation reverse current of the diode I_0 is a fundamental property of the cell type or the semiconductor. Further quantities are the applied voltage U across the diode. e_0 is the elementary charge, k_B the Boltzmann constant and T the temperature. Conversely, the voltage is

$$U = \frac{k_B \cdot T}{e_0} \cdot \ln\left(1 - \frac{I_{cell} - I_{photo}}{I_0}\right) \tag{12.4}$$

Typical voltages of individual cells are in the volt range and direct current is generated. Therefore, a PV system includes an inverter that generates the required alternating voltage for grid feed-in (Quaschning 2023).

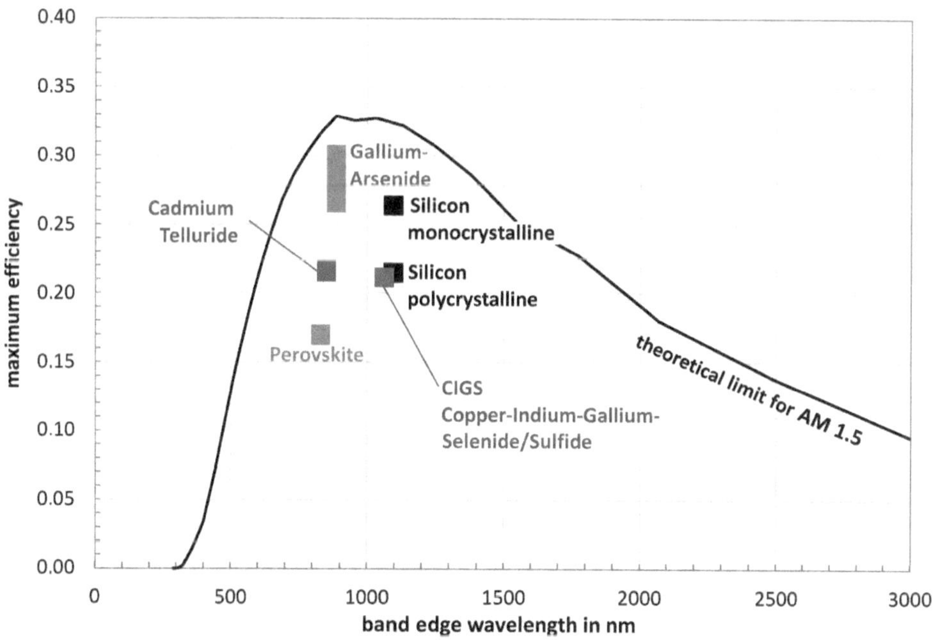

Figure 12.2 Maximum conversion efficiency for photovoltaics, theoretical limit and achieved values of real cells for AM 1.5 (based on Rühle 2016)

12.2 Important PV Variants

A large number of suitable semiconductors are known with which photovoltaics would in principle be possible. However, the technologies available on the market are limited to a few semiconductors. This is historically due to the fact that an early focus was placed on a few promising technologies and investments were concentrated there. This also reflects technical reasons. For a semiconductor to be used, it must have some key properties that were only known early on for a few materials.

The following are concepts that are available on the market and enable a sufficient EROI. A good overview of relevant technologies and potential future developments is provided by the National Renewable Laboratory (NREL) on its website (*https://www.nrel.gov/pv/cell-efficiency.html*). There, the historical and current efficiencies of individual cells from research are presented. Furthermore, historical and current best efficiencies of commercially manufactured modules can be found there.

Requirements for photovoltaics

Efficiency In order to justify the efforts for installation, mounting, connection to the electrical grid and maintenance, the energy yield must be sufficiently high. This requires a corresponding efficiency. Modules available on the market today always have a module efficiency of more than 10%.

Yield The energy yield of the module must be sufficiently high compared to the expenditure for manufacturing and installation, thus enabling an EROI well above 3. The achievable financial yield must also be sufficiently high. This limits the maximum possible energy payback time of the modules.

Lifetime In principle, PV modules require a considerable amount of energy during manufacturing. They have a high CED. Therefore, the modules must be operated for a long time in order to return their invested energy with a sufficiently high EROI. Modules must be designed and manufactured to be resistant to common external influences, and all components must be designed for a sufficiently long service life. A significant risk here is mechanical damage caused by storms, hail or falling branches. With regard to service life, the necessary seals and generally UV-sensitive plastics are critical. The intended service life is 30 years or longer.

Material The chemical elements, semi-finished products and components required for manufacturing must be available on the market in sufficient quantities and at acceptable costs over the long term.

Processes The technical requirements for some of the fundamental manufacturing processes are very high. It is a major challenge for plant manufacturers to develop and produce equipment and machinery that can deliver these processes at the required quality and homogeneity with high throughput. In particular, the transition from the laboratory (demonstrator with 1 cm^2) to module size (prototype) represents a major hurdle that not all concepts manage to overcome.

Toxic substances During operation, the module must not release any toxic substances.

End of life In the interest of closed-loop use of the valuable materials employed, the modules must be disassemblable at end of life. This requirement is technically often directly at odds with service life.

Silicon

Silicon-based modules account for the largest share of all installed PV modules, worldwide and in Germany (Table 12.1). There are a number of variants depending on the manufacturing process and the crystal structure of the silicon used. Silicon exhibits good efficiency with very good service life. Many of the necessary processes were first established for the manufacture of components for microelectronics and have since been optimized with regard to PV.

The production of silicon for PV cells is very energy-intensive and comprises the following steps:

1. First, the reduction of quartz sand to silicon takes place in electric arc furnaces using carbon in the chemical reaction occurring at approximately 2000 °C $SiO_2 + C \Longrightarrow Si + CO_2$. The temperature near the arc is sufficient to keep quartz sand and silicon in a liquid state. Carbon dioxide is emitted during the production of silicon.
2. Conversion of the pure silicon into $SiCl_4$ or a substance with similar technical properties
3. Next, silicon is deposited from the gas phase at approximately 800 °C as high-purity crystalline silicon. For this purpose, the $SiCl_4$ decomposes at electrodes at high temperature. The Cl_2 can be extracted from the reactor as a gas, while the Si deposits as a solid.
4. If the purity is insufficient, the silicon can be inductively remelted under high vacuum at 1500 °C again to outgas further foreign atoms.
5. Drawing of silicon as a crystal from a melt at approximately 1400 °C
6. Sawing of wafers (the future cells) from the ingot

Subsequently, several additional thermal processes are required (diffusion, plasma coating, solder reflow, etc.), during which the cells are each heated to up to 1000 °C.

Table 12.1 Common single-layer photovoltaic systems: Efficiency is specified for the following applications: in the laboratory | maximum efficiency of commercial systems | typical field efficiency (NREL))

Semiconductor	Band edge	Efficiency	TRL
Silicon single crystal	1.13 EV or 1100 nm	26.7 % | 24 % | 19 %	TRL 9
Silicon polycrystalline	1.13 EV or 1100 nm	22.3 % | 20 % | 17 %	TRL 9
Gallium arsenide (GaAs)	1.42 EV or 875 nm	27.5 % | 25.1 % | -	TRL 8
Cadmium telluride (CdTe)	1.5 EV or 825 nm	22.1 % | 17 % | 16 %	TRL 9
CIGS – copper indium gallium (selenium or sulfur)	1.5 EV or 830 nm	22.6 % | 19 % | 17 %	TLR 9
Perovskite	Depending on composition	24.2 % | 11.7 % | -	TLR 6

Cadmium Telluride (CdTe)

From CdTe thin-film solar cells are constructed (Table 12.1). For this purpose, a glass pane is used as a substrate and coated as homogeneously as possible with the necessary functional layers by vapor deposition. For CdTe, the required temperature is approximately 800 °C. CdTe has a solid market segment. The cells exhibit a comparatively very low embodied energy, have a good efficiency (also under diffuse light) and a good service life. They are easy to recycle.

Copper Indium Gallium Selenide and Similar Layer Sequences (CIGS)

This is a group of semiconductors in which copper, indium, usually gallium, and either sulfur or selenium form the core of the cell (Table 12.1). The modules have good efficiency even for diffuse light, the efficiency is not very temperature-sensitive, and they have a good service life. Recycling is fundamentally possible, similar to CdTe.

Similar to CdTe, the cell is built up in a vacuum by vapor deposition on a glass pane. In this process, the semiconductor must be eutectic-melted at approximately 600 °C as an intermediate step. The major technical challenge here is handling the difficult selenium and achieving the required very high homogeneity of the processes over the large module area.

Gallium Arsenide (GaAs)

This semiconductor exhibits a very good efficiency (Table 12.1). It has so far been used in niche applications where high performance justifies the high costs. Due to material costs, only thin-film modules are financially attractive. For this purpose, the cells are built on GaAs crystals in a special process and then removed. The challenge is to provide the required crystal structure in the necessary quality over large areas.

Perovskites

This is a large group of materials that have by now achieved comparable efficiency in the laboratory to established cells (Table 12.1). A key challenge for future use is the stabilization of the semiconductors over a long period of time. Therefore, these systems are not yet ready for market.

Organic Semiconductors

It is possible to produce cells from organic dyes. The only module listed at NREL so far has an efficiency of 8 %. Since organic dyes age and decompose under light, the service life of the cells and modules is significantly limited. For a time, the research approach was to apply these substances inexpensively to plastic films using printing processes, thereby producing flexible but very short-lived solar cells with low efficiency. The extent to which this variant is energetically justified remains unclear.

12.3 EROI of Photovoltaics

The EROI depends on the one hand on the amount of electricity that a module or a power plant actually provides, and on the other hand on the energy input required for this module or power plant. Both points are briefly discussed.

Useful energy The total electricity generated by a module or an installation is determined in particular by the following aspects:

- The actual efficiency of the module

- The actually utilized share of the generated electricity: Outside the system boundaries of the module or the installation, it is decided whether the generated electricity is used, i.e., fed into the grid or used on-site.
- The actual service life of the module: The service life can be shortened by lack of maintenance, storms or other natural events, vandalism, insolvency of the operator, or the end of financial subsidies.
- The installation of the module influences how much solar radiation falls on the module and can thus be harvested. The latitude determines the annual cycle of solar irradiation. The weather influences how much radiation can reach the ground and whether it is predominantly direct or diffuse. The tilt angle and orientation define the actual intensity on the module. Vegetation or buildings can cast shadows.
- The ambient temperature and any wind influence the cell temperature of the modules and thus the efficiency. Silicon in particular favors high intensity at low temperature.
- Dust, leaves, snow, and other deposits determine what fraction of the intensity can reach the cells. Maintenance has an effect on performance. Scratches on the surface also reduce the performance of the modules through scattering.

These factors can be summarized as

$$W_{el} = \int_0^{\Delta t} \eta(t,T) \cdot \tau_{glass}(t) \cdot \psi(t) \cdot I(t) \cdot dt \qquad (12.5)$$

Here is $\eta(t, T)$ the time-dependent and temperature-dependent efficiency of the cell, $\tau_{glass}(t)$ the time-dependent transmission of the module including the effects of dust, leaves or snow, $\psi(t)$ is the time-resolved capacity factor of the generated electricity and $I(t)$ the actually usable intensity reaching the module. This can be estimated by

$$W_{el} = \eta \cdot \tau_{glass} \cdot \psi \cdot Q_{in}^* \qquad (12.6)$$

Here η, τ_{glass} and ψ are each mean values over the lifetime and Q_{in}^* is the radiation that has entered the module over its lifetime. This incoming radiation is given in databases either for flat ground or already for an optimal module orientation (see Section 12.4).

Invested energy Important factors are:

- The energy input for the production of the cells
- The design, the components used and the manufacturing process for the module from the cells
- The design, the components used and the manufacturing process for the frame on which the module is mounted
- Energy inputs for the electrical connection and rectifier

- Maintenance and upkeep of the modules: all repairs, cleaning of surfaces, replacement of components

In the literature, widely diverging values for the EROI of PV systems can be found. Here it is necessary to look very carefully at what the system boundaries of each study were, which inputs thus became part of the invested energy, and how precisely the life cycle of the modules was taken into account. General conclusions from a single literature source cannot simply be applied to a specific project. Rather, it must be checked whether reliable data for a comparable site already exist. The fundamental method for PV systems is described in (Fthenakis et al. 2011). A discussion of fundamental problems in studies on the EROI of PV systems and the presentation of a methodical and clean assignment of data can be found in Raugei et al. References to relevant studies are also found there (Raugei et al. 2017).

Small or large PV systems

Photovoltaics consists of many small modules. These are ideally scalable and interchangeable. This allows systems of any size to be built. This is advantageous for small systems, since a fully functional system for the generation of electrical energy is available with just a few elements. On the other hand, many advantages of other energy technologies are thus foregone. Even very large power plants consist of the same set of elements. Only a very large number of identical components are built and deployed side by side. In wind power and thermal power plants, size is a decisive factor for the efficiency of a system. This applies only to a limited extent to photovoltaics. Photovoltaics is therefore ideally suited for off-grid solutions, for small systems and for private electricity generation. There are many possible applications in which the electricity is consumed directly and therefore no storage or grid feed-in is required. These include, for example, irrigation pumps or machines and systems that can be operated in accordance with supply availability. For other purposes and many off-grid solutions, small storage units are useful or necessary (see Chapter 13).

12.4 Installation of PV Systems

The yield of a PV system depends directly on the location of installation. The more light falls on the cells over the period of use, the higher the generated electricity. This section discusses relevant influencing variables from the general to the specific site.

Local Intensity Outside the Atmosphere

The solar incoming radiation still above the atmosphere defines the weather-independent potential of solar incoming radiation for a location on earth. This figure makes it possible to see what incoming radiation could exist at a location at a given time without weather effects:

$$I(\theta,\delta,\psi) = I_{sol} \cdot \left(\frac{\overline{d}}{d}\right)^2 \cdot (\sin(\theta) \cdot \sin(\delta) + \cos(\theta) \cdot \cos(\delta) \cdot \cos(\psi)) \quad (12.7)$$

In this equation I_{Sol} = 1366 W m^{-2} is the solar constant, d is the current and $\bar{d}$ the mean distance between earth and sun. The angles denote with θ the longitude of the location. The angle δ describes the current distance between the equator and the point at which the sun's radiation falls perpendicularly on the ground: On 21 June in the northern summer this is +23° 45' and on 21 December in the northern winter –23° 45'. The angle ψ defines the local time of day. This is 0 when the sun reaches its maximum height above the horizon, and it advances by 15° with each hour. The first term of the bracket depends on the position of the earth in its orbit, i.e. the date. The second term describes the earth's rotation and thus the time of day.

For the influence of the eccentricity of the earth's orbit, there are precise approximation functions that accurately describe the orbit of the earth around the sun. One approximation is

$$\left(\frac{\bar{d}}{d}\right)^2 \cong 1 + 0.034 \cdot \cos\left(\frac{2 \cdot \pi \cdot n}{365.25}\right) \tag{12.8}$$

Here n is the day of the year. The intensity of solar incoming radiation thus varies by ±3.4 % over the year. Key statements about the variation of incoming radiation above the atmosphere are:

- The solar incoming radiation incorporates as properties of the earth the tilt of the earth's axis at 23° 45' and the eccentricity of the earth's orbit, roughly approximated by Equation 12.8, as inputs.
- The annual variation of solar incoming radiation at a location depends only on the longitude. Further north or south, the annual total is lower. At the poles it is approximately a factor of 4 lower than at the equator.
- The diurnal variation depends on the cosine of the time of day (and more weakly on the season).
- At the equator, the available solar incoming radiation is distributed particularly evenly over the year. The daily mean is between 400 and 450 W m^{-2}.
- Near the poles, the seasonal variation is particularly high. North or south of the polar circles, the sun does not shine at all for some periods of the year. The maximum daily mean of solar incoming radiation at the poles is each slightly higher than the maximum at the equator at 550 W m^{-2}.

Influence of the Atmosphere

In clear weather, the intensity available at the ground depends on the length of the path that the radiation travels through the air. We follow the light through the atmosphere before describing the intensity relative to the ground.

Atmospheric mass This is expressed by the relative air mass (Air Mass) AM specified. This air mass is 0 outside the atmosphere. When the sun shines exactly perpendicular to the ground, it has (under clear sky conditions) the value 1 and increases with the angle γ of the current solar elevation above the horizon:

$$AM = \frac{1}{\sin(\gamma)} \tag{12.9}$$

The solar elevation γ is the angle between the horizon and the current position of the sun above the horizon (Figure 12.3). With the air mass AM, the scattering of light by air and particles increases. The scattering at the molecules of the air depends almost exclusively on AM, while the scattering at particles is strongly influenced by the current particle loading of the atmosphere. In this process, a portion of the scattered radiation always reaches the ground as diffuse radiation and is also utilized. A mean estimate for the irradiated intensity at sea level as a function of the air mass is

$$I_{radiation} = I_{sol} \cdot 1.1 \cdot 0.7^{AM^{0.678}} = I_{sol} \cdot \exp\left(0.0953 - 0.3567 \cdot AM^{0.678}\right) \tag{12.10}$$

Numerical values of this function and observed values are given in Table 12.2. Due to the curvature of the earth's surface, a certain intensity still reaches the ground even at 0° solar elevation. The intensity $I_{radiation}$ refers to a surface perpendicular to the direction of propagation of the direct solar radiation, not to the ground.

Table 12.2 Air Mass, Solar Elevation and Intensity at Ground Level Under Clear Sky: the Zenith Angle Is 90° – γ.

Air mass AM	Solar elevation γ	Zenith angle	$I_{radiation}$	Observed values for $I_{radiation}$	I_{ground}
0	90°	0°	1366 W	1366 W	1366 W
1	90°	0°	1052 W	1130 W to 840 W	1052 W
1.09	67°	23°	1030 W	1110 W to 800 W	948 W
1.15	60°	30°	1015 W	1100 W to 780 W	879 W
1.5	42°	48°	940 W	1050 W to 680 W	629 W
2	30°	60°	849 W	970 W to 560 W	425 W
2.9	20°	70°	721 W	880 W to 430 W	24 W
5.6	10°	80°	477 W	660 W to 200 W	83 W
38	0°	90°	22 W	–	0 W

The intensity related to the horizontal ground depends on the angle of incidence. It equals the radiation intensity when the radiation falls perpendicularly from above. Otherwise, the radiation intensity is distributed over a larger ground area and I_{ground} is therefore always lower. The intensity on a horizontal surface at ground level depends on the solar elevation with

$$I_{ground} = I_{radiation} \cdot \sin(\gamma) \tag{12.11}$$

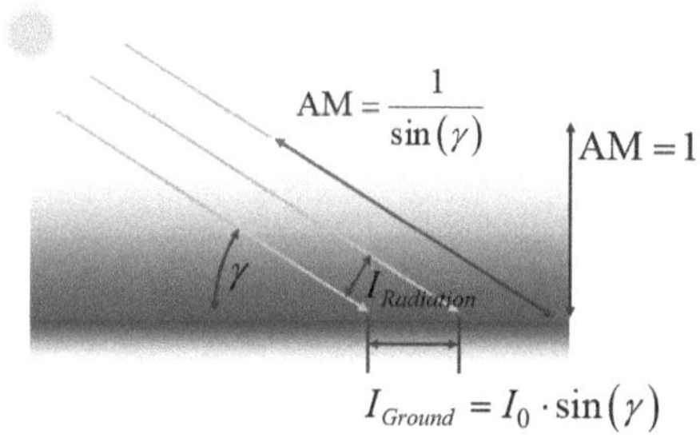

Figure 12.3 Essential quantities as a function of the local solar elevation γ

Diffuse radiation At ground level, not only the directed radiation arrives directly from the sun, but also a – depending on the weather – considerable proportion as diffuse radiation. In the case of heavy cloud cover, this diffuse radiation can come almost homogeneously from all directions. Even in actually clear skies, there is a diffuse component that depends strongly on the aerosol content of the air as well as on the current position of the sun (Quaschning 2023).

Climate, weather and PV potential on site

Section 12.4 so far was about the influence of clear air. How much sunlight is actually available on site depends on the current weather and, in the long term, on the regional climate. The greatest influence on the radiation actually usable at ground level is exerted by clouds, since these can absorb almost all of the radiation depending on their density and thickness. These weather effects can be well described regionally by the climate or by long-term averaged data on solar irradiation at ground level. Reliable databases containing this information are now available.

Good global data is provided by the Global Solar Atlas (*https://globalsolaratlas.info*). This database contains a number of further relevant pieces of information down to a resolution of approximately 1 km. The relevant information consists of mean annual values (Figure 12.4):

- **Global Horizontal Irradiation (GHI):** the total irradiation related to flat ground
- **Diffuse Horizontal Irradiation (DHI):** the diffuse component of irradiation related to flat ground
- **Direct Normal Irradiation (DNI):** the direct irradiation of the sun – its intensity
- **Optimum Angle:** the optimum orientation of stationary PV modules
- **Global Irradiation at Optimum Tilt (GTI):** the total irradiation at optimum orientation
- Mean temperature at 2 m above ground

With this information, it is already possible to assess quite well what the mean annual yield of a photovoltaic system in a region will be:

- Clearly recognizable in the data are regions with high cloud cover, such as the area near the equator. There, usable annual amounts in the range of 1500 to 1800 kWh $m^{-2}\,a^{-1}$ are typical.
- In the desert regions between the equator and the tropics, values of up to 2700 kWh $m^{-2}\,a^{-1}$ are reached. This is already quite close to the theoretically possible maximum.
- In the far south of Europe, usable annual amounts of up to 1800 kWh $m^{-2}\,a^{-1}$ are still possible.
- In Germany, the value of usable irradiation varies between 1300 kWh $m^{-2}\,a^{-1}$ in the Alpine foothills and under 900 kWh $m^{-2}\,a^{-1}$ on the North Sea coast.

These data take into account all on-site effects in their mean value. They provide no information about the annual cycle. Many regions show strong seasonal variations in cloud cover, aerosol and dust content of the atmosphere, and other factors that determine what intensity reaches the ground and whether this intensity is direct radiation or diffuse radiation. These seasonal influences of the local climate are significant in different regions at different times of year. Often particularly important are the seasonal changes in the position of the major wind zones as well as the monsoon.

For the EU, the PVGIS collection is another very good data source. Detailed datasets for locations in the EU can be retrieved there (Huld et al. 2012). For Germany, there is additionally the solar roof potential of the DLR (*https://eosolar.dlr.de*). In addition, there are further tools from individual federal states, which in some cases allow the planning of installations on individual rooftops.

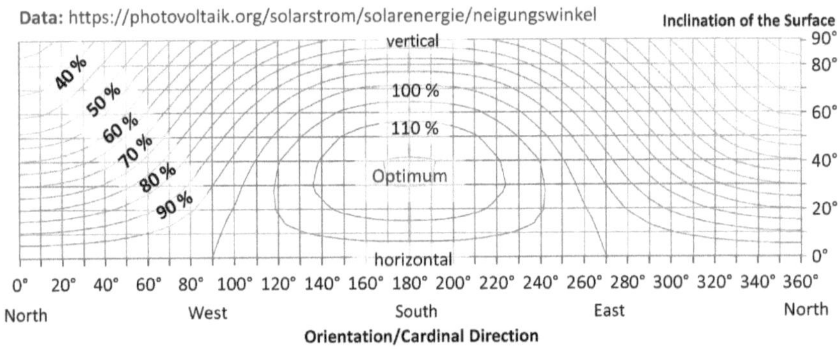

Figure 12.4 dependence of the mean annual irradiation on an unshaded surface (e.g., a PV module) for central Europe at approximately 50° north: A horizontal surface was used as the reference (100%) (data source: photovoltaik.org)

Orientation of the modules

In order to achieve the highest possible output, i.e., to use as little module area as possible for a given yield, the modules should be positioned as perpendicular as possible to the solar irradiation. Since the position of the sun changes throughout the day, this cannot be done statically. Tracking the module according to the position of the sun is repeatedly discussed, but this requires considerable mechanical and electrical effort. The associated energy costs do not appear to justify the effort compared to the resulting benefit.

This leaves, for optimization purposes, the preferred orientation of the module toward the south and installation at a tilt angle that is essentially determined by the latitude as well as, where applicable, local climate-related variations in the annual cycle of irradiation:

- At the equator, this should be 0°, with the module arranged (nearly) parallel to the ground. To facilitate rainwater runoff and cleaning, an angle of 10° is often used.
- In the favorable regions of the near-equatorial deserts, the angle should vary between 10° and 20°.
- In southern Europe, the angle is approximately 28° to 30°.
- In Germany, the optimal angle is approximately 38°. It varies slightly depending on the region.

Further details on optimizing the installation are discussed in Quaschning (Quaschning 2023).

Annual cycle of usable radiation

Figure 12.5 shows the annual cycle of monthly sunshine hours for Hesse and the development of annual sunshine hours since 1951. Sunshine hours are a first indicator of the mean monthly yield. The sunshine hours do not account for the fact that the sun is higher in the sky in summer and therefore a higher output reaches the modules at that time. This is an effect of the lower atmospheric mass AM. In addition, more shading occurs in winter. The data are derived from a small number of weather stations and can vary considerably locally: local weather or shading from mountains can contribute to this. The number of sunshine hours has increased significantly from the earlier to the later period shown. This effect is more pronounced in summer.

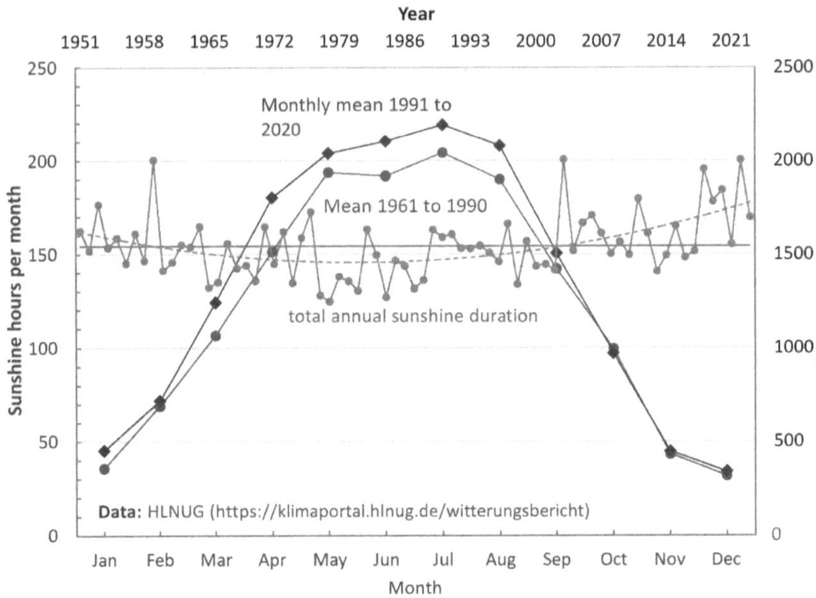

Figure 12.5 Annual cycle of monthly sunshine hours for Hesse and development of sunshine hours from 1951 to the present (data source: HLNUG)

Land Use

The goal of a photovoltaic system is to capture sunlight. This creates either a competing use with other forms of land use or a potential for shared use:

- Photovoltaic systems are ideally mounted on buildings, as this rarely creates a competing use.
- Installation on unused areas generally does not represent a competing use. Such areas are often part of modern transport infrastructure, such as noise barriers or dead zones between carriageways. Small residual areas between transport surfaces that are neither used for agriculture nor connected to any of the surrounding ecosystems are also suitable.
- Photovoltaic systems on agricultural land must initially be viewed critically. In order to occupy these soils, agricultural land must be made available elsewhere: a natural ecosystem is sacrificed at another location, and the products generated there are transported using fossil fuels: a systemic and global assessment of the environmental impacts and energy input is necessary here.
- In principle, the combination of extensive grazing or specialty crops and photovoltaics is sensible in regions with seasonal drought stress (as is now also the case in Germany): the photovoltaic installation significantly reduces the evaporation rate

and plants can thrive overall more effectively. A relevant example of well-suited specialty crops is viticulture.

When large areas are covered with modules, the entire area cannot be fully occupied:

- Access routes are needed to be able to maintain or clean the modules. This area then remains unused.
- If the modules are installed at an angle, they must be positioned at a certain distance from one another to avoid excessive mutual shading at low sun angles. Shading reduces the output of the system when the solar elevation is lower than the angle that allows complete irradiation of all modules.

Shading Figure 12.6 shows a typical open-area geometry and the limiting angle up to which no shading occurs. Here we define a geometric land use factor that describes how closely the modules are positioned:

$$\alpha_{mod} = \frac{h}{d} \tag{12.12}$$

Here, h is the total height of the modules and d their spacing. The shadow angle ω can be calculated from these data (Quaschning 2023).

$$\omega = \arctan\left(\frac{\alpha_{mod} \cdot \sin\delta}{1 - \alpha_{mod} \cdot \cos\delta}\right) \tag{12.13}$$

Figure 12.7 shows the curves of this equation for a wide range of tilt angles:

- 10° is the minimum angle. Then rain can still wash dust and other contaminants off the module.
- For Germany, a tilt angle of 38° and orientation towards the south is optimal (Figure 12.4). Under this orientation, the area utilization should be $\alpha_{Mod} < 0.35$ in order to avoid shading even in winter.
- In regions closer to the equator, modules can be installed with a lower inclination and higher area utilization.

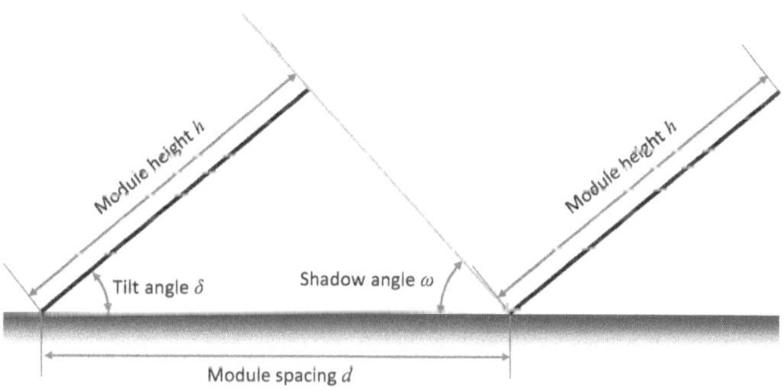

Figure 12.6 Geometry of module rows and shadow angle below which shading occurs

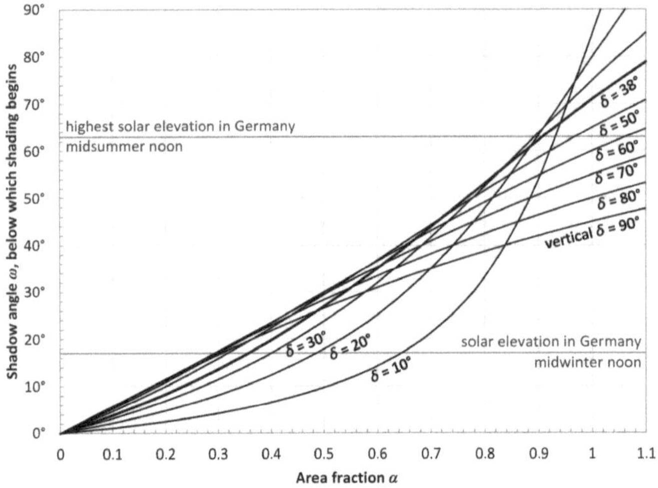

Figure 12.7 Dependence of the shadow angle on the area fraction and on the tilt angle

Intermittency of photovoltaics

Photovoltaics does not generate constant power. One of the essential characteristics of photovoltaics is the variation in power output due to the constantly changing solar irradiance: the diurnal variation of irradiance proceeds slowly. Often, additional strong fluctuations occur due to passing clouds. Since photovoltaics generates electricity, the power provided by a system follows every fluctuation in the intensity reaching the system without any time delay:

- The maximum possible intensity is subject to seasonal variations. These are small near the equator and very pronounced near the poles of the earth.
- The diurnal course of the sun means that light is only available during the day. Only then can electricity be generated.
- The weather influences the actually generated intensity, in particular through clouds, dust on the module, and snow. Even light cirrus clouds in the stratosphere or contrails from aircraft noticeably reduce the intensity available at ground level. Fair-weather convection of cumulus clouds leads to strong fluctuations in the irradiated intensity at ground level.

If photovoltaics is part of the electrical grid, small-scale effects such as contrails or fair-weather convection can be compensated by systems at another location. Large-scale weather structures, such as frontal troughs or high-fog situations, cannot easily be compensated even across a grid spanning a few hundred km. In such cases, significantly less electrical power from photovoltaics is simply available.

In island locations, for private use, or when photovoltaics generates a relevant share of electricity, storage of electricity is therefore essential in order to compensate for such fluctuations.

Photovoltaics for Germany

Germany has an area of 357 578 km² and a mean electricity demand of currently about 60 GW. It is located at approximately 50° north latitude. Modern photovoltaic systems achieve an efficiency of 20% or more. We calculate for the case of ideal PV weather (Figure 12.8).

What land area must Germany cover with photovoltaics in order to generate its electricity demand on 21 June?

To determine this, we need the intensity related to the horizontal surface. The sun is then at an angle of $\gamma = 40° + 23°$ in the sky. Thus

$$\mathrm{AM} = \frac{1}{\sin\gamma} = \frac{1}{\sin(63\,°)} = 1.122$$

From the air mass, the intensity of the direct solar radiation follows as

$$I = I_{sol} \cdot \exp\left(0.0953 - 0.3567 \cdot \mathrm{AM}^{0.678}\right)$$
$$I = 1366\,\frac{\mathrm{W}}{\mathrm{m}^2} \cdot \exp\left(0.0953 - 0.3567 \cdot 1.122^{0.678}\right) = 1022\,\frac{\mathrm{W}}{\mathrm{m}^2}$$

The required area is

$$A = \frac{P_{el}}{I_{ground} \cdot \eta_{PV}} = \frac{P_{el}}{I_0 \cdot \sin\gamma \cdot \eta_{PV}} = \frac{60 \cdot 10^9\ \mathrm{W}}{1022\,\frac{\mathrm{W}}{\mathrm{m}^2} \cdot \sin(63\,°) \cdot 0.20} = 329 \cdot 10^6\ \mathrm{m}^2 = 329\ \mathrm{km}^2$$

This is approximately 1/1000 of the country's land area.

What land area must Germany cover with photovoltaics in order to generate its electricity demand on December 21?

To determine this, we need the intensity related to the horizontal surface. The sun is then at an angle of $\gamma = 40° - 23°$ in the sky. Thus

$$\mathrm{AM} = \frac{1}{\sin\gamma} = \frac{1}{\sin(17\,°)} = 3.420$$

From the air mass, the intensity of the direct solar radiation follows as

$$I = 1366\,\frac{\mathrm{W}}{\mathrm{m}^2} \cdot \exp\left(0.0953 - 0.3567 \cdot 3.420^{0.678}\right) = 661\,\frac{\mathrm{W}}{\mathrm{m}^2}$$

The required area is

$$A = \frac{60 \cdot 10^9\ \mathrm{W}}{661\,\frac{\mathrm{W}}{\mathrm{m}^2} \cdot \sin(17\,°) \cdot 0.20} = 1552 \cdot 10^6\ \mathrm{m}^2 = 1552\ \mathrm{km}^2$$

This is approximately 0.5% of the country's land area.

What minimum land area must Germany cover in order to generate its electricity demand on an annual average?

For this calculation, the mean annual irradiation is required. This is 1100 kWh m^{-2}: This value corresponds to 3.96 GJ m^{-2} or an annual average intensity at ground level of 126 W m^{-2}. The required area thus follows from

$$A = \frac{P_{el}}{I_{ground} \cdot \eta_{PV}} = \frac{60 \cdot 10^9\ \mathrm{W}}{126\,\frac{\mathrm{W}}{\mathrm{m}^2} \cdot 0.20} = 2381 \cdot 10^6\ \mathrm{m}^2 = 2381\ \mathrm{km}^2$$

This is approximately 0.75% of the country's land area. If we also account for losses from storing electricity, this required area can quickly increase by a factor of two. This would correspond to approximately 10% of today's settlement and transport area. Thus, photovoltaics could largely be accommodated on buildings.

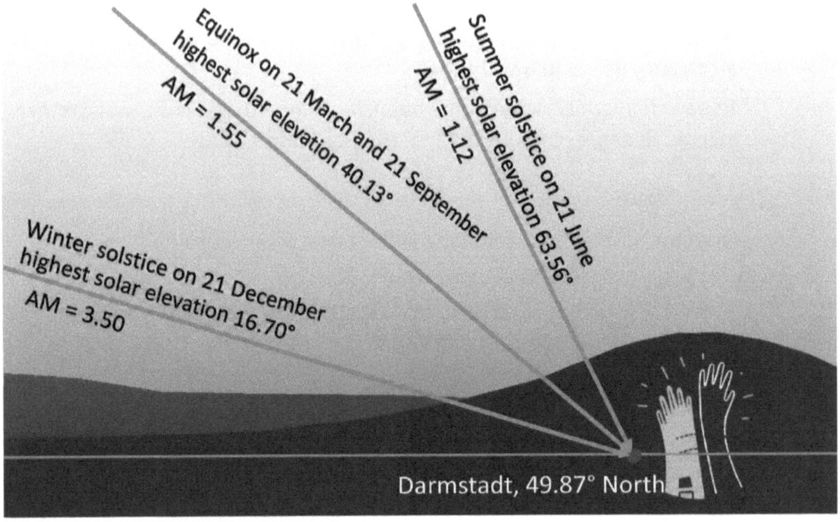

Figure 12.8 Sun position for Darmstadt in winter, in summer, and at the equinox

What maximum power does photovoltaics provide if the most recently determined area is installed?

This can be determined simply by the ratio of areas. On June 21 under ideal conditions, this would be 435 GW, and on December 21 under ideal weather conditions, 92 GW. The challenge for the intensive use of photovoltaics in Germany arises from the very large fluctuations in generated power indicated here.

In summer, a maximum of approximately 7 times the required power is generated. For photovoltaics to achieve a socially acceptable EROI, this power must then also be utilized, i.e., stored, for example. Storage infrastructure will be required that on the one hand can absorb such very large power, while on the other hand absorbs hardly any power for large parts of the year. This makes it clear that a mix of different technologies is needed here, which together generates power more evenly throughout the year.

Biomass instead of photovoltaics?

What fraction of Germany's land area would need to be used to provide this power from biomass?

This can be determined by various methods. A simple approach is to compare efficiencies. With intensive forestry or agriculture, a maximum conversion efficiency into usable biomass of approximately 1% is conceivable. This biomass could be used in state-of-the-art steam turbine power plants with an efficiency of 40%. Thus, the efficiency for electricity generation, without accounting for the inputs for harvesting and provision, is 0.4%. Comparing this with the efficiency for photovoltaics, 45 times the area would be required, or 119,000 km^2.

Another approach is to use the yield from forestry as a basis. This is reported at up to 8600 kg ha^{-1} of hardwood (*http://www.tfz.bayern.de*). From this, with a heating value of 16 MJ kg^{-1}, a mean usable heat flux of 0.436 W m^{-2} follows. With 40% as the efficiency for electricity generation, an area of 343,800 km^2 or nearly the entire area of Germany would need to be used. Both estimates are lower bounds. The actual land requirement will be higher, since the total energy input for agriculture or forestry and transport to the power plants must also be included. Biomass alone is not an option.

12.5 Thermoelectrics

In principle, temperature differences and electricity can be linked and made usable in suitable materials through special physical effects. If an electric field is present in a special circuit, which is usually composed of several different materials, a temperature gradient arises. In the Peltier effect, a flowing current in a suitable material generates a temperature gradient. This effect is used, for example, for cooling electronic components or hotel refrigerators. In the reverse Seebeck effect, a temperature gradient in a suitable circuit generates a voltage gradient. A usable current can then flow due to this voltage.

In principle, a temperature gradient can thus generate a voltage and thereby cause a current to flow. In doing so, the temperature gradient is reduced, i.e., quasi consumed. In order to generate electricity in a steady state, heat must flow so that the generating gradient is maintained. The conversion efficiency is theoretically limited by the Carnot efficiency. In practice, the efficiencies of today's systems are significantly lower.

Thermoelectrics can be used to harvest heat flows that cannot be utilized otherwise. However, the following must always be examined:

- Would it not be energetically more favorable to prevent the (waste) heat flow from arising in the first place?
- Does the energetic benefit justify the energetic input, i.e., does the system have a sufficient EROI?
- Cannot more energy be provided with the same energetic input by other means?
- Is the heat flow better utilized in a district heating network?

For cost reasons, thermoelectrics are ultimately only used in extreme situations where another form of energy supply is impossible or becomes too expensive. It is not a method by which relevant quantities of electricity can be provided.

References

DLR – Deutsches Zentrum für Luft- und Raumfahrt: Solardach-Potenzial der DLR. *https://eosolar.dlr.de*

Fthenakis, V./Frischknecht, R./Raugei, M./Kim, H. C./Erik Alsema, E./Michael Held, M./Wild-Scholten, M. de: Methodology Guidelines on Life Cycle Assessment of Photovoltaic Electricity. International Energy Agency. 2011. *https://iea-pvps.org/wp-content/uploads/2020/01/rep12_11.pdf*

Global Solar Atlas. *https://globalsolaratlas.info*

HLNUG – Hessisches Landesamt für Naturschutz, Umwelt und Geologie: Witterungsbericht.*https://klimaportal.hlnug.de/witterungsbericht*

Huld, T./Müller, R./Gambardella, A.: A new solar radiation database for estimating PV performance in Europe and Africa. Solar Energy 86, 2012, pp. 1803–1815. *https://joint-research-centre.ec.europa.eu/photovoltaic-geographical-information-system-pvgis_en*

Kleidon, A.: Thermodynamics of the Earth System. Cambridge University Press, 2016

NREL – National Renewable Energy Laboratory: Best Research-Cell Efficiency Chart. *https://www.nrel.gov/pv/cell-efficiency.html*

photovoltaik.org: Photovoltaik Neigungswinkel: Tabelle, Rechner & Ausrichtung. *https://photovoltaik.org/solarstrom/solarenergie/neigungswinkel*

Quaschning, V.: Regenerative Energiesysteme. Technologie – Berechnung – Klimaschutz. 12. Auflage. Carl Hanser Verlag, München 2023

Raugei, M./Sgouridis, S./Murphy, D./Fthenakis, V./Frischknecht, R./Breyer, C./Bardi, U./Barnhart, C./Buckley, A./Carbajales-Dale, M./Csala, D./Wild-Scholten, M. de/Heath, G./Jæger-Waldau, A./Jones, C./Keller, A./Leccisi, E./Mancarella, P./Pearsall, N./Siegel, A./Stolz, P.: Energy Return on Energy Invested (ERoEI) for photovoltaic solar systems in regions of moderate insolation. A comprehensive response. In: Energy Policy 102, 2017, pp. 377–384

Rühle, S.: Tabulated values of the Shockley-Queisser limit for single junction solar cells. In: Solar Energy 130, 2016, pp. 139–147

13 Storing and Transporting Energy

Storage and transport of energy are central elements of all energy systems. The transformation of an energy system also requires a transformation of storage and transport routes: Until now, fossil fuels have been kept in above-ground and underground storage facilities, in many forms of tanks, as well as in pipelines or ships, and burned when needed. In the transition to a predominantly electrical energy system, it must be examined whether fuels will continue to represent the most favorable option for energy storage and transport in the future, or whether other technologies are needed.

Electricity from wind and photovoltaics is sometimes available in abundance and sometimes regionally unavailable for extended periods. Here we need ways to connect generation and use over time:

- Starting from electricity, transport and storage as electricity or as another form of work is energetically particularly straightforward. In the ideal case, no entropy is generated when converting electricity into work and vice versa. The conversion would be lossless in the ideal case. Even real conversion methods still achieve considerable efficiency (see Table 17.17 in Section 17.10).
- The conversion of electricity into heat succeeds almost without losses.
- The conversion of electricity into chemical energy carriers and fuels is associated with relevant losses (see Chapter 10 and Table 17.17 in Section 17.10).
- When converting energy from fuels or when converting heat into electricity, substantial losses are then unavoidably incurred (see Chapter 3 and fundamentals of thermodynamics in Linow 2022). Storing as fuel or heat is therefore energetically very unfavorable.

In any case, the storage and transport of energy require additional infrastructure and further resources. Furthermore, losses in the system are unavoidable. Therefore, the socially usable EROI of the energy system is reduced. Storing and transporting are thus activities that should only be carried out to a very limited extent if an energy system is to have high social utility. At the same time, it is necessary to use the infrastructure for electricity generation as intensively as possible in order to keep its EROI high. Trade-offs and adjustments are necessary here (Palmer/Floyd 2020).

13.1 Storing Energy As Work

As technical options, in addition to electricity, potential energy and kinetic energy are available to avoid conversion via heat. This limits the technical options to a few relevant ones:

- Kinetic energy: as rotating masses
- Potential energy: pumped-storage hydropower plants as well as some ideas for lifting large masses
- Electricity: in capacitors or batteries

Kinetic energy The storage of kinetic energy can be achieved via flywheels or large gyroscopes. An optimal storage density is achieved with extremely fast-rotating gyroscopes that rotate in a vacuum. In this way, kinetic energy can be stored for a few minutes. Such concepts can primarily serve the short-term stabilization of power grids: electrical energy is fed in via an external field and also extracted again. Prototypes are in use, but overall this is a niche technology for very specific applications, particularly since the material input is high and the stored energy remains small.

Potential energy There are some niche ideas that each combine high material expenditure with low stored energy or appear very immature. On a large industrial scale and relevant today are only pumped-storage hydropower plants.

Pumped-Storage Hydropower Plants

These are essentially hydropower plants that can also be operated in reverse: additionally, large electrically driven pumps are available that can pump the water back into an upper reservoir. They do this when there is a surplus of electricity in the grid. The water is then stored in the upper reservoir until electricity is needed again. This water can then be used to generate electrical power via turbines and generators. The water is collected in the lower reservoir so that it is available for the next storage cycle. Originally, pumped-storage hydropower plants were designed as a fast-responding means of stabilizing the grid. Pumped-storage hydropower plants can absorb power from the grid or feed it back into the grid very quickly, much faster than is possible with fossil-fueled power plants. They are a technically simple, fully mature (TRL 9), energetically cost-effective, and very long-term investment. Larger plants can provide approximately 1 GW of power over a period of typically 8 hours. The following example determines the need for pumped-storage hydropower plants to be able to bridge one week of cold dark doldrums for Germany.

A major obstacle to new plants is the current remuneration structure for electricity, because the operator must, in order to act economically, purchase and store electricity significantly more cheaply than they sell it. The taxes and levies, which are sometimes due multiple times, make the required price difference large and the financial risk is high for the operators. If more storage capacity is to be created in Germany, it must be taken into account that operators of storage facilities neither consume nor generate electricity in the first place. At least these shares would therefore have to be assessed differently from the storage cycle loss. A second aspect is the location for further projects: preference is given to two large open areas that are close to each other but separated by a sufficiently large difference in elevation: one area for the upper reservoir on the mountain and one for the lower reservoir in the valley. The capacity of the pumped-storage hydropower plant depends linearly on the usable difference in elevation. If a large difference in elevation can be realized, the reservoirs may be able to be smaller.

Storage requirement for Germany

The greatest challenge for the transition of the electricity supply to photovoltaics and wind is the required storage capacity for electrical energy in order to be able to meet the possible full demand at any time. In winter, in the worst case, weather conditions can occur in which it is simultaneously almost windless and high fog covers the sky for up to a week (cold dark doldrums). In that case, the electrical energy demand would have to be covered entirely from storage. The consumption of electrical energy for Germany amounts to approximately E_{BRD} = 525 TWh per year (2023).

Pumped-storage hydropower plants are established at suitable locations in the low or high mountain ranges, where a sufficiently large usable head between closely situated lower and upper reservoirs is possible: typical for such pumped-storage hydropower plants is a realizable difference in elevation of Δz = 275 m. The reservoirs then have a mean depth of h_{PSH} = 30 m. Modern turbines for hydropower plants achieve an efficiency of 93 % for the generation of electrical power.

What is the mean electrical power demand for Germany?

The mean electrical power is

$$\overline{P_{De}} = \frac{W_{el}}{\Delta t} = \frac{525\ \text{TWh}}{1\ \text{a}} = \frac{525 \cdot 10^{12}\ \text{Wh}}{365 \cdot 24\ \text{h}} = 59.93 \cdot 10^{9}\ \text{W}$$

If we now assume that during a dark doldrums period the electrical demand is not reduced – e.g. because there are no suitable price signals for this – then the further steps follow. We also assume that the electricity during the dark doldrums period comes only from the pumped-storage hydropower plants.

What mass flow of water generates this electrical power?

Using the assumed typical data for pumped-storage hydropower plants, the result is

$$\dot{m}_{PSH} = \frac{\overline{P_{De}}}{\eta \cdot g \cdot \Delta z} = \frac{59.93 \cdot 10^{9}\ \text{W}}{0.93 \cdot 9.81\ \frac{\text{m}}{\text{s}^2} \cdot 275\ \text{m}} = 23.89 \cdot 10^{6}\ \frac{\text{kg}}{\text{s}} \cong 23\,900\ \frac{\text{m}^3}{\text{s}}$$

In the following, we assume a longer dark doldrums period of one week.

What is the area requirement for the required energy storage?

What is needed is the volume that flows at the calculated mass flow rate in one week. This is

$$V_{PSH} = \frac{\dot{m}_{PSH}}{\rho_{water}} \cdot \Delta t_{DC} = \frac{23.89 \cdot 10^6 \, \frac{\text{kg}}{\text{s}}}{996 \, \frac{\text{kg}}{\text{m}^3}} \cdot 10 \cdot 24 \cdot 3600 \text{ s} = 20.72 \cdot 10^9 \text{ m}^3 = 20.72 \text{ km}^3$$

With the specified mean depth of the basins, the area is then

$$A_{PSH} = \frac{V_{PSH}}{h_{PSH}} = \frac{20.72 \cdot 10^9 \text{ m}^3}{30 \text{ m}} = 690 \cdot 10^6 \text{ m}^2 = 690 \text{ km}^2$$

This is the area of the upper reservoirs. The lower reservoirs require a comparable area. In total, this amounts to approximately 0.4% of Germany's land area, or slightly more than the area of Lake Constance.

The construction of such a storage capacity would be cost-intensive and the infrastructure would be clearly visible in the landscape. However, it would certainly be technically feasible, especially if even greater head heights are utilized and deeper basins are constructed. Independently of this, the societal question must be assessed as to whether such a large amount of capital, labor, material, and energy should be committed to demand-oriented electricity generation.

Pumped-storage hydropower in the Rhine-Main region

In the Rhine-Main region, the Feldberg north of Frankfurt would be an ideal location: to the south, the mountain descends with a good gradient almost 600 m into the Main plain. At the same time, the summit area offers several square kilometers of space for one or more upper reservoirs. The water supply is secured by precipitation on the mountain alone. Only at the start would the facility need to be flooded once with water from the Main. Considerably smaller and lower, but still a genuinely favorable location, is the Melibokus south of Darmstadt. Here, a height difference of around 300 m could still be utilized for a large upper reservoir.

The infrastructure would significantly and visibly alter the shape of the mountains over a wide area. Their value as a recreational and forest area would be reduced. The residential areas on the southern slope of the Feldberg (Königstein, Bad Homburg vor der Höhe, Oberursel) and also on the western slope of the Melibokus would lose residential value. They are currently high-priced residential areas for wealthy Frankfurt residents.

The upper reservoir and the resulting infrastructure would simultaneously enable straightforward development of the Feldberg or the Melibokus as wind energy sites. The Feldberg is one of the best sites for wind turbines in Hesse, with particularly high and consistent wind speeds, and the Melibokus would also be very well suited.

The benefit for the region from such centrally located, energetically favorable energy storage facilities and wind farms would be high. Due to the growth of data centers in Frankfurt, electricity demand is increasing significantly. There is therefore a need for additional generation capacity. At the same time, data centers require particularly high electrical supply security, to which these facilities can make a substantial contribution. Nevertheless, political resistance would be considerable, and electricity suppliers therefore strive to find remote, often less favorable locations.

13.2 Storing Electricity Directly

Currently, it is batteries or capacitors in which electricity can be stored directly.

Capacitors This technology is relevant for niche applications where fast charging and discharging is the primary concern. For long-term storage of large quantities of electricity, the technology is too complex.

Batteries and accumulators Batteries were at the very beginning of electrification and have accompanied the second industrial revolution throughout. On the one hand, this is absolutely established technology; on the other hand, intensive work is being done on new concepts with which high energy density can be achieved using widely available metals.

High-performance batteries are the prerequisite for digital mobile devices and electrically powered vehicles. In both applications, the weight of the battery is the primary concern, which is why lithium-ion batteries have prevailed here. The extent to which sufficient lithium and electrode materials will be available at market-compatible prices in the future to enable envisaged plans for complete electromobility is being intensively discussed (Lundaev et al. 2023; König 2023).

Requirements for Batteries

Batteries for storing fluctuatingly generated electricity, for stabilizing the electrical supply grid, and for compensating for periods in which no electricity can be generated, are stationary. This entails special requirements:

- There must be the possibility of storing very large amounts of electricity.
- There is a very high power level during charging and discharging.
- There must be a high cycle efficiency and a high cycle count to enable a long service life, as well as low maintenance requirements. If a battery is to compensate for a day-night cycle, one cycle per day or 4,000 cycles in ten years will occur. Capital goods of this magnitude have projected service lives in the range of 30 years or more, so that 12,000 cycles are targeted.
- The requirements for mass or form factor are low.

- High safety – even when integrated into very large storage systems – requires low flammability, the ability to contain and quickly suppress accidents, and a low proportion of toxic substances. The electricity supplied and the required voltage should be within a manageable range, and they should behave benignly with respect to overload and necessary protection.
- Also important is the use of metals that are available in the long term and a high recycling rate for all materials used, so that the storage systems can be used sustainably at socially acceptable costs.

An overview of possible battery technology with regard to this application is given in Table 13.1 and Table 13.2. All battery types share common properties that make large-scale deployment of a storage system for a large city or a region difficult for this application:

- Only some types exhibit cycle counts that suggest they can be built and used as storage systems for a relevant period of time in the future.
- The energy density that can be stored requires very large installations in which considerable quantities of raw materials are bound over the long term.
- The ESOI is low, so that considerable energy is required for the provision and long-term operation of these storage systems. This has implications for the energy and raw material price.

For many battery types, large-scale processes for recycling the bound raw materials are currently lacking. Complete recycling is one of the fundamental prerequisites for the large-scale use of batteries over a longer period of time.

Table 13.1 Typical Electrical Parameters for Electricity Storage (Based on Barnhart et al. 2013 and IEC 2011; Updated): Efficiency η_{store}, Maximum Discharge, Number of Cycles n_{Zy}, Energy Density W_{el}/m and Power Density P_N/V

Storage	η_{store}	Discharge	n_{cycle}	W_{el}/m in MJ kg^{-1}	P_N/V in kW m^{-3}
Pumped storage	85 %–90 %	~ 100 %	> 25 000	0.0007–0.007	0.2–2
Compressed air (diabatic)	40 %–75 %	Plant	> 25 000	–	0.2–2
Battery (PbA)	70 %–80 %	~ 80 %	< 1000	0.1–0.16	40–400
Battery (Li-Ion)	85 %–95 %	~ 80 %	< 4000	0.3–1.0	2000–10 000
Battery (Na-Ion)	90 %	~ 80 %	> 1000	ca. 0.4	> 1000
Battery (NiMH)	65 %–75 %	-	< 1200	0.15–0.3	500–3000
Battery (NaS)	70 %–80 %	~ 80 %	< 4000	0.36–0.9	> 2000

Table 13.2 Key Figures on Materiality and ESOI for Electricity Storage (Based on Barnhart et al. 2013 and IEC 2011): Number of Cycles n_{Zy}, Specific Embodied Energy ε_{Zy}, Storage Stability Δt_S and ESOI

Storage	n_{Zy}	ε_{Zy}	Material	Δt_S	ESOI
Pumped hydro storage	> 25 000	0.033		Long	700
Compressed air (diabatic)	> 25 000	0.045		Long	800
Battery (PbA)	< 1500	0.0105		Short	5
Battery (Li-Ion)	< 10 000	0.00735	Critical	Long	32
Battery (Na-Ion)	> 1000			Long	
Battery (NiMH)	< 1200	–	Critical		–
Battery (NaS)	< 4750	0.0069			20

How many storage units are needed?

In electrically powered vehicles, storage units of approximately 100 kWh, i.e. 360 MJ, are used today. To store the current average weekly electricity demand of Germany, $36 \cdot 10^{15}$ J or $10 \cdot 10^{9}$ kWh are required. This corresponds to the storage capacity of 100 000 000 electric passenger cars. Since Germany currently has almost 50 000 000 passenger cars, a considerable part of this storage demand could be realized in car batteries.

13.3 Storing Energy As Heat

Energy can either be stored directly as heat, or the energy is used to produce a fuel. Fuels can be a wide variety of chemical compounds, but nuclear fuels would also be conceivable.

Fuel The term Power-to-X encompasses many storage options for electricity. Originally, only Power-to-Gas was meant, i.e. the storage of electricity in the chemical energy of hydrogen or methane. Here it is understood as the conversion of electricity into a chemical energy carrier, i.e. a fuel. The technical fundamentals, the most important options and key figures were already presented in Chapter 10.

Since the cycle efficiency of these methods inevitably remains low, this cannot represent a universal storage method. The use of these methods will be limited to necessary areas where other direct storage options are not suitable. Today these are aircraft and ocean-going ships.

13.3.1 Thermal Storage

In the future, there will be a need for suitable technologies to store heat directly (and not as a fuel). Solar thermal power plants can only absorb heat from solar radiation during the day. If they are to generate electricity into the night or if another process is to operate during the night, part of this heat must be stored appropriately. Residential buildings can be supplied from thermal storage units, either during the night or over longer periods of time. Direct storage as heat also includes the storage of cold – i.e. the ability to absorb heat at low temperature.

The essential requirements for a thermal storage system depend on the specific application. The requirements include the following:

- The stored heat quantity or the heat quantity available after a specified period of time Q in the storage
- The minimum temperature $T_{out,min}$, at which this heat is to be provided
- The transferable heat flow $\dot{Q}$ when charging or discharging the storage and thus the rate at which heat can be supplied or removed
- The permissible heat loss from the storage over time
- The number of cycles over which the storage operates

These requirements can be translated into possible technical solutions based on fundamental properties:

Insulation The heat loss from a storage increases with the storage temperature. Therefore, with regard to the effort required for insulation, it is more favorable to store heat at lower temperatures. In general, gases have the lowest thermal conductivity when convection can be prevented. Foam-like or fibrous material in which the gas fills the interstices is ideal here. The required quality of the insulation depends on (i) the time over which storage is intended, (ii) the permissible losses over time, (iii) the temperature difference between inside and outside, and (iv) the surface area of the storage. In principle, good insulation can be achieved over a long period of time, but this requires a comparatively high expenditure of material. The ideal shape of a storage is spherical in order to minimize the surface area.

Storage size Large storage units have a significantly more favorable ratio between the volume of the storage medium and the surface area: the volume increases with the third power and the surface area with the second power. For large storage units, the heat losses related to the stored heat are lower, or the specific effort for insulation decreases.

Specific storage capacity The amount of heat that can be stored in a volume depends, for sensible heat, on the density ρ and the specific heat capacity c_p of the storage material, and for latent heat on the density and the specific enthalpy of the phase transition Δh_P, which is utilized:

$$Q_S = \underbrace{V_S \cdot \rho \cdot c_p \Delta T}_{sensible} + \underbrace{V_S \cdot \rho \cdot \Delta h_P}_{latent} \tag{13.1}$$

The choice of material is decisive here.

Temperature The maximum temperature in the storage may place very high demands on the material used:

- The material must remain thermally stable. It must not decompose or oxidize. It must not undergo chemical reactions and must not exhibit a high diffusion rate into adjacent components.
- All components must remain mechanically stable, i.e., their temperature-dependent mechanical properties must not change too much.
- All components must be resistant to thermal cycling. This means they must not build up stresses during the intended temperature changes that could cause them to rupture. They must remain mechanically stable over many cycles. Therefore, crack formation must be kept to a minimum.
- All material combinations must remain stable in the long term even at maximum temperature. They must not damage each other.
- The availability of materials nearby and low costs for the storage material must be ensured.

Storage for Sensible Heat

Table 13.3 lists some typical materials for storage units. They are arranged in ascending order according to the temperature range. In principle, higher temperatures can still be achieved using special ceramics, carbides (silicon carbide SiC) or refractory metals (tungsten, tantalum), but then at very high costs very quickly. The storage capacity is calculated as the sensible heat that can be stored between 20 °C and maximum temperature. All inputs for the real storage system increase with temperature. Established technologies exist with air preheaters for blast furnaces and glass melting furnaces, which can be scaled fairly easily for other applications (Sterner/Stadler 2017).

Latent Heat Storage

When the heat is to be provided at a relatively low temperature, such as for building heating or industrial processes, it is advantageous to store it not via sensible heat but via latent heat: in this case the thermal storage can be operated almost isothermally (i.e. at constant temperature), thereby avoiding many mechanical problems caused by thermal expansion. Solid-liquid transitions are often used, in which the specific enthalpy of the phase transition utilized mostly lies in the range between 100 kJ kg^{-1} and 330 kJ kg^{-1}. The volume-related storage capacity is thus comparable to that of water and sensible heat. This is also the case because most suitable salt solutions have a density between 1000 kg m^{-3} and 2000 kg m^{-3}.

Table 13.3 Typical Storage Materials for Sensible Heat, the Density ρ, the Specific Heat Capacity c_p, the Volumetric Heat Capacity $\rho \cdot c_p$, the Maximum Volumetric Storage Capacity q_V and the Usable Temperature Range T_{use} As Well As the Typical Application

Material	ρ in kg m^{-3}	c_p in J kg^{-1} K^{-1}	$\rho \cdot c_p$ in MJ m^{-3} K^{-1}	q_V in MJ m^{-3}	T_{use}	Application
Water	1000	4200	4.2	336	0–100 °C	Building heating
Molten salt	2200–2600	1.3	2.9–3.4	870–1020	150–450 °C	Solar thermal power plants
Concrete	1900–2300	0.88	1.7–2.1	820–1000	< 500 °C	Structural components
Rock	2500	0.7	1.75	1370	< 800 °C	Ground storage
Iron	7800	0.5	3.9	3040	< 800 °C	Solar thermal storage
Brick	1400–1900	0.84	1.2–1.6	1200–1600	< 1000 °C	
Fireclay	2000	1.0	2.0	2400	<1200 °C	Air preheaters in the glass and steel industry

The specific enthalpy of vaporization of some common substances is significantly higher than the enthalpy of fusion. For water, the enthalpy of fusion is 330 kJ kg^{-1} and the enthalpy of vaporization is 2500 kJ kg^{-1}. Liquid-solid phase transitions are nevertheless technically preferred, since the specific volume of the storage material hardly changes in the process. For gas-liquid storage, suitable and large pressure vessels are required that can accommodate the gas phase.

A number of well-defined systems are now available for operating temperatures between 0 °C and approximately 300 °C, so that this technology can be implemented for a wide variety of purposes (Sterner/Stadler 2017).

Thermochemical Storage

Similar to latent heat storage thermochemical storage systems also use a reversible reaction that can absorb or release a large amount of energy within a narrow temperature window. In principle, the storage densities are comparable to or higher than those of the other two methods. However, the effort required to provide the storage system is overall higher (Sterner/Stadler 2017).

Heat storage for Darmstadt

This example is intended to give a sense of the size of such storage systems. At the end of 2018, Darmstadt had a stock of 79,854 apartments, in which n_{Da} = 161,857 inhabitants live (*https://www.darmstadt.de*). However, no data are available for the heating energy demand in Darmstadt. Therefore, we use the values for Germany: In 2016, the N_{DE} = 82,180,000 inhabitants of Germany together spent Q_{DE} = 462 TWh of energy for space heating in their households, or related to the living area Q/A = 132 kWh m^{-2} (*https://www.umweltbundesamt.de*).

What is the per capita heating energy demand and the average living area?

The per capita heating energy demand for one year is

$$Q_{cap} = \frac{Q_{De}}{N_{De}} = \frac{462\ \mathrm{TWh}}{82.18 \cdot 10^6} = \frac{462 \cdot 10^9\ \mathrm{kWh}}{82.18 \cdot 10^6} = 5622\ \mathrm{kWh}$$

and the living area follows with

$$A_{cap} = \frac{Q_{cap}}{Q/A} = \frac{5622\ \mathrm{kWh}}{132\ \frac{\mathrm{kWh}}{\mathrm{m}^2}} = 43\ \mathrm{m}^2$$

What is the heat demand for Darmstadt in one year?

Provided we assume that Darmstadt approximately reflects the German average, then

$$Q_{Da} = N_{Da} \cdot Q_{cap} = 161\ 857 \cdot 5622\ \mathrm{kWh} = 0.910\ \mathrm{TWh} = 3.276 \cdot 10^{15}\ \mathrm{J}$$

Darmstadt is wealthier than average, but is currently growing very rapidly, so this is likely to balance out with regard to apartment size.

How large would a thermal storage unit need to be for Darmstadt?

What is meant here is a thermal storage unit that can fully supply all apartments in Darmstadt with heat, where this is a simple water storage unit that absorbs sensible heat.

The volume results, e.g. under the assumption that the heat is stored in the range between 10 °C and 60 °C, as

$$V_{Da} = \frac{Q_{Da}}{\rho_{water} \cdot c_{p,water} \cdot \Delta T} = \frac{3.276 \cdot 10^{15}\ \mathrm{J}}{996\ \frac{\mathrm{kg}}{\mathrm{m}^3} \cdot 4.20\ \frac{\mathrm{kJ}}{\mathrm{kg \cdot K}} \cdot 50\ \mathrm{K}} = 15.66 \cdot 10^6\ \mathrm{m}^3$$

This corresponds to a storage volume of 97 m^3 per resident or approximately the residential volume. With an area of 1 km^2 the storage unit would need to be 16 m deep.

What area would need to be covered with solar thermal collectors to be able to provide the required heat?

For Darmstadt, the annual solar incoming radiation is approximately q_{in} = 1100 kWh m^{-2} (*https://globalsolaratlas.info*). Let us assume that the solar thermal system has an efficiency of approximately 50 % and the storage unit has a cycle efficiency of 80 %. Then the required area is

$$A_{solar} = \frac{Q_{Da}}{q_{in} \cdot \eta_{solar} \cdot \eta_S} = \frac{0.910 \text{ TWh}}{1100 \frac{\text{kWh}}{\text{m}^2} \cdot 0.5 \cdot 0.8} = 2.07 \cdot 10^6 \text{ m}^2$$

These dimensions are impressive. They can be influenced by significantly reducing the demand for heat. This works (i) through significantly less heating in heatable rooms, (ii) fewer rooms that are heated, and (iii) smaller living areas per person, as well as (iv) lower energy losses from heated rooms. The first three approaches significantly reduce comfort and interfere with the expectations of all residents. The fourth approach is being intensively pursued in new buildings, but not in existing buildings.

Since such a storage unit as a tank would be enormous and therefore energetically and financially very expensive, it will certainly not be feasible. However, such a thermal storage unit would be conceivable as a medium-depth geothermal system, in which the heat is stored underground at a depth of approximately 1000 m. Even for this, the effort due to the many required drillings would be impressive. At the same time, a risk assessment would first need to clarify what environmental impacts can arise from cyclic thermal expansion and other effects.

13.3.2 Compressed Air Storage

The storage of mechanical work in the form of compressed air is a method that was first used in railways and has been used for many years at several locations for the storage of electricity. It can be regarded as an available technology with TRL 7. Ambient air is electrically compressed to a high degree and stored in a large vessel as the storage medium. Often the vessel is a large cavern underground, e.g. in a salt dome. The enthalpy difference stored in the air between the storage unit (index S) and the environment (index 0) is

$$\Delta H = m_{air} \cdot (u_S - u_0 + p_S \cdot v_S - p_0 \cdot v_0) \tag{13.2}$$

The air in the cavern must not have a very high temperature, as otherwise moisture from the rock passes into the air or other reactions can occur with the salt. Therefore, the temperature of the air in the cavern is often not much higher than that of the surroundings. To achieve this, the air must be cooled simultaneously during compression. Since the specific internal energy u of an ideal gas depends only on temperature, it follows for the case where the temperature in the cavern approximately equals the ambient temperature and thus the two internal energies are nearly identical,

$$\Delta H = m_{air} \cdot (p_S \cdot v_S - p_0 \cdot v_0) \tag{13.3}$$

In addition, the equation of state of the ideal gas can be used here:

$$\Delta H = m_{air} \cdot (R_{air} \cdot T_S - R_{air} \cdot T_0) \cong 0 \tag{13.4}$$

The equation suggests that no energy is stored. This appears contradictory, because mechanical work is nevertheless stored:

- Ambient air is highly compressed and cooled to approximately ambient temperature in the process. This process is an isothermal compression, in which the heat removed approximately equals the mechanical power supplied.
- If work is to be extracted from the storage unit, the air must be reheated during expansion in the turbine. So that this process also proceeds approximately isothermally, approximately the energy must be supplied as heat that is to be extracted as mechanical work via the turbine.

Today this is technically solved, e.g. by dissipating the heat during compression to the surroundings. If the compressed air is to be used, it is fed to a gas turbine: the air from the storage unit is first heated by burning natural gas and then expands in the turbine, which drives a generator. In this way, the efficiency of the gas turbine is higher, since the energy-intensive compression of the incoming air of the gas turbine is eliminated. For this form of diabatic compressed air storage, an additional fuel supply, e.g. for natural gas, is required.

In principle, the heat extracted during compression of the air can be stored separately from the compressed air. When the stored compressed air is extracted, the stored heat is then supplied back to it. This concept of adiabatic compressed air storage does not require an additional fuel supply, but has not yet been implemented on a large technical scale (TRL 6).

13.4 Storage, EROI and Curtailing

Renewable energy is only available in a highly variable manner. Thus, when using renewable energy sources, too little or too much is usually generated for the current demand. Today, fossil energy carriers are available to replace renewable energy during times when too little renewable energy can be generated. However, this will foreseeably no longer be the case.

In times when more renewable energy is generated than is needed or stored, the renewable power plants must be curtailed. Today this is also done when conventional power plants cannot be ramped down quickly enough. In principle, it needs to be assessed for the future whether, in the event of an oversupply of renewable energy, it is better to curtail or store it.

Curtailing means that the fraction φ of the energy $E_{pot,pp}$ that could be generated by a power plant in a given period was not generated, i.e., the energy actually generated with curtailing is

$$E_{use,pp} = (1 - \varphi) \cdot E_{pot,pp} \tag{13.5}$$

This directly reduces the EROI of a renewable energy technology accordingly. The loss in EROI depends on the extent of curtailing – whether only one power plant is shut down or larger parts of the renewable infrastructure. This energy wasted through curtailing can instead be used to fill a storage unit and thus be useful at a later point in time. In the case where all the energy from curtailing is stored, it holds that

$$E_{use,s} = \eta_s \cdot \varphi \cdot E_{pot,pp} \tag{13.6}$$

The total energy provided then consists of the directly used fraction and the stored fraction:

$$E_{use,tot} = E_{use,pp} + E_{use,s} = (1 - \varphi) \cdot E_{pot,pp} + \eta_s \cdot \varphi \cdot E_{pot,pp} \tag{13.7}$$

In this way, the EROI of the power plant follows (see Chapter 5):

$$\mathrm{EROI}_{pp} = \frac{E_{use,tot}}{E_{invest}} = \frac{(1 - \varphi) \cdot E_{pot,pp} + \eta_s \cdot \varphi \cdot E_{pot,pp}}{E_{invest,pp} + E_{invest,s}} \tag{13.8}$$

From the EROI of the power plant, the invested energy of the power plant follows:

$$E_{invest,pp} = \frac{E_{pot,pp}}{\mathrm{EROI}_{pp}} \tag{13.9}$$

From the ESOI of the storage unit for the power plant, the invested energy for the storage unit follows:

$$E_{invest,s} = \frac{E_{use,s}}{\mathrm{ESOI}_s} \tag{13.10}$$

In the best case, a storage unit is only needed for the fraction φ of the energy. This gives the EROI of the entire infrastructure (Pellow et al. 2015) as

$$\mathrm{EROI}_{tot} = \frac{E_{use,pp} + E_{use,s}}{E_{invest,pp} + \varphi \cdot E_{invest,s}} = \frac{(1 - \varphi) \cdot E_{pot,pp} + \eta_s \cdot \varphi \cdot E_{pot,pp}}{\frac{1}{\mathrm{EROI_{pp}}} + \frac{\varphi}{\mathrm{ESOI}_s}} \tag{13.11}$$

This equation represents the simplest possible case. It can be extended for the analysis of more complex relationships. In Figure 13.1 this equation is illustrated for two variants that are typical for wind energy as a renewable energy source:

1. For storage units with a cycle efficiency of 0.85, the curves are shown for four ESOI values covering the entire range from batteries to pumped-storage hydropower plants. In this case, the EROI drops by only a factor of two even for the worst storage variants when all the energy is passed through the storage unit. This cycle efficiency is characteristic of pumped-storage hydropower plants or batteries.

2. For storage units with a cycle efficiency of 0.25, the curves for four identical ESOI values are shown as in the previous case. In this case, the EROI decreases significantly faster with the storage rate. This cycle efficiency is characteristic of power-to-X.

Figure 13.1 shows that the necessary storage technology to compensate for fluctuations will have a significant impact on the price of electricity. The extent depends critically on what fraction of the electricity is to be intermediately stored, and thus ultimately on how strongly electricity is to be provided in a demand-oriented manner in the future.

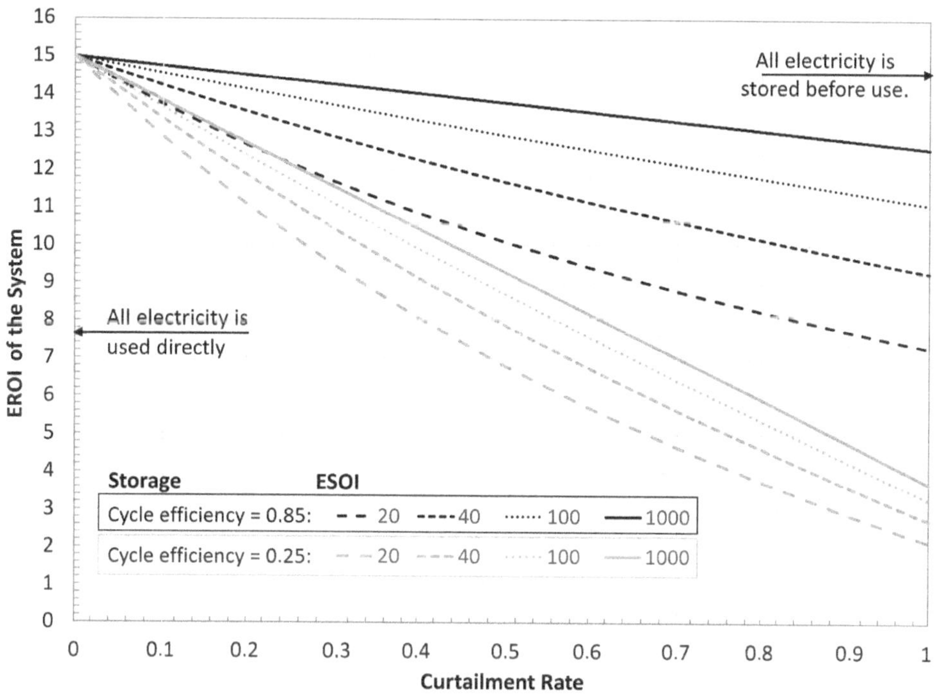

Figure 13.1 Effect of the storage rate φ and the ESOI on the overall EROI of generation infrastructure and storage

13.5 Transporting Energy

Energy can also be transported safely and in larger quantities in the future via two routes, either as electricity in corresponding lines or as a transportable energy carrier or fuel. Independently of this, there is the possibility of not transporting the energy to the point of application, but instead relocating particularly industrial processes to where suitable energy is available at low cost. The energy is then transported as embodied grey energy bound in products. This is a particularly obvious option for very energy-intensive processes.

Energy Carriers

The conversion of electricity into energy carriers is the subject of Chapter 10. The associated high losses were addressed there; numerical values are provided in Table 17.17 in Section 17.10. Chapter 10 also already discussed the effort required for the compression of gaseous fuels: while this is relatively low for methane compared to the energy content of the methane, it is impressively high for hydrogen. For this reason too, it is quite unclear whether there will be an extensive hydrogen pipeline network or even overseas transport by tanker ships in the future. The cost structure is absolutely not comparable to that for LNG (liquefied natural gas).

Electricity

The direct transport of electricity in lines avoids the critical losses of conversion into energy carriers. However, losses also occur here, now due to the electrical resistance of the lines used, in the case of alternating current due to field effects in the immediate vicinity of the line, and due to transformers during voltage conversion.

Resistance The electrical resistance of a conductor is a function of the length l, the cross-sectional area A and the temperature-dependent specific electrical resistance of the conductor material:

$$R_{wire} = \frac{l_{wire}}{A_{wire}} \cdot \sigma_{wire}(T) \cong \frac{l_{wire}}{A_{wire}} \cdot \sigma_{20°\mathrm{C,cond}} \cdot (1 + \alpha \cdot |\vartheta - 20\ °\mathrm{C}|) \qquad (13.12)$$

Conductivity is usually specified at 20 °C and the temperature dependence can be described linearly over a certain temperature range using a coefficient α.

This resistance and the associated heat generation ultimately limits the current-carrying capacity of electrical lines. This leaves the voltage used to transport electricity as the only free parameter. Conversely, above 400,000 V, optical phenomena and noise increasingly occur on AC lines. Therefore, the backbone of the grid is designed for a typical 400,000 V. Additional losses then arise from dielectric field effects of the alternating field with its surroundings. This is why overhead lines are particularly simple and cost-effective to design.

Grid structure and expansion

The electrical grid in Central Europe has four levels:

1. The 400-kV grid transports large amounts of power over long distances, connecting generation with consumption. Typically, line losses amount to approximately 1% per 100 km.
2. The typical 110-kV grid branches off from the 400-kV grid and supplies cities or large industrial consumers.

3. The medium-voltage grid (mostly 35 kV) is the distribution network across the area.
4. The 230-V grid (400-V three-phase current) serves for residential connections.

Between these levels there are transformers that step the current up or down to the required voltage. To transport large amounts of power, several cables are connected in parallel to form a single line.

An additional requirement for the electrical grid is its high reliability. It must also function when individual lines fail for a certain period of time. For this purpose, multiple paths are generally provided at least between nodes of the higher levels, each of which can individually transport at least a typical demand.

Historically, the electrical grid was designed to run from a few power plants to many users. In the meantime, it is being massively restructured to integrate a very large number of small generators (PV, small wind farms) into the grid. Power no longer flows steadily from large power plants into the grid; instead, power is distributed much more widely across the grid.

Grid expansion in Germany

At full expansion of the currently planned wind power capacity, the German Bight will eventually have 120 GW of rated capacity available. The key question of this example is how this offshore wind power can be transported approximately 600 km southward, where the greater demand exists today. The EROI of North Sea wind power is approximately 15, when the system boundary of the calculation is the connection to the power grid on the mainland.

Overhead lines with a voltage of 380 kV form the backbone of the European power grid. These lines transport high power levels over long distances: typically, each cable of these lines has a steel core of 40 mm^2 and an aluminum sheath (through which the major portion of the current flows) of 240 mm^2. Under normal weather conditions, such a cable can carry approximately 1000 A. The typical 400-kV line south of Frankfurt has a total of 18 bundles arranged in four levels, with each bundle consisting of four such conductors.

Main natural gas pipelines operate at a pressure of approximately 80 bar and the gas flows at a maximum flow velocity of 10 m s^{-1}. The large supply lines have a typical inner diameter of 1000 mm. They are buried in the ground and therefore have a constant temperature of 10 °C.

How many overhead lines are needed to transport 120 GW?

One cable transports a maximum of

$$P_{power\ line} = U \cdot I_{power\ line} = 400\ \text{kV} \cdot 1000\ \text{A} = 400\ \text{MW}$$

Thus, 300 cables or 5 overhead lines, each with $4 \cdot 18$ cables, are required from north to south.

What is the maximum amount of gas that a natural gas main pipeline can transport?

The question is what power (as potential heat of combustion based on the lower heating value) can be transported through a natural gas main pipeline at maximum. The volumetric flow rate of natural gas in a pipeline is

$$\dot{V}_{pipe} = A_{pipe} \cdot c_{NG} = \frac{\pi}{4} \cdot d_{pipe}^2 \cdot c = \frac{\pi}{4} \cdot (1\ \mathrm{m})^2 \cdot 10\ \frac{\mathrm{m}}{\mathrm{s}} = 7.854\ \frac{\mathrm{m}^3}{\mathrm{s}}$$

at 80 bar and 12 °C and thus at standard conditions

$$\dot{V}_N = \dot{V}_{pipe} \cdot \frac{T_N}{T} \cdot \frac{p}{p_N} = 7.854\ \frac{\mathrm{m}^3}{\mathrm{s}} \cdot \frac{273.15\ \mathrm{K}}{285\ \mathrm{K}} \cdot \frac{80\ \mathrm{bar}}{1.013\ \mathrm{bar}} = 594.5\ \frac{\mathrm{Nm}^3}{\mathrm{s}}$$

This corresponds to a transmission power of

$$\dot{Q}_{pipe} = \dot{V} \cdot H_{I,methane} = 594.5\ \frac{\mathrm{Nm}^3}{\mathrm{s}} \cdot 36.32\ \frac{\mathrm{MJ}}{\mathrm{Nm}^3} = 21.59 \cdot 10^9\ \mathrm{W}$$

How many natural gas pipelines can be supplied?

From 120 GW of wind power, a quantity of gas can be produced of

$$\dot{Q}_{windgas} = \eta_{el \to CH_4} \cdot P_{wind} = 0.55 \cdot 120\ \mathrm{GW} = 66 \cdot 10^9\ \mathrm{W}$$

The following number of pipelines is required for transport:

$$N_{pipe} = \frac{\dot{Q}_{windgas}}{\dot{Q}_{pipe}} = \frac{66\ \mathrm{GW}}{21.59\ \mathrm{GW}} = 3.06$$

This should be feasible with 3 pipelines.

What power is available in the south?

As a heat flow, the already determined 66 GW are available – provided that the pipeline does not require some of it for intermediate compressors or that some is lost through leakages.

For the reconversion of methane to electricity, either a gas turbine with an efficiency of approximately 38 % can be used

$$P_{el \to CH_4 \to el} = \eta_{el \to CH_4} \cdot \eta_{CH_4 \to el} \cdot P_{wind} = 0.55 \cdot 0.38 \cdot 120\ \mathrm{GW} = 25.04\ \mathrm{GW}$$

or for a modern CCGT power plant with 60 % it is

$$P_{el \to CH_4 \to el} = \eta_{el \to CH_4} \cdot \eta_{CH_4 \to el} \cdot P_{wind} = 0.55 \cdot 0.60 \cdot 120\ \mathrm{GW} = 39.60\ \mathrm{GW}$$

What amount of energy is stored in the pipeline?

At any given time, a large amount of gas is stored in a gas pipeline:

$$V_{pipe} = A \cdot l = \frac{\pi}{4} \cdot d^2 \cdot l = \frac{\pi}{4} \cdot (1\ \mathrm{m})^2 \cdot 600\ 000\ \mathrm{m} = 471\ 200\ \mathrm{m}^3$$

The gas volume contained therein is at maximum

$$V_{N,pipe} = V_{pipe} \cdot \frac{T_N}{T} \cdot \frac{p}{p_N} = 471\ 200\ \mathrm{m}^3 \cdot \frac{273.15\ \mathrm{K}}{285\ \mathrm{K}} \cdot \frac{80\ \mathrm{bar}}{1.013\ \mathrm{bar}} = 35.66 \cdot 10^6\ \mathrm{Nm}^3$$

Thus the stored chemical energy is

$$Q_{pipe} = V_{N,pipe} \cdot H_{I,CH_4} = 35.66 \cdot 10^6\ \mathrm{Nm}^3 \cdot 36.32\ \frac{\mathrm{MJ}}{\mathrm{Nm}^3} = 1.295 \cdot 10^{15}\ \mathrm{J}$$

No energy is stored in the electrical power line.

How does energy transport via synthetic natural gas and subsequent reconversion to electricity change the EROI?

The EROI of the system, neglecting the energy costs of the gas infrastructure, amounts to 3.13 (gas turbine with 38 %) to 4.95 (CCGT power plant).

This exercise compares two discussed variants and highlights the advantages of both options. The direct transport of electricity produces lower losses than transport as natural gas. However, the transmission losses of electricity of approximately 6 % are not taken into account here. On the other hand, transport as gas has the advantage of compensating for fluctuations by using the pipeline as a storage medium.

References

Barnhart, C. J./Dale, M./Brandt, A. R./Benson, S. M.: The energetic implications of curtailing versus storing solar- and wind-generated electricity. In: Energy and Environmental Science 6, 2013, pp. 2804–2810

Carbajales-Dale, M./Barnhart, C. J./Benson, S. M.: Can we afford storage? A dynamic net energy analysis of renewable electricity generation supported by energy storage. In: Energy Environ. Sci. 7, 2014, p. 1538

IEC – International Electrotechnical Committee (Hrsg.): Electrical Energy Storage. White Paper. Genf 2011. *https://basecamp.iec.ch/download/iec-white-paper-electrical-energy-storage*

König, M.: Entwicklung der Versorgungssituation und des Energiebedarfes mineralischer Rohstoffe im Kontext der globalen Energiewende. Masterthesis. Hochschule Darmstadt 2023. *https://doi.org/10.48444/h_docs-pub-394*

Linow, S.: Angewandte technische Thermodynamik. Carl Hanser Verlag, München 2022

Lundaev, V./Solomen, A. A./Le, T./Lohrmann, A./Breyer, C.: Review of critical materials for the energy transition, an analysis of global resources and production databases and the state of material circularity. In: Minerals Engineering 203, 2023, 108282

Palmer, G./Floyd, J.: Energy, storage and civilization. A systems approach. Springer, Cham 2020

Pellow, M. A./Emmott, C. J. M./Barnhart, C. J./Benson, S. M.: Hydrogen or batteries for grid storage? A net energy analysis. In: Energy and Environmental Science 8, 2015, pp. 1938–1952

Sterner, M./Stadler, I.: Energiespeicher. Bedarf, Technologien, Integration. Springer, Wiesbaden 2017

Zapf, M.: Stromspeicher und Power-to-Gas im deutschen Energiesystem. Rahmenbedingungen, Bedarf und Einsatzmöglichkeiten. Springer, Wiesbaden 2017

14
Treating Greenhouse Gases

This chapter deals with the handling of greenhouse gases emitted by our energy system and our technology. To this end, we first look at the handling of other emissions from the combustion of fossil fuels, in order to get an impression of the energy input required for flue gas cleaning. Based on this, we discuss the possibilities of extracting carbon dioxide as the central greenhouse gas from the flue gas or from the atmosphere. The focus is on the energy costs and on possible contributions to achieving the Paris climate targets or keeping the earth system within planetary boundaries (see Chapter 8). Other greenhouse gases are only briefly touched upon. Finally, we look at methods and the possible role of hard geoengineering. The goal is to understand where the energy challenge lies when it comes to separating and depositing carbon dioxide and other greenhouse gases, or reducing their impact on our earth by other means.

14.1 Conventional Flue Gas Cleaning

Every combustion reaction leads to the emission of pollutants. Not all typical pollutants, such as dust, sulfur, carbon monoxide or nitrogen oxides occur in all reactions. This section summarizes, as a benchmark, the input and energy costs for established techniques for flue gas cleaning (see box on conventional flue gas cleaning).

Conventional flue gas cleaning

The emissions of dust, sulfur, and nitrogen oxides lead to massive health consequences and economic damage, which is why their release has been increasingly regulated by law and reduced. These regulations were enforceable due to the impressive health and macroeconomic benefits (Romanello et al. 2022) as well as the very low associated costs.

Dust

When burning organic material (wood, straw, dung) and solid fossil fuels (coal, lignite, peat), some constituents of the fuel form residues that are released to the environment as dust with the flue gas. The amount of dust that can be released from systems without any flue gas cleaning is impressive, though rarely encountered today: historical steam locomotives give an indication and can still make this actively tangible today, although they have a rather low power output. Impressive are images of coal-fired battleships at full speed.

Fly ash consists of sedimentary rock present in the fuel and inorganic constituents of plants. Depending on the speed at which the flue gases are drawn from the combustion zone and other particularities of the plant, larger or smaller proportions remain as ash in the combustion chamber or are transported as fly ash with the flue gas into the environment.

Fly coke is unburned coal dust that forms under stronger draft, i.e., high velocity of the flue gas stream, and certain firing systems. Fly coke represents a loss of efficiency of the combustion system.

There are many reasons to reduce dust emissions:

- Dust negatively affects other activities in the vicinity. Contamination can complicate or impede other technical processes.
- Dust covers plants and thus leads to reduced photosynthesis, lower yields, reduced growth, and the die-off of vegetation.
- Dust is directly harmful to health; it contains unpleasant or toxic components such as arsenic, cadmium, chromium, copper, mercury, lead, selenium, or vanadium, which poison humans, animals, and soils.
- Dust is harmful to the respiratory tract. In particular, fine particulate matter has a multitude of direct and indirect effects on the health of humans and animals (Romanello et al. 2022).

For this reason, there are today strict limit values for dust emissions, also because the separation of dust is technically easy to manage. If less toxic dust can continue to be used as a commodity (aggregate in the construction industry), it disappears from a societal energy balance. Dusts that are so toxic that they must be permanently disposed of represent a long-term burden: what is relevant here is whether such dusts are deposited in a long-term secure, inaccessible, and cost-free manner, or whether they burden future generations as a source of poison — also in energetic terms — forever (perpetual burdens).

Sulfur

Sulfur is a constituent of organic material. During the formation of fossil energy carriers in geological processes, it is either bound directly in the fuel or it is present as hydrogen sulfide (H_2S). Gasoline, diesel, kerosene, and natural gas are cleaned of sulfur, so that no flue gas cleaning is necessary.

Coal is usually burned directly with its sulfur. In combustion, sulfur or hydrogen sulfide reacts with oxygen to form sulfur dioxide (SO_2). In this reaction, additional usable heat is released. Sulfur dioxide is a pungently smelling toxic gas. It causes the yellow coloration of flue gas. Sulfur dioxide reacts further with water to form sulfuric acid (H_2SO_4):

- Hydrogen sulfide (H_2S) is highly toxic.
- Special weather conditions can lead to extreme sulfur concentrations and smog in regions with high industrial density and high emissions.
- Sulfur dioxide is harmful to health and can lead to death directly or in the long term through respiratory diseases during smog events.
- Sulfuric acid destroys a great many materials, especially buildings.
- Sulfur is transported over long distances by the wind and causes soil acidification far away: close to the source, acid wastelands form (e.g., near Nickel in Russia), or damage to ecosystems occurs at greater distances (forest dieback or acidification of bodies of water).

Sulfur can easily be bound with lime and then removed from the flue gas as gypsum. During flue gas scrubbing, other substances such as halogens (chlorine, bromine, and fluorine) are simultaneously bound.

The specific energy consumption in the power plant for this process is 0.8 to 1.2 kWh per 1000 m^3 of flue gas. The energy input for the provision of limestone must be additionally accounted for. In return, the gypsum can be sold as a commercial product and is today consumed entirely by the construction industry.

Nitrogen oxides

In combustion reactions, nitrogen oxides form via three independent pathways:

Fuel NO Nitrogen atoms already bound in the fuel react with the oxygen in the air to form NO. The formation reaction of fuel NO proceeds largely independently of the maximum temperature of combustion and other parameters. Therefore, this form of NO can only be reduced either by using a different fuel or by secondary measures, i.e., downstream flue gas cleaning.

Thermal NO In the region of the highest temperature of combustion, above approximately 1000 °C, nitrogen and oxygen can react with each other. The amount of NO formed increases rapidly with the maximum temperature of combustion.

Prompt NO Here the chemical reaction by which nitrogen and oxygen react with each other proceeds via the mediation of reactive carbon radicals. As with thermal NO, the reaction rate here also increases significantly with temperature. Prompt NO forms preferentially in the rich zone and thermal NO in the lean zone. Thus, formation can be effectively reduced by staged combustion in suitable burners. NO is of particular importance in engine combustion, since the compression of the air in the cylinder generates a very high combustion temperature. Here, primary measures achieve good reduction rates when they are planned and taken into account.

If primary measures are insufficient, the emission can be further reduced by a downstream catalytic reaction. In most cases, a selective catalytic reduction (SCR) is used. In addition, ammonia must be available as a reaction partner for the nitrogen oxides at the catalyst. The ammonia can either be added directly or be generated from urea in the flue gas.

urea is therefore added as an additive in motor vehicles, as it is less hazardous to health than ammonia (Ad-Blue). The energy input is defined by the additional infrastructure required, the pressure drop across the catalysts, and the provision of ammonia or uric acid.

Other substances

All fossil fuels began as dead plant material and therefore contain a large number of additional elements. During the refining of crude oil, these substances are mostly carried through to the end and accumulate in the last fractions. Therefore, these fractions are quite highly enriched with sulfur and heavy metals.

In the case of coal there is no purification or separation prior to use as a fuel; therefore, coal contains all the heavy elements that were deposited in it during its formation. Typical flue gas cleaning removes many of these substances. Only a few regularly slip through, in particular mercury.

During the combustion of solids, especially those with low calorific value, high moisture content, and a high proportion of halogens, extremely toxic compounds such as dioxins or furans are formed. This applies in particular to the combustion of waste. Therefore, particular effort must be expended here on flue gas cleaning. The filter dusts and other substances from the flue gas cleaning of waste incineration plants are therefore extremely toxic and must be deposited separately.

Technical information on the processes is provided by Krauß as well as the VDI guidelines (Krauß 2009).

Emission limit values

Pollutants from technical sources are the largest environmental cause of disease and premature death (Landrigan et al. 2018). The relevant burdens here vary greatly by region. In addition, climate change will further alter these regional burdens (Romanello et al. 2022). To limit these effects, limit values have developed over recent decades, which have often arisen from a balancing of the socially desirable right to health, the technically feasible reductions, and economic interests in externalizing these costs.

Terms More specifically, the statutory requirements have emerged from a process aimed at achieving a public health objective, for which the best available technology (BAT) is intended for reduction purposes. In this way, what is technically achievable is established by law as a requirement. The public health objective can be expressed as the number of still tolerable deaths (measured as Years of Life Lost, YOLL) as well as serious health damage (measured as Disability Adjusted Life Years, DALY). BAT is understood to mean at least a TRL 8. This refers to a technique that is available on the market and can be used at reasonable cost. Thus, binding limit values have been developed for a number of pollutants, such as for dust (in particular from coal combustion, waste incineration, and diesel engines), carbon monoxide, sulfur, nitrogen oxides, organic gases with methane (Volatile Organic Compounds, VOC) or without methane (non-Methane VOC, NMVOC) or mercury. Not all of these limit values are always enforced. BATs are not always updated with technological progress. Not all critical substances are regulated with due caution (example: PFAS).

The technology required to reduce emissions can become quite large and thus influences the costs for fuels or for plants in which emissions are produced. From a technical perspective, combustion and the associated emissions have thereby become manageable. This statement holds as long as the exhaust gas cleaning system is switched on and functioning (i.e., only to a very limited extent in the case of internal combustion engine passenger cars).

Reference oxygen Limit values for emissions always refer to the dry flue gas at standard conditions. The water vapor is not taken into account and is condensed out beforehand during flue gas measurement. The emission limit values always refer to a specific, predefined oxygen fraction in the flue gas, which is regulated together with the limit value. In order to compare actually measured values of an emission r_{meas} with the limit values, these are first converted to the specified reference oxygen concentration:

$$r_{ref} = \frac{21 - O_{2,ref}}{21 - O_{2,meas}} \cdot r_{meas}$$

Here $O_{2,ref}$ is the reference oxygen concentration in % given in the law or the applicable regulation, and $O_{2,meas}$ is the measured oxygen content in %.

Typical emissions of a power plant

The Staudinger 5 power plant in Großkrotzenburg serves as an example. It has a rated electrical output of 510 MW. In 2013, it emitted 2,560,000 t of carbon dioxide, 1,710 t of nitrogen oxides, 589 t of sulfur dioxide, and 29.8 kg of mercury. The power plant uses hard coal, petroleum coke, and sewage sludge as fuel. To simplify the calculation, we assume that the fuel is pure carbon with $H_{l,C}$ = 32.8 MJ kg^{-1}. In 2007, the power plant had a specific emission of m_{CO2}/W_{el} = 1.090 kg CO_2 per kWh of electricity generated.

What are the efficiency and the capacity factor of the power plant?

Based on the given specific CO_2 emission per kWh, we can determine the efficiency. For this purpose, we use the definition of the lower heating value and the definition of the thermal efficiency:

$$\eta = \frac{W_{el}}{Q} = \frac{W_{el}}{m_{fuel} \cdot H_{I,fuel}}$$

From each carbon atom in the fuel, one CO_2molecule is produced, so the amounts of substance of carbon in the fuel and carbon dioxide in the flue gas are identical:

$$\frac{m_C}{M_C} = n_C = n_{CO_2} = \frac{m_{CO_2}}{M_{CO_2}}$$

Substituting yields

$$\eta = \frac{W_{el}}{m_C \cdot H_{I,C}} = \frac{1}{H_{I,C}} \cdot \frac{W_{el}}{m_{CO_2}} \cdot \frac{M_{CO_2}}{M_C} = \frac{1}{32.8 \frac{MJ}{kg}} \cdot \frac{1}{1.090 \frac{kg}{kWh}} \cdot \frac{44.01 \frac{kg}{kmol}}{12.01 \frac{kg}{kmol}} = 0.369$$

The actual efficiency is different, since we have used neither the actual lower heating value of the fuel mixture nor its carbon content. Both are somewhat lower.

The capacity factor is the ratio of the actually delivered electricity to the maximum possible electricity. Again, from the mass of CO_2 and the specific CO_2emission per kWh, the actually generated electricity follows as

$$W_{el,2013} = m_{CO_2} \cdot \frac{W_{el}}{m_{CO_2}} = 2.56 \cdot 10^9 \text{ kg} \cdot \frac{1}{1.090 \frac{kg}{kWh}} = 8.455 \cdot 10^{15} \text{ J}$$

The maximum possible work over one year is

$$W_{el,\max} = P_N \cdot 1 \text{ a} = 510 \text{ MW} \cdot 365 \cdot 24 \cdot 3600 \text{ s} = 16.08 \cdot 10^{15} \text{ J}$$

The utilization factor is therefore

$$\psi_{2013} = \frac{W_{el,2013}}{W_{el,\max}} = \frac{8.455 \cdot 10^{15} \text{ J}}{16.08 \cdot 10^{15} \text{ J}} = 0.526$$

What is the mercury content (mass fraction) of the fuel?

The mercury is not separated in the flue gas cleaning system. The actual mass of fuel converted amounts to (if it were pure carbon)

$$m_C = m_{CO_2} \cdot \frac{M_C}{M_{CO_2}} = 2.56 \cdot 10^9 \text{ kg} \cdot \frac{12.01 \frac{kg}{kmol}}{44.01 \frac{kg}{kmol}} = 698.6 \cdot 10^6 \text{ kg}$$

Thus the mass fraction of mercury in the fuel is (since the fuel consists almost entirely of carbon and since mercury is present almost completely in the flue gas)

$$\mu_{Hg} = \frac{m_{Hg}}{m_C} = \frac{29.8 \text{ kg}}{698.6 \cdot 10^6 \text{ kg}} = 42.66 \cdot 10^{-9}$$

What is the sulfur content of the fuel?

We cannot answer this, since a large portion of the sulfur is separated in the flue gas scrubber and only the residual emission is given.

What is the nitrogen oxide concentration (volume fraction) in the flue gas?

For this purpose, we determine the volume of NO and of the total flue gas.

Nitric oxide First, the specific gas constant for NO is needed:

$$R_{NO} = \frac{R}{M_{NO}} = \frac{8314 \frac{J}{kmol \cdot K}}{30.01 \frac{kg}{kmol}} = 277.0 \frac{J}{kg \cdot K}$$

From this, for standard conditions p_N = 101 325 Pa and T_N = 273.17 K

$$V_{NO,2013,N} = \frac{m_{NO,2013} \cdot R_{NO} \cdot T_N}{p_N} = \frac{1\,710\,000 \text{ kg} \cdot 277.0 \frac{J}{kg \cdot K} \cdot 273.15 \text{ K}}{101\,325 \text{ Pa}} = 1.277 \cdot 10^6 \text{ Nm}^3$$

Flue gas The flue gas from the stoichiometric reaction of the fuel with air consists only of the carbon dioxide and the nitrogen from the air. The oxygen from the supplied combustion air is also present in the CO_2 . The volume of carbon dioxide is

$$V_{CO_2,2013,N} = \frac{m_{CO_2,2013} \cdot R_{CO_2} \cdot T_N}{p_N} = \frac{2.56 \cdot 10^9 \text{ kg} \cdot 188.9 \frac{J}{kg \cdot K} \cdot 273.15 \text{ K}}{101\,325 \text{ Pa}} = 1.304 \cdot 10^9 \text{ Nm}^3$$

Thus the volume of nitrogen for the case of stoichiometric combustion, in which exactly one O atom is supplied for each C atom_2 required for combustion,

$$V_{N_2,2013,N} = \frac{0.79}{0.21} \cdot V_{CO_2,2013,N} = \frac{0.79}{0.21} \cdot 1.304 \cdot 10^9 \text{ Nm}^3 = 4.906 \cdot 10^9 \text{ Nm}^3$$

The total flue gas volume at 0 % reference oxygen is

$$V_{fluegas,2013,N} = \frac{1}{0.21} \cdot V_{CO_2,2013,N} = \frac{1}{0.21} \cdot 1.304 \cdot 10^9 \text{ Nm}^3 = 6.210 \cdot 10^9 \text{ Nm}^3$$

We still need to relate this flue gas volume to the applicable reference oxygen of 6 %:

$$V_{6\%,2013,N} = V_{fluegas,2013,N} \cdot \frac{21-0}{21-6} = 6.210 \cdot 10^9 \text{ Nm}^3 \cdot \frac{21-0}{21-6} = 8.694 \cdot 10^9 \text{ Nm}^3$$

Thus the volume ratio for nitrogen oxide is $y_{NO} = 14.70 \cdot 10^{-6}$ or, on a mass basis, 196 mg Nm^{-3}.

14.2 Greenhouse Gases

Among the existing limit values for emissions from power plants, greenhouse gases are only included to a very limited extent: methane and other hydrocarbons (VOC) have a greenhouse effect, but restrictions on their emission have mostly been introduced for other reasons (harmful to health, ozone-depleting, explosive). Carbon dioxide is not taken into account.

By now there are a large number of approaches and methods aimed at reducing greenhouse gas emissions. Some of these have the desired effect, but many do not (Plötz et al. 2024). These mostly involve financial incentives or increases in the efficiency of technical processes. It is now clear that the actual goal of timely decarbonization cannot be achieved by these means alone and within the framework of current policy (Climate Action Tracker).

In the following, we therefore discuss framework conditions and energy inputs required to keep the earth system within a (climatically) habitable range, going beyond these approaches. We orient ourselves by the planetary boundary. This states that the mean temperature of the earth should remain in the long term below +1 °C above the pre-industrial value and that the radiative forcing of all greenhouse gases combined should be less than 1 W/m^2 (Chapter 8). Both values have already been significantly exceeded. The following possibilities are discussed.

Carbon Capture and Storage (CCS) This refers to the separation of a portion of the carbon dioxide emissions from the flue gas stream of large point sources such as fossil-fuel power plants and cement works, and the subsequent geological storage of this carbon dioxide. In this way, carbon dioxide emissions could be reduced in the future. This is often linked to the goal of continuing to use fossil fuels until they are exhausted (Section 14.3).

Carbon Capture and Usage (CCU, CCUS) Here, the captured carbon dioxide is used to manufacture new products. The carbon is then intended to be stored in these products for a certain period of time (Section 14.3).

Negative CO_2 Emissions (NET) This involves the active removal of carbon dioxide from the atmosphere in order to store it geologically over the long term. In this way, the concentration of carbon dioxide in the atmosphere is to be actively reduced. Most nations have, within the framework of the Paris Climate Agreement, pledged such negative emissions on a large scale from 2050 onwards (Section 14.4).

Solar Radiation Management (SRM) These are methods by which the incoming solar radiation into the earth system is to be reduced by means of geoengineering. The goal would be to adjust the radiation balance of the earth so that the mean temperature of the atmosphere is appropriate (Section 14.5). These methods explicitly do not include a reduction in greenhouse gas concentration, but address only the radiative forcing.

Is natural gas really less harmful to the climate than coal?

This example estimates under what conditions natural gas is less harmful to the climate than coal. For this purpose, the emissions of two power plants of equal rated electrical output and with an assumed equal utilization factor of $\psi = 0.85$ are compared.

A modern coal-fired power plant, comparable to Maasvlakte, with a rated output of 1000 MW, hard coal as fuel and an electrical efficiency of 45 % is used as the coal power plant. The coal here is Australian hard coal. The rail distance from the coal mine to the port in Australia is 450 km, the sea route from Australia to Rotterdam is 25,000 km, and the distance along the Rhine from Rotterdam to Groß-Krotzenburg is 520 km.

A CCGT power plant with two gas turbines each with a rated output of 400 MW, a downstream steam turbine with 200 MW and an overall efficiency of 60 % is used as the natural gas power plant. This could be, for example, the Trianel power plant in Hamm.

Approximately 10 % of the natural gas is used for the transport and purification of the gas. Depending on the infrastructure, transport losses of between 5 % and 15 % occur.

The energy input for the transport of the coal is taken from Table 17.13 in Section 17.7, where the fuel in each case is fossil diesel. Approximately 20 % of the extracted crude is used for the extraction, processing and provision of the diesel.

What is the required fuel mass for both power plants and for one year?

Both power plants generate the same amount of electricity per year of

$$W_{el} = P_N \cdot \psi \cdot \Delta t = 1.00 \cdot 10^9 \text{ W} \cdot 0.85 \cdot 365 \cdot 24 \cdot 3600 \text{ s} = 26.01 \cdot 10^{15} \text{ J}$$

Coal power plant The coal power plant requires a heat input of

$$Q_{coal} = \frac{W_{el}}{\eta_{coal}} = \frac{26.01 \cdot 10^{15} \text{ J}}{0.45} = 57.80 \cdot 10^{15} \text{ J}$$

For this, a mass of hard coal (as pure carbon) is required of

$$m_{coal} = \frac{Q_{coal}}{H_{I,coal}} = \frac{57.80 \cdot 10^{15} \text{ J}}{25 \cdot 10^6 \frac{\text{J}}{\text{kg}}} = 2312 \cdot 10^6 \text{ kg}$$

CCGT The CCGT power plant requires as heat

$$Q_{NG} = \frac{W_{el}}{\eta_{CCGT}} = \frac{26.01 \cdot 10^{15} \text{ J}}{0.60} = 43.35 \cdot 10^{15} \text{ J}$$

and thus

$$m_{NG} = \frac{Q_{NG}}{H_{I,NG}} = \frac{43.35 \cdot 10^{15} \text{ J}}{50 \cdot 106 \frac{\text{J}}{\text{kg}}} = 867 \cdot 10^6 \text{ kg}$$

What is the required extracted mass of natural gas for one year?

We can only estimate the extracted mass, since the extraction losses are not specified. For a share of 80 % of the extracted natural gas that reaches the power plant, we obtain

$$m_{NG,ex} = \frac{m_{NG}}{\eta_{transport}} = \frac{867 \cdot 10^6 \text{ kg}}{0.8} = 1084 \cdot 10^6 \text{ kg}$$

How much diesel mass is required for the transport of the coal?

The total amount of diesel follows from the individual contributions listed below.

For transport by rail, an energy input of

$$Q_{rail} = e_{rail} \cdot l_{rail} \cdot m_{coal} = 0.25 \frac{\text{kJ}}{\text{kg} \cdot \text{km}} \cdot 450 \text{ km} \cdot 2312 \cdot 10^6 \text{ kg} = 260 \cdot 10^{12} \text{ J}$$

and thus

$$m_{Diesel,rail} = \frac{Q_{rail}}{H_{I,Diesel}} = \frac{260 \cdot 10^{12} \text{ J}}{43.2 \cdot 10^6 \frac{\text{J}}{\text{kg}}} = 6.02 \cdot 10^6 \text{ kg}$$

For transport across the ocean in a bulk carrier, the following applies

$$Q_{ocean} = e_{ocean} \cdot l_{ocean} \cdot m_{coal} = 0.16 \frac{\text{kJ}}{\text{kg} \cdot \text{km}} \cdot 25\,000 \text{ km} \cdot 2312 \cdot 10^6 \text{ kg} = 9248 \cdot 10^{12} \text{ J}$$

$$m_{Diesel,ocean} = \frac{Q_{ocean}}{H_{I,Diesel}} = \frac{9248 \cdot 10^{12} \text{ J}}{43.2 \cdot 10^6 \frac{\text{J}}{\text{kg}}} = 214.1 \cdot 10^6 \text{ kg}$$

For transport by inland vessel, the following applies

$$Q_{inland} = e_{inland} \cdot l_{inland} \cdot m_{coal} = 0.36\ \frac{\text{kJ}}{\text{kg} \cdot \text{km}} \cdot 520\ \text{km} \cdot 2312 \cdot 10^6\ \text{kg} = 433 \cdot 10^{12}\ \text{J}$$

$$m_{Diesel,inland} = \frac{Q_{inland}}{H_{I,Diesel}} = \frac{433 \cdot 10^{12}\ \text{J}}{43.2 \cdot 10^6\ \frac{\text{J}}{\text{kg}}} = 10.02 \cdot 10^6\ \text{kg}$$

The total amount of diesel required for transport is then $m_{Diesel} = 230 \cdot 10^6$ kg.

The required amount of crude oil for the provision of the diesel is thus

$$m_{Diesel,ex} = \frac{m_{Diesel}}{0.80} = \frac{230 \cdot 10^6\ \text{kg}}{0.8} = 288 \cdot 10^6\ \text{kg}$$

This crude oil is completely combusted.

What is the total emitted amount of substance of carbon dioxide?

Calculated for both power plants over one year and taking into account the inputs for the transport of the raw materials

Coal For the coal power plant, the coal and the diesel must be taken into account. For the coal,

$$n_{\text{CO2,coal}} = n_{\text{C,coal}} = \frac{m_{coal}}{M_{\text{C}}} = \frac{2312 \cdot 10^6\ \text{kg}}{12.01\ \frac{\text{kg}}{\text{kmol}}} = 192.5 \cdot 10^6\ \text{kmol}$$

Diesel is more difficult to calculate, because we must take into account that a diesel molecule, as a longer-chain hydrocarbon, contains a number of carbon atoms. Here, heavy diesel or even heavy fuel oil is more likely to be used, so that the number of carbon atoms is typically closer to $N = 20$. The molar mass of such a hydrocarbon is then approximately 283 kg/kmol. We obtain approximately

$$n_{\text{CO}_2,Diesel} = N_{C\ in\ Diesel} \cdot n_{Diesel} = N_{Diesel} \cdot \frac{m_{Diesel}}{M_{Diesel}} = 20 \cdot \frac{288 \cdot 10^6\ \text{kg}}{283\ \frac{\text{kg}}{\text{kmol}}} = 20.40 \cdot 10^6\ \text{kmol}$$

This is only an estimate, since ultimately the crude oil used influences the emission.

In total, the amount of substance of emitted carbon dioxide resulting from the provision of coal for the power plant is $n_{\text{CO2},coal} = 213 \cdot 10^6$ kmol.

NG For the CCGT power plant, one molecule of methane reacts to form one molecule of CO_2, and thus

$$n_{\text{CO2,NG}} = n_{NG} = \frac{m_{NG,ex}}{M_{methane}} = \frac{1084 \cdot 10^6\ \text{kg}}{16.04\ \frac{\text{kg}}{\text{kmol}}} = 67.58 \cdot 10^6\ \text{kmol}$$

What quantity of natural gas can escape into the atmosphere as methane?

When is the same global warming potential as that of the coal power plant reached over the next 20 years? The coal power plant emits a mass of carbon dioxide per year of

$$m_{\text{CO2,coal-pp}} = n_{\text{CO2,coal-pp}} \cdot M_{\text{CO}_2} = 213 \cdot 10^6\ \text{kmol} \cdot 44.01\ \frac{\text{kg}}{\text{kmol}} = 9.37 \cdot 10^9\ \text{kg} = 9.37\ \text{Mt}$$

The CCGT power plant emits

$$m_{\text{CO2,CCCT}} = n_{\text{CO2,CCGT}} \cdot M_{\text{CO}_2} = 68 \cdot 10^6 \text{ kmol} \cdot 44.01 \frac{\text{kg}}{\text{kmol}} = 2.99 \cdot 10^9 \text{ kg} = 2.99 \text{ Mt}$$

The difference in released CO_2 mass between coal and natural gas is

$$\Delta m_{\text{CO}_2} = m_{\text{CO2,coal-pp}} - n_{\text{CO2,CCGT}} = 9.37 \cdot 10^9 \text{ kg} - 2.99 \cdot 10^9 \text{ kg} = 6.38 \cdot 10^9 \text{ kg}$$

The global warming potential of methane over 20 years is 86. Thus, the maximum mass of methane that may be released is

$$\Delta m_{\text{CH}_4} = \frac{\Delta m_{\text{CO}_2}}{GWP_{methane}} = \frac{6.38 \cdot 10^9 \text{ kg}}{86} = 74.2 \cdot 10^6 \text{ kg}$$

Of the extracted mass, this is a maximum fraction of

$$\Delta \varphi_{loss} = \frac{\Delta m_{\text{CH}_4}}{m_{NG,ex}} = \frac{74.2 \cdot 10^6 \text{ kg}}{1084 \cdot 10^6 \text{ kg}} = 0.068$$

This means that the use of natural gas causes less global warming than coal only if very little natural gas is released. We have determined a limit of 7% here. Our approach is not very precise and we have not included some effects. Losses due to venting of natural gas during normal pipeline operation, losses during extraction through leakages, leakages in the transport infrastructure, and boil-off losses with liquefied natural gas (LNG) quickly add up to values above 10%. Thus, natural gas is no better than coal with regard to greenhouse gas emissions. The result obtained here is confirmed in much more detailed global analyses (Tanaka et al. 2019; Howarth 2024). Whether CCS makes a difference here is discussed in Section 14.3.

14.3 Carbon Capture and Storage (CCS) and Carbon Capture and Usage (CCU)

The CCS process consists of two steps: the separation of carbon dioxide from a gas stream and the geological storage. It will often be necessary to transport the carbon dioxide from the point of separation to the storage site. In the CCU process, the carbon dioxide is to be used as a raw material to manufacture a product that, in the best case, stores the contained carbon for some time (CCUS).

Separation of Carbon Dioxide

The goal of the separation process is to demix the flue gas, i.e., to reverse the mixing of flue gas and carbon dioxide (see Section 3.3). Figure 14.1 shows the minimum specific energy input for the complete separation of carbon dioxide as a function of the molar fraction of carbon dioxide in the gas. Relevant numerical values are:

- Approximately 500 kJ/kg for complete extraction from air
- Approximately 200 kJ/kg for complete extraction from the flue gas of gas turbines or biomass power plants
- Approximately 150 kJ/kg for complete extraction from the flue gas of coal power plants

- Approximately 100 kJ/kg for complete extraction from the flue gas of cement plants

This allows an efficiency to be established for the evaluation of real processes (Michaelidis 2021):

$$\eta_{cap} = \frac{w_{\min}}{w_{real}} \tag{14.1}$$

In many real processes, only very little work w_{cap}, but a large amount of heat q_{cap} is used. This heat can then be converted into work via a Carnot efficiency (or a real appropriate efficiency):

$$\eta_{cap} = \frac{w_{\min}}{w_{real}} = \frac{w_{\min}}{w_{cap} + \left(1 - \frac{T_0}{T_{cap}}\right) \cdot q_{cap}} \tag{14.2}$$

Typical values of technical processes for the separation of CO_2 are provided by Table 14.1. Efficiencies are well below 20%, with no expectation that these will improve significantly in the future.

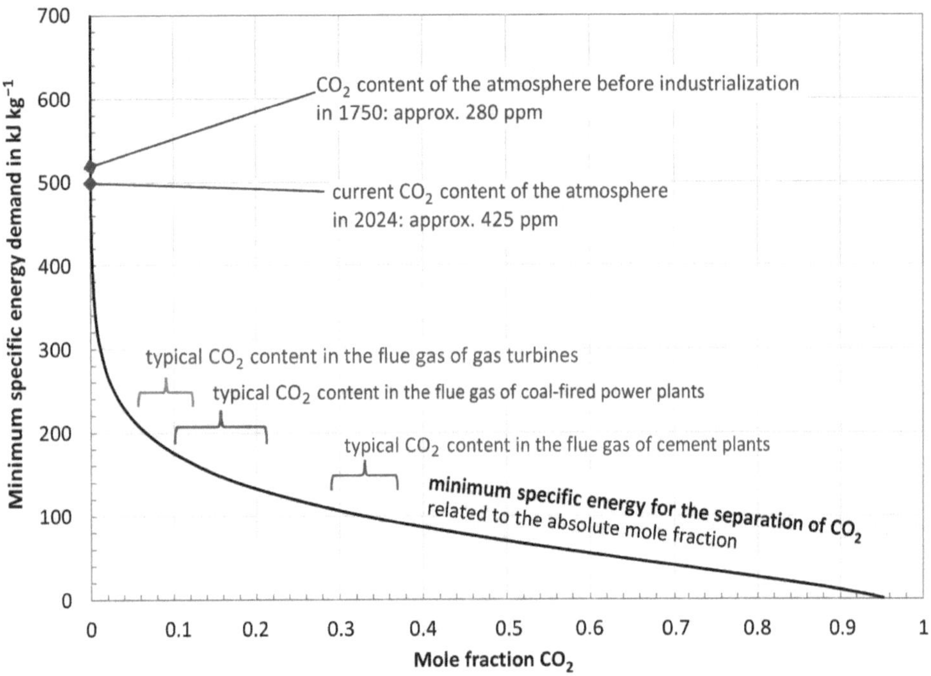

Figure 14.1 Specific minimum (ideal) energy input for the extraction of carbon dioxide from a gas at 300 K ambient temperature

Table 14.1 Input for the Separation of Carbon Dioxide Relative to 1 kg CO_2: Required Heat, Work, Temperature and Efficiency η_{cap} (Source: Michaelidis 2021)

	Work	Heat	Temperature	η_{cap}
Amine-based (flue gas)	0.4 MJ	2.0 MJ to 4.3 MJ	110 °C	8.3 % to 12.8 %
Ammonia-based (flue gas)	0.4 MJ	1.53 MJ	110 °C	15.5 %
Liquefaction				5 %

Transport of Carbon Dioxide

Larger mass flows of CO_2, as could be separated from large power plants or cement factories, are most economically transported via pipelines. Due to the special material properties of CO_2, it is necessary that the pressure is approximately 100 bar and does not fall below 80 bar: the critical point of carbon dioxide is at 74 bar and 31 °C, therefore the pressure must always be higher to prevent the formation of CO_2 wet steam (Onyebuchi et al. 2018).

For transport, the CO_2 must be compressed from ambient pressure to typically 100 bar. Using the ideal gas approximation, which is only limitedly applicable here, the specific work of single-stage compression starting from standard conditions is 400 kJ kg^{-1}. A calculation with real substance data yields 312 kJ kg^{-1} for a two-stage ideal compression (Linow 2022). Taking into account the efficiencies of real compressors, a specific required work of 400 kJ kg^{-1} is realistic. This work depends critically on the pressure at which the CO_2 can be transferred from the capture stage to the compression stage.

The energy required for the transport of fluids in pipelines depends critically on the dimensions of the pipeline and the transport velocity (Cerbe/Lendt 2016). Since CO_2 is conveyed in a supercritical state, it has a comparatively high density. Friction in the pipeline becomes relevant. The actual energy demand is only determined during the construction of a real pipeline and the trade-offs made between investment costs (diameter) and operating costs (flow velocity). Theoretical values are in the range of 100 kJ kg^{-1} to 200 kJ kg^{-1} per 1000 km of distance.

Geological Storage of Carbon Dioxide

Figure 14.2 illustrates typical geological conditions, i.e. pressure and temperature of geologically suitable conditions. Old natural gas and oil fields, volcanic rock, unexploited coal seams, and very deep water-bearing strata are discussed as geological storage sites. It is important that the CO_2 does not escape again through the overlying layers and, in the best case, becomes chemically bound in the reservoir. This limits the mass flow rate at which carbon dioxide can be injected (Mühlbauer et al. 2024).

The compression of CO_2 from conditions in a pipeline to the pressure in the subsurface typically requires about 100 kJ kg^{-1} (Linow et al. 2022, p. 112).

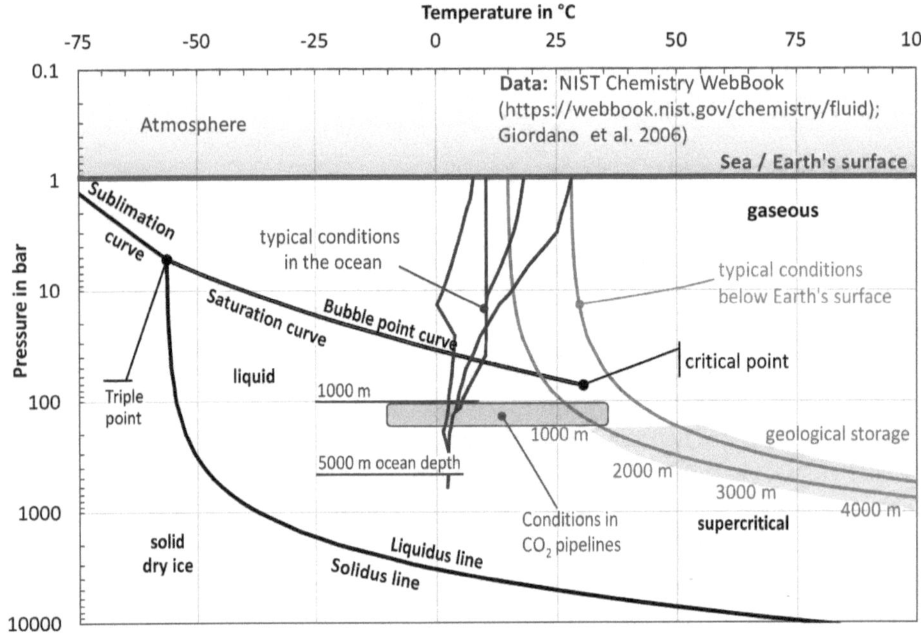

Figure 14.2 Storage of carbon dioxide: shown are typical pressure-temperature profiles in the ocean and in the subsurface of the continents, as well as the corresponding phases of pure carbon dioxide (data sources: NIST Chemistry WebBook, Giordano et al. 2006).

Total Input for CCS

We use a key figure that relates the original benefit from burning the fossil fuel to the input for CCS. Here, work values are to be compared with each other consistently. Therefore, the original benefit is the generated electricity related to the emitted mass of CO_2 and the input is the total work for the CCS process:

$$\varpi = \frac{w_{use}}{w_{CCS}} = \frac{H_{I,fuel} \cdot \frac{M_{Br}}{M_{CO_2}} \cdot \eta_{el}}{w_{cap} + \left(1 - \frac{T_0}{T}\right) \cdot q_{cap} + w_{comp} + w_{pipe} + w_{inj}} \tag{14.3}$$

Input for CCS

For natural gas as fuel

The specific benefit in generating electricity from natural gas is

$$w_{el} = H_{I,NG} \cdot \frac{M_{NG}}{M_{CO_2}} \cdot \eta_{pp} \cong 50\ \frac{\text{MJ}}{\text{kg}} \cdot \frac{16.04\ \frac{\text{kg}}{\text{kmol}}}{44.01\ \frac{\text{kg}}{\text{kmol}}} \cdot 0.40 = 7.29\ \frac{\text{MJ}}{\text{kg}}$$

The input required to remove this CO_2 from the atmosphere again results from two processes: About 50 % of the CO_2 can be captured directly from the flue gas (these are currently real values, even if supposedly 90 % should be captured). The other part must be extracted from the air via DACCS. Assuming that the CO_2 is transported 1000 km from the power plant to the storage site, the following applies to the direct share

$$w_{dir} = \mu_{dir} \cdot \left(w_{cap} + \left(1 - \frac{T_0}{T}\right) \cdot q_{cap} + w_{comp} + w_{pipe} + w_{inj}\right)$$

$$w_{dir} = 0.50 \cdot \left(0.4\ \frac{\text{MJ}}{\text{kg}} + \left(1 - \frac{283\ \text{K}}{383\ \text{K}}\right) \cdot 3.0\ \frac{\text{MJ}}{\text{kg}} + 0.40\ \frac{\text{MJ}}{\text{kg}} + 0.2\ \frac{\text{MJ}}{\text{kg}} + 0.10\ \frac{\text{MJ}}{\text{kg}}\right) = 0.94\ \frac{\text{MJ}}{\text{kg}}$$

In addition, there is the share from DACCS. This can be extracted directly from the air at the storage site. Remarkably, the inputs for DACCS are otherwise comparable to those of CO_2 flue gas cleaning:

$$w_{ind} = \mu_{ind} \cdot \left(w_{cap} + \left(1 - \frac{T_0}{T}\right) \cdot q_{cap} + w_{comp} + w_{inj}\right)$$

$$w_{ind} = 0.50 \cdot \left(0.4\ \frac{\text{MJ}}{\text{kg}} + \left(1 - \frac{283\ \text{K}}{383\ \text{K}}\right) \cdot 3.0\ \frac{\text{MJ}}{\text{kg}} + 0.40\ \frac{\text{MJ}}{\text{kg}} + 0.10\ \frac{\text{MJ}}{\text{kg}}\right) = 0.84\ \frac{\text{MJ}}{\text{kg}}$$

Thus the relative benefit is

$$\varpi = \frac{w_{use}}{w_{CCS}} = \frac{w_{use}}{w_{dir} + w_{ind}} = \frac{7.29\ \frac{\text{MJ}}{\text{kg}}}{0.94\ \frac{\text{MJ}}{\text{kg}} + 0.84\ \frac{\text{MJ}}{\text{kg}}} = 4.1$$

This means that 25 % of the work generated with the methane must be expended for CCS.

For coal as fuel

The specific benefit in generating electricity from coal is

$$w_{el} = H_{I,C} \cdot \frac{M_{coal}}{M_{CO_2}} \cdot \eta_{pp} \cong 32.5\ \frac{\text{MJ}}{\text{kg}} \cdot \frac{12.01\ \frac{\text{kg}}{\text{kmol}}}{44.01\ \frac{\text{kg}}{\text{kmol}}} \cdot 0.40 = 3.55\ \frac{\text{MJ}}{\text{kg}}$$

The benefit is significantly lower, even if we use the lower heating value of pure carbon here and not that of hard coal. The input for CCS is identical to that for natural gas and as the relative benefit we obtain

$$\varpi = \frac{w_{use}}{w_{CCS}} = \frac{w_{use}}{w_{dir} + w_{ind}} = \frac{3.55\ \frac{\text{MJ}}{\text{kg}}}{0.94\ \frac{\text{MJ}}{\text{kg}} + 0.84\ \frac{\text{MJ}}{\text{kg}}} = 2.0$$

This means that 50 % of the generated work must be expended for CCS.

These key figures are directly linked to the EROI. They suggest that the use of fossil fuels with CCS does not represent an economically viable or socially acceptable option.

Carbon Capture and Utilisation (CCU)

The utilisation of the captured CO_2 to produce new products (synthetic fuel, plastics, basic materials for the chemical or construction industry) contributes to emission reduction only if the carbon remains bound in these products in the long term. If these products bind the carbon for only weeks (fuel or single-use plastics) to a few years (other plastics) and return it to the atmosphere at the end of their life, this is irrelevant with regard to a reduction in CO_2 concentration. Only long-term storage over centuries would be significant as Carbon Capture, Usage and Storage (CCUS). It is not yet foreseeable whether the costs of producing such products will allow large quantities (several Gt per year) of CO_2 to be bound in this way. Short-term methods of CCUS that bind CO_2 over decades are suitable as intermediate storage. In their assessment, however, intergenerational equity must be taken into account in every case, since further negative emissions are foreseeably necessary.

14.4 Negative Greenhouse Gas Emissions

Negative emissions of carbon dioxide are an essential part of almost all scenarios with which the Paris climate targets are still to be achieved. They are included in many commitments of nation states to these climate targets (pledges) – including those of the European Union and Germany (Shukla et al. 2022; Erlach et al. 2022; BVerfG 2021). The required negative emission is foreseeably impressively high over very many years. It corresponds approximately to today's carbon dioxide emissions. Concrete values depend on how quickly we succeed globally in reducing further emissions of greenhouse gases.

Although the international community has agreed to limit global warming to well below +2 °C, the current commitments of states to reduce their emissions amount to approximately +3 °C of warming (Climate Action Tracker). The assumption that negative emissions will in the future simply and without great cost or effort solve all climate problems can in the worst case be used as a justification to reduce greenhouse gas emissions even less or not at all.

Order of magnitude of the required negative emissions

In the IPCC scenarios according to which the Paris climate targets could still be achieved, it is already anticipated today that, in addition to nearly complete emissions reduction by 2050, we will also need substantial annual negative emissions by the year 2050. The pledges with which Germany intends to achieve its climate targets also include massive negative emissions from 2050 onwards.

Exact figures depend on how quickly it will be possible in the coming years to halt the emission of CO_2 from fossil sources (and other greenhouse gases), and to what extent the 1.5 °C temperature threshold will be temporarily exceeded, with the associated risks (Schleussner et al. 2024).

By the year 2050, more than 10 Gt of CO_2 will need to be removed from the atmosphere by technical means, and by the year 2100 an order of magnitude of 1000 Gt. For comparison: current global CO_2 emissions amount to approximately 40 Gt per year, to which Germany contributes approximately 0.7 Gt per year.

The 1000 Gt of CO_2 correspond to 125 t for every living person, if we distribute this equally among all. The 1000 Gt CO_2 correspond to approximately 67 % of total historical emissions from the burning of fossil fuels to date. We would therefore need to reverse approximately 67 % of our historical emissions over the next 75 years, as well as all emissions that we still allow ourselves from now on.

In fact, this task could be considerably larger. In order to keep the climate within a temperature range that is favorable for most of our civilizations and to stabilize sea levels, it would be necessary to get below 350 ppm in the foreseeable future (planetary boundary). To implement this, we would have to compensate for the interim losses in reflectivity caused by sea ice and glaciers: clean ice reflects sunlight very well, whereas dirty or wet ice caps and glaciers absorb much more heat, and open ocean absorbs almost all of the incoming solar radiation.

In addition, we must offset the effect of other highly potent greenhouse gases (methane and nitrous oxide, refrigerants such as R134a and R410, SF6, etc.) through additional negative emissions of CO_2. When the term greenhouse gas neutrality is used today — to be achieved in the EU by 2050 — this inevitably entails negative emissions of its own: the emissions of all other greenhouse gases occurring after 2050 must be offset by the removal of CO_2 from the atmosphere.

What is 1 Gt of carbon dioxide?

One gigaton (1 Gt) of carbon dioxide is something we cannot really imagine. The mass 1 Gt denotes 1 000 000 000 000 kg. This mass as water occupies one cubic kilometer, since water has a density of 1000 kg per m^3 and one billion cubic meters corresponds to one cubic kilometer. Lake Müritz in Mecklenburg is Germany's second largest lake and has a volume of 0.7 km^3, i.e., the water in Lake Müritz weighs 0.7 Gt.

Carbon dioxide is a gas under ambient conditions (i.e., 1 bar pressure and 25 °C). Under these conditions, 1 Gt of carbon dioxide occupies a volume of 567 km^3. This would be enough to cover Germany to a depth of approximately 1.5 m.

At higher pressure, CO_2 occupies a smaller volume. Above 74 bar, we can no longer distinguish between liquid phase and gaseous phase. This is referred to as supercritical. CO_2 should always be present in pipelines or geological reservoirs in a supercritical state, i.e., at pressures above 74 bar. Supercritical carbon dioxide at a temperature of 25 °C and a pressure of 400 bar has the same density as liquid water. This roughly corresponds to the conditions at a depth of 1500 m below ground. In that case, 1 Gt again corresponds to the volume of one cubic kilometer. To store 1 Gt geologically, the cap rock over an area of 1000 km^2 may need to be lifted by 1 m to accommodate it at depth (and if there are no pores in the rock).

Carbon dioxide is produced by the combustion of carbon with oxygen. Oxygen contributes 72.8 % to the mass of CO_2 and carbon the remaining 27.2 %: thus 1 Gt of CO_2 contains a mass of 0.272 Gt of carbon. We can think of this carbon as coal. High-quality pure coal loosely piled has a density of approximately 900 kg/m^3. The carbon from 1 Gt of carbon dioxide thus corresponds, for example, to a pyramid with a base length of 1220 m and a height of 610 m.

If we distribute this coal onto freight trains, 75 m^3 fit into a four-axle, 12.5 m long freight wagon of the FALNS type. Per freight wagon this amounts to approximately 67.5 t. We therefore require 4.49 million of these freight wagons. This corresponds to 112 000 freight trains of 40 wagons each, or a single freight train with a length (of the wagons) of 56 100 km. This train would go around the earth 1.4 times.

14.4.1 What Are Good/useful NET Methods?

Criteria arise directly from the goal (binding carbon dioxide in the long term) and the requirement to do this as reliably, simply, using few resources, and in a globally equitable manner as possible (Linow et al. 2022). Table 14.2 contains a proposal for the evaluation of the criteria:

- CO_2 is bound sufficiently long-term, i.e., over geological time periods or well beyond 1000 years.
- A large amount of CO_2 is bound: a significant quantity each year and equally large in total over the long term.
- Considerably more CO_2 is bound than is released after accounting for all required processes. This is evaluated as the degree of binding:

$$\varepsilon_{CO_2} = \frac{m_{\mathrm{CO2,stored}}}{\sum_{processes} m_{\mathrm{CO2,leaked}}} \tag{14.4}$$

- Few critical resources are required for implementation (water, raw materials, land area, agriculturally usable area, soils).
- There are little or no harmful effects on existing ecosystems and their biodiversity.
- All other planetary boundaries are respected. Competing uses between the method and other essential goals of our sustainable development must remain resolvable.
- As little energy as possible is required for the implementation of the processes, and this energy is provided entirely from renewable sources.
- Food security, health, peace, and development opportunities (SDGs) are maintained. The efforts and changes of the method can be distributed equitably and shouldered jointly at the international level (Fyson et al. 2020).
- We can begin implementation immediately. The technologies exist.
- Storage sites for CO_2 are easy to monitor and require little effort for their security.
- Knowledge about the technologies is freely accessible and their use is not restricted by patents.
- Where possible, there are further positive effects from the application through the improvement of soils, water bodies, living conditions, and ecosystems.

Completely irrelevant is whether the method is purely technical or includes biological elements: the required orders of magnitude of interventions in the natural systems of the earth remain impressive, and all methods are technical, since nonc proceeds on its own.

An essential aspect here is that energy, land area, soil, water, and biodiversity are fundamentally not depreciable in the economic sense. These resources must therefore be evaluated and protected with great care. It is already foreseeable that negative emissions represent a heavy burden for the near future. It is therefore our task to ensure that these burdens do not crush people in the near future or even destroy their living conditions (Hansen et al. 2017; BVerfG 2021). For the application of these technologies, we therefore need appropriate long-term tolerable limits in the sense of sustainable development. It is foreseeable that the actually implementable order of magnitude may be significantly lower than the questionable pledges of the signatory states of the Paris Climate Agreement (Deprez et al. 2024).

Table 14.2 Assessment Categories and Simple Traffic Light Scheme for NET Methods

Category	Relevant (+++)	Suitable (++)	Critical (+)	Hazardous (Z)
Storage duration	Over geological time periods	300 to 1000 years	100 to 300 years	Less than 100 years
Storage quantity	Over 500 Gt possible in 100 years	50 Gt to 500 Gt in 100 years	Under 50 Gt in 100 years	Zero or negative
Degree of binding	Above 10	3 to 10	Below 3	Zero or negative
Raw materials	Low input required / renewable raw materials / readily available raw materials	Moderate input use of other infrastructures (transport)	High input unavoidable losses long-term binding	High losses raw materials are insufficient
Soils	Clearly positive effect	Essentially neutral effect or positive and negative impacts roughly balance out	Negative but tolerable impact	Destructive effect
Ecosystems	Clearly positive effect	Essentially neutral effect or positive and negative impacts roughly balance out	Negative but tolerable impact	Destructive effect

Category	Relevant (+++)	Suitable (++)	Critical (+)	Hazardous (Z)
Planetary boundaries	All planetary boundaries are safely respected.	Efforts are needed to remain within planetary boundaries.	Substantial impacts on the ability to comply with planetary boundaries	no
Energy demand	Less than 0.1 MJ per kg CO_2	Between 0.1 MJ and 1.0 MJ per kg CO_2	Between 1.0 MJ and 10 MJ per kg CO_2	More than 10 MJ per kg CO_2 use of fossil fuels
Food	Clearly positive effect: higher long-term yields or security of supply	Essentially neutral effect or positive and negative impacts roughly balance out	Negative but tolerable impact	Destructive effect
People, SDGs	Local benefit in poor regions / increasing resilience / creating fulfilling work	Essentially neutral effect or positive and negative impacts roughly balance out	Negative but tolerable impact	Expropriation, deprivation of rights, displacement hardly avoidable
Technology Readiness Level (TRL)	8 to 9 established and immediately deployable	6 to 7 low remaining development risks: use foreseeable in < 10 years	3 to 5 significant development risks: use not clearly foreseeable	Less than 3 science fiction, real technical implementation not foreseeable
Risks and monitoring	Risks are known and low	Risks are known and manageable	Risks are little known or high	Unknown or unacceptable risks
Open source	Free	Minor restrictions on a few components	Core components not free	Monopoly, oligopoly

14.4.2 Technical Methods for CCS

Carbon dioxide has been released by our technical processes. It can be removed from the air again by other technical processes.

DA-CCS Carbon dioxide can be extracted directly from the air in order to store it geologically (Direct Air Capture with Carbon Storage, DACCS). Suitable separation processes are used for this purpose. The energy and material input required is quite high.

The potential of DACCS depends on the available energy and the resources needed for the manufacture and operation of the plants (Herzog 2024; Madhu et al. 2021).

The accounting of all processes from separation to storage of CO_2 is necessary to determine the total energy demand:

- Thus, wet-chemical processes for DACCS require approximately as much energy as could be made usable at the time of burning the fossil fuel that released this carbon dioxide into the atmosphere.
- Dry processes for DACCS in combination with heat pumps or solar thermal energy are energetically considerably more favorable, but still demanding. Therefore, the energy demand for negative emissions must be taken into account in models for the global future energy supply.

BE-CCS Plants bind CO_2 from the air during growth. The biomass produced in this way can then be burned in power plants for the provision of heat or electricity, or converted into coal. Approximately 50% to 80% of the CO_2 produced in the process could be separated from the flue gas by energy-intensive processes and stored geologically (Bioenergy with Carbon Capture and Storage, BECCS). The potential of BECCS is assessed very differently. For 1 Gt of negative emissions per year by means of BECCS, approximately 300,000 km^2 of agricultural or forestry land is required (Creutzig et al. 2021). This land cannot then be used for other purposes, which is why BECCS is in direct competition with agriculture and forestry, the preservation of biodiversity, and natural terrestrial ecosystems.

In scenarios for achieving the Paris climate targets, it is sometimes assumed that more than 5,000,000 km^2 of land will be used for BECCS alone. This is more than 30 times the agricultural area of Germany. Such scenarios are not compatible with other goals such as the protection of biodiversity or food security (Shukla et al. 2022; Popp et al. 2017; Heck et al. 2018).

In the case of BECCS, the energy input is typically so high that the energy made usable by burning the biomass is almost entirely required for the CCS, meaning no additional benefit is generated. The main obstacle to BECCS or even massive reforestation is the immense land requirement and the associated necessary land acquisition and displacement of population groups with insufficiently secure land rights.

Lime Quicklime is produced from limestone in an energy-intensive process. The carbon dioxide generated in this process could be captured from the flue gas and stored geologically (CCS). This lime is then to be applied to land or water surfaces to bind carbon dioxide there. This is another process associated with high energy costs.

14.4.3 Technical Interventions in Biological Processes

In the earth system, two natural cycles transport carbon between the atmosphere, the sea, vegetation, and the subsurface:

1. In the organic cycle, carbon is fixed in plants as biomass through photosynthesis. A portion of this biomass can be deposited in the soil, as soil formation. A small fraction of the biomass can be geologically incorporated in such a way that it eventually becomes deposits of fossil fuels again over the long term.
2. In the inorganic cycle, carbon is bound long-term through stable compounds in water (e.g., bicarbonate) and in the form of minerals (carbonates).

Some processes proceed quickly, others slowly. If we want to remove CO_2 from the atmosphere, we must influence these two cycles in such a way that carbon is bound as quickly as possible and released again as slowly as possible.

Possibilities for enhancing the organic cycle via photosynthesis begin with plants taking up CO_2 and converting it into organic material through photosynthesis. This organic material can then be directed in such a way that it persists over the long term, thereby binding CO_2. Photosynthesis uses only an exceptionally small fraction of the sun's energy, and climate change is increasingly impairing the productivity of many regions, so that these methods remain limited in their effect.

Whether future binding of CO_2 on land is still possible must also be assessed against current developments: the naturally occurring binding that should take place is barely happening in the global sum in terrestrial ecosystems (Piyu Ke 2024). The causes include biodiversity loss due to human destruction of terrestrial ecosystems in combination with the climate heating we are further accelerating. In particular, the hottest regions contribute disproportionately to the emission of carbon dioxide from soils.

Ocean Oceans contain approximately 60 times as much CO_2 as the atmosphere and could bind considerably more through slowly occurring processes. However, the required processes proceed too slowly to have a decisive effect now, and existing processes are currently weakening (Ditlevsen 2023).

There are two approaches by which these processes could be technically accelerated:

1. Fertilizing the ocean with iron, nitrogen, or phosphorus – depending on which element is lacking
2. Ensuring that nutrient-rich water rises from the depths of the ocean

In both cases, the aim is for far more algae to grow and bind CO_2 for their growth. If these algae, or the small creatures that feed on them, then sink rapidly to the ocean floor, they would remove the CO_2 from the atmosphere and bind it over the long term. The potential could be quite considerable, but the expected impacts on ocean ecosystems are not foreseeable today, while the required energy and technical input will be large (Masson-Delmotte et al. 2021).

Terrestrial ecosystems Large, old, undisturbed, diverse forests are very good and – if left undisturbed – long-term stores of carbon. In addition, they often enable, as a biotic pump, the rainfall necessary for them in the interior of continents.

Harvested timber from forests can store carbon if it is used for a long time, e.g., as a building material in structures. Worldwide, forests are increasingly threatened by livestock farming, mining, drought, storms, and fire. In addition, forests are often transformed by these processes or by use into CO_2 sources. For this reason, this option is extremely limited and does not represent a secure long-term store (Ke et al. 2024).

Massive and large-scale reforestation is in direct competition with the supply of food for people. Since ecosystems already exist on the designated areas, this often conflicts with the protection of biodiversity.

It is estimated that a maximum of approximately 1.5 Gt to 3 Gt of carbon dioxide per year can be bound in this way, which is more than an order of magnitude too little to achieve the goal by this route alone (IPCC WG I, 2021; cf. however McKay et al. 2022 on long-term stability).

Peatlands A very large amount of carbon is bound in peatlands. Under favorable natural conditions, it remains stored there over the long term. Currently, peatlands are threatened globally by drainage. Even drained peatlands remain important greenhouse gas sources for a long time. Therefore, their protection and restoration are central measures for reducing CO_2 and CH_4 emissions. Restored peatlands can become a carbon sink if peat formation resumes there. Storage processes in peatlands proceed slowly; therefore, approximately 1 Gt of carbon dioxide per year is cited as the global potential (Masson-Delmotte et al. 2021).

Coasts It is being discussed that natural coastal ecosystems function as carbon sinks. It would therefore be sensible to preserve seagrass meadows and mangrove forests, and to re-establish them where they have been destroyed. It would also be urgently necessary to do so for other reasons such as coastal protection and biodiversity. The quantities of carbon that these ecosystems actually store have not been clarified. Currently, the contribution is estimated at less than 1 Gt CO_2 per year (Masson-Delmotte et al. 2021).

Soils Currently, soils are a relevant source of CO_2emissions (Piyu Ke 2024). In addition, at the current rate of soil loss, it is quite foreseeable that no soils will remain available for agriculture (Wilkinson 2007). Conversely, healthy, deep, old soils of temperate and polar latitudes each store impressive amounts of carbon.

The development and stabilization of agriculturally used soils therefore have an exceptionally high priority. This would allow yields to be significantly improved and secured over the long term. Increasing the carbon content, here as organic material or as biochar (see the following discussion), is the foundation and requires adapted farming practices for this purpose. In this way, more than 2 Gt CO_2 per year can be sequestered (Masson-Delmotte et al. 2021).

Biochar Organic material can be converted into charcoal in special kilns and then incorporated into agriculturally used soils. This charcoal is stable over several centuries. It significantly improves the water and nutrient retention of the soil – and it would be a carbon sink.

During the carbonization of plant material, approximately 50% of the bound carbon is released as CO_2. Usable heat is released during the process. Therefore, the origin of the biomass is relevant for the climate impact. Organic material that would otherwise decompose is well suited, while forests felled specifically for production show a harmful balance. The estimates of the potential vary particularly widely for this method. Approximately 2 Gt of carbon dioxide per year appears realistic (Masson-Delmotte et al. 2021).

All of these options represent technical interventions in ecosystems and natural nutrient cycles. They partly require large-scale technologies for their implementation at the required scale. These biological processes can together bind approximately 5 Gt to 15 Gt CO_2 per year, with the carbon then remaining securely bound for decades to millennia (Masson-Delmotte et al. 2021). All of these processes have further important positive effects on biodiversity, nutrient and water cycles. They require significant changes to current forms of land use. Ultimately, they represent forms of climate adaptation.

14.4.4 Technical Interventions in Geological Processes

A further option is offered by the acceleration of geological processes that bind CO_2.

Enhanced weathering In the long term, CO_2 is bound by geological processes: temperature and moisture influence how quickly rocks weather. In particular, volcanic rocks react with CO_2 during their weathering and bind it geologically over the long term. The weathering of volcanic rock can be accelerated by grinding it into powder and then spreading it in forests, on agricultural land or in the ocean (Enhanced Rock Weathering, ERW). The capacity of seawater to buffer acid is improved and the pH value is increased. This counteracts the acidification of the oceans (Chapter 8). Some types of rock improve soil properties, while others require caution regarding where and how they are used. Basalt can replace part of the mineral fertilizer and makes liming unnecessary.

Enhanced weathering can bind several tens of Gt of carbon dioxide per year. For this, sufficient quantities of silicate rock must be ground and spread. For the sequestration of 1 t of carbon dioxide, between 1 t and 3 t of rock are required depending on the type of rock (approx. 1 t for dunite or approx. 3 t for basalt). To bind 1 Gt of carbon dioxide, between 0.4 km^3 and 1.2 km^3 of rock will need to be processed. The major challenge here is the energy required for processes at a sufficient scale.

The method does not compete with agriculture for land. With suitable rocks, it can support agriculture.

14.4.5 Assessment of the Approaches

Based on the brief introduction of the methods included here and taking into account the state of knowledge, the assessment results from Table 14.3. Some of the methods have limited benefit for negative emissions, but are useful for other reasons.

Table 14.3 Assessment of the Methods for Negative Emissions Discussed Here Using the Criteria from Table 14.2

Category	DACCS	BECCS	CaCCS	Ocean	Land	Bog	Soil	Char	ERW
Duration	+++	+++	+++	+++	a)	a)	a)	++	+++
Quantity	++	++	++	+	a)	a)	a)	+	++
ε_{CO2}	+++	++	++	(?)	+++	+++	+++	+	+++
Raw materials	+	+	++	+	+++	+++	+++	++	++
Ground	++	Z	++	++	+++	++	+++	+++	+++
Ecosystem	++	Z	++	++ (?)	++ (?)	+++	+++	+	++
Planet	++	Z	+	+	++	+++	+++	++	++
Energie	+	+	+	++	++	+++	++	++	+
Nahrung	++	Z	++	++	+	+	+++	+++	+++
SDGs	++	Z	++	++	b)	b)	++	+++	++
TRL	+++	+ c)	+ c)	+	+++	+++	+++	+++	+++
Risk	+++	+	++	+	+++	+++	+++	+++	++

a) critically depends on future climate: the hotter, the shorter
b) critically depends on local implementation
c) capture from flue gas has not yet been demonstrated at large scale and over the long term.

14.5 Geoengineering

Negative emissions are actually a form of geoengineering, just as the massive burning of fossil fuels represents an originally unintended form of geoengineering. In both cases, the transmission of the atmosphere in the infrared spectral range is altered. Geoengineering in the narrower sense refers to modifying and controlling short-wave solar incoming radiation (Solar Radiation Management, SRM). It is a way to compensate, to a certain extent, for the symptoms of elevated greenhouse gas concentrations in the atmosphere. Possible ideas are the subject of this section. They are currently being discussed very critically. What ultimately interests us is the energy input, since this would have to be factored into an assessment as part of the use of fossil fuels.

The incoming power of the sun depends on the current shape of the Earth's orbit and the distance between the Earth and the Sun. These parameters cannot be changed. What can be changed are two things:

- The reflectivity of the Earth itself can be increased. Options exist at the Earth's surface or in the atmosphere. This results in less radiation being absorbed overall in the earth system.
- The Earth can be shaded by objects located between the Sun and the Earth. These objects are thus outside the earth system. However, they should be in an orbit around the Earth or the Sun in which they can cast their shadow on the Earth as permanently as possible.

Table 14.4 summarizes the key aspects of the methods presented below.

Increasing the Reflection of the Earth's Surface

The reflectivity of the Earth's surface can be increased locally. In this case, the measure acts in the region that has been modified, not globally:

- On a small scale, this is already being applied today in cities, for example through white roofs or light-colored ground surfaces. Especially in cities, this can be one of the many measures that enable climate adaptation there. However, settlements cover only a small fraction of the Earth's surface.
- Desert soils could be covered with lighter-colored material. However, this requires a massive movement of material, the destruction of ecosystems, and regular maintenance of this technology.
- Strongly absorbing tundra and taiga could be converted back into mammoth steppe: this has a significantly higher reflectivity and stabilizes the permafrost beneath it. To achieve this, the population of large wild animals would have to be increased back to levels as before the ice age: bison, horses, moose, etc. The former biodiversity with mammoths, rhinoceroses, deer and lions cannot be restored at the required speed (Zimow 2012).

To produce substantial effects, significant portions of the land surface would have to be altered considerably. This is technically associated with the use of immense resources.

Increasing the Reflectivity of the Atmosphere

Changes in the troposphere are difficult, as it is very dynamic and any measure disappears within days. The stratosphere is different. There, reflective particles could remain for many months. By introducing scattering particles, the reflectivity for incoming sunlight is reduced. Originally, the analysis was limited to sulfur dioxide droplets (typical of larger volcanic eruptions and their climate effects), but many other variants are now being discussed: titanium dioxide, sapphire (Al_2O_3), diamond and others (Vattioni et al. 2024). The optimal reflectivity at high residence time can be set via the diameter of the particles. In particular, the use of sulfur dioxide is extremely cost-effective and can be implemented quickly. It is quite conceivable that a nation state would use this method to make the climate more pleasant for itself and its region.

Effects of SRM

These methods each result in a reduction of the solar radiation available at the earth's surface: it becomes somewhat darker. This directly leads to some changes in the earth system:

- The evaporation rate is reduced, i.e., the precipitation rate decreases and the earth becomes drier overall. This can also lead to a shift and change in climate and precipitation overall.
- Less light reduces the yield of photosynthesis. This directly affects agriculture, i.e., the nutrition of people and the availability of biomass for other uses.
- The combination of both effects results in a restriction of the area in which vegetation can exist.

For this reason, geoengineering is a measure with high risk for all human societies. Any decision in favor of it depends not only on technical feasibility, but also on possibilities for correction (Lenzi et al. 2018).

Termination Shock Many of the measures only act relatively short-term. If they are discontinued, the earth's climate very quickly transitions to its actual state, determined by the greenhouse gas concentration. Within months, the mean temperature of the earth can rise by several degrees, with unforeseeable and massive consequences. These scenarios are therefore referred to as termination shock. In particular, if the measures cannot be maintained due to other global crises, this shock can be fatal (Parker/Irvine 2017).

Energy input In addition to ethical considerations, questions of energy costs are also decisive for a possible application of these technologies. Large mass flows must be transported into the stratosphere or into space. If no naturally occurring raw materials are used, the sometimes high input for producing the particles in the required size must be added.

Table 14.4 Overview of Geoengineering Measures and Their Possible Contribution to Changes in Radiation Drivers, Influence As Maximum Possible Change in Radiative Forcing ΔF

Measure	ΔF in W m^{-2}	Start	Lifetime	Control	Limits
Space sunshades	-3.7 or more	Unclear	Approx. 20 a	no	Available energy, emissions from rockets
Increasing aerosols in the stratosphere to increase reflectivity	-3 or more	5 a to 10 a	Approx. 3 a	yes	Saturation of particle number
Increasing reflectivity of tropospheric clouds (mechanical)	-3	5 a to 10 a	Days to weeks	yes	Pollution of the troposphere and the earth
Increasing reflectivity of desert soils	-2	Approx. 10 a	Approx. 10 a	no	Change of surface
Increase reflectivity of land surfaces through grassland	Relevant, indirect	Now	Unclear	no	Available land surface (primarily tundra)
Increase reflectivity of settlements and cities	-0.15	Approx. 10 a	Approx. 70 a	yes	Affects only small land area

14.6 Outlook

This chapter is necessary in the context of this book. It shows that it is not possible without massive consequences to implement the transformation of our energy system as a side note or only at some point in the distant future and then hope that everything will somehow sort itself out. Energy cannot be discounted, which means that the task grows ever larger the longer it is left unaddressed and could soon become unsolvable.

Regardless of this, it has by now become clear that overshooting our climate targets and leaving the cleanup as a task for future generations entails massive risks of its own. It is quite possible that the cleanup cannot return our earth to a state we require – on the state of the discussion (Schleussner et al. 2024).

References

BVerfG – Bundesverfassungsgericht: Beschluss des Ersten Senats vom 24. März 2021, 1 BvR 2656/18, 1 BvR 96/20, 1 BvR 78/20, 1 BvR 288/20, 1 BvR 96/20, 1 BvR 78/20 (Klimaschutz). *https://www.bundesverfassungsgericht.de/SharedDocs/Entscheidungen/DE/2021/03/rs20210324_1bvr265618.html*

Cerbe, G./Lendt, B.: Grundlagen der Gastechnik: Gasbeschaffung – Gasverteilung – Gasverwendung. 8. Auflage. Carl Hanser Verlag, München 2016

Climate Action Tracker. *https://climateactiontracker.org*

Creutzig, F./Erb, K.-H./Haberl, H./Hof, C./Hunsberger, C./Roe, S.: Considering sustainability thresholds for BECCS in IPCC and biodiversity assessments. In: GCB Bioenergy 13, 2021, pp. 510–515

Deprez, A./Leadley, P./Dooley, K./Williamson, P./Cramer, W./Gattuso, J.-P./Rankovic, A./Carlson, E. L./Creutzig, F.: Sustainability limits needed for CO2 removal. In: Science 383, 2024, pp. 484–486

Erlach, B./Fuss, S./Geden, O./Glotzbach, U./Henning, H.-M./Pittel, K./Renn, J./Rens, S./Sauer, D. U./Schmidt, C. M./Spiecker, I./ Stemmler, C./Stephanos, C./Strefler, J.: Was sind negative Emissionen, und warum brauchen wir sie? (Kurz erklärt!). Deutsche Akademie der Naturforscher Leopoldina e. V. 2022.*https://doi.org/10.48669/ESYS_2022-2*

Fyson, C. L./Baur, S./Gidden, M./Schleussner, C.-F.: Fair-share carbon dioxide removal increases major emitter responsibility. In: Nature Climate Change 10, 2020, pp. 836–841

Giordano, V. M./Datchi, F./Dewaele, A.: Melting curve and fluid equation of state of carbon dioxide at high pressure and high temperature. In: J. Chem. Phys. 125, 2006, 054504

Hansen, J./Sato, M./Kharecha, P./Schuckmann, K. von/Beerling, D. J./Cao, J./Marcott, S./Masson-Delmotte, V./Prather, M. J./Rohling, E. J./Shakun, J./Smith, P./Lacis, A./Russell, G./Ruedy, R.: Young people's burden: requirement of negative CO_2 emissions. In: Earth System Dynamics 8, 2017, pp. 577–616

Heck, V./Gerten, D./Lucht, W./Popp, A.: Biomass-based negative emissions difficult to reconcile with planetary boundaries. In: Nature Climate Change 8, 2018, pp. 151–155

Herzog, H./Morris, J./Gurgel, A./Paltsev, S.: Getting real about capturing carbon from the air. In: One Earth 7, 2024, pp. 1477–1488

Howarth, R. W.: The greenhouse gas footprint of liquefied natural gas (LNG) exported from the United States. In: Energy Science and Engineering, Volume 12, Issue 11, November 2024, pp. 4843–4859

Ke, P./Ciais, P./Sitch, S./Li, W./Bastos, A./Liu, Z./Xu, Y./Gui, X./Bian, J./Goll, D. S./Xi, Y./Li, A./Sullivan, M. O./Goncalves de Souza, J./Friedlingstein, P./Chevallier, F.: Low latency carbon budget analysis reveals a large decline of the land carbon sink in 2023. In: National Science Review, Volume 11, 2024, nwae367

Krauß, K.: Kraftwerkstechnik. Springer, Berlin 2009

Landrigan, P. J./Fuller, R./Acosta, N. J. R./Adeyi, O./Arnold, R./Basu, N. N./Baldé, A. B./Bertollini, R./Bose-O'Reilly, S./Boufford, J. I./Breysse, P. N./Chiles, T./Mahidol, C./Coll-Seck, A. M./Cropper, M. L./Fobil, J./Fuster, V./Greenstone, M./Haines, A./Hanrahan, D. et al.: The Lancet Commission on pollution and health. In: The Lancet 392, 2018, pp. 462–512

Lenton, T. M./Vaughan, N. E.: The radiative forcing potential of different climate geoengineering options. In: Atmos. Chem. Phys. Discuss. 9, 2009, pp. 2559–2608

Lenzi, D./Lamb, W. F./Hilaire, J./Kowarsch, M./Minx, J. C.: Weight the ethics of plans to mop up carbon dioxide. In: Nature 561, 2018, pp. 303–305

Linow, S./Bijma, J./Gerhards, C./Hickler, T./Kammann, C./Reichelt, F./Scheffran, J.: Kurzimpuls – Perspektiven auf negative CO_2-Emissionen. Diskussionsbeiträge der Scientists for Future 12, 2022. (DOI: 10.5281/zenodo.7392348)

Madhu, K./Pauliuk, S./Dhathri, S./Creutzig, F.: Understanding environmental trade-offs and resource demand of direct air capture technologies through comparative life-cycle assessment. In: Nature Energy 6, 2021, pp. 1035–1044

Masson-Delmotte, V./Zhai, P./Pirani, A./Connors, S. L./Péan, C./Berger, S./Caud, N./Chen, Y./Goldfarb, L./Gomis, M. I./Huang, M./Leitzell, K./Lonnoy, E./Matthews, J. B. R./Maycock, T. K./Waterfield, T./Yelekçi, O./Yu, R./Zhou, B. (Hrsg.): Climate change 2021. The Physical Science Basis. Contribution of Working Group I to the Sixth Assessment Report of the Intergovernmental Panel on Climate Change. Cambridge University Press 2021. *http://www.ipcc.ch/report/sixth-assessment-report-working-group-i*

McKay, D. I. A./Staal, A./Abrams, J. F./Winkelmann, R./Sakschewski, B./Loriani, S./Fetzer, I./Cornell, S. E./Rockström, J./Lenton, T. M.: Exceeding 1.5 °C global warming could trigger multiple climate tipping points. In: Science 377, 2022

Michaelides, E. E.: Thermodynamic analysis and power requirement of CO2 capture, transportation, and storage in the ocean. In: Energy 230, 2021, 120804

Mühlbauer, A./Kleiner, D./Breyer, C.: Techno-economic insights and deployment prospects of permanent carbon dioxide sequestration in solid carbonates. In: Energy & Environmental Science 17, 2024, pp. 8756–8775

NIST Chemistry WebBook, SRD 69. Thermophysical Properties of Fluid Systems. *https://webbook.nist.gov/chemistry/fluid*

Onyebuchi, V. E./Kolios, A./Hanak, D. P./Biliyok, C./Monovic, V.: A systematic review of key challenges of CO2 transport via pipelines. In: Renewable and Sustainable Energy Reviews 81, 2018, pp. 2563–2583

Parker, A./Irvine, P. I.: The risk of termination shock from solar geoengineering. In: Earth's Future 6, 2017, pp. 456–467

Plötz, P./Rohde, C./Repenning, J./Auf der Maur, A./Becker, L./Braungardt, S./Deurer, J./Dünnebeil, F./Friedrichsen, N./Heidt, C./Hennenberg, K./Hermann, H./Jöhrens, J./Kasten, P./Köppen, S./Lutz, C./Scheffler, M./Thamling, N./Wünsch, M.: Quantifizierung der Treibhausgaswirkung von staatlichen Begünstigungen in Deutschland. Bericht zum Vorhaben Wissenschaftliche Unterstützung Klimapolitik und Maßnahmenprogramm (14-BE-2203). Karlsruhe, Berlin, 2024.*https://www.oeko.de/fileadmin/oekodoc/THG-Wirkung_staatliche-Beguenstigungen.pdf*

Popp, A./Calvin, K./Fujimori, S./Havlik, P./Humpenoder, F./Stehfest, E./Bodirsky, B. L./Dietrich, J. P./Doelmann, J. C./Gusti, M./Hasegawa, T./Kyle, P./Obersteiner, M./Tabeau, A./Takahashi, K./Valin, H./Waldhoff, S./Weindl, I./Wise, M./Kriegler, E./Lotze-Campen, H./Fricko, O./Riahi, K./Vuuren, D. P. van: Land-use futures in the shared socio-economic pathways. In: Global Environmental Change 42, 2017, pp. 331–345

Romanello, M./Di Napoli, C./Drummond, P./Green, C./Kennard, H./Lampard, P./Scamman, D./Arnell, N./Ayeb-Karlsson, S./Ford, L. B./Belesova, K./Bowen, K./Cai, W./Callaghan, M./Campbell-Lendrum, D./Chambers, J./Daalen, K. R. van/Dalin, C./Dasandi, N./Dasgupta, S./et al.: The 2022 report of the Lancet Countdown on health and climate change: health at the mercy of fossil fuels. In: The Lancet 400, 2022, pp. 1619–1654

Schleussner, C. F./Ganti, G./Lejeune, Q./Zhu, B./Pfleiderer, P./Prütz, R./Ciais, P./Frölicher, T. L./Fuss, S./Gasser, T./Gidden, M. J./Kropf, C. M./Lacroix, F./Lamboli, R./Martyr, R./Maussion, F./McCaughey, J. W./Meinshausen, M./Mengel, M./Nicholls, Z./Quilcaille, Y./Sanderson, B./Seneviratne, S./Sillmann, J./Smith, C. J./Steiert, N. J./Theokritoff, E./Warren, R./Price, J./Rogeli, J.: Overconfidence in climate overshoot. In: Nature 634, 2024, pp. 366–373

Shukla, P. R./Skea, J./Slade, R./Al Khourdajie, A./Diemen, R. van/McCollum, D./Pathak, M./Some, S./Vyas, P./Fradera, R./Belkacemi, M./Hasija, A./Lisboa, G./Luz, S./Malley, J. (Hrsg.): Climate Change 2022: Mitigation of Climate Change. Contribution of Working Group III to the Sixth Assessment Report of the Intergovernmental Panel on Climate Change. Cambridge University Press 2022.*http://www.ipcc.ch/report/sixth-assessment-report-working-group-3*

Tanaka, K./Cavalett, O./Collins, W. J./Cherubini, F.: Asserting the climate benefits of the coal-to-gas shift across temporal and spatial scales. Nature Climate Change 9, 2019, S. 389–396

Vattioni, S./Käslin, S. K./Dykema, J. A./Beiping, I./Sukhodolov, T./Sedlacek, J./Keutsch, F. N./Peter, T./Chiodo, G.: Microphysical interactions determine the effectiveness of solar radiation modification via stratospheric particle injection. In: Geophysical research Letters, Volume 51, Issue 19, 16 October 2024, e2024GL110575

Vaughan, N. E./Lenton, T. M.: A review of climate geoengineering proposals. In: Climatic Change 109, 2009, pp. 745–790

VDI 3677-1: Filternde Abscheider – Oberflächenfilter

VDI 3677-3: Filternde Abscheider – Heißgasfiltration

VDI 3678-1: Elektrofilter – Prozessgas- und Abgasreinigung

VDI 3927-1: Abgasreinigung – Minderung von Schwefeloxiden, Stickstoffoxiden und Halogeniden aus Abgasen von Verbrennungsprozessen (Rauchgasen)

Wilkinson, B. H./McElroy, B. J.: The impact of humans on continental erosion and sedimentation. In: GSA Bulletin 119, 2007, pp. 140–156

Wunderlin, E./Chiodo, G./Sukhodolov, T./Vattioni, S./Visioni, D./Tilmes, S.: Side effects of Sulfur-based geoengineering due to absorptivity of sulfate aerosols. In: Geophysical Research Letters, Volume 51, Issue 4, 28 February 2024, e2023GL107285

Zimov, S. A./Zimov, N. S./Tikhonov, A. N./Chapin, F. S.: Mammoth steppe. A high-productivity phenomenon. In: Quaternary Science Reviews 57, 2012, pp. 26–45. *https://doi.org/10.1016/j.quascirev.2012.10.005*

15 What Comes Next?

In the preceding chapters, methods were presented with which solution approaches for the major, wicked, and interconnected problems of ensuring a long-term energy supply, avoiding dangerous climate change, and dealing with increasingly scarce resources can be evaluated thermodynamically. The insight gained so far is that there is no simple solution and that any approach to shaping the future will be difficult and full of obstacles. This chapter first summarizes the findings on possible energy technologies before addressing the obstacles and opportunities of transforming our energy system.

15.1 Energy Technologies

In Table 8.5 evaluation options for individual energy technologies were proposed. Based on this scheme, Table 15.1 for the direct conversion of work into electricity and Table 15.2 for the conversion of heat into electricity or the use of heat contain an assessment of the relevant technologies. Included are all technologies that are expected to provide significant shares of the required energy in the future. The assessments are to be regarded as preliminary in accordance with standard scientific methodology. They start from the question of whether these are essential technologies. Some of the technologies will occupy a relevant niche function. The assessments are to be understood as a summary of the aspects of the individual technologies presented here and as an incentive to explore individual aspects in greater depth.

The criteria and the assessments are at the same time a summary of the contents of this book. Their core question is: Will the yield justify the input? Or, more precisely, the following questions are at stake:

- Is the planned technology fundamentally in accordance with the principles of thermodynamics, in particular the 2nd law?
- Has a robust EROI been determined for the project? Is this EROI at least above 10 to 1, as is required for supplying a modern society and its economy?

- Have all external energy subsidies been taken into account? This includes energy cannibalism as well as impacts that impair or destroy ecosystem services. These are services that must therefore be provided from the generated energy in the future.
- Can the energy that can be generated be easily delivered to consumers? Is there a process by which the energy is suitably delivered to the consumer, and how costly is this in terms of energy, resources to be deployed, and finances?
- Can the energy be generated according to demand? Can the generated energy be stored, e.g. as electricity? To what extent can generation and consumption be synchronized?
- Is the energy provided in such a way that it can be easily transported, stored, and used? Are there relevant safety aspects?
- Were the expenditures for transport, storage, and safety included in the calculation of the EROI?
- Do the benefits of using the technology sufficiently outweigh the risks? Have the risks been sufficiently well understood and described?
- Are there resources or elements that could hinder large-scale or global use of the technology because they are not available in sufficient quantities, or because their extraction would be too costly or would entail excessively high environmental impacts?
- Will the energy source provide sufficient energy as net energy to make a relevant contribution to society and the economy?
- Are sufficient resources and capital available to realize the plan?

In Table 15.1 and Table 15.2 there is in each case a simple point-based assessment that enables an initial comparison.

Table 15.1 Assessment of Energy Technologies, Part 1 – Electricity: Application of the Assessment Criteria from Table 8.5 (+++: High or Central, ++: Medium or Contribution, +: Low or Niche, 0: No Contribution, Negative or Impossible)

Aspect	Wind off-shore	Wind on-shore	Hydro-power	Tidal	Ocean current	PV
Potential, long-term total capacity	+++	+++	+	+	0	+++
Heat for heating	++	++	++	++	++	++
Heat for processes	++	++	++	++	0	++
Electricity	+++	+++	+++	+++	0	+++

Aspect	Wind off-shore	Wind on-shore	Hydro-power	Tidal	Ocean current	PV
Fuel for ships, aircraft, etc.	+ + a)	+ + a)	+ + a)	+ + a)	+ + a)	+ + a)
Technical difficulties	+ + +	+ + +	+ + +	+ + +	+	+ + +
Technology readiness level	+ + + 9	+ + + 9	+ + + 9	+ + + 8	+ 2	+ + + 8
Energy costs b)	+ + +	+ + +	+ + +	+ +	+	+ + +
Social costs c)	+ + +	+ + +	+ d)	+ +		+ + +
Acceptance	+ +	+ +	+ +	+ +		+ + +
Risk	+ + +	+ + +	+ +	+ + +		+ + +
Backyard	+ +	+ +	+ + +	+ +	+	+ + +
Technical complexity	+ +	+ + +	+ + +	+ +	+	+ +
Efficiency	+ + + d)	+ + + d)	+ + + d)	+ + + d)	+ + + d)	+ + d)
Intermittency	+ +	+ +	+ + +	+ + +	+ + +	+ +
EROI	+ + +	+ +	+ + +	+ +	+	+ +
Storage	+ +	+ +	+ + +	+ +	+ +	+ +
Resource requirements	+ +	+ +	+ + +	+ +	+	+ +
Recycling	+ + +	+ + +	+ + +	+ +	+	+ +
Points e)	10	8	10	5	–4	9

a) synthetic fuels produced from electricity
b) depending on the location or at suitable sites
c) depending on the location: typically, interventions in ecosystems, loss of fisheries, the climate impact via methane and reduced albedo, corruption and national debt, loss of arable land and displacement are rated as such.
d) related to the maximum possible conversion: for wind and water the Betz limit; for photovoltaics the Shockley-Queisser limit
e) for the evaluation the following were counted: +++ = +1 / ++ = 0 / + = –1 / 0 = –2.

Table 15.2 Assessment of Energy Technologies, Part 2 – Heat: Application of the Assessment Criteria from Table 8.5 (+++: High or Central, ++: Medium or Contribution, +: Low or Niche, 0: No Contribution, Negative or Impossible, HP: Heat Pump, ST: Solar Thermal, ST_{conc}: solar Thermal with Concentrator, GT: Geothermal)

Aspect	Bio-mass	HP	ST	ST_{conc}	H_2 (green)	GT	Nu-clear fission	Fu-sion
Potential, long-term total capacity	+	++	+++	+++	+++	+++	+	0
Heat for heating	+++	++	+++	++	+++	+++	++	++
Heat for processes	+++	+	+	+++	+++	+	++	++
Electricity	++	n. a.	n. a.	+++	++	n. a.	+++	+++
Fuel for ships, aircraft, etc.	+++	n. a.	n. a.	+ [a]	++ [a]	n. a.	++ [a]	++ [a]
Technical difficulties	++	+++	+++	+++	++	+++	++	0
Technol-ogy readiness level	+++ 9	+++ 7	+++ 9	+++ 9	++ 6	+++ 9	+++ 7 [e]	+ 2
Energy costs [b]	+++	+++ [c]	+++	+++	+	+++ [c]	+	+
Social costs	++ [d]	+++	+++	+++	++	+++	+ [e]	[e]
Accep-tance	++	++	+++	++	+++	+++	+	+
Risk	++	+++	+++	+++	+++	+++	+	+
Backyard	++	+++	+++	++	++	+++	+	+
Technical complexity	+++	+++	+++	++	+ [f]	+++	+	+
Efficiency	+ [g]	+++	+++	++	+++	+++	++	++

Aspect	Bio-mass	HP	ST	ST_{conc}	H_2 (green)	GT	Nu-clear fission	Fu-sion
Intermit-tency	+ + +	+ + +	+ +	+ +	+ + +	+ + +	+ + +	
EROI	+	n. a.	+	+	+	+ +	0 [e)]	+
Storage	+ + +	n. a.	+ + +	+ + +	+ +	+ + +	+ +	+ +
Resource require-ments	+ +	+ + +	+ + +	+ +	+ [f)]	+ + +	+	+
Recycling	+ + +	+ + +	+ +	+ +	+ + +	+ +	0 [e)]	0 [e)]
Points [h)]	6	10	11	6	4	13	-9	-9

a) synthetic fuels produced with electricity
b) depending on the location or at suitable sites
c) depending on the costs of the required electricity
d) depending on the location: typically, interventions in ecosystems, loss of fisheries, the climate impact through methane and reduced albedo, corruption and national debt, loss of arable land and displacement are rated this way.
e) taking into account the unresolved final disposal and the impressive energetic and societal efforts required after shutting down the reactors, an energetic credit from future generations is concealed here – energy that these generations will have to provide without benefiting from it themselves.
f) electrolysis membranes and iridium supply
g) compared to the yield of photovoltaics on the same area
h) for the evaluation, the following were counted: +++ = +1 / ++ = 0 / + = –1 / 0 = –2

15.2 Transformation Is Complex

Shaping our future through ourselves and in our own interest, rather than letting it be shaped by avoidable catastrophes, is the central and urgent task of our time (by design, not by disaster). Engineers should, can, and must make a significant contribution to this. It should also have become clear by now that there will be no purely technical solutions, but that technology can only provide a contribution.

There is no going back to any dreamed-of past. Our modernity, our innovations, and our legacy drive our world forward. Legacy here means all the changes in the earth system that we or our ancestors have set in motion and that continue to accumulate, like our legacy burdens (Schlaudt 2024). We cannot reject this legacy. It is a part of our world. All we can do is accept it, understand it as precisely as possible, nurture our ethical compass, and try to shape a good life for all of us here under these evolving conditions. It is not a simple legacy, because dealing with it requires constant vigilance, critical observation, and considerable resources that we must now and in the future expend solely on managing this legacy. In the best case, this is what the word Anthropocene means. In this sense, the transformation of our energy system is not a project with a clear conclusion. It will be a balancing act between what is possible in the earth system, what is technically feasible, and what remains ethically justifiable. At best, it is a mixture of adaptation and management of super-wicked problems.

Goals for Engineers

As engineers, we are accustomed to receiving a good and clear definition of objectives. The complete definition of the goal itself is already a relevant part of the project, and as in Chapter 2 shown, the target definition evolves with the learning curve in the project. At the same time, good targets are SMART. This formulation is associated with five concrete requirements for targets, which are provided in Table 15.3. At the core, the SMART formulation of problems is about getting into an orderly course of action very quickly and, in particular, not being pushed into a state of shock paralysis by an unclear, uncertain, daunting assignment.

Table 15.3 When Are Aims SMART?

No.	Aspect	Meaning
1	Specific	The target or targets must be clearly understandable and unambiguously formulated. The scope for interpretation should be reduced as much as possible, so that all parties involved in the process have understood and can use the same target definition.
2	Measur-able	Targets and important aspects of the targets should be measurable. This makes it clearly recognizable whether a target has been achieved. It becomes possible to describe progress unambiguously and to identify deviations from the plan. This also establishes that these measurable quantities are to be tracked. In addition, it often makes sense to define early on how the key figures are to be measured precisely.
3	Accepted	All parties involved in the process have a fundamental common consensus on the target itself. They do not need to have a common consensus on all details. Many things will be negotiated in the course of the process. This point is also defined as *assignable*, i.e., assignable to a responsible person.
4	Realistic	The target must be fundamentally achievable. This is often a question of the resources provided for working on the project.

No.	Aspect	Meaning
5	Time-bound	A clear deadline must be set for the project by which it is to be completed. What is essential in project management is to ensure that the project progresses quickly, especially in the initial phases. There are very good methods available today to ensure this, up to and including competition between multiple solution approaches or teams for a single target.

However, it is not possible to formulate super-wicked problems as a SMART project. This creates a first obstacle. We feel overwhelmed and therefore ignore or suppress such problems. We are not prepared for this. Our comfort zone is clearly not here. But it cannot be helped – if we want to shape things, then we must face the entire entangled issue. This works much better when we seek help: we do not have to approach this solely from an engineering perspective. We can bring in all the other disciplines and perspectives and work together with them on problem descriptions and solution pathways.

15.2.1 Transformation or Disruption?

In many industrial segments, a technology transition can occur rapidly. This sudden transition takes place in a free market when the new technology meets the necessary requirements for the required performance and at the same time brings obvious advantages over a previously dominant technology. If a technology transition proceeds unimpeded, it can happen so quickly that many established market players are caught off guard by the process and some disappear from the market. This process is a disruption of the existing market (Christensen 1997).

The model of disruption can only be applied to a limited extent to the provision of energy, since many of the prerequisites for a disruption are not present in the energy market. This market is obstructed by state actors, a close entanglement between state, economic and personal interests, rent payments and subsidies (Coady et al. 2015). It is neither a free market nor a pure capital goods market.

Independently of these power-political aspects, technical aspects are of great importance, which at least significantly slow down any disruption, strongly influence every transition, and ultimately require a certain degree of state control:

- Every technology transition in the energy sector is a transformation of the energy supply itself, i.e., the energy supply of society must always function with sufficient stability and reliability.
- The technology transition presupposes the availability and deployment of resources to build and maintain the required new energy infrastructure.

- Even maintaining a steady state, i.e., a uniform energy supply using a particular technology, requires constant effort. In the case of fossil energy carriers, these efforts increase over time (Court/Fizaine 2017), thus increasingly binding energy and resources themselves that can then no longer be deployed for other purposes. Independently of the technology transition, the price of energy therefore rises.
- The new technology must be built up alongside the existing one. In doing so, it must either be immediately market-competitive or be supported through subsidies until it is comparable to the old one.
- All components of the new energy technology must be developed; large-scale implementation can only take place at TRL 8.
- Learning curves are required: different technical solution approaches must be tried out in the field. Business models must be created and designed. Environmental impacts must be understood and reduced where necessary.
- Social acceptance for changing technical framework conditions must be developed in parallel.
- The old technology must continue to be maintained and operated until the new one can take over its function, so that a seamless transition is achieved. Sudden premature shutdown causes at least supply bottlenecks. In particular, the old technology may only be shut down incrementally – always only as much as the new technology can already reliably take over. This must even be observed when energy from the old technology becomes significantly more expensive than from the new one.
- The sector of a technology becoming obsolete must preserve its viability with shrinking revenues and an increasingly difficult cost structure. It must function reliably at least as long as it is needed. It must not simply become insolvent as long as the security of energy supply still partially depends on it.

The core of these arguments is that several technologies must coexist simultaneously and each individual one requires special technical attention. The new technologies must mature rapidly and be expanded quickly so that they can take on their role. To do so, they simultaneously require large amounts of energy and resources. The old technologies must be carefully maintained and can only be shut down incrementally. In the case of fossil energy carriers, energy and resources are also required for this. This directly influences the costs of the energy provided.

The supply of energy is a core element of public services. It therefore belongs to the most important state responsibilities. States cannot expect their energy system to simply transform on its own. It is a central responsibility of every good government to pursue a fact-based and future-oriented energy policy that provides all other actors with a clear and long-term reliable framework. Engineers can at least persistently demand this.

Expectations of new technology

A new technology for the provision of energy has fundamental technical and financial disadvantages compared to established technology:

1. The technology starts with initial, rather experimental installations. Not all technical problems have yet been identified or even solved, i.e., it often functions less reliably.
2. It does not have a widely spread network of similar facilities, but is first tried out in a few unconnected locations. Thus, it does not have an established infrastructure or experienced experts.
3. Social procedures for dealing with it are lacking. Users embark on an adventure.
4. If this technology is to provide a relevant share of energy, a great deal of capital, resources, and energy will be required for its development.

If the already established technology has additionally succeeded in externalizing its own costs effectively, in acquiring rent payments (Coady et al. 2015) or in becoming part of the power structure, then a technology change is even more difficult or impossible. The externalization of costs can take the form, for example, of societal acceptance of health consequences, illness, or premature death caused by this technology (Romanello et al. 2022). Rent payments include subsidies for an established technology (Coady et al. 2015).

In public discourse, expectations of new, especially renewable energy technology are very high and are often deliberately driven even higher:

- It should immediately be more reliable and cheaper than all existing and long-established technologies.
- It is granted no learning curve. It should work flawlessly right now.
- It should exhibit at least the identical performance as the established technologies.
- It should in particular not impose its own or new requirements on its use.
- It should, unlike established technology, receive no subsidies.
- It should, unlike established technology, not be visible.
- Unlike old technology, it must not devastate landscapes, force child labor, or produce unpleasant images.
- It should be deployable at any time, cost-neutrally, within a short period.

These expectations cannot be met by a developing technology in its early stage, and they are not met by the currently established technologies either (Smil 2017). Our fossil-based energy technologies have each grown over long periods of time into large sectors in an environment barely characterized by technical competition with other technologies. The fossil technologies had and continue to have a great deal of freedom for their massive impacts.

Essential to the low energy price of fossil technology is its often massive subsidization. The subsidization of fossil fuels is generally accepted or is often not understood as such. In 2015, the total global subsidies for fossil fuels amounted to 5,300 billion

US dollars, or 6.5% of global gross domestic product, with coal accounting for the largest share of these subsidies (Coady et al. 2015). Not included in this figure are the externalized costs, such as the consequences of industrial global warming. In addition, strong personal interests of important political actors are tied to established technology (Auzanneau 2018).

To what extent the asymmetric expectations compiled above, directed specifically at renewable energy technology, also originate from marketing for established fossil technologies is an interesting question, since these expectations directly support established technology and prevent fossil fuels from facing competition. The task for engineers is to understand these connections, communicate them, and advocate for acceptance of the necessary changes.

15.2.2 Speed of Transformation

How quickly can the required new renewable technologies be expanded? There are different possible speeds. Extreme positions are the following:

Directed transformation A massive immediate transition, in which, as in a war economy, all resources are subordinated to this goal: renewable energy technology, including the required grid structure and storage, is expanded very rapidly to a sufficient extent for future needs. This path can achieve a complete transition within a few years. It depends on the energy supply from fossil sources during the transition phase. In order to extract the required resources and build the infrastructure, planning must be greatly accelerated. This is achieved through appropriate regulations and new distribution procedures (ideas on this, Herrmann 2022).

This approach carries a risk: if the targeted infrastructure is too large, i.e., is intended to provide too much energy, then the impacts of the rapidly required resource extraction may themselves lead to unintended and destructive effects in the earth system. Modesty in objectives is necessary here.

Market-oriented transformation If consideration is to be given to the interests of the fossil energy sectors and the geopolitical ambitions of the main producing countries, this means waiting until the fossil fuels themselves fall into a substantial crisis and then create space for something new.

Also, fully accommodating societal expectations of low energy prices, avoiding changes to the landscape through the development of renewable energy technology or the restructuring of the electrical supply network, as well as avoiding any further inconveniences, means waiting. If the energy supply environment changes through crises to such an extent that societal interest in a reliable energy supply after the end of fossil fuels dominates the discussion and such reasons recede into the background, action can be taken. This approach has a high potential for a complete collapse of the energy system, since the required future technology will not be available in time when the established one is exhausted.

Between the two limiting cases there is a range – so we hope – characterized by enabling the required transformation sufficiently quickly and in a socially acceptable manner, without exceeding technical boundary conditions and planetary boundaries. Requirements for this space of possibilities can be developed from the framework of planetary boundaries and the SDGs (Kenny et al. 2010, Sgouridis et al. 2016):

- The impacts arising from the process of technology change should themselves not exceed planetary boundaries.
- The energy supply should be continuously ensured overall without major disruptions. Adequate living conditions for all should always be made possible.
- The future energy supply should be provided at a level that ensures the continued functioning of societies at a sufficient level. What this means in concrete terms must be technically assessed and socially negotiated. With a view to the goal of adequate living conditions for all, this energy must not be too low, as it is intended to enable social structures of a certain complexity.
- Fossil fuels should emit at most the carbon dioxide budget considered safe into the atmosphere. This amounts to the 1.5 °C target. This mass of carbon dioxide limits the timeframe by which the technology change must be completed and fossil energy technology is shut down: from today's perspective (2025), this budget is completely exhausted.
- The future energy technology must have a sufficient EROI. During the transition, fossil technology must also still have a sufficient EROI to at least ensure supply. Therefore, only those fossil fuels that have a particularly high EROI should be used.
- Energy cannibalism should be avoided. This means, among other things, that renewable energy technologies should contain as little energy from fossil fuels as possible in their CED.
- The resources required for the provision of renewable energy technology and bound within it should be extracted at a rate that remains within the framework of planetary boundaries.
- The resources bound in renewable energy technology must be sufficiently low to continue allowing other uses.
- The most complete possible recycling of all material resources, especially all metals, should be pursued.

15.2.3 Material-Technical Limits

These limitations have two concrete implications for our transformation. They define a maximum possible growth rate for renewable energy technology and implicitly a maximum possible global generation (at least during the transition phase). In addition, the space of possibilities in which such a transition can be technically designed to produce only minor social disruptions is shrinking rapidly. On the one hand, the maximum carbon dioxide budget has been reached, and on the other hand, the available quantity of fossil energy carriers with sufficient EROI is rapidly decreasing, so that ever larger shares of fossil energy carriers are being used in the provision of these energy carriers instead of being available for other purposes. If, with the end of fossil energy carriers, only a significantly lower generatable power from renewable energy carriers is available, this leads to scarcity, supply bottlenecks, price turbulences and disruptions in the economic and social fabric.

Scenarios that explore this space of possibilities are mostly based on simple models, but require many assumptions (Kenny et al. 2010; Sgouridis et al. 2016). These models focus on the energetic framework conditions, not on financial aspects. The justification for this choice arises from the physical framework conditions themselves, which can be expressed in energy, but not in capital. This underlying argument is presented in Chapter 4 to Chapter 6.

Energy cannibalism

Energy cannibalism means that an inherently environmentally friendly technology requires many resources from an existing, e.g. fossil, infrastructure for its rapid build-up. If the technology grows too fast, the inputs for its growth are greater than the actually low-emission energy it could provide. For a specific technology, the energy provided over a period of time from all its real power plants is (Kenny et al. 2010)

$$E_{tech} = \sum_i P_{N,i} \cdot \psi_i \cdot \Delta t = \overline{P_{N,tech}} \cdot \overline{\psi_{tech}} \cdot \Delta t$$

The generation of all power plants *i* with individual rated capacities $P_{N,i}$ and individual service lives ψ_i can be aggregated into a total installed rated capacity and mean service life. The energy payback time, i.e. the time until which this technology only returns the energy invested in it, is given in years:

$$\Delta t_{pb} = \frac{E_{inv}}{E_{prod}}$$

Here E_{invest} is the energy invested over the life cycle and E_{prod} is the energy generated over the life cycle. From this, an energy demand for the growth of the specific technology can be calculated:

$$E_w = \frac{E_{inv}}{E_{prod}} \cdot \left(\overline{P_{N,tech}} \cdot \overline{\psi_{tech}} \cdot \Delta t\right) \cdot r = \Delta t_{pb} \cdot E_{tech} \cdot r$$

If the generated energy $E_{tech} > E_w$ is, then the technology generates net energy. Conversely, for $E_{tech} < E_w$ the energy technology must be energetically subsidized by other technologies: it cannibalizes energy from other technologies. Thus, the maximum non-cannibalizing annual growth rate of an energy technology is (Kenny et al. 2010)

$$r = \frac{1}{\Delta t_{pb}} = \frac{E_{prod}}{E_{invest}}$$

This key figure can equally be applied to the emission of greenhouse gases. This analysis is significant for two reasons:

1. This establishes an upper growth rate, in case the new energy technology is to multiply only from itself, or it must not generate any additional greenhouse gas emissions. The growth rate depends on the technology, but also on the location and operating conditions including curtailment.
2. The maximum growth rate can only be realized if the existing old fossil energy technology continues to operate. If the old technology simultaneously shrinks in its output, because this is planned or because fewer fossil energy carriers are available, this correspondingly reduces the growth rate, since the new technology is already needed for supply, i.e. it passes on part of its energy yield to society.

The second point shows that an energy transition must be started early if no disruption or interruption of the energy supply is desired. If the energy transition begins too late or is disrupted or stopped in the interim, then later (especially when fossil energy sources are no longer available) severe restrictions on energy supply will have to be accepted for several years or decades. This unnecessary waiting or prevention of an essential transformation is often justified by arguing that a transformation would be too burdensome for the stability of society. As this argument shows, waiting or preventing is itself quickly causally responsible for a sharp decline in the energy available to society. Waiting or preventing can lead to the collapse of the social order or civilization through the unprepared reduction of available energy (Tainter 1990).

15.2.4 Usable Energy Flows

Only a portion of the earth's energy flows is usable for human societies. Figure 15.1 illustrates this in summary. For power plants, only a fraction of these flows is truly accessible, and only a certain portion of this power can actually be fed into the grid by the power plants. A further portion is lost as transmission loss and during storage. A relevant portion of the generated energy must be used to compensate for the services of damaged or destroyed ecosystems. A balance must be found here between human needs, the energy flows accessible with high EROI, and what our earth requires in order to continue contributing as many valuable ecosystem services as possible.

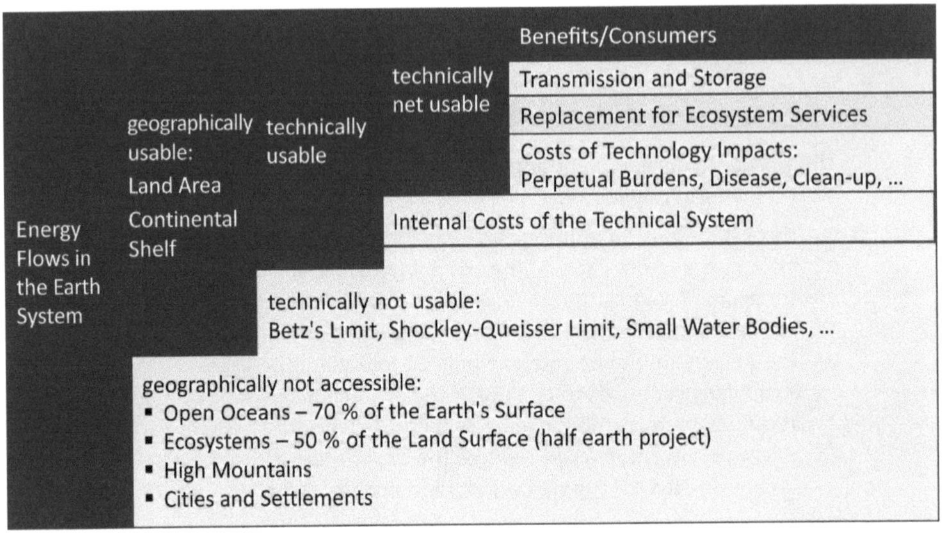

Figure 15.1 Schematic representation of the actual potential for utilizing the earth's energy flows (Moriarty/Honnery 2016): the individual contributions depend on many decisions

15.3 Energy Scenarios

A number of relevant scenarios and analyses exist regarding how a functioning regenerative energy supply can be achieved and designed. The value of such analyses arises in particular from the fact that they identify obstacles, approach the limits of what is technically possible, or explore the framework conditions through which a transformation of our energy system can be shaped. None of the scenarios is a simple technical guide, nor should they be understood as such. Not all of the assumptions made always align with what is technically feasible or truly desirable. At the same time, these scenarios provide encouragement that it is indeed possible to redesign our energy system within a short period of time.

Transformation scenarios

This is a non-exhaustive introduction to relevant studies, methods, and scenarios.

Global

In retrospect, the study by Jacobsen et al. is important for the USA: Jacobsen et al. argue that it would be possible to completely convert the energy supply of the USA to electricity and, in terms of energy carriers, to hydrogen by the year 2050 (Jacobsen et al. 2015). Many of the assumptions used were subsequently critically questioned or assessed as very optimistic (Clack et al. 2017). This initiated an important learning curve that makes new scenarios considerably more robust. Moriarty and Honnery as well as Rye and Jackson are pessimistic regarding the scope of what is possible (Moriarty/Honnery 2016; Rye/Jackson 2019).

Europe

Ram et al. is ambitious and relevant (Ram et al. 2022). The underlying model maps generation, distribution, and key users in the European grid over an entire year, resolved on an hourly basis. This also captures critical moments (Bogdanov et al. 2023).

This looks quite different in an analysis in which the existing level of ambition was used as a basis. Starting from a scenario process and the measures developed, a significant gap emerges (Frantziskaki et al. 2019).

Germany

The Deutsche Energie Agentur (DENA) published a study in 2018. It continues to accompany the transformation intensively and provides relevant data and publications. The Wuppertal Institut has developed an ambitious plan for how Germany can be CO_2-neutral by 2035 (Kobiela et al. 2020). The Bundesverband der deutschen Industrie (BDI) has updated its 2018 study (BDI 2021). The Fraunhofer ISE (Institut für Solare Energiesysteme) regularly publishes a new edition of its analysis and scenarios (Thelen et al. 2024).

The scenarios cited here

Bogdanov, D./Oyewo, A. S./Breyer, C.: Hierarchical approach to energy system modelling. Complexity reduction with minor changes in results. In: Energy 273, 2023, 127213

BDI – Bundesverband der Deutschen Industrie: Klimapfade 2.0. Ein Wirtschaftsprogramm für Klima und Zukunft. 2021.*https://bdi.eu/publikation/news/klimapfade-fuer-deutschland*

Clack, C. T./Qvist, S. A./Bazillian, M./Brandt, R. M./Caldeira, K./Davis, S. J./Diakov, V./Handschy, M. A./Hines, P. D. H./Jaramillo, P./Kammen, D. M./Long, J. C. S./Morgan, M. G./Reed, A./Sivaram, V./Sweeney, J./Tynan, G. R./Victor, D. G./Weyant, J. P./Whitacre, J. F.: Evaluation of a proposal for reliable low-cost grid power with 100 % wind, water, and solar. In: PNAS 114, 2017, pp. 6722–6727

DENA – Deutsche Energieagentur. *https://www.dena.de*

Frantziskaki, N./Hölscher, K./Holman, I. P./Pedde, S./Jaeger, J./Kok, J./Harrison, P. A.: Transition pathways to sustainability in greater than 2 °C climate futures of Europe. In: Regional Environmental Change 19, 2019, pp. 777–789

Jacobson, M. Z./Delucchi, M. A./Cameron, M. A./Frew, B. A.: Low-cost solution to the grid reliability problem with 100 % penetration of intermittent wind, water, and solar for all purposes. In: PNAS 112, 2015, pp. 15 060–15 065

Kobiela, G./Samadi, S./Kurwan, J./Tönjes, A./Fischedick, M./Koska, T./Lechtenböhmer, S./März, S./Schüwer, D.: CO2-neutral bis 2035: Eckpunkte eines deutschen Beitrags zur Einhaltung der 1,5-°C-Grenze. Wuppertal Institut 2020

Moriarty, P./Honnery, P.: Can renewable energy power the future? In: Energy Policy 93, 2016, pp. 3–7

Ram, M./Bogdanov, D./Satymov, R./Lopez, G./Mensah, T./Sadovskaia, K./Breyer, C.: Accelerating the European renewable energy transition. Lappeenranta University of Technology 2022

Rye, C. D./Jackson, T.: A review of EROEI-dynamics energy-transition models. In: Energy Policy 122, 2019, pp. 260–272

Thelen, C./Nolte, H./Kaiser, M./Jürgens, P./Müller, P./Senkpiel, C./Kost, C.: Wege zu einem klimaneutralen Energiesystem: Bundesländer im Transformationsprozess. Fraunhofer ISE 2024

What the German scenarios have in common is that they justify and highlight the framework conditions that legislators must create in order to enable the respective transformation. These studies open up a broad focus, as they include energy-intensive sectors of industry and identify possible technical changes therein. Many studies share the characteristic that they implicitly make optimistic assumptions about the future: the pathways presented are seen as fundamentally possible. The required resources, primary energy sources, and framework conditions are available in sufficient quantities and mostly at approximately today's conditions. This is methodologically acceptable, but must be taken into account when interpreting the findings. Such studies are also to be understood politically. They are intended to show what is considered acceptable and what is not from the perspective of those who commissioned them.

The following key theses emerge from the studies:

1. A reduction of CO_2emissions by more than 80 % appears foreseeable for Germany and other industrialized nations and achievable without major social disruptions. Many technical problems need to be solved along the way, and clear and timely groundwork by legislators is required. However, these reductions and the associated necessary social adjustments do not yet present themselves as wicked.
2. It appears advantageous to advance available technical options for a future energy infrastructure and to integrate them into a future system as much as possible. Experience can often only be gained with large-scale installations, and not every path or approach will succeed. Therefore, a certain degree of fault tolerance is necessary. This must at the same time be explained repeatedly. However, this does not mean "technology openness", because there are not enough resources available for ideas that are foreseeably unsuitable or harmful.
3. The transition to a fully regenerative energy supply requires far-reaching adjustments that have strong societal impacts. This complete transition seems barely possible to many actors today, as it obviously creates wicked problems.
4. The disruption of the market for fossil fuels is regarded in many studies as a conventional technical change. In fact, it is doubtful whether this *"creative destruction"* in the Schumpeterian sense will be accepted as such by those affected, and whether they will not instead cling by all means to their business model of burning all extractable fossil fuels as quickly and completely as possible until the bitter end.

5. Effective measures are systemic, are large-scale, and often go far beyond what seems conceivable today. Here, a lack of imagination with regard to future impacts is just as obstructive as the will of key actors to entrench the status quo as much as possible and to write off and externalize all impacts in an inappropriate manner.
6. International cooperation appears necessary and helpful in order to implement these goals globally. However, it is not a prerequisite for advancing the energy transition, the necessary adaptation to industrial global warming, and the reduction of dependence on raw material imports for one's own country.
7. For Germany as a country very poor in resources and internationally interconnected, international cooperation is on the one hand essential; on the other hand, for strategic and economic reasons, it is necessary to strive for a high degree of possible energy autarky, if necessary together with the nearest neighbors. The supply of fossil fuels is foreseeably not guaranteed in the future under current conditions for a number of reasons. The growth in global demand combined with the drying up of resources with high EROI (Court/Fizaine 2017) can lead to massive geopolitical disruptions (Diamond 2019). Freeing oneself from these problems as early as possible is one of the core arguments for the rapid implementation of the energy transition.
8. The path of transition is at best arduous. It will not be easy politically, technically, or socially, and it will be felt by everyone through higher energy prices and the necessary changes in energy use. What level of supply will be possible in the long term cannot be decided today; however, it may be significantly lower than today.
9. Evidence-based policy is needed. No society can afford to install or support populist, fact-denying, ignorant politics. This is a challenge for all of us.
10. The role of engineers is to contribute the technical solutions and at the same time to repeatedly bring decision-making and communication back to a fact-based discourse.

The role of the key figures, approaches, and calculation methods presented in this book is to support and guide this process by bringing the core energy question to the forefront of every solution idea and illuminating it intensively. The core questions are always: What relevant contribution does a solution idea deliver at what cost? For whom is the measure meaningful and beneficial? In doing so, waiting and inaction is always one of these solution approaches.

15.4 Managing Crises

The goal of this book is to open up the space of possibilities for a responsible and caring shaping of our future on our planet. A core thesis is that we must now rapidly transform our energy system so that we succeed in adequately controlling global heating, resource availability, and further planetary boundaries.

Currently there is no broad social consensus on what the truly decisive drivers and framework conditions of our future can be. There is only a feeling that we are in a crisis. Dealing with crises is difficult. A crisis can completely overwhelm and destroy the individual or even a society, often because the crisis is disavowed for too long or because it simply becomes too large at some point.

We can try to describe how to deal with severe shocks and massive crises. Typically, we go through a series of phases: denial → anger → bargaining → grief, depression, and suffering → emotional acceptance → experimentation → integrating into our lives. Currently, many of us remain stuck in the first two steps. Only with acceptance of the crisis does it become possible to face it in an appropriate manner (cf. Bendell 2018). It is obvious that successfully dealing with a crisis will lead to changes and adaptations in the personal and societal sphere. This was already discussed in Section 8.3.2 on resilience. The willingness to change and the acceptance of the crisis go hand in hand.

Table 15.4 compiles relevant factors for dealing with crises at a national or societal level and from a point of acceptance (Diamond 2019). These factors are analyzed in Table 15.4 with regard to the unfolding crisis of our core topics. If it succeeds in holding together the core values of the Federal Republic and European integration in Germany and Central Europe, then important prerequisites for an orderly handling of this unfolding crisis are in place.

Table 15.4 Factors for Managing Crises and Their Relevance to a Transformation of Our Energy System Under the Harsh Framework Conditions of Planetary Boundaries (Based on Diamond 2019, Expanded)

No.	Factors influencing the management of a crisis	Concrete connection to the energy transformation
1	Social consensus that one's own society is in a crisis: At least large parts of the population must agree that the situation constitutes a crisis that threatens the community in its continued existence or cohesion.	The supply of primary energy sources and other resources is threatened (costs, availability). Climate heating will massively change life here. It will severely burden the regional and global environment and trigger further crises in many regions, each of which will make it more difficult to supply the society.

No.	Factors influencing the management of a crisis	Concrete connection to the energy transformation
2	Acceptance that society is responsible for taking active action: We are responsible and capable of helping ourselves. Self-pity, blaming others, or seeing oneself as a victim does not help – on the contrary, it is ideal for turning the crisis into a catastrophe.	No help is coming from outside. We can, we are allowed to, and we should act ourselves.
3	Building a fence: which areas belong to this problem and which do not? At the national level, this means determining which institutions, rules, and sectors are to be changed and which remain the same.	The supply of energy and resources affects many basic living conditions, but by no means all of them. Some rules and institutions urgently need to be adapted, others do not. A discussion on sufficiency is necessary here: what are the necessary conditions for living a good life? Should everyone lead an adequate life or should a few lead a wasteful one?
4	Seeking and obtaining material and financial support from other nations	This is only possible to a very limited extent, as all regions are simultaneously affected in their own way. Cooperation at the European level is central, as this is a common problem.
5	Using other nations as role models and understanding how they successfully address this problem	There are a number of initial experiences, failed attempts and successful approaches (as a starting point: Tainter 1990 and Diamond 2011).
6	Shared identity	In this context, these include, for example, good education for all, the appreciation of engineers and their work, as well as the ability – often perceived as boorish – to address problems directly and without regard for a person's reputation or status.
7	Honest national self-assessment: Open and free shared discourse in media that are tied to facts: on the radio, on television, on open digital platforms, in newspapers, but not in the closed opinion bubbles of so-called "social media".	The culture of debate is increasingly dominated by bullshit. This must be clearly named. Courageous, fact-oriented media need our support in remaining so. Engineers play a responsible role here, which includes: promoting energy literacy, identifying bullshit, and naming and evaluating what is possible from our technical perspective.

No.	Factors influencing the management of a crisis	Concrete connection to the energy transformation
8	Historical experience from previous national crises	Most countries have successfully overcome several crises. In particular, facing the problem, understanding the outside perspective, and making an international consensus part of one's own culture were successful approaches.
9	Dealing with national failure: what happens/happened after major disasters?	Europe has shown, after the end of the Second World War, that it can pick itself up and start completely anew. This capability is needed again here.
10	National flexibility: persisting in the same old ways or the possibility of change; the ability to further develop principles	What is hindering is the tendency toward ready-made solutions. What is needed here is room for maneuver and space for experimentation rather than a corset of rigid and prefabricated rules.
11	National core values: what defines the community and cannot be changed?	One of the core values needed here is 'German Engineering' good engineering: this is the capability to invent and build new, good, reliable and beautiful solutions to all technical problems. Many further non-negotiable core values are connected to human rights for all.
12	Freedom from geopolitical demands: key factors are the attitudes of (powerful) neighbors, military and political framework conditions, and available (financial) resources.	The energy transition liberates us in the longer term from the vulnerability to blackmail by current and future suppliers of primary energy carriers. A common European stance can only be helpful in this regard. Today, Europe possess the resources to implement a rapid transformation of the energy supply while maintaining a high level of supply security.

It is foreseeable that these intertwined crises are only just beginning to unfold in their full magnitude. Accepting this situation and shaping how we deal with it requires new patterns of thinking. The declining EROI of all fossil technologies for providing usable energy (Rye/Jackson 2019), climate change and the associated losses in all areas, as well as the scarcity and rising cost of necessary resources, will have an impact on the space of possibilities for economic and social action (Rye/Jackson 2019). With the end of fossil energy carriers, there will be a shift in the significance, the scope of influence, and the available resources of regions or states (Goldthau et al. 2019). These changes will strongly influence what is technically possible, desirable and feasible.

Good technical solutions are designed for a surprising future. They are flexible in their application, they are available to the whole of society, they are not tools of oppression, and they can cope with constraints or crises. The attitude prevalent in many public discussions of immediately dismissing every new approach or placing it under a ban on thinking testifies to an impressive lack of imagination and knowledge on the part of those who presume to do so: it is never sufficient to explain what cannot be done. The central question is what can be done in the future, and for that we must argue.

15.5 Getting Started

Getting started is hard: Am I allowed to do this? Should I do this? Can I do this? Am I alone? What will others think? Should I wait for someone else to do something first? Do I have the time for it?

Getting started is easy: I am allowed to try something new! I can break out of the daily routine! I am allowed to make mistakes! I come into contact with other people and ideas that are new to me!

The challenges of securing our energy supply, minimizing and managing the impacts of global warming, and making use of increasingly scarce and expensive resources – which are coming our way regardless – will demand our attention whether we are ready for them or not. They will require engineers to apply many new as well as proven approaches: a thoroughly solid and broadly based course of study prepares one to tackle a wide variety of problems with the appropriate tools. Knowledge of earlier technical solutions is helpful when new technology is to be designed to be resilient, flexible and sufficient. The aim of this collection of tools is to provide, in a compact and usable form, some new and some proven tools that may be important in the future but are currently barely or not at all taught in bachelor's degree programs. These are tools intended to provide the necessary orientation amid the emerging changes.

The great personal challenge that I, as the author, am living through – and which will be of significance for us as practicing engineers, but also as a society – is to find a balance from which meaningful and future-oriented action remains possible for us. The facts are serious and the situation is urgent. Therefore, the following cheap escape routes often present themselves:

- Turning away ("I don't care ...")
- Anger ("XYZ are to blame for everything, so let's just get rid of them or do away with them ...")
- Despair ("It's already too late ...") or
- Cynicism ("They deserved it anyway ...")

A more extensive compilation of such disruptive and obstructive narratives is provided by Lamb et al. and *https://www.leolinne.com/-discourses-of-climate-delay* (Lamb et al. 2020). It is a constant challenge for everyone to continue thinking and analyzing clearly, and to continue not losing hope that one's own contribution counts and that it is worthwhile to become part of the solution right now.

The clear and realistic voice of engineers is needed now: it is our task to explain what is technically possible, what is technically necessary, and what raises ethical or social problems that technology alone must not or cannot solve. To fulfill this role, it is important to think in systems, and not to regard the system boundary of our technology as the limit of our responsibility, but rather to understand our beautiful, living earth as our system (and our irreplaceable home).

References

Auzanneau, M.: Oil power and war. A dark history. Chelsea Green, White River Junction 2018

Bendell, J.: Deep Adaption. A Map for Navigating Climate Tragedy. IFLAS Occasional Paper. 2018

Christensen, C.: The innovators dilemma. When new technologies cause great firms to fail. Harvard Business School Press, Boston 1997

Coady, D./Parry, I./Sears, L./Baoping, S.: How large are global energy subsidies? IMF working papers. International Monetary Fund 2015. *https://www.imf.org/external/pubs/ft/wp/2015/wp15105.pdf*

Conway, E. M./Oreskes, N.: Merchants of Doubt. How a Handful of Scientists Obscured the Truth on Issues from Tobacco Smoke to Global Warming. Bloomsbury 2012

Court, V./Fizaine, F.: Long-term estimates of the energy-return-on-investment (EROI) of coal, oil, and gas global productions. In: Ecological Economics 138, 2017, pp. 145–159

Diamond, J.: Kollaps. Warum Gesellschaften überleben oder untergehen. S. Fischer, Frankfurt 2011

Diamond, J.: Krise. Wie Nationen scheitern oder sich erneuern. S. Fischer, Frankfurt 2019

Goldthau, A./Westphal, K./Bazilian, M./Bradshaw, M.: How the energy transition will reshape geopolitics. In: Nature 569, 2019, pp. 29–31

Herrmann, U.: Das Ende des Kapitalismus. Warum Wachstum und Klimaschutz nicht vereinbar sind – und wie wir in Zukunft leben werden. Kiepenheuer & Witsch 2022

Jackson, T.: The Post-growth Challenge. Secular Stagnation, Inequality and the Limits to Growths. In: Ecological Economics 156, 2019, pp. 236–246

Jess, A.: What might be the energy demand and energy mix to reconcile the world's pursuit of welfare and happiness with the necessity to preserve the integrity of the biosphere? In: Energy Policy 38, 2010, pp. 4663–4678

Kenny, R./Lwa, C./Pearce, J. M.: Towards real energy economics: Energy policy driven by life-cycle carbon emission. In: Energy Policy 38, 2010, pp. 1969–1978

Lamb, W. F./Mattioli, G./Levi, S./Roberts, J. T./Capstick, S./Creutzig, F./Minx, J. C./Culhane, T./Steinberger, J. K.: Discourses of climate delay. In: Global Sustainability 3, 2020

*Romanello, M./Di Napoli, C./Drummond, P./Green, C./Kennard, H./Lampard, P./Scamman, D./Arnell, N./Ayeb-Karlsson, S./Ford, L. B./Belesova, K./Bowen, K./Cai, W./Callaghan, M./Campbell-Lendrum,

D./Chambers, J./Daalen, K. R. van/Dalin, C./Dasandi, N./Dasgupta, S./et al.: The 2022 report of the Lancet Countdown on health and climate change: health at the mercy of fossil fuels. In: The Lancet 400, 2022, pp. 1619–1654

Schlaudt, O.: Zugemüllt Eine müllphilosophische Deutschlandreise. C. H. Beck München 2024

Sgouridis, S./Csala, D./Bardi, U.: The sower's way. Quantifying the narrowing net-energy pathways to a global energy transition. In: Environmental Research Letters 11, 2016, 094009

Smil, V.: Energy and Civilization. A History. MIT Press, Cambridge 2017

Tainter, J.: The collapse of complex societies. Cambridge University Press 1990

16
Constants, Symbols and SI Prefixes

Constants

Symbol	Value	Meaning
c_0	$= 299\,792\,458$ m s^{-1}	Speed of light in vacuum
c_1	$= 142 \cdot 10^{-16}$ W m^2	Constant of Planck's law
c_2	$= 0.014387$ m K	Constant of Planck's law
c_W	$= 2.89777 \cdot 10^{-3}$ m K	Constant in Wien's displacement law for radiation
$c_{W,S}$	$= 3.00292 \cdot 10^{-3}$ m K	Constant in Wien's displacement law for entropy S
e_0	$= 1.602 \cdot 10^{-19}$ C	Elementary charge
g	$= 9.81$ m s^{-2}	Gravitational acceleration (at the earth's surface)
I_{sol}	$= 1366$ W m^{-2}	Solar constant, intensity above the earth's atmosphere
h	$= 6.626 \cdot 10^{-34}$ J s	Planck's constant
k_B	$= 1.381 \cdot 10^{-23}$ J K^{-1}	Boltzmann constant
N_A	$= 6.02214 \cdot 10^{23}$ mol^{-1}	Avogadro constant – number of molecules in one mol
p_N	$= 101\,235$ Pa	Pressure at standard conditions
R	$= 8314$ J kmol^{-1} K^{-1}	Universal gas constant
σ	$= 5.67 \cdot 10^{-8}$ W m^{-2} K^{-4}	Stefan-Boltzmann constant
T_N	$= 273.15$ K	Temperature at standard conditions

Formula Symbols – Latin Letters

Symbol	Unit	Meaning
$a(\lambda)$	m^{-1}	Absorption rate
A	m^2	Area
AM	–	Relative air mass
B	€	Monetary value, benefit
c	$m\ s^{-1}$	Velocity
c_p	$J\ kg^{-1}\ K^{-1}$	Specific heat capacity of a substance at constant pressure
c_v	$J\ kg^{-1}\ K^{-1}$	Specific heat capacity of a substance at constant volume
C	€	Monetary value, cost
C	–	Concentration of a greenhouse gas as volume fraction
COP	–	Coefficient of performance – performance factor of a heat pump
d	m	Distance
d	m	Diameter
e	$J\ kg^{-1}$	Specific energy
e_{net}	–	Relative specific net energy
E	J	Energy
E_{net}	J	Net energy
EROI	–	Energy returned on Energy invested
ESOI	–	Energy saved on Energy invested
f	–	Function
ΔF	$W\ m^{-2}$	Radiative forcing
g	–	Supply efficiency of an energy carrier
G	–	Geometry factor
h	–	Fractional factor of energy
h	$J\ kg^{-1}$	Specific enthalpy (intensive caloric state variable)
H	J	Enthalpy (extensive caloric state variable)
H_I	$J\ kg^{-1}$	Lower heating value
H_S	$J\ kg^{-1}$	Upper heating value or calorific value
$H_€$	$J\ €^{-1}$	Energy value
I	$W\ m^{-2}$	Intensity of radiation
I_0	$W\ m^{-2}$	Incoming intensity at ground level
I_{ground}	$W\ m^{-2}$	Incoming intensity at ground level, related to the horizontal surface
I	A	Electric current
I_{photo}	A	Induced photocurrent

Symbol	Unit	Meaning
k	$W\,m^{-2}\,K^{-1}$	Heat loss of a solar thermal collector
k	–	Pipe roughness
$k(\lambda)$	–	Extinction coefficient or imaginary part of the refractive index
K	€	Capital
kea	$J\,kg^{-1}$	Specific cumulative energy demand
CED	J	Cumulative energy demand
l	m	Length
L	–	Air requirement
L_{min}	–	Minimum air requirement
L^*	m	Characteristic length in convection
m	kg	Mass
$\dot{m}$	$kg\,s^{-1}$	Mass flow rate
M_i	$kg\,kmol^{-1}$	Molar mass of substance i
M	$W\,m^{-2}$	Emission of thermal radiation
n	kmol	Amount of substance
$\dot{n}$	$kmol\,s^{-1}$	Molar flow rate
$n(\lambda)$	–	Refractive index
N	–	Number
N_{cycle}	–	Number of cycles of a storage system
p	Pa	Pressure
P	W	Power
P_N	W	Rated power
q	$J\,kg^{-1}$	Specific heat
q_A	%	Flue gas loss from the Siegert formula
Q	kJ	Heat
$\dot{Q}$	W	Heat flux
r	–	Rate of change
r	m	Radius
r	m	Roughness of a surface
r_i		Volume fraction of substance i
R_i	$J\,kg^{-1}\,K^{-1}$	Specific gas constant of the gas i

Symbol	Unit	Meaning
Re	–	Reynolds number
s	$J\,kg^{-1}\,K^{-1}$	Specific entropy
$s(\lambda)$	m^{-1}	Scattering rate
S	$J\,K^{-1}$	Entropy
$\dot{S}$	$W\,K^{-1}$	Entropy flux
t	s	Time
T	K	Absolute temperature
TLR	–	Technology readiness level
u	$J\,kg^{-1}$	Specific internal energy
U	J	Internal energy (extensive caloric state variable)
U	V	Electric voltage
v	$m^3\,kg^{-1}$	Specific volume
v	s^{-1}	Frequency
V	m^3	Volume
V_N	${m_N}^3$	Volume at standard conditions
$\dot{V}$	$m^3\,s^{-1}$	Volumetric flow rate
w	$J\,kg^{-1}$	Specific work
W	J	Work
W_{Diss}	J	Dissipated work
W_p	J	Pressure work
W_V	J	Volume work
x	tbd	Growing quantity
x_0	tbd	Initial value of a growing quantity
y_i	–	Mole fraction of substance *i*
$y(x)$	tbd	Time integral over a (growing) quantity *x*
z	m	Height
z	–	Growth rate (with exponential growth)
z	–	Concentrator ratio

Formula Symbols – Greek Letters

Symbol	Unit	Meaning
α	–	Wind shear exponent
α	–	Area utilization factor for PV
$\alpha(\lambda)$	–	Absorptivity of a body
δ	rad	Angle between equator and sun position
Δ	–	Difference
ϕ	rad	Angle
γ	–	Quality of a (solar thermal) reflector
γ	rad	Solar elevation above the horizon
δ	rad	Installation angle of (PV) modules
ε	–	Emissivity of a surface
$\varepsilon(T)$	–	Emissivity of a body in the stefan-boltzmann formula
$\varepsilon(\lambda, T)$		Emissivity of a body in the Planck formula
ε	–	Energy efficiency ratio of a refrigeration machine
ε_{cycle}	–	Stored energy related to the energy bound in the storage
η	–	Efficiency
η_{Betz}		Maximum efficiency of a wind power plant according to Betz
η_{Carnot}	–	Carnot efficiency
η_F	–	Combustion efficiency
θ	rad	Longitude
κ	–	Concentration ratio of a solar thermal concentrator
κ	–	Isentropic exponent
λ	–	Pipe friction factor
λ	–	Excess air ratio
λ	$K\ W^{-1}\ m^{-2}$	Climate sensitivity
λ	m	Wavelength of electromagnetic radiation
λ_{max}	m	Wavelength of the maximum of blackbody radiation
μ	–	Mass fraction
υ	–	Kinematic viscosity
Π	–	Pressure ratio
ϖ	–	Energetic benefit after CCS
ρ	$kg\ m^{-3}$	Density (intensive state variable)
$\rho(\lambda)$	–	Reflectivity of a surface
ψ	–	Capacity factor of a power plant

Symbol	Unit	Meaning
ψ	rad	Time of day as angle
κ	–	Concentrator ratio
κ	–	Isentropic exponent
$\tau(\lambda)$	–	Transmissivity of a material
ξ	–	Pressure loss coefficient
ω	rad	Shadow angle

Prefixes for Units

The naming of quantities in the SI system is illustrated using the example of the joule as follows:

- Kilo: 1 kJ = 1000 J = 10^3 J
- Mega: 1 MJ = 1 000 000 J = 10^6 J
- Giga: 1 GJ = 1 000 000 000 J = 10^9 J
- Tera: 1 TJ = 1 000 000 000 000 J = 10^{12} J
- Peta: 1 PJ = 1 000 000 000 000 000 J = 10^{15} J
- Exa: 1 EJ = 1 000 000 000 000 000 000 J = 10^{18} J
- Zeta: 1 ZJ = 1 000 000 000 000 000 000 000 J = 10^{21} J

Special prefixes for concentrations as particles per particle

- ppt: 1/1000 = 10^{-3}
- ppm: 1/1 000 000 = 10^{-6}
- ppb: 1/1 000 000 000 = 10^{-9}

17 Appendix: Data on the Core Topics of the Book

This appendix collects useful data on the core topics of the book that can be used for one's own practical work:

- Table 17.1 provides an overview of the elements and their use.
- Table 17.2 contains fundamental data on important technical gases.
- Table 17.3 shows the current composition of the earth's atmosphere and Table 17.4 the greenhouse effect of important (mostly technical) gases.
- Table 17.5 summarizes relevant units of energy.
- Table 17.6, Table 17.7 and Table 17.8 contain data on relevant technical fuels.
- Table 17.9 shows the land requirement and the nitrogen input for energy crops
- Table 17.10 and Table 17.11 show important properties of metals and plastics. Table 17.12 contains data on the energy input for processing.
- Table 17.13 shows typical values for the specific energy input of transport.
- Table 17.14 contains data on the land use intensity of energy infrastructure.
- Table 17.15 summarizes data on the energy input for the provision of energy carriers.
- Table 17.16 contains typical efficiencies of fundamental processes and Table 17.17 summarizes conversion efficiencies based on these. Table 17.18 focuses on power-to-X as a storage option.
- Table 17.19 shows numerical values that illustrate the energy cliff and Table 17.20 shows typical EROI values of current energy infrastructure.
- Table 17.21 shows data on the reserves of important metals and is a supplement to Table 17.1.
- Table 17.22 contains data and references on the energy requirements for adequate living conditions.
- Section 17.14 lists the most important IEC standards for energy infrastructure.

17.1 The Elements

The elements, their molar mass M, melting and boiling temperature, concentration in the earth's crust μ_k, typical extracted ore grade μ_{Erz}, annual production volume $\dot{m}_i$, CED and CO_2 equivalent of provision as pure material. The share of the element in the earth's crust provides a measure of abundance and availability, and the ore grade indicates the effort required for processing the extracted rock. CED and $CO_{2,eq}$ reflect the current technical input for the provision of the pure metal.

Table 17.1 The Elements

Name		M in kg $kmol^{-1}$	T_m in K	T_b in K	μ_k in ppm	μ_{Erz} in %	$\dot{m}_i$ in kg a^{-1}	CED in MJ kg^{-1}	$CO_{2,eq}$ in kg kg^{-1}
Aluminum	Al	26.982	933	2750	8.23 %	32	$57 \cdot 10^9$	210	13
Antimony	Sb	121.75	904	1905	0.320	13	$190 \cdot 10^6$	140	13
Argon	Ar	39.948	83.0	87.2	3.5	-	-		
Arsenic	As	74.92	885		1.8				
Barium	Ba	137.33	1002	4450	2.2	0.45	140 000		
Beryllium	Be	9.0122	1560	2750	120	0.42	$7 \cdot 10^6$	6400	400
Lead	Pb	207.2	601	2025	14	8	$4 \cdot 10^9$	27	1.9
Boron	B	10.811	2320	4000	10				
Bromine	Br	79.90	266	332	2.4				
Cadmium	Ca	112.414	594	1038	120	0.2	$20 \cdot 10^6$	13	0.7
Caesium	Cs	132.905	302	951	3				

Name		M in kg kmol^{-1}	T_m in K	T_b in K	μ_k in ppm	μ_{Erz} in %	$\dot{m}_i$ in kg a^{-1}	CED in MJ kg^{-1}	$CO_{2,eq}$ in kg kg^{-1}
Cerium	Ce	140.12	1072	9750	64	1.5	$50 \cdot 10^6$	510	33
Chlorine	Cl	35.453	172	239	0.57 %				
Chromium	Cr	51.996	2133	2950	120	32	$23 \cdot 10^9$	300	16
Dyspro-sium	Dy	162.5	1670	2855	5.2		$1 \cdot 10^6$	1400	90
Iron	Fe	55.847	1811	3136	0.51 %		$2.3 \cdot 10^{12}$	45	3
Erbium	Er	167.26	1795	3135	3.5		500 000	2900	190
Eu-ropium	Eu	151.96	1092	1850	1.9		590 000	20 000	1300
Fluorine	F	18.9984	53.5	85	585				
Gadolin-ium	Gd	157.25	1585	3540	6.4		$2.3 \cdot 10^6$	1300	86
Gallium	Ga	69.72	303	2500	19	0.0025	78 000	3100	210
Germa-nium	Ge	72.59	1211	3110					
Gold	Au	196.967	1337	3130	0.0018		$2.4 \cdot 10^6$	250 000	15 000
Hafnium	Hf	178.49	2485	4885	4.2	16	90 000	1100	69
Helium	He	4.0026	3.5	4.22	0.008				
Holmium	Ho	164.93	1744	2968	1.3		10 000	1900	120
Indium	In	114.82	430	2346	0.250	0.005	600 000	2900	180
Iodine	I	126.904	387	457	0.45				

Name		M in kg kmol^{-1}	T_m in K	T_b in K	μ_k in ppm	μ_{Erz} in %	$\dot{m}_i$ in kg a^{-1}	CED in MJ kg^{-1}	$CO_{2,eq}$ in kg kg^{-1}
Iridium	Ir	192.22	2718	4740	0.000035		3000	45 000	2900
Potas-sium	K	39.098	336	1032	2.9 %			150	9.5
Calcium	Ca	40.078	1112	1763	4.5 %	40			
Cobalt	Co	58.933	1766	3185	24	0.55	$72 \cdot 10^6$	130	8.3
Carbon	C	12.011	3810	7200	200				
Krypton	Kr	83.80	115.8	119.8	0.0001				
Copper	Cu	63.546	1357	2845	58	2.6	$16 \cdot 10^9$	60	3.7
Lan-thanum	La	138.906	1194	3715	39		$35 \cdot 10^6$	290	19
Lithium	Li	6.941	454	1607	20	0.6	$18 \cdot 10^6$	590	28
Lutetium	Lu	174.967	1937	3668	0.530		10 000	48 000	3000
Magne-sium	Mg	24.305	922	1364	2.33 %	0.15	$570 \cdot 10^6$	320	36
Man-ganese	Mn	54.938	1518	2334	950	0.15	$10 \cdot 10^9$	65	4
Molybde-num	Mo	95.94	2892	4900	1.3		$200 \cdot 10^6$	380	34
Sodium	Na	22.989	371	1155	2.36 %				
Neon	Ne	20.179	24.5	27.1	0.005				
Neody-mium	Nd	144.24	1290	3341	41		$22 \cdot 10^6$	85	5.5

Name		M in kg kmol^{-1}	T_m in K	T_b in K	μ_k in ppm	μ_{Erz} in %	$\dot{m}_i$ in kg a^{-1}	CED in MJ kg^{-1}	$CO_{2,eq}$ in kg kg^{-1}
Nickel	Ni	58.693	1728	3190	85	1	$1.4 \cdot 10^9$	170	11
Niobium	Nb	92.906	2740	5020	18	1.7	$62 \cdot 10^6$	1700	110
Osmium	Os	190.2	3310	5300	0.42		50	62 000	4000
Palladium	Pd	106.4	1826	3240	15		190 000	160 000	8500
Phosphorus	P	30.974	317	995	1500				
Platinum	Pt	195.09	2045	4100	0.005		180 000	270 000	15 000
Plutonium	Pu	244	913	3505	–				
Praseodymium	Pr	140.908	1205	3785	9.1		63 000	650	42
Promethium	Pm	145	1353		–				
Mercury	Hg	200.59	235	630	0.058		$1.3 \cdot 10^6$	130	8.8
Radium	Ra	226	973		$9 \cdot 10^{-7}$				
Radon	Rn	222	202	211	$4 \cdot 10^{-13}$				
Rhenium	Re	186,207	3453	5920	0.0007	0.5	52 000	12 000	770
Rhodium	Rh	102,906	2230	3980	0.00037	0.5	16 000	560 000	30 000
Ruthenium	Ru	101,07	2525	4430	0.001				

Name		M in kg kmol^{-1}	T_m in K	T_b in K	μ_k in ppm	μ_{Erz} in %	$\dot{m}_i$ in kg a^{-1}	CED in MJ kg^{-1}	$CO_{2,eq}$ in kg kg^{-1}
Rubidium	Rb	85,4678	313	964	90				
Samarium	Sm	150,4	1345	2064	6.9		$3 \cdot 10^6$	1200	77
Oxygen	O	15.9994	54.4	90.2	46.1 %				
Scandium	Sc	44.956	1813	3550	20		$2 \cdot 10^6$	59 000	3800
Sulfur	S	32.066	388	718					
Selenium	Se	78.96	494	958	0.05	11	$1.5 \cdot 10^6$	72	4.5
Silver	Ag	107.868	1234	2435	0.075	0.005	$21 \cdot 10^6$	1500	100
Silicon	Si	28.086	1684	3540	28.2 %				
Nitrogen	N	14.0067	63.2	77.3	19				
Strontium	Sr	87.62	1043	1650	370				
Tantalum	Ta	180.948	3252	5640	1.8	0.3	$1.2 \cdot 10^6$	4300	260
Tellurium	Te	127.6	723	1261	0.001	4	200 000	160	7.5
Terbium	Tb	158.925	1631	3500	0.97		250 000	9000	580
Thallium	Tl	204.38	570	1733	0.61	0.007	10 000	2400	150
Thorium	Th	232	2028	5067	9.6				
Thulium	Tm	168.934	1819	2220	0.470		50 000	26 000	1700
Titanium	Ti	47.867	1943	3565	0.54 %	16	$200 \cdot 10^6$	580	39

Name		M in kg $kmol^{-1}$	T_m in K	T_b in K	μ_k in ppm	μ_{Erz} in %	$\dot{m}_i$ in kg a^{-1}	CED in MJ kg^{-1}	$CO_{2,eq}$ in kg kg^{-1}
Uranium	U	238	1406	4422	1.9	0.2	$40 \cdot 10^6$	1100	81
Vana-dium	V	50.942	2191	3680	170	1.2	$54 \cdot 10^6$	3700	250
Hydro-gen	H	1.0079	14		0.14 %				
Bismuth	Bi	208.98	544	1838	0.02	0.42	$7.3 \cdot 10^6$	150	10
Tungsten	W	183.85	3660	5890	1	1.7	$58 \cdot 10^6$	540	34
Xenon	Xe	131.3	161.3	164.9	$3 \cdot 10^{-5}$	–	–	–	–
Ytter-bium	Yb	173.04	1098	1407	2.9		50 000	5600	360
Yttrium	Y	88.9059	1775	3610	29		$8.7 \cdot 10^6$	1500	93
Zinc	Zn	65.38	692	1182	70				
Tin	Sn	118.69	505	2890	2.1	2	$310 \cdot 10^6$	230	13
Zirco-nium	Zr	91.22	2125	4681	165				

17.2 Fundamental Material Properties

Table 17.2 Data of Technically Relevant Gases: the Specific Heat Capacity c_p and the Isentropic Exponent κ in the Standard State at p_N = 101 325 Pa, T_N = 273.15 K

Name, refrigerant, formula	c_p in kJ kg^{-1} K^{-1}	M in kg $kmol^{-1}$	R in J kg^{-1} K^{-1}	κ
Air, R729	1.004	28.96	287.2	1.400
Nitrogen, R728, N_2	1.039	28.01	296.8	1.400
Oxygen, R732, O_2	0.915	32.00	259.8	1.400
Water, R718, H_2O	1.858	18.02	461.5	1.33
Hydrogen, R702, H_2	14.200	2.016	4125	1.41
Carbon monoxide, CO	1.040	28.01	296.8	1.40
Carbon dioxide, R744, CO_2	0.8169	44.01	188.9	1.30
Ammonia, R717, NH_3	2.056	17.03	488.2	1.31
Methane, R50, CH_4	2.156	16.04	518.3	1.32
Ethane, R170, C_2H_6	1.729	30.07	276.5	1.20
Propane, R290, C_3H_8	1.667	44.10	189.6	1.13
Helium, R704, He	5.238	4.003	2077	1.66
Argon, R740, Ar	0.5203	39.95	208.1	1.66
Krypton, Kr	0.248	83.80	99.2	1.66

17.3 Atmosphere and Greenhouse Gases

Table 17.3 Current Composition of the Dry atmosphere, Greenhouse Effect (GHG), Volume Fraction r, Mass Fraction μ and Absolute Mass of Gases

Gas, formula	GHG	r	μ	mass
Total atmosphere		1	1	$5285 \cdot 10^{15}$ kg
Nitrogen, R728, N_2	–	0.78084	0.7552	$3991 \cdot 10^{15}$ kg
Oxygen, R732, O_2	–	0.20942	0.2314	$1223 \cdot 10^{15}$ kg

Gas, formula	GHG	r	μ	mass
Argon, R740, Ar	–	0.00934	0.0129	$68.1 \cdot 10^{15}$ kg
Neon, Ne	–	$18.180 \cdot 10^{-6}$	$12.7 \cdot 10^{-6}$	$67 \cdot 10^{12}$ kg
Helium, R704, He	–	$5.240 \cdot 10^{-6}$	$0.724 \cdot 10^{-6}$	$3.8 \cdot 10^{12}$ kg
Krypton, Kr	–	$1.140 \cdot 10^{-6}$	$9.76 \cdot 10^{-6}$	$51 \cdot 10^{12}$ kg
Xenon, Xe	–	$0.087 \cdot 10^{-9}$	$0.394 \cdot 10^{-9}$	$2.08 \cdot 10^{9}$ kg
Carbon dioxide, R744, CO_2	yes	$425 \cdot 10^{-6}$	$646 \cdot 10^{-6}$	$3.413 \cdot 10^{15}$ kg
Methane, R50, CH_4	yes	$1.90 \cdot 10^{-6}$	$1.05 \cdot 10^{-6}$	$5.6 \cdot 10^{12}$ kg
Hydrogen, H_2	yes	$500 \cdot 10^{-9}$	$35 \cdot 10^{-9}$	$0.18 \cdot 10^{12}$ kg
Nitrous oxide, laughing gas, N_2O	yes	$340 \cdot 10^{-9}$	$517 \cdot 10^{-9}$	$2.73 \cdot 10^{12}$ kg
Carbon monoxide, CO	–	$50\text{–}200 \cdot 10^{-9}$	$50\text{–}200 \cdot 10^{-9}$	
Dichlorodifluoromethane, R12, CCl_2F_2	yes	$490 \cdot 10^{-12}$	$2050 \cdot 10^{-12}$	$11 \cdot 10^{9}$ kg
Trichlorofluoromethane, R11, CCl_3F	yes	$220 \cdot 10^{-12}$	$1040 \cdot 10^{-12}$	$5.5 \cdot 10^{9}$ kg
Chlorodifluoromethane, R22, $CHClF_2$	yes	$160 \cdot 10^{-12}$	$4800 \cdot 10^{-12}$	$2.5 \cdot 10^{9}$ kg
Carbon tetrachloride, R10, CCl_4	yes	$75 \cdot 10^{-12}$	$398 \cdot 10^{-12}$	$2.1 \cdot 10^{9}$ kg
Trichlorotrifluoroethane, R113, $C_2Cl_3F_3$	yes	$67 \cdot 10^{-12}$	$434 \cdot 10^{-12}$	$2.3 \cdot 10^{9}$ kg
1,1-Dichloro-1-fluoroethane, R141b, $CCl_2F\text{-}CH_3$	yes	$25 \cdot 10^{-12}$	$100 \cdot 10^{-12}$	$530 \cdot 10^{6}$ kg
1-Chloro-1,1-difluoroethane, R142b, $CClF_2\text{-}CH_3$	yes	$22 \cdot 10^{-12}$	$76 \cdot 10^{-12}$	$400 \cdot 10^{6}$ kg
Sulfur hexafluoride, SF_6	yes	$12 \cdot 10^{-12}$	$60 \cdot 10^{-12}$	$320 \cdot 10^{6}$ kg
Bromochlorodifluoromethane, Halon 1211, $CBrClF_2$	yes	$2.8 \cdot 10^{-12}$	$23 \cdot 10^{-12}$	$120 \cdot 10^{6}$ kg
Bromotrifluoromethane, Halon 1301, $CBrF_3$	yes	$3.4 \cdot 10^{-12}$	$17 \cdot 10^{\ 12}$	$92 \cdot 10^{6}$ kg

Table 17.4 Greenhouse Gas Effect of Important Gases: Given Are the Half-Life $\Delta t_{0.5}$ of the Gas in the Atmosphere, the Global Warming Potential (GWP) for 20 Years, 100 Years and 500 Years, the Concentration r for Pre-Industrial Conditions and Today, As Well As the Growth Rate of the Atmospheric Concentration z (IPCC, AR6, WG1).

Gas, formula	$\Delta t_{0.5}$	GWP 20	GWP 100	GWP 500	r(1750)	r(2024)	z
Carbon dioxide, R744, CO_2		1	1	1	$280 \cdot 10^{-6}$	$425 \cdot 10^{-6}$	+0.004
Methane, CH_4	12.4 a	86	34	10	$0.70 \cdot 10^{-6}$	$1.9 \cdot 10^{-6}$	+0.004
Hydrogen, H_2		35	11			$0.50 \cdot 10^{-6}$	
Nitrous oxide, N_2O	273 a	273	273	130	$0.27 \cdot 10^{-6}$	$0.34 \cdot 10^{-6}$	+0.003
Dichlorodifluoromethane, R12, CCl_2F_2	100 a	10 800	10 200		-	$490 \cdot 10^{-12}$	-0.005
Trichlorofluoromethane, R11, CCl_3F	45 a	7020	5250	2900	-	$220 \cdot 10^{-12}$	-0.005
Chlorodifluoromethane, R22, $CClF_2H$	12 a	5280	1760		-	$220 \cdot 10^{-12}$	-0.01
Carbon tetrachloride, R10, CCl_4	32	3810	2200	658	-	$72 \cdot 10^{-12}$	-0.01
Trichlorotrifluoroethane, R113, $CClF_2$-CCl_2F	93	6860	6520	2830	-	$67 \cdot 10^{-12}$	-0.01
1,1-Dichloro-1-fluoroethane, R141b, CCl_2F-CH_3	9.4	2710	860	246	-	$25 \cdot 10^{-12}$	0
1-Chloro-1,1-difluoroethane, R142b, $CClF_2$-CH_3	18	5510	2300	658	-	$22 \cdot 10^{-12}$	0
Sulfur hexafluoride, SF_6	3200 a	17 500	23 500		-	$12 \cdot 10^{-12}$	+0.04

Gas, formula	$\Delta t_{0.5}$	GWP 20	GWP 100	GWP 500	r(1750)	r(2024)	z
Bromochlorodifluoromethane, Halon 1211, $CBrClF_2$	16	4920	1930	552	–	$2.8 \cdot 10^{-12}$	-0.02
Bromotrifluoromethane, Halon 1301, $CBrF_3$	72	8320	7200	2750	–	$3.4 \cdot 10^{-12}$	0
1,1,1,2-Tetrafluoroethane, R134a, CF_3-CH_2F	13.4 a	3790	1550	436	–	$130 \cdot 10^{-12}$	+0.10
Difluoromethane, R32, CF_2H_2	5.4 a	2693	771	220	–	$30 \cdot 10^{-12}$	+0.21
Carbon tetrafluoride, R14, CF_4	50000 a	5300	7380	10 587	–	$90 \cdot 10^{-12}$	+0.01
Hexafluoroethane, R116, C_2F_6	10000 a	8210	11 100		–	$5.2 \cdot 10^{-12}$	+0.02
Nitrogen trifluoride, NF_3	500 a	12 800	16 100		–	$3.6 \cdot 10^{-12}$	+0.12

17.4 Units of Energy

A large number of units of energy are used. These can be converted into each other, but always with the risk that information about the quality of the energy is lost (see Chapter 5).

Table 17.5 Units of Energy and Conversion Factors

Unit	Conversion to SI system	Notes
Joule, J	$J = kg \cdot m^2 \cdot s^{-2}$	Unit in the SI system
Kilowatt hour (kWh)	1 kWh = 1000 W · 3600 s 1 kWh = 3.6 MJ	Typical for electricity
BTU (British Thermal Unit)	1 BTU = 1.0548 kJ	Energy required to heat one british pound of water by one degree Fahrenheit
Calorie (cal)	1 cal = 4.1868 J	Calorie: the energy required to heat 1 g of water from 0 °C to 1 °C
Barrel Barrel of Oil Equivalent (BOE)	1 BOE = 5 800 000 btu 1 BOE = 6117 MJ	Barrel is an obsolete unit of volume of approx. 159 l. The energy content applies to a specific grade of crude oil, i.e., not every barrel of crude oil contains 1 BOE.
Tonne of oil Ton of Oil Equivalent (TOE)	1 TOE = 41 870 MJ 1 TOE = 39.683 MBTU 1 TOE = 7.11 BOE or 7.33 BOE or 7.4 BOE	The following are possible as the unit of measurement "tonne": ▪ Short ton with 1 tn. sh. = 2000 lb = 907.2 kg (i.e., 2000 pounds) ▪ Metric tonne with 1 to = 1000 kg ▪ Long ton with 1 tn. l. = 160 st. = 2240 lb. = 1016 kg (i.e., 160 stone = 2240 pounds)
Oil unit (OE)	1 OE = 10 000 kcal 1 OE = 41.87 MJ	
Hard coal unit (SKE)	1 SKE = 0.7 OE 1 SKE = 7000 kcal = 29.31 MJ	

17.5 Fuels

Table 17.6 Data of Relevant Gaseous Fuels (GWI Worksheets): Lower Heating Value H_I, Upper Heating Value H_S, Density ρ As Well As the Volumetric Heating Value at Standard Conditions $H_{I,N}$

Fuel, formula	H_I in MJ kg^{-1}	H_S in MJ kg^{-1}	ρ in kg Nm^{-3}	$H_{I,N}$ in MJ Nm^{-3}
Methane, CH_4	50.00	55.50	0.7175	35.9
Ethane, C_2H_6	47.62	51.90	1.355	64.5
Propane, C_3H_8	46.35	50.35	2.011	93.2
Butane, C_4H_{10}	45.75	49.50	2.70	123.5
Hydrogen, H_2	119.96	141.80	0.0899	10.8
Carbon monoxide, CO	10.11	10.11	1.2505	12.5
Ammonia, NH_3	18.65	22.5	0.7714	14.4
Acetylene, C_2H_2	48.32	50.02	1.171	56.6
Dimethyl ether, CH_3-O-CH_3	28.4	37.6	1.97	55.9
Natural gas H	50.5	55.5	0.72–0.83	36.2
Natural gas L	40.7	45.1	0.80–0.83	33.2
Town gas	34.5	38.9	0.497	17.1

Natural gas H is almost pure methane with small admixtures of ethane, propane, N_2.
Natural gas L has an admixture of approximately 10 % nitrogen.
Town gas, coke oven gas, or blast furnace gas contain hydrogen and carbon dioxide in somewhat varying proportions.

Table 17.7 Data of Relevant Liquid Fossil and Biological Fuels: Heating Value H_I, Calorific Value H_S and Density ρ

Fuel, formula	H_I in MJ kg^{-1}	H_S in MJ kg^{-1}	ρ in kg m^{-3}	Composition/comments
Mineral oil	42 ± 2		800	Primarily C_xH_y (impurities)
Heavy oil/bitumen	40		800	Primarily long-chain C_xH_y (impurities)
Oil shale (Estonia)	12			Limestone, precursor stages to mineral oil

Fuel, formula	H_I in MJ kg^{-1}	H_S in MJ kg^{-1}	ρ in kg m^{-3}	Composition/comments
Oil sands (USA, Canada)	6			Rock, long-chain CxHy (impurities)
Gasoline	40.1–41.8	46.7	720–775	For use in gasoline engines
Kerosene	42.6–43.5	46.20	750–845	For use in aircraft engines and the generation of light
Diesel	43.40	44.80	820–845	Use in diesel engines
Methanol, CH_3OH	19.93	22.7	790	
Ethanol, C_2H_5OH	26.70	29.7	790	As a substitute for gasoline
Isopropanol, C_3H_7OH	30.45	33.6	790	

Table 17.8 Data of Relevant Solid Fossil and Biological Fuels: Lower Heating Value H_I, Higher Heating Value H_S and Density ρ

Fuel	Water content	H_I in MJ kg^{-1}	H_S in MJ kg^{-1}	ρ in kg m^{-3}	Composition/comments
Anthracite	< 1 %	32.50	32.50		Almost pure carbon
Hard coal	< 1 %	25 ± 5	25 ± 5	800	Carbon (impurities)
Lignite	50 %	7 ± 2			Water, carbon, rock/sand
Peat		5–10			Dried on site
Peat (dry)		15	22		
Hardwood	1–50 %	16–19		1000	
Hardwood		16 ± 1	21 ± 1	700–900	Air-dried
Softwood	15–50 %	21–23		450–550	Air-dried
Charcoal	< 1 %	28–32		450	
Harvest residues	5–60 %	12–15			Air-dried
Straw	7–15 %	17–18	18.4		Air-dried
Manure	10–20 %	8–14			Air-dried
Paraffin, wax	< 1 %	41.50	46.00		Long-chain C_xH_y

Table 17.9 Land Intensity and Nitrogen Demand for Energy Crops (Fritsche et al. 2017)

Plant	Land intensity in $m^2\ GJ^{-1}\ a^{-1}$	Yield in GJ $ha^{-1}\ a^{-1}$	Nitrogen demand in kg N $GJ^{-1}\ a^{-1}$	Nitrogen demand in kg N $ha^{-1}\ a^{-1}$	Share N	Product
Sugar-cane	23	435	0.11	48	0.6 %	Ethanol
Oil palm	30	333	0.44	147	2.4 %	Palm oil
Soy	20	500	3.90	1950	21.5 %	Beans only for oil
Miscant-hus	42	238	0.21	50	1.2 %	
Switch-grass	65	154	0.30	46	1.7 %	
Sweet Sorghum (millet)	61	164	0.39	64	2.1 %	
Grain Sorghum (millet)	16	625	1.00	625	5.5 %	
Willow	53	189	0.09	17	0.5 %	Via cellulose
Birch	68	147	0.16	24	0.9 %	
Poplar	72	139	0.16	22	0.9 %	
Corn	49	204	0.49	100	2.7 %	
Rapeseed	16.5	606	1.40	848	7.7 %	Seeds only for oil
Sugar beet	19	526	0.46	242	2.5 %	

17.6 Embodied Energy

For the extraction of non-energy raw materials, the same basic considerations apply as for energy raw materials. The number of processes for initial processing are already much greater, and up to the primary forming of semi-finished products from the pure material, many complex processes can take place. In Table 17.10 relevant key parameters for important inorganic materials and in Table 17.11 for plastics are compiled. The density provides an orientation for the weight of components, and the temperature range indicates the application range of the material or material group (not all nickel-base steels can be used in air up to 1200 °C, but some are designed for this application). The price offers a rough orientation. The cumulative energy demand (CED) is defined in Chapter 5. It covers the entire process from extraction to the semi-finished product, as indicated here. The CED and the CO_2-equivalent refer to typical semi-finished products, i.e., sheets, bars or wires for metals and primary-formed glass, as well as powder for ceramics. The recycling potential refers in each case to a pure, uncontaminated batch of sufficient mass. These data also serve as an illustration of the sources, since for many further questions values can be obtained from Ashby (Ashby 2009), the GEMIS software (Fritsche et al.), the PROBAS database, but also from databases for LCA software such as GABI.

Embodied energy refers to the energy that was typically expended for the provision of raw materials, the production of semi-finished products or the manufacture of products. Precise data are only determined by an LCA. The data given in Table 17.10 are more of a basis for initial estimates.

Table 17.10 Properties of Typical Engineering materials, Their Density ρ, the Temperature Range of Use T_{use} and Some Relevant Properties

Material	ρ in kg m^{-3}	T_{use} in °C	Price in € kg^{-1}	CED in MJ kg^{-1}	$CO_{2,eq}$ in kg kg^{-1}	Recycle	Properties
Structural steel	7800–7900	-70–360	0.50	57–72	1.8	+ + +	Inexpensive, strong, robust
Alloyed steel	7800–7900	-70–660	0.50–1.00	60–83	2.0	+ + +	Very firm and robust
Stainless steel	7400–8100	-270–850	1.00–8.00	83–115	5.0	+ + +	Durable, temperature-resistant depending on alloy
Chromium	7100–8000	-270–120		ca. 480	24.0	+ +	Alloy component
Aluminum[a]	2500–2950	-270–180	1.00–6.00	235–335	13.0	+ + +	Lightweight, corrosion-resistant
Magnesium	1730–1950	-40–300	2.50–11.00	300–500	35	+ + +	Very lightweight
Titanium[a]	4360–4840	-40–500	21.00–28.0	750–125(	46	+ + +	Highly corrosion-resistant and strong, robust
Nickel[a]	7650–9300	-200–120	4.00–30.00	40–690	12	+ + +	Highly corrosion-resistant and strong, robust
Zinc[a]	5500–7200	-45–120	0.90–2.90	50–145	4.1	+ + +	Corrosion-resistant
Copper	8930–8940	-270–180	1.50–2.00	100–130	3.7	+ + +	Ductile, very good thermal conductivity

Material	ρ in kg m^{-3}	T_{use} in °C	Price in € kg^{-1}	CED in MJ kg^{-1}	$CO_{2,eq}$ in kg kg^{-1}	Recy-cle	Properties
Brass	7800–8800	-270–220	1.40–2.00	100–120	3.6	+ + +	Very good wear resistance
Bronze	8500–9000	-270–200	3.50–5.50	110–120	4.2	+ + +	Corrosion-resistant
Gold	19 300			260 000	17 000	+ + +	Very good conductivity
Alu-minum oxide Al_2O_3	3700–3800	-270–170	4.00–12.00	150–200	2.8	+	Very good ceramic properties
Soda-lime glass	2200–2500	-270–250	0.60–1.00	20–25	0.8	+ + +	Inexpensive, transparent

[a] and their alloys

Table 17.11 Properties of Plastics: Their Density ρ, Temperature Range T_{use}

Material[a]	ρ in kg dm^{-3}	T_{use} in °C	Price in € kg^{-1}	CED in MJ kg^{-1}	$CO_{2,eq}$ in kg kg^{-1}	Recy-cle	Properties
Polyethylene (PE)[a]	0.94–1.40	-40–100	1.00–4.00	105–115	2.8	+ [b]	Thermoplas-tic, durable
Polypropy-lene (PP)[a]	0.89–0.92	-40–120	ca. 1.00	75–85	3.1	+ [b]	Thermoplas-tic, durable and rigid
Polystyrene (PS)[a]	1.04–1.05	-18–100	2.00–3.80	100–110	3.8	-	Thermoplas-tic, transparent[b]
Acrylonitrile butadiene styrene (ABS)	1.00–1.21	-18–90	1.50–2.80	95–104	3.8	+ [b]	Durable and rigid, easily dyeable
Polyamide (PA)/Nylon	1.10–1.42	-80–120	2.90–11.50	110–120	8.0	+ [b]	Tensile strength, hard

Material[a]	ρ in kg dm^{-3}	T_{use} in °C	Price in € kg^{-1}	CED in MJ kg^{-1}	$CO_{2,eq}$ in kg kg^{-1}	Recy-cle	Properties
Polyacryl (PMMA)	1.16–1.22	-50–100	ca. 2.00	97–105	6.8	–	Transparent, easily dyeable, UV-resistant
Polycarbon-ate (PC)	1.14–1.21	-40–120	ca. 4.00	120–130	6.0	+ [b]	Transparent, rigid
Poly-oxymethy-lene (POM)	1.39–1.43	-30–110	2.70–4.00	115–121	4.0	+ [b]	Abrasion-resistant, low friction, waterproof
Polytetraflu-oroethylene (PTFE)	2.14–2.20	-270–250	ca. 15.00	180–195	6.0	+ [b]	Thermoplas-tic, low friction, extremely resistant
Polyurethane (PU)	1.12–1.24	-30–80	ca. 5.50	110–118	3.7	+ [b]	Flexible, resistant
Polyvinyl chloride (PVC)	1.30–1.58	-20–70	ca. 1.00	77–83	2.5	+ [b]	Corrosion-resistant, durable
Polyester (PET)	1.19–1.81	-20–160	1.25–2.50	89–95	3.0	+ [b]	Thermoplas-tic, rigid
Silicone	1.1–2.3	-100–300	7.00–17.00	175–190	7.9	–	Flexible, low friction
Phenolic resins/ Bakelite	1.24–1.32	-20–160	0.90–1.20	94–100	3.6	–	Flame-retardant, durable, hard and rigid
Cellulose acetate (CA)[e]	0.98–1.40	-10–60	3.00–4.00	120–126	3.8	+ [b]	Transparent
Celluloid[e]	1.35–1.40	[d]	ca. 4.00	117–130	–	+	

[a] including additives, fillers, dyes, functional substances, flame retardants, ...

[b] applies only to pure polymers, not to mixtures

[c] releases fluorine as a decomposition product above 300 °C

[d] highly flammable

[e] from biological material

Table 17.12 Typical Energy Input for Forming of Metals

Process	For material	Energy in MJ kg^{-1}	$CO_{2,eq}$ in kg kg^{-1}
Casting	Metals	8–12	0.4–0.6
Rolling, forging	Metals	3-5	0.15–0.25
Extrusion, rolling films	Metals	10–20	0.5–1.0
Wire drawing	Metals	20–40	1.0–2.0
Powder processes (sintering)	Metals	20-30	1-1.5
Extrusion	Polymers/Plastics	3.1–5.4	0.15–0.30
Molding	Polymers/Plastics	11–27	0.55–1.4
Powder processes	Ceramics	20–30	1-1.5
Forming	Glass	2-4	0.1–0.2

17.7 Transport

Table 17.13 Typical Energy Input for Transport (Source: Ashby 2009)

Mode of transport	Energy input
Ship, long distances across oceans: large container ships, bulk carriers, oil tankers (heavy fuel oil or diesel)	0.16 kJ kg^{-1} km^{-1}
Ship, shorter distances near the coast: smaller vessels such as feeders (heavy fuel oil or diesel)	0.27 kJ kg^{-1} km^{-1}
Ship, transport on rivers and canals (diesel)	0.36 kJ kg^{-1} km^{-1}
Railway (diesel)	0.25 kJ kg^{-1} km^{-1}
Van (parcel delivery service, diesel)	2.50 kJ kg^{-1} km^{-1}
Truck up to 14 t payload (diesel)	1.5 kJ kg^{-1} km^{-1}
Truck up to 32 t payload (diesel)	0.94 kJ kg^{-1} km^{-1}
Truck up to 40 t payload (diesel)	0.82 kJ kg^{-1} km^{-1}
Truck up to 55 t payload (diesel)	0.71 kJ kg^{-1} km^{-1}
Car (gasoline)	2.2–3.0 kJ kg^{-1} km^{-1}
Car (LPG)	3.9 kJ kg^{-1} km^{-1}
Car (hybrid)	1.55 kJ kg^{-1} km^{-1}
SUV (gasoline)	4.8 kJ kg^{-1} km^{-1}
Aircraft, long-haul (kerosene)	6.5 kJ kg^{-1} km^{-1}
Airplane, short-haul (kerosene)	11–15 kJ kg^{-1} km^{-1}
Helicopter (Eurocopter, kerosene)	55 kJ kg^{-1} km^{-1}

17.8 Land Use

The intensity of land use for fossil and renewable energy technologies is shown in Table 17.14 It should be noted here that renewable energies and in particular biomass can mostly be used repeatedly on a given land area and depending on the type of land management, whereas fossil energies always require new land areas once old deposits are exhausted. Fossil energies leave behind, after the extracted energy is depleted, areas without soils or with poisoned or severely degraded soils. These areas therefore exhibit significantly lower yields or biodiversity in the future and in the long term than existed before the start of extraction. Such severely disturbed areas require very long periods of time for regeneration (Fritsche et al. 2017). The data in Table 17.14 thus represent a rough estimate for a short period of time (< 100 years). For a long-term assessment, the type of land management and measures actually taken for regeneration after use would need to be included.

Table 17.14 Intensity of Land Use for Energy Provision (Source: Fritsche et al. 2017)

Primary energy	Secondary energy	Land intensity in $m^2\ GJ^{-1}$		Note
		Typical	Range	
Coal, underground	Electricity	0.72	0.6–2.0	
Coal, surface mining	Electricity	18	0.6–30	Highest values for mountaintop removal
Natural gas	Electricity	0.72	0.4–3.6	
Crude oil	Fuel	1.44	0.4–2.2	
Uranium 235	Electricity	0.36	0.3–3.6	Nuclear power plants
Wind	Electricity	3.6	1.0–4.7	Large wind turbines, onshore wind farms
Water, large dams	Electricity	36	12–560	
Photovoltaics	Electricity	36	30–54	Freestanding, not on buildings
Solar thermal	Electricity	54	28–70	With concentrator
Biomass	Electricity	1800	1600–2900	Biomass from arable farming
Biomass	Fuel	1260	790–2000	Low values for corn and sugarcane, high values for soy and cellulose
Biomass, harvest residues	Fuel	0.4	–	No secondary effects considered such as loss as fertilizer

17.9 Provision of Energy Carriers

Before energy carriers can be used, they must be provided. The appropriate handover point for the analysis is the tank, the bunker, and the storage facility. This data focuses on direct losses. The embodied energy in the infrastructure is not taken into account unless explicitly stated.

Solid fuels Coal is mostly stored simply as a pile. Wood, dung, and straw are present as stacks, bales, or clamps. Wood chips and other solid fuels are stored in special storage rooms. Waste should be stored in suitable rooms for hygienic reasons. This is not always the case.

Liquid fuels For a large number of liquid energy carriers, there are reliable figures for their provision on the European market as well-to-tank (Table 17.15). For other regions, these figures deviate somewhat.

Gaseous fuels For gaseous fuels, a large number of forms exist: some are unpressurized (gasometer), others are not (cavern, high-pressure tank). Here the appropriate transfer point is the meter at the consumer. The following factors must be distinguished:

- All losses at the source due to leakages
- All unused but extracted components
- Inputs for compression and transport in pipelines: the gas must be brought to the pressure in the pipeline and recompressed approximately every 100 km (onshore) or less frequently (subsea) to avoid pressure losses.
- Inputs for liquefaction during transport as liquefied natural gas (LNG) at –162 °C or as liquefied hydrogen (LH2) at –253 °C / propulsion of tankers / boil-off losses of the boiling liquid / vaporization upon landing in order to have gaseous fuel again
- Venting and burning (flaring) during maintenance work in the network
- Leakages and diffusion losses in the network
- Common accidents and incidents

Electricity Here there are fundamentally no tanks or anything similar. The input here is the share of electricity that is fed into the grid by the generator but does not reach any consumer. These are the following losses:

- Electrical resistance in the lines: to keep this particularly low, electricity is transmitted at very high voltages over longer distances (110 kV or 400 kV in Europe).
- Transformation of the voltage of electricity between the individual voltage levels of the electrical grid
- At high voltage, additional losses also occur, since the strong electric field interacts with the air or significantly more so with the ground in the case of underground cables.

The sum of these losses varies with the utilization of the grid. It is below 6 %.

Table 17.15 Energy Input for the Provision of Liquid Fuels (Prussi et al. 2020): *Well to Tank* As Total Energy Input e_{tot} and the Fossil Energy Input e_{fos} Alone

Fuel	Production	e_{tot} in MJ MJ^{-1}	e_{fos} in MJ MJ^{-1}	$CO_{2,eq}$ in kg MJ^{-1}
Gasoline	Conventional crude oil	0.18	0.18	0.014
Diesel	Conventional crude oil	0.21	0.21	0.015
Diesel	Rapeseed oil	1.2 to 1.3	0.4 to 0.5	0.060
Diesel	Soy	2.5 to 2.7	0.4 to 0.6	0.070 to 0.080
Diesel	Palm oil	1.2 to 1.3	0.2 to 0.3	0.040 to 0.060
Diesel	Wood from forestry to synthetic diesel	1.2	–	0.010
Natural gas	Compressed, transport by pipeline	0.16 to 0.29	0.16 to 0.29	0.013 to 0.023
Liquefied gas	Liquefied, transport by ship	0.22 to 0.33	0.22 to 0.33	0.018 to 0.022
Biogas	Compressed, from corn	1.26	0.29	0.041
Liquefied gas	Electricity, power-to-methane	1.05	–	0.0032
Ethanol	Sugarcane in Brazil	2.1	0.15	0.026
Ethanol	Sugar beet or wheat in the EU	0.9 to 1.7	0.2 to 0.9	0.022 to 0.095
Ethanol	Wood from forestry	1.9	0.25	0.021
Dimethyl ether (DME)	Wood from forestry	1.1	–	0.010
Methanol	Wood from forestry	1.1	–	0.010

17.10 Efficiencies

Energy conversion takes place in steps. Table 17.16 contains efficiencies of basic components. Table 17.17 contains conversion efficiencies for processes in which multiple components act. Both tables offer only a selection of data. For lighting, the actual specification is lumens per watt (electrical). The lumen is defined as 683 W of radiation at 555 nm. This allows a rough conversion into an efficiency.

Table 17.16 Efficiencies η of Fundamental Processes and Aggregates

Unit	Process	η	Details
Turbomachinery		Efficiency at the operating point, often strong variation with deviation from the operating point	
Pump for liquids	Shaft work → volume flow	< 0.90	
Compressor for gases	Wave work → volume flow rate	< 0.8	Pressure difference and volume flow rate determine the operating point.
Water turbine	Flowing water → wave work	< 0.9	Ususally wide operation range (Quaschning 2023)
Savonius rotor	Wind → wave work	< 0.15	Vertical small wind turbines: < 25 % of what is technically possible
Windmill	Wind → wave work	< 0,30	Middle ages and museum: < 50 % of what is technically possible
Darrieus rotor	Wind → wave work	< 0,40	Vertical wind turbines: < 66 % of the technically possible
Three-blade wind turbine	Wind → wave work	< 0.55	Typical wind turbines: < 90 % of what is technically possible
Radiation			
Photovoltaics, single layer	Light (sun) → electricity	up to 0.25	Depending on cell type and design

Unit	Process	η	Details
Photovoltaics, multilayer	Light (sun) → electricity	up to 0.6	Depending on cell type and design, suitable for special applications only
Incandescent lamp	Electricity → light	< 0.06	Depending on filament temperature, often significantly lower
Gas discharge (sodium vapor)	Electricity → light	up to 20 %	up to 150 lm/W
LED	Electricity → light	up to 50 %	up to 350 lm/W
Electrical systems		Operating point of rotating machines depends on the rotational frequency	
Generator	Shaft work → electricity	up to > 0.95	Depending on operating point
Electric motor	Electricity → shaft work	up to > 0.97	Large variation possible
Inverter	Direct current → alternating current	0.95–0.99	
Rectifier	Alternating current → direct current	0.95–0.99	
Transformer	Change voltage	up to > 0.99	$U_{in} \cdot I_{in} = \eta \cdot U_{out} \cdot I_{out}$

Table 17.17 Efficiencies of Energy Conversion: the Product Provided, the Process Used for It, the Efficiency and, Where Applicable, Further Explanations

Product	Process	Efficiency	Comment
Shaft work from coal	Coal → boiler → Newcomen steam engine	< 0.005	First steam engine from the year 1713
Shaft work from coal	Coal → boiler → Watt steam engine	< 0.04	Low-pressure steam engine by James Watt, ca. 1770
Shaft work from coal	Coal → boiler → three-cylinder expansion steam engine	0.10–0.15	Ships, steam locomotives, ca. 1900
Shaft work from coal	Coal → boiler → steam turbine	0.25	Ship propulsion 1914
Electricity from coal	Coal → boiler → steam turbine → generator	0.40–0.47	State of the art today
Electricity from crude oil products	Crude oil → boiler → steam turbine	0.20–0.30	Obsolete usage
Electricity from diesel/ natural gas	Diesel → diesel engine → generator	0.30–0.40	Emergency power generators
Electricity from natural gas	Natural gas → stationary gas turbine → generator	0.30–0.40	Gas power plant
Electricity from natural gas (CCGT)	Natural gas → stationary gas turbine and boiler → generator	0.50–0.60	The waste heat of the gas turbine generates superheated steam.
Electricity from uranium	Reactor → steam turbine	0.35	Boiling water and pressurized water reactors
Hydrogen from electricity	Electricity → electrolysis → H_2	0.68 ± 0.02	The reference here is the lower heating value of the hydrogen.
Hydrogen, transport only	Transport of H_2 in a pipeline a few 100 km	0.92 ± 0.01	Maintenance losses, diffusion losses

Product	Process	Effi-ciency	Comment
Hydrogen, compression only	H_2 (standard conditions) → compressor → H_2 (80 bar, ambient temperature)	0.85 ± 0.02	Electric drive
Hydrogen, compression only	H_2 (standard conditions) → compressor → H_2 (80 bar, ambient temperature)	0.75 ± 0.03	Drive with hydrogen gas turbine
Methane from electricity	Electricity → electrolysis → carbonization → methane (standard conditions)	0.55 ± 0.05	The reference here is the lower heating value of the methane.
Methane from electricity	Electricity → electrolysis → carbonization → methane (80 bar)	0.50 ± 0.05	The reference here is the heating value of methane.
Synthetic gasoline or synthetic diesel from electricity (e-fuels)	Electricity → electrolysis → carbonization → Fischer-Tropsch → refinery → gasoline and diesel and naphtha	0.39 ± 0.02 (gasoline and diesel)	Naphtha as by-product with $\eta = 0.09$
Passenger car drive, electric	Power grid → battery → drive	0.80	Plug-to-wheel
Passenger car drive, gasoline	Filling station → tank → drive	0.25	Tank-to-wheel
Truck drive, diesel	Filling station → tank → drive	0.33	Tank-to-wheel
Battery storage, cycle efficiency	electricity → battery → Electricity	0.80	Na-ion battery
Wood stove	Wood → space heating	0.1–0.4	Simple stove
Coal stove	Coal → space heating	0.3–0.6	Simple stove
Space heating	Fuel → space heating	0.9	Heating water circuit, flow temperature > 50 °C
Space heating	Fuel → space heating	1.1	Heating water circuit, flow temperature < 40 °C

Table 17.18 Reference Values for Power-to-X As Electricity Storage: Transmission Losses and Expenditures for the Compression of Hydrogen Are Not Taken into Account.

Conversion	Cycle efficiency	Boundary condition
Electricity → heat → steam → electricity	< 0.32	High-temperature storage with $\eta = 0.8$
Electricity → hydrogen → electricity	~ 0.25	Gas turbine power plant
Electricity → hydrogen → electricity	~ 0.40	CCGT power plant
Electricity → methane → electricity	~ 0.23	Storage at 80 bar/gas turbine power plant
Electricity → methane → electricity	~ 0.35	Storage at 80 bar/CCGT power plant
Electricity → ammonia → hydrogen → electricity	~ 0.15	Gas turbine power plant/storage of liquid ammonia
Electricity → methanol → electricity	~ 0.20	Gas turbine power plant/storage of liquid methanol

17.11 EROI

Table 17.19 contains numerical values for the functions in Figure 5.6. Table 17.20 gives an overview of EROI values. The values deviate in some cases significantly from literature values, since a factor of 0.3 was applied throughout to convert fuels into electricity or other usable forms of work. For hydrogen, 0.65 is used for the efficiency of conversion from electricity and 0.3 for the reconversion into electricity.

Table 17.19 Relationship Between EROI, Invested and Delivered Energy: e_{use} Is the Gross Delivered Energy Relative to the Invested Energy. e_{invest} Is the Invested Energy Relative to e_{use}.

EROI	Gross e_{use}	Net e_{use} - e_{invest}	e_{invest}	χ(EROI)
0	None	-∞	∞	-∞
0.01	0.01	-0.99	100	-99
0.1	0.1	-0.9	10	-90
0.2	0.2	-0.8	5.0	-40
0.5	0.5	-0.5	2.0	-10
1	1	0	1.00	0
1.1	1.1	0.1	0.91	0.091
1.5	1.5	0.2	0.67	0.33
2	2.0	1.0	0.50	0.50
3	3.0	2.0	0.33	0.67
4	4.0	3.0	0.25	0.75
5	5.0	4.0	0.20	0.80
6	6.0	5.0	0.17	0.83
7	7.0	6.0	0.14	0.86
8	8.0	7.0	0.13	0.88
9	9.0	8.0	0.11	0.89
10	10	9.0	0.10	0.90
15	15	14	0.067	0.93
20	20	19	0.05	0.95
30	30	29	0.033	0.97
50	50	49	0.02	0.98
100	100	99	0.01	0.99

Table 17.20 Typical EROI Values for Relevant Energy Carriers and Infrastructure: the Reference Is Always the Generated Electricity, i.e., the Quality of the Primary Energy for Fuels Is Taken into Account accordingly.

Type	EROI	Comments
Coal	3 to 33	Without perpetual liabilities/without climate adaptation
Coal with Carbon Capture and Storage (CCS)	0 to 2	Without perpetual liabilities
Conventional crude oil	5 to 33	Without climate adaptation etc.
Unconventional crude oil: deep sea, heavy oil	1 to 5	Without climate adaptation etc.
Crude oil from fracking	0 to 3	Without climate adaptation etc.
Crude oil from tar sands	0 to 3	Without climate adaptation etc.
Natural gas: on-site use	5 to 33	Without climate adaptation etc.
Natural gas: costly transport, LNG	3 to 6	Without climate adaptation etc.
Fuel from biomass: favorable locations	1.5 to 3	E.g. from Brazil / without biodiversity and soil loss
Fuel from biomass: temperate climate	Approx. 1	E.g. Germany, USA / without biodiversity and soil loss
Natural gas from fracking	1 to 4	Without climate adaptation etc.
Hydrogen: grey or brown	0.7 to 7	Without perpetual liabilities / without climate adaptation
Hydrogen: blue or turquoise	0.5 to 5	Without perpetual liabilities / without further necessary climate adaptation
Hydrogen: pink	1 to 5	Without perpetual liabilities
Hydrogen: green	< 1 to 5	
Hydropower: large	20 to 100	Without biodiversity loss and further consequences
Hydropower: small	5 to 25	Without biodiversity loss and further consequences
Wind	5 to 25	
Photovoltaics	1 to 20	Curtailing, location, shading, ...
Solar thermal: concentrated	2 to 7	Favorable location
Nuclear fission	10 to 30	Without perpetual liabilities
Nuclear fusion	Significantly less than 2	Without perpetual liabilities

17.12 Range of Important Metals

Table 17.21 Availability of Relevant Metals (Sources: Lundaev et al. 2023; Astra 2025): Data Are the Extracted Mass m_{ext}; Reserves m_{res}; Resources m_{sou}; Point in Time for Peak Extraction t_{peak} Based on Reserves and, Where Applicable, Resources; Static Range t_{stat} Based on reserves)

Raw material	m_{ext}	m_{res}	m_{sou}	t_{peak}	t_{stat}
Aluminum	$1.6 \cdot 10^9$ t	$12 \cdot 10^9$ t	$30 \cdot 10^9$ t	2060–2080	> 2200
Antimony	$7.7 \cdot 10^6$ t	$2.9 \cdot 10^6$ t	–	2026	2038
Beryllium	22 000 t	–	100 000 t	Until 2070	> 2200
Cadmium	$1.4 \cdot 10^6$ t	$0.69 \cdot 10^6$ t	$5.7 \cdot 10^6$ t	2034–2089	2038
Chromium	$0.99 \cdot 10^9$ t	$0.57 \cdot 10^9$ t	$12 \cdot 10^9$ t	2027–2061	2035
Dysprosium	$0.02 \cdot 10^6$ t	$0.5 \cdot 10^6$ t	$4.0 \cdot 10^6$ t	2076–2110	> 2200
Fluorine	$170 \cdot 10^6$ t	$160 \cdot 10^6$ t	$240 \cdot 10^6$ t	2035–2043	2060
Gallium	5060 t	$1.5 \cdot 10^6$ t	$14 \cdot 10^6$ t	2079–2102	> 2200
Germanium	4613 t	–	36 000 t	Until 2085	–
Gold	170 000 t	64 000 t	132 000 t	2036–2039	2044
Indium	$15 \cdot 10^3$ t	$21 \cdot 10^3$ t	$356 \cdot 10^3$ t	2029–2063	2042
Cobalt	$3.2 \cdot 10^6$ t	$7.6 \cdot 10^6$ t	$25 \cdot 10^6$ t	2039–2057	2074
Copper	$0.79 \cdot 10^9$ t	$0.88 \cdot 10^9$ t	$5.6 \cdot 10^9$ t	2038–2079	2063
Lithium	$1.0 \cdot 10^6$ t	$22 \cdot 10^6$ t	$89 \cdot 10^6$ t	2046–2057	> 2200
Magnesium	$41 \cdot 10^6$ t	$4.6 \cdot 10^9$ t	$7.8 \cdot 10^9$ t	2097–2101	> 2200
Manganese	$0.70 \cdot 10^6$ t	$1.5 \cdot 10^9$ t	$17 \cdot 10^9$ t	2048–2115	2099
Molybdenum	$9.2 \cdot 10^6$ t	$16 \cdot 10^6$ t	$25 \cdot 10^6$ t	2040–2046	2074
Neodymium	$0.61 \cdot 10^6$ t	$16 \cdot 10^6$ t	$62 \cdot 10^6$ t	2074–2099	> 2200
Nickel	$77 \cdot 10^6$ t	$95 \cdot 10^6$ t	$300 \cdot 10^6$ t	2036–2056	2058
Niobium	$1.9 \cdot 10^6$ t	$17 \cdot 10^6$ t	–	After 2089	2180
Platinum	$7.2 \cdot 10^3$ t	$31 \cdot 10^3$ t	$42 \cdot 10^3$ t	2057–2068	> 2200
Praseodymium	$0.19 \cdot 10^6$ t	$4.7 \cdot 10^6$ t	$20 \cdot 10^6$ t	2075–2104	> 2200
Rhenium	1400 t	2500 t	11 000 t	2031–2035	2064
Selenium	$0.11 \cdot 10^6$ t	$0.1 \cdot 10^6$ t	–	After 2035	2052
Silver	$1.3 \cdot 10^6$ t	$0.53 \cdot 10^6$ t	$1.2 \cdot 10^6$ t	2030–2040	2023
Tantalum	44 000 t	123 000 t	–	After 2043	2083
Titanium	$190 \cdot 10^6$ t	$450 \cdot 10^6$ t	$599 \cdot 10^6$ t	2056–2061	2107

Raw material	m_{ext}	m_{res}	m_{sou}	t_{peak}	t_{stat}
Vanadium	$2.7 \cdot 10^6$ t	$24 \cdot 10^6$ t	$63 \cdot 10^6$ t	2073–2096	> 2200
Tungsten	$3.9 \cdot 10^6$ t	$4.6 \cdot 10^6$ t	–	After 2072	2080
Zinc	$570 \cdot 10^6$ t	$250 \cdot 10^6$ t	$1900 \cdot 10^3$ t	2029–2068	2041
Tin	$19 \cdot 10^6$ t	$5.5 \cdot 10^6$ t	$15.4 \cdot 10^6$ t	2031–2034	2042
Zircon	$26 \cdot 10^6$ t	$52 \cdot 10^6$ t	–	2050 –	2107

17.13 Decent Living Conditions

The concept of decent living conditions for all people and everywhere is a concretization of the SDGs. The focus is on a society in which everyone can lead a dignified life and participate in social life. This gives rise to requirements for the resources and energy needed. Table 17.22 summarizes the minimum requirements. From this, bound resources can be determined (Kikstra et al. 2021; Vélez-Henao/Pauliuk 2023). The resulting energy demand for use in Table 17.22 varies with regional conditions.

Table 17.22 Material Indicators for Decent Living Conditions of a Person (Source: Kikstra et al. 2021 and Vélez-Henao/Pauliuk 2023): "Minimum" Lists the Specific Requirement. the Average Energy Demand Includes Only Operation. a Household Is Assumed to Be Inhabited by 2 to 8 People.

Aspect	Minimum	Energy demand
Shelter/housing	15 m^2 living space	
Heating	up to 18 °C	0–45 MJ/d
Cooling	From 26 °C/60 % humidity (sleeping area only)	0–4 MJ/d
Nutrition	9 MJ/d (2100 kcal/d) 0.2 stove 0.2 refrigerator	1.7–14 GJ/d 1.2 MJ/d 1.2 MJ/d
Clothing and footwear	2.4 kg/a + 0.9 kg/a	1.7 GJ/a + 0.2 GJ/a
Hygiene	65 l/d water 0.75 kg/d waste	150 MJ/d 440 MJ/a
Hot water	30 l/d	3-9 MJ/d
Education	2.5 m^2 facility	Approx. 3 GJ/a
Health	2.5 m^2 facility	Approx. 3 GJ/a
Communication	0.8 mobile phone 0.25 laptop and 0.25 router	
Mobility	23 km/d	2.3-60 MJ/d
Public facilities	4.7 m^2	
Space for gatherings		

17.14 Standards for Energy Infrastructure

Steam Turbines

IEC 60045-1: Steam turbines – Part 1: Specifications

Generators (Rotating Machines)

IEC 60034-3: Rotating electrical machines – Part 3: Specific requirements for synchronous generators driven by steam turbines or combustion gas turbines and for synchronous compensators

IEC 60034-22: Rotating electrical machines – Part 22: AC generators for reciprocating internal combustion (RIC) engine driven generating sets

IEC 60034-33: Rotating electrical machines – Part 33: Specific technical requirements for hydro generators

Fuel Cells

IEC 62282-3-200: Fuel cell technologies – Part 3-200: Stationary fuel cell power systems – Performance test methods

IEC 62282-3-201: Fuel cell technologies – Part 3-201: Stationary fuel cell power systems – Performance test methods for small fuel cell power systems

IEC 62282-4-102: Fuel cell technologies – Part 4-102: Fuel cell power systems for industrial electric trucks – Performance test methods

IEC 62282-6-400: Fuel cell technologies – Part 6-400: Micro fuel cell power systems – Power and data interchangeability

Tidal Power Plants

IEC TS 62600-100: Marine energy – Wave, tidal and other water current converters – Part 100: Electricity producing wave energy converters – Power performance assessment

IEC TS 62600-101: Marine energy – Wave, tidal and other water current converters – Part 101: Wave energy resource assessment and characterization

IEC TS 62600-200: Marine energy – Wave, tidal and other water current converters – Part 200: Electricity producing tidal energy converters – Power performance assessment

IEC TS 62600-201: Marine energy – Wave, tidal and other water current converters – Part 201: Tidal energy resource assessment and characterization

Photovoltaics

IEC 61215-1: Terrestrial photovoltaic (PV) modules – Design qualification and type approval – Part 1: Test requirements

IEC 61215-1-1: Terrestrial photovoltaic (PV) modules – Design qualification and type approval – Part 1-1: Special requirements for testing of crystalline silicon photovoltaic (PV) modules

IEC 61215-1-2: Terrestrial photovoltaic (PV) modules – Design qualification and type approval – Part 1-2: Special requirements for testing of thin-film Cadmium Telluride (CdTe) based photovoltaic (PV) modules

IEC 61215-1-3: Terrestrial photovoltaic (PV) modules – Design qualification and type approval – Part 1-3: Special requirements for testing of thin-film amorphous silicon based photovoltaic (PV) modules

IEC 61215-1-4: Terrestrial photovoltaic (PV) modules – Design qualification and type approval – Part 1-4: Special requirements for testing of thin-film Cu(In,GA)(S,Se)2 based photovoltaic (PV) modules

IEC 61829: Photovoltaic (PV) array – On-site measurement of current-voltage characteristics

IEC 61853-1: Photovoltaic (PV) module performance testing and energy rating – Part 1: Irradiance and temperature performance measurements and power rating

IEC 62670-3: Photovoltaic concentrators (CPV) – Performance testing – Part 3: Performance measurements and power rating

Solar Thermal Energy

IEC 62862-3-2: Solar thermal electric plants – Part 3-2: Systems and components – General requirements and test methods for large-size parabolic-trough collectors

IEC 62862-4-1: Solar thermal electric plants – Part 4-1: General requirements for the design of solar tower plants

Wind Turbines

IEC 61400-1: Wind turbines – Part 1: Design requirements

IEC 61400-2: Wind turbines – Part 2: Small wind turbines

IEC 61400-12-1: Wind energy generation systems – Part 12-1: Power performance measurements of electricity producing wind turbines

References

Ashby, M. F.: Materials and the Environment. Eco-Informed Material Choice. Elsevier, Amsterdam 2009

Ashby, M. F.: Materials and sustainable development. Butterworth Heinemann, Oxford 2016

Ashby, M./Johnson, K.: Materials and Design. The art and science of material selection in product design. Butterworth Heinemann, Oxford 2014

Astra, M.: The State of Material Circularity for Critical Metals. Hochschule Darmstadt 2025

Barnhart, C. J./Dale, M./Brandt, A. R./Benson, S. M.: The energetic implications of curtailing versus storing solar- and wind-generated electricity. In: Energy and Environmental Science 6, 2013, pp. 2804–2810

Carroll, A. R.: Geofuels. Energy and the Earth. Cambridge University Press 2015

Fritsche, U. R./Berndes, G./Cowie, L. A./Dale, V. H./Kline, K. L./Johnson, F. X./Langeveld, H./Sharma, N./Watson, H./Woods, J.: Global land outlook. Working paper. Energy and land use. UNCCD & IRENA 2017

Fritsche, U./Greß, H.-W.: Der nichterneuerbare kumulierte Energieverbrauch und THG-Emissionen des deutschen Strommix im Jahr 2017 sowie Ausblicke auf 2020 bis 2050. IINAS GmbH, Darmstadt 2018. *http://iinas.org/tl_files/iinas/downloads/GEMIS/2018_KEV_THG_Strom-2017_2020-2050.pdf*

Fritsche, U./Rausch, L./Simon, K.-H.: Globales Emissions-Modell integrierter Systeme (GEMIS). *https://iinas.org/arbeit/gemis*

GWI – Gas- und Wärme-Institut Essen e. V.: GWI-Arbeitsblätter. Verbrennungskennwerte | Gaseigenschaften | Berechnung. Vulkan Verlag, Essen 2013

Hall, C. A. S.: Energy Return on Investment. A Unifying Principle for Biology, Economics, and Sustainability. Springer 2017

IEC – International Electrotechnical Committee (Hrsg.): Electrical Energy Storage. White Paper. Genf 2011. *https://basecamp.iec.ch/download/iec-white-paper-electrical-energy-storage*

Kikstra, J. S./Mastrucci, A./Min, J./Riahi, K./Rao, N. D.: Decent living gaps and energy needs around the world. In: Environmental Research Letters 16, 2021, 095006

Linow, S.: Angewandte technische Thermodynamik. Carl Hanser Verlag, München 2022

Lundaev, V./Solomon, A. A./Le, T./Lohrmann, A./Breyer, C.: Review of critical materials for the energy transition, an analysis of global resources and production databases and the state of material circularity. In: Minerals Engineering 203, 2023, 108282

Prussi, M./Yugo, M./De Prada, L./Padella, M./Edwards, R./Lonza, L.: JEC Well-to-Tank report v5. EUR 30269 EN. Publications Office of the European Union, Luxembourg 2020

Smil, V.: Energy and civilisation. A history. The MIT Press, Cambridge 2017

Umweltbundesamt: Prozessorientierte Basisdaten für Umweltmanagement-Instrumente (PROBAS). *http://www.probas.umweltbundesamt.de/php/index.php*

Vélez-Henao, J. A./Pauliuk, S.: Material requirements of decent living standards. In: Environmental Science & Technology 57, 2023, pp. 14 206–14 217

Index